ANNUAL REVIEW OF EARTH AND PLANETARY SCIENCES

ANNUAL REVIEW OF EARTH AND PLANETARY SCIENCES

FRED A. DONATH, *Editor*
University of Illinois—Urbana

FRANCIS G. STEHLI, *Associate Editor*
Case-Western Reserve University

GEORGE W. WETHERILL, *Associate Editor*
University of California—Los Angeles

VOLUME 3

1975

ANNUAL REVIEWS INC.   4139 EL CAMINO WAY   PALO ALTO, CALIFORNIA 94306

ANNUAL REVIEWS INC.
Palo Alto, California, USA

International Standard Book Number 0-8243-2003-4
Library of Congress Catalog Card Number 72-82137

REPRINTS

The conspicuous number aligned in the margin with the title of each article in this volume is a key for use in ordering reprints. Available reprints are priced at the uniform rate of $1 each postpaid. Effective January 1, 1975, the minimum acceptable reprint order is 10 reprints and/or $10, prepaid. A quantity discount is available.

PREFACE

The range of topics appropriate for a volume entitled *Annual Review of Earth and Planetary Sciences* is vast, but dwarfed by the problem of selecting those few that will lead to broad and balanced coverage. Nevertheless, there assembles each year a group of patient optimists known as the Editorial Committee, who have rashly expressed a willingness to come to grips with this task. It takes both optimism and patience to face the earth, the sun, the planets, and their many biographers, and to annually attempt to wring from them an inventory of events ripe for review. Optimism is required of those who would set forth balanced presentation as a goal. Patience is needed to achieve broad coverage in the only possible way—over a cycle of several years. To patience and optimism must be added good fortune in seeking out qualified, often uniquely qualified, scientists willing to share their special knowledge and perspectives, to give their time to this endeavor, and to produce completed manuscripts when they are due.

Despite its optimism and its patience, the Editorial Committee has a grievous failing—it is not perfect. It may overlook topics ready for review, thus contributing to lack of balance and to diminished breadth. Sometimes it can find no author willing to review a complex and rapidly changing subject. Occasionally an invitation is accepted by a scientist who later, for one reason or another, defaults, and this is a major cause of imbalance and reduced scope. Yet, despite such failings and with or without hoped-for chapters, each volume must follow an ineluctable course through deadlines imposed by the closely coordinated production of the 20 other volumes of the Annual Review series.

Each January, the Editorial Committee meets to select topics and suggest authors for the chapters of the volume of *Annual Review of Earth and Planetary Sciences* that will appear two years hence. Invitations to prospective authors are normally sent out during the spring, and a final listing of both topics and authors is ordinarily confirmed by midsummer. Manuscripts are due the following summer; publication follows approximately 10 months after the manuscript due date. The Editorial Committee welcomes suggestions for topics that would be timely for review, as well as suggestions for appropriate authors. It will give careful attention to each suggestion received. Perhaps with the help of its readers, *Annual Review of Earth and Planetary Sciences* can continue to improve both the balance and breadth of its coverage.

THE EDITORS

SOME RELATED ARTICLES APPEARING
IN OTHER ANNUAL REVIEWS

From the *Annual Review of Astronomy and Astrophysics*, Volume 12 (1974)

Comets, Brian G. Marsden
Surface Geology of the Moon, Farouk El-Baz
Nucleo-Cosmochronology, David N. Schramm
On the Origin of the Light Elements, Hubert Reeves

From the *Annual Review of Biochemistry*, Volume 43 (1974)

Membrane Receptors, Pedro Cuatrecasas
Methods of Gene Isolation, Donald D. Brown and Ralph Stern
The Molecular Organization of Membranes, S. J. Singer

From the *Annual Review of Ecology and Systematics*, Volume 5 (1974)

The Ecology of Secondary Succession, Henry S. Horn
The Ecology of Mangroves, Ariel E. Lugo and Samuel C. Snedaker
The Ecology of Macroscopic Marine Algae, A. R. O. Chapman
Methods of Comparing Classifications, F. James Rohlf
Frequency-Dependent Selection, Francisco J. Ayala and Cathryn A. Campbell
Closed Ecological Systems, Frieda B. Taub
Equilibrium Theory of Island Biogeography and Ecology, Daniel S. Simberloff
Desert Ecosystems: Higher Trophic Levels, Imanuel Noy-Meir
Continental Drift and Vertebrate Distribution, Joel Cracraft
Biogeographical Considerations of the Marsupial-Placental Dichotomy, Jason A.
    Lillegraven
The Measurement of Species Diversity, Robert K. Peet

From the *Annual Review of Fluid Mechanics*, Volume 7 (1975)

Experiments in Granular Flow, K. Wieghardt
Experiments in Rotating and Stratified Flows: Oceanographic Application,
    T. Maxworthy and F. K. Browand
Hydrodynamics of Large Lakes, G. T. Csanady

From the *Annual Review of Materials Science*, Volume 4 (1974)

Some Defect Structures in Crystalline Solids, B. G. Hyde, A. N. Bagshaw, Sten
    Andersson, and M. O'Keeffe

CONTENTS

ANNUAL REVIEWS INC. is a nonprofit corporation established to promote the advancement of the sciences. Beginning in 1932 with the *Annual Review of Biochemistry,* the Company has pursued as its principal function the publication of high quality, reasonably priced Annual Review volumes. The volumes are organized by Editors and Editorial Committees who invite qualified authors to contribute critical articles reviewing significant developments within each major discipline.

Annual Reviews Inc. is administered by a Board of Directors whose members serve without compensation.

Annual Reviews are published in the following sciences: Anthropology, Astronomy and Astrophysics, Biochemistry, Biophysics and Bioengineering, Earth and Planetary Sciences, Ecology and Systematics, Entomology, Fluid Mechanics, Genetics, Materials Science, Medicine, Microbiology, Nuclear Science, Pharmacology, Physical Chemistry, Physiology, Phytopathology, Plant Physiology, Psychology, and Sociology (to begin publication in 1975). In addition, two special volumes have been published by Annual Reviews Inc.: *History of Entomology* (1973) and *The Excitement and Fascination of Science* (1965).

Edward Bullard

THE EMERGENCE OF PLATE TECTONICS: A PERSONAL VIEW

✖10030

Edward Bullard

Department of Geodesy and Geophysics, University of Cambridge,
Cambridge CB3 0EZ, Great Britain

THE PROBLEM

In the early nineteenth century it was clear that the continents were very old and had a complex history. Ideas about the time scale were vague but the bulk of well-informed opinion spoke of millions or hundreds of millions of years or, more vaguely, of there being "no vestige of a beginning, no prospect of an end" (Hutton 1795). Of course there were many, particularly among English clergyman-geologists, who still tried to fit earth history into the time since 4004 B.C. allowed by Bishop Usher, but the effort was by then clearly hopeless (Rudwick 1972). The view of Hutton and of Lyell that the time scale was long enough for processes observable today to have caused the changes that had occurred in the past was widely accepted by the 1830s.

The geologists of the nineteenth century were enormously successful; they devised methods of elucidating the history of particular areas in great detail and setting them in an ordered time sequence (a time scale without dates). They were much less successful in giving a global picture of events; in fact, the subject tended to fragment into local monographs. It is only in our own day that the pendulum has swung back and the center of interest shifted to the relation of local phenomena to global processes.

This review attempts to describe the course of this change of emphasis. The central theme is the relation of the continents to each other and to the floor of the ocean. In the absence of knowledge about the ocean floor a variety of hypotheses could be entertained. The oceans could be considered as like the continents but eroded from material originally above sea level, or as flooded low lying continents, or as subsided continents. Alternatively they could be regarded as essentially different from the continents but as of the same range of age and with an equally complicated history. All of these views are certainly incorrect and of all the possible hypotheses the one that seemed to many the least likely has obtained a nearly universal acceptance. This is the hypothesis of sea-floor spreading and continental drift. The editor has asked me to tell the story of this change in ideas from a personal

1

point of view. This seems to me appropriate since a revolution in scientific opinion is not only a question of new data and new theories, it has also its emotional and psychological aspects.

BEFORE WEGENER

Reading the literature with hindsight reveals a number of passages that can be regarded as prefiguring the idea of mobile continents. A passage from Bacon (1620) is frequently quoted but this appears to me to be merely a comment on the similar shape of the west coasts of Africa and South America, with no implication of movement. The first undoubted reference seems to me to be in Snider's (1858) book *La création et ses mystères dévoilé*. This book is a rather cranky attempt to square geology with Holy Writ and, even when first published, can have contained little of interest to serious students of geology. It does, however, contain a diagram showing how well the west coast of Africa fits the east coast of South America. The diagram (Figure 1) is a rough sketch and the fit has been obtained by distorting the coastlines; in fact the coastlines do not fit, the good fit is between the continental edges (that is, the outer edge of the continental shelves). The distortion is not trivial

Figure 1   Snider's reassembly of the continents [a redrawn version from Hallam (1973)].

as the Argentine continental shelf is 400 km wide. In fairness to Snider it should be said that he could have had only the vaguest idea of where the edge of the shelf lay, and could not have known that it is the true edge of the continent. Snider's diagram was reproduced by Pepper (1861) in an admirable book on minerals and metallurgy intended primarily for schoolboys. By a strange chance the page heading is "The drift theory," but the text shows that the drift theory referred to concerns the formation of coal seams from driftwood.

So far as I know, the writings of Snider and Pepper had no influence on scientific opinion. The same can be said of the other nineteenth and early twentieth century references to continental movement. A number of these are listed by Du Toit (1927) and Meyerhoff (1968); they will not be discussed here. The best known is Taylor's (1910); his movements are primarily away from the pole and are not closely related to later work.

WEGENER AND THE DISPUTES OF 1920–1940

The source of the controversies of the 1920s and 1930s is two lectures given in 1912 by Alfred Wegener (1880–1930) in Frankfurt and Marburg. These were published (Wegener 1912a,b) and later appeared as a book (Wegener 1915). Wegener brought together a great collection of evidence from diverse fields of earth science to show that the continents had formerly been collected together into a single land mass, Pangea. The main arguments were: (a) that the continents would fit together as a jigsaw puzzle; (b) that not only the shapes but the geological structures fit; (c) that longitude determinations showed Greenland to be moving away from Europe; (d) that the distributions of past climates (particularly the Permo-Carboniferous ice age), salt deposits, and corals were incompatible with the present positions of the continents, but compatible with the former existence of Pangea; and (e) that the distributions of animals and plants were also consistent with the idea of Pangea.

The publication of Wegener's book was followed by 20 years of controversy which was ended more by the exhaustion of the possible arguments and of the contestants, than by any firm conclusion. It is interesting to consider why Wegener's arguments did not carry conviction, since it is now clear that many of them are, in principle, sound. The reasons were, in part, associated with the nature of Wegener's presentation. He argues too hard and was often accused (e.g. by Lake 1922) of advocating a cause rather than seeking truth.

Wegener laid great stress on the fitting together of the continents which, he tells us, was the original source of his ideas. The crucial illustration, reproduced here as Figure 2, is a mere sketch in which the continents are substantially distorted. For example Alaska and Asia are left in contact at the Bering Straits, which is impossible if the Atlantic is closed and the continents left intact. India is elongated to about three times its present length to fill the Indian Ocean, and it is supposed that most of it was crushed to form the Himalayas and the mountains of central Asia. North and South America are joined by a large continental area, whose origin is not clear. The whole argument was much weakened by Wegener's belief that the southern limit of the moraines of the Quaternary glaciation were continuous in his

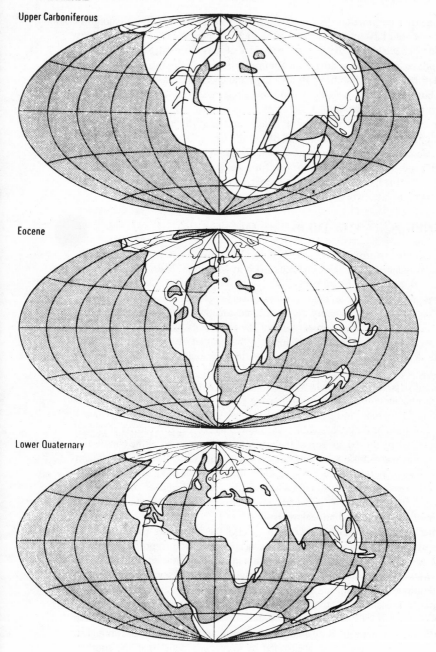

Figure 2   Wegener's reassembly of the continents, redrawn by Hallam (1973).

reconstruction and indicated that the separation of Greenland from Europe had occurred 50,000 to 100,000 years ago. This implied a motion of Greenland away from Europe at the almost incredible speed of 30 m/yr. The longitude measurements that were supposed to support this conclusion were never convincing, as they involved a comparison of results from the old method of lunars with modern observations.

The argument from the distribution of the Permo-Carboniferous ice age in the southern continents was a strong one, but in an age when travel was by ship, most geologists resident in the northern hemisphere had not seen the evidence and were fairly easily able to explain it away, doubt it, or ignore it. Paleontological arguments are usually not, in themselves, convincing; for example, if the Mesozoic reptiles of Europe and North America are similar but the Tertiary mammals are different, this is compatible with separation of the continents in the Cretaceous, but it is also consistent with the former existence of a land bridge which foundered into the Atlantic. The evidence does not point uniquely to relative horizontal movement of the continents.

It is easy to see why there was such strong opposition to Wegener in the 1920s and 1930s. If weak or fallacious arguments are mixed with strong ones, it is natural for opponents to refute the former and to believe that the whole position has been refuted. There is always a strong inclination for a body of professionals to oppose an unorthodox view. Such a group has a considerable investment in orthodoxy: they have learned to interpret a large body of data in terms of the old view, and they have prepared lectures and perhaps written books with the old background. To think the whole subject through again when one is no longer young is not easy and involves admitting a partially misspent youth. Further, if one endeavors to change one's views in midcareer, one may be wrong and be shown to have adopted a specious novelty and tried to overthrow a well-founded view that one has oneself helped to build up. Clearly it is more prudent to keep quiet, to be a moderate defender of orthodoxy, or to maintain that all is doubtful, sit on the fence, and wait in statesmanlike ambiguity for more data (my own line till 1959).

More sanguine defenders of orthodoxy may be driven by similar motives to quite violent and logically indefensible attacks on the innovators. Examples of this class of argument are Chamberlain's (1928) approving quotation from an unnamed colleague: "If we are to believe Wegener's hypothesis we must forget everything which has been learned in the last 70 years and start all over again". Termier (1925) also commented upon it as "...a beautiful dream, the dream of a great poet. One tries to embrace it, and finds that he has in his arms a little vapour or smoke...." [Termier was, of course, a Frenchman; Du Toit (1937) says, a little inappropriately, that this remark "reveals the gloomy spirit of its author".] Another geologist quoted, but not named, by Gevers (1950) said that in Wegener's views he saw "only a drunken sialic upper crust hopelessly floundering on the sober sima". Such irrational reactions are not unique to earth scientists. There is a celebrated story, whose origin I cannot at the moment trace, about a pure mathematician who was so incensed by the attempts, popular 60 years ago, to refine the definition of a function that he referred to "certain young men who have introduced into analysis new and extraordinary functions whose sole purpose is to make a mock of definitions that have been found adequate by generations of mathematicians" (I cannot quote the exact words).

The main arguments for and against Wegener's position were conveniently brought together in the published account of a symposium held in New York in 1926 at a meeting of the American Association of Petroleum Geologists (van Watershoot van der Gracht 1928). The most systematic attack came from Charles Schuchert. He attempted to show that Wegener's fit of the continents around the Atlantic is illusory. He does this by moving plasticine models of the Americas over the surface of the globe and gets gaps of 1200 miles. The illustrations are so bad that it is difficult to trace the reason for this extraordinary and quite false result. It is probably mainly due to leaving Africa fixed relative to Europe. It is only fair to say that Wegener had also done this and that his diagram suffered some distortion to get as good a fit as he did. It is interesting that Schuchert so desired to refute Wegener that he did not see what an excellent fit could be obtained by separating Africa from Europe by widening Tethys at its eastern end. In 1926 this would have been an important discovery and might greatly have affected the development of ideas. As it was he quoted a friend who said that the fit of Africa and South America was "made by Satan" to vex geologists. He even reproduced a diagram from Behm (1923) showing a widened Tethys and a good fit (I have not seen Behm's book). He remarks that Behm's map is distorted, which it is, but misses the hint of how to close the Atlantic. The whole story of the fits is an illustration of the sloppy way in which new ideas can be treated by very able men when their only object is to refute them.

The main features of Wegener's work that contemporaries found unacceptable were: (a) the late date ascribed to the separation of Greenland from Europe, (b) Washington's (1923) statement that the rocks on the two sides of the Atlantic do not match, (c) the improbability that Pangea had survived intact till the Cretaceous and then broken up, (d) the absence of a plausible mechanism driving the movements, and (e) the impossibility of the strong continents being buckled into mountains by pushing through the weaker ocean floor.

The absence of a mechanism may have been unfortunate, but it was not strictly an argument against drift. We believe many things of which we do not know the cause; for example, no one doubts that there have been ice ages; it was known that cholera was caused by drinking dirty water many years before the role of bacteria was understood; and even today the precise mechanism producing the undoubted correlation between smoking and lung cancer is not clear.

The arguments about mechanism were rendered somewhat inconclusive by the almost complete ignorance of mechanics of several of the contestants. Baily Willis (1944), for example, wrote:

I confess that my reason refuses to consider "continental drift" possible....when conclusive negative evidence regarding any hypothesis is available, that hypothesis should, in my judgement, be placed in the discard, since further discussion of it merely incumbers the literature and befogs the minds of fellow students.... Now, it is a well established principle of mechanics that any floating object moving through air, water, or a viscous medium creates behind it a suction of the same order as the pressure developed in front of it. This law applies equally to airplanes, ships, rafts, and drifting continents (if there are any). The pressure which could raise the Andes must, therefore, have been approximately equaled by the suction and tension in the rear. Sections of the continent must have been sucked off.

In 1929, J. W. Gregory gave a presidential address to the Geological Society of London. In this he supported an elaborate system of sunken continents beneath the oceans which bobbed up when needed. To believe this he felt he had to dispose of isostasy. He wrote: "If the ocean surface does not conform to a regular ellipsoid or spheroid, if it sags down in mid-ocean owing to the lateral attraction of the water toward the land, or sinks with variations in the specific gravity of the water, ... then the slight differences in the attraction of the ocean floor may be due to the depth being overestimated and not to the higher density of the floor". He ends: "If isostasy be so stated that it is inconsistent with the subsidence of the ocean-floors, so much the worse for that kind of isostasy". (He adds a footnote to say that Harold Jeffreys "fully agrees" with the last sentiment.) Gregory was not a crank, he was among the most knowledgeable and influential geologists of his day. He was the first geologist to see the Kenya rift valley and his two books about it greatly interested me as a young man. It is odd that he could not restrain himself from writing nonsense about things of which he knew nothing. His presidential address is well worth reading both for the discussion of the need for transatlantic connections and for the world view of a distinguished geologist in the 1920s. Perhaps the most remarkable feature is the absence of any appeal for more information about the oceans. Gregory seems very satisfied by what he has; maybe it was convenient to keep the ocean as *terra incognita* about which anything could be assumed.

So much for Wegener's opponents of the 1920s. His most effective supporter was A. L. Du Toit (1878–1948) who was a distinguished South African geologist. In 1923 he spent some months in South America with the express purpose of comparing the rocks with those of South Africa (Du Toit 1927). He was deeply impressed by the similarity of the stratigraphy. Of South America he wrote, "To a visitor from South Africa the resemblances to that country are simply astounding.... I had great difficulty in realising that this was another continent and not some portion of one of the southern districts of the Cape." In fact, the evidence for former connections between the continents is much easier to see in the southern hemisphere than it is in the northern. This is largely due to the general tendency of the Caledonian orogeny in the northern hemisphere to run parallel to the Atlantic margins while in the southern continents the split has crossed tectonic lines more nearly at right angles. The Permo-Carboniferous glaciations of the southern hemisphere also provided strong arguments. Du Toit's unrivaled knowledge of South African geology [R. A. Daly is said by Gevers (1950) to have called him the "world's greatest field geologist"] enabled him to write a book (Du Toit 1937) which had a great influence on subsequent opinion. The biographical notice by Gevers (1950) gives a most interesting account of the development of Du Toit's views and their relation to other aspects of his work. It is strange that Gevers felt it necessary, even in a laudatory obituary notice, to half dissociate himself from heresy. He summed up Du Toit's work in these words: "... notwithstanding the zealous and valiant efforts of du Toit and others, there has in recent years been a marked regression of opinion away from continental drift. . . . The greater the indignation to which the orthodox minds are roused by revolutionary heresies, the greater the amount of un-sympathetic attention. This again stimulates the zeal and ardour of the heretics and their disciples. . . . Of late, however, the obstacles to smooth continental drifting are being more strongly felt in many quarters previously sympathetic."

The supporter of Wegener who came nearest to modern views was Arthur Holmes (1929, 1944). His ideas are most conveniently discussed later, in the context of plate tectonics. Daly (1926) flirted with the idea of large movements but does not mention the matter in a later book (1942).

THE RE-EMERGENCE OF CONTINENTAL MOVEMENT: 1945–50

On the whole Wegener's opponents had the best of the prewar arguments. During the 1930s and 1940s it was unusual and a little reprehensible to believe in continental drift. It is easy now to see that what was needed was not further disputes about the old arguments, which had been demonstrated not to carry conviction, but new evidence. The new evidence, when it came, was of two kinds. Paleomagnetism provided virtually direct proof of movement, and the study of the ocean floor provided an entirely new insight into the relations of continents to oceans and into the geometry of the movement.

It is here that my personal experience begins. I migrated from physics to geophysics in 1931, shortly after Wegener's death, and spent some years learning the techniques of applying physics to the earth and trying to understand the modes of thought of geology. My initial idea was that geophysics should be used to solve specific geological problems, the paradigm being the applications to prospecting for oil. Gradually I realized that, important as such applications were, I was more interested in major problems of earth structure and history. I gradually came to see how weak the underpinning of geological hypotheses was, and I began to look around for experimental projects that were really significant and yet within the capabilities of our small group. We were fortunate in some of our choices, such as gravity in East Africa and Cyprus, and heat flow in England, Persia, and South Africa, but our real piece of good fortune was to turn to the study of the geology of the ocean floor.

MARINE GEOLOGY: 1936–1960

The way in which we came to study the ocean floor is worth recounting. The initial marine project was for B. C. Browne to measure gravity over the continental margin to the west of the British Isles, using the pendulum apparatus of Vening Meinesz and a submarine of the British navy. At about the time this was being planned, in 1936 I think, I met Richard Field of the Geology Department at Princeton. Field was a remarkable man; he was in a large degree the founder of marine geology. He explained to me that what was wrong with geology was that it studied only the dry land and that you could not expect to have sensible views about the earth if you studied only one third of its surface. The critical problem was to study the ocean floor starting from the land and working outwards into the ocean. The idea was simple and must have occurred to many people, but in Field it was combined with the burning zeal of an Old Testament prophet. He would not take no for an answer, he would not stop talking, he had no doubts, he was embarrassing and sometimes a nuisance, and yet he struck the match that set earth science alight. For some accounts of his life see Hess (1962a) and Bullard (1962); for his own views see Field (1938). He invited me to the United States in 1937 (and, I suspect, paid my fare), he drove me hither and thither in his car, tried to teach me geology, and sent me out to sea with the Coast and Geodetic Survey and with Maurice Ewing. After a few days he was preaching to the converted, but he did produce a sense of urgency that was new to me in science. By similar methods he recruited Maurice Ewing and Harry Hess (the latter carried his marine enthusiasm to the point of being simultaneously a professor of geology and an admiral).

Marine geology has many threads going far back into the past (Deacon 1971), but its development in the 1950s was largely the work of the groups at Columbia, the Scripps Institution, Woods Hole, Princeton, and Cambridge. Of course there were also many other groups such as Hans Petersson's Albatross expedition and the work of Gaskell and Swallow in H.M.S. *Challenger*. It is remarkable how, after the vigorous start of marine geology in the voyage of the original *Challenger* (1873–1876), the subject had been allowed almost to collapse and had to await the arrival of a new generation with new ideas about the amounts of money it was appropriate to spend on research [on this, but in a more general context, see Bullard (1975) and other papers in the same symposium].

Before the war the continental shelf was the main area of interest. Enormous and unexpected thicknesses of sediment were discovered. These have proved of immense economic importance and it is of interest that they were first found, not by an oil company, but by groups of young men interested in the earth. The paper by Ewing, Crary & Rutherford (1937) describes the beginning of a worldwide investigation which combines in the happiest way great scientific interest and a start towards the discovery of enormous resources. My own first publications described work with T. F. Gaskell in the western approaches to the English Channel (Bullard & Gaskell 1938, 1941).

In the years immediately before the war both Ewing's group and that at Cambridge turned their thoughts to using the seismic method to measure the thickness of sediments in the deep ocean. This was not as easy as it now appears; work on the continental shelf involved putting the explosives and the instruments on the sea floor which was, at that time, a new and complicated procedure. Just before the war it occurred to Gaskell and me (Bullard & Gaskell 1941) that the instruments could be in the water near the surface of the sea. Some trials were made in the late summer of 1939. After the war this, with the explosions also near the surface, became the usual method of operation.

After the war a great effort was put into work in the deep sea, the work of Ewing at Columbia and of Hill at Cambridge being particularly significant. At first the only equipment available in surface ships was the echo sounder, the corer, the dredge, and the seismic gear. The Meinesz pendulum equipment could also be used, but only in a submarine. Soon a great discovery was made, or rather gradually became apparent; the oceans are quite different from the continents. At sea all the hard rocks are basalts; all the hills are volcanoes; the sediments are usually less than 1 km thick and often much thinner; the Moho is at a depth of about 10 km compared to 30 km under the continents; and, most important of all, the rocks recovered were all Cretaceous or younger. The fact of the youth of the ocean floor took some time to sink in. It is not unexpected that the sea floor should be covered with young sediments, and it was only gradually realized that we had not only recent sediments, but also many samples of all ages back to the middle Cretaceous and, before that, nothing. When this was realized great efforts were made to find older rocks by dredging on fault scarps, but none were found; it seemed clear that the history of the oceans went back to 100 m.y., but no further. The contrast with the continents and their great areas of 2000 to 3000 m.y. old rocks was a critical discovery and showed how prescient Field had been in believing that the study of the oceans was what was missing from geology.

The oceans were formed recently, but the results of the gravity measurements and the shallow depth of the Moho made it certain that they were not sunken continents covered with lavas and recent sediments. A whole range of speculation on paleogeography was excluded. If the paleontologists needed land bridges between now-separated continents they must fit them into gaps between surveys which became narrower and narrower. I suppose the only candidates left today are Lomonosov Ridge in the Arctic Ocean and Broken Ridge sticking out from Australia into the Indian Ocean and ending in the deep sea.

It is not practicable in a brief account to trace all the ideas of the 1950s to their sources in the literature and, instead, I shall use a series of review articles that I wrote. These give convenient summaries and, while they are inevitably colored by my own interests and beliefs, they do, I think, give a fair picture of the development of ideas. The first of these reviews was written just before the war (Bullard 1939): it opens with a sceptical statement about our ability to decide, using information then available, between theories of the history of the oceans; it mentions permanent oceans, continental foundering, and continental drift as possibilities. The rarity of deep water sediments on land suggests that oceans are not converted to continents. The key is "to study the form and nature of the ocean floor and, if possible, its structure....The first requirement...is to develop instruments and methods for collecting information about the submerged rocks." A project is suggested for a detailed survey of part of the mid-Atlantic Ridge using moored buoys, to determine "whether the topography consists of submarine volcanoes, of undenuded fold mountains, or of fault scarps". The rapid rate of accumulation shown by cores of deep ocean sediment, obtained by Piggot, is said to suggest impossibly great thicknesses of sediment if the oceans are old. Hopes are expressed that the seismic method may show how much sediment is present in the deep sea and the work on the continental shelf is described. Surface waves and the nature of the rocks from islands and from dredging are said to suggest the absence of granite in the oceans. The gravity results of Vening Meinesz and the locations of earthquakes in the East Indies are discussed and it is suggested, following Visser (1936), that there is an inclined thrust plane along which the continent and the island arc are thrust over the ocean. The review concludes that "the difficulties lie in organisation and finance rather than in technique".

This review is interesting in that it emphasizes a major problem. How much of the ocean floor is truly oceanic and how much is continental fragments? In particular, what is the nature of the mid-ocean ridge? Daly (1942) and Holmes (1929) both thought that the ridge was a sialic, continental feature. If this were so, large pieces must have broken from the continents in the course of separation and no fit is to be expected. Studies of the propagation of surface waves, showing that the Atlantic had a structure intermediate between that of a continent and the Pacific Ocean, seemed to confirm this view. Detailed studies at sea gradually dispelled this idea and left only minor continental fragments in the oceans, such as the Seychelles and Rockall Bank. The difficulties with the surface waves were resolved by Ewing & Press (1956).

Reviews written during the war (Bullard 1940a,b) add nothing to that of 1939, except the statement that the continental shelf had been built outwards; this also occurs in Bullard & Gaskell (1941). We should have realized that this view was incompatible with the results of dredging which showed Mesozoic rocks near the outer ends of the submarine canyons off the east coast of the United States.

In 1954 the Royal Society held a discussion about the floor of the Atlantic Ocean. The time was well chosen; the first deep-sea seismic results were available giving thin sediments and a shallow Moho. There were also a few measurements of heat flow.

G. M. Lees, in the best 1920s manner, contrasted "geophysical conceptions" with "geological evidence", and maintained that foundered continents were buried beneath the lavas and sediments of the ocean floor. I replied (Bullard 1954) with what now seems remarkable restraint. My main theme was the importance of establishing beyond doubt the continuity of the continental and oceanic Mohos, which Lees had denied and which would exclude his hypothesis of buried continents beneath the oceans. I also stressed the desirability of showing whether the Hercynian structures of western Europe and Newfoundland ran out beyond the continental edge. I must have had in mind the idea that they might have been truncated by the separation of North America from Europe, but I did not mention it; I was still sitting on the fence. I did not go any further in a review published two years later (Bullard 1956).

A most interesting paper was published by Hess (1954) in the Proceedings of the Royal Society Symposium. In it he discusses the nature of the mid-Atlantic Ridge and considers various possibilities. One of them is the existence of a rising convection current in the mantle beneath the ridge: a large rising mass of basalt breaks through to the ocean floor and carries blocks of peridotite with it. The ridge stands high because of the high temperature of the rising basalt, because of the serpentization of the basalt and because of "lifting" by the upward convection current. It is easy to see here the germ of the idea of sea-floor spreading which is first clearly stated nine years later (see below). There is, however, no mention of plates or of any general outward motion of the crust; the argument is directed exclusively to phenomena near the ridge crest. Rather oddly, there is no mention of the very similar speculations of Holmes (1929, 1944) which are discussed in a later section of this review.

The 1954 symposium also contains a remarkable paper by Rothé (1954). He shows that the earthquakes of the deep Atlantic are concentrated along the crest of the mid-Atlantic Ridge and that they run around South Africa into the Indian Ocean, which suggests that the ridge is connected to the ridges of the Indian Ocean. His diagram is reproduced in Figure 3. Shortly after this, Ewing & Heezen (1956) stressed the worldwide continuity of the ridge and of its remarkable central valley along which the earthquakes lie. These discoveries were the first clue from which plate tectonics developed, but before describing this we turn back to the vindication of Wegener by paleomagnetism.

Paleomagnetism

The study of the magnetization of rocks has led to the discovery of a surprising variety of phenomena which have been of critical importance for the understanding of the relative motions of portions of the earth's surface. It is impossible to describe the history of these discoveries in detail here. Excellent accounts have been given by Cox & Doell (1960) and Irving (1964).

Knowledge of rock magnetism goes back to classical times [the references have been collected by Gilbert (1600)], but nothing that is relevant to our purpose was done till the work of Bruhnes (1906), who discovered reversely magnetized rocks. He and his successors [see Koenigsberger (1938) for references] showed that: (a) modern lavas are magnetized in the direction of the present field, (b) Tertiary lavas are magnetized roughly in the direction of the present field or in the reverse direction, and (c) older lavas are often magnetized in directions making large angles with the present field. The results were quite complicated and some hindsight is involved in putting the conclusions so simply. It is remarkable that such fascinating and obviously significant phenomena should have been so widely ignored for so many years. I first heard of them in 1939 from Basil Schonland who was interested in the work of Gelletich (1937) on the Pilansburg dykes in South Africa. We discussed the results a good deal, largely in an effort to explain them without assuming reversals of field. We did not, so far as I remember, read the earlier literature or propose any program of work. We were both occupied with other things, I with heat flow, he with thunderstorms.

Towards the end of the war Patrick Blackett and I spent much time discussing the shape of science after the war (Bullard 1975). One of the topics we discussed was the origin of the earth's magnetic field. Blackett decided to test the hypothesis that a rotating body is spontaneously magnetized. For this purpose he developed a very sensitive magnetometer. When the theory collapsed he turned his attention to the use of this instrument for measuring the magnetization of rocks (Blackett 1956). In this he was joined by a number of graduate students, among whom was Keith Runcorn. They embarked on a large program of measurement and soon had very striking results confirming and greatly extending the prewar French and German work.

The results raised many questions which had to be answered before a convincing interpretation could be made:

(a) Were the rocks really behaving as fossil compasses and indicating the direction

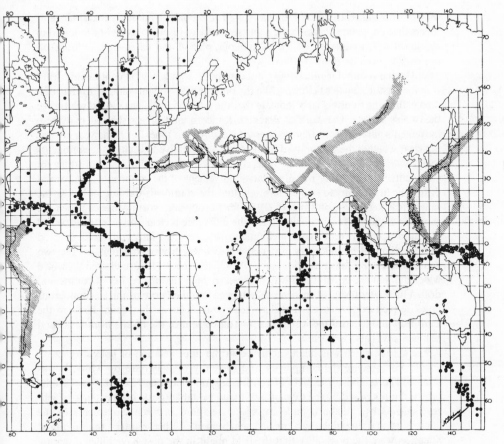

Figure 3   Rothé's map of the earthquake epicenters of the Atlantic, showing their alignment along the axis of the mid-ocean ridges (from Rothé 1954).

of the field in the past? The fold and conglomerate tests of Graham (1949) showed that a large class of rocks, particularly basalts and red sandstones, were satisfactory.

(*b*) Does reversed magnetization [of which striking examples had been found by Hospers (1951, 1953, 1954) in Iceland] always indicate a reversed field at the time of magnetization? This was for a long time a controversial issue which was greatly complicated by a lava, found by Nagata (1951, 1953) on the Haruna volcano in Japan, which magnetized itself backwards on cooling through its Curie point. It gradually became clear that such rocks are extremely rare. The universal normal magnetization of recent rocks (except for Nagata's rock) also suggests the rarity of self-reversal. The matter was settled 12 years later by the demonstration that the reversals were simultaneous in widely separated places (Cox, Doell & Dalrymple 1963a,b, McDougall & Tarling 1963) and that the sediments of the ocean floor also showed reversals (Harrison & Funnell 1964). The proof by Ninkovich et al (1966) and others that the sediments show the same sequence of reversals as had been found in the lavas left no further doubt. A bibliography of the extensive literature on reversals is given by Bullard (1968).

(*c*) Do the results for older rocks indicate polar wandering or continental drift? If in the past the earth's field was, like the present field, roughly similar to that of a dipole, then the results clearly indicate that the magnetic pole has moved relative to the earth's surface. The work of Blackett, Runcorn, and others showed that these movements gave motions that progressed steadily with time. For some years it was not clear whether the pole moved in the same way relative to each of the continents (polar wandering) or whether different tracks were obtained from different continents, which would indicate relative movement of the continents (continental drift). The first results to suggest strongly that the continents did not all follow the same track were those of Runcorn (1956) for Europe and North America which are shown in Figure 4. In a statistical sense the difference is clear and is in the expected direction but at the time many, including myself, were not fully convinced. I feared that there might be unknown systematic errors which were different for the two continents. For me full conviction came when Blackett et al (1960) collected the world wide data and presented it in a way that clearly indicated the reality of continental drift. A little later Irving (Runcorn 1962) showed that the Permian and Carboniferous poles derived from Australian rocks were about as far as they could be from the European and North American poles. The agreement of the results from lavas and sediments was also important.

The clarity which was finally achieved in the interpretation of paleomagnetism should not obscure the complexity and difficulty of the route by which it was attained. To establish the facts of continental movement and field reversal in the face of doubts raised by the existence of many unstable rocks, the existence of self-reversing rocks and the complexity of the relations between movements of the continents and the pole is a major achievement. Other matters that required attention were the possibility that the field might in the past have differed greatly from that of a dipole (Cox & Doell 1961), that the magnetic poles might have departed greatly from the poles of rotation, and that the earth's radius might have changed. All this necessarily took some years to elucidate; the surprising thing is that by 1960 the case was substantially complete. Runcorn (1962) gives a good account of the state of knowledge at that time.

In a review lecture given at the first International Oceanographic Congress in 1959 (Bullard 1961) I said, for the first time, that I thought that the paleomagnetic evidence gave "a strong case for relative movements of the continents." I discussed what should be done to find confirmation in the oceans and suggested again the study of structural lines in western Europe and Newfoundland to determine if they are truncated by the continental edges. I also returned to the question of the arrangement of the sediments on the shelf, about which I had been in error in 1939, and suggested that a sedimentary basin may have been cut in half by the separation of Europe and North America. I said that "the mid-Atlantic ridge and particularly its central valley might be regarded as the place where the Atlantic is at present

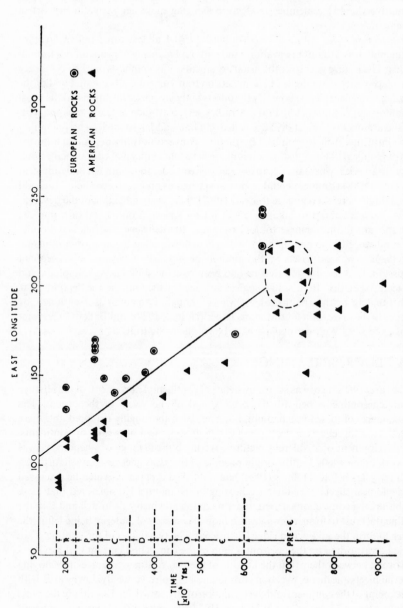

Figure 4 Variation of the longitude of paleomagnetic poles for Europe and North America. The longitude can be measured because the pole was then near the west coast of the Pacific [from Runcorn (1962), by permission from Academic Press].

widening." The published account still shows some concessions to the possibility of alternative views; I remember the lecture as being more aggressive in support of movement.

In a lecture given in June 1963 (Bullard 1964) I at last came out in favor of continental drift without reservations and without balancing of probabilities for and against. The lecture goes over the usual arguments, with emphasis on the occurrence of displacements of hundreds of kilometers on transcurrent faults on land and of the even larger displacements shown by the offsets of the magnetic lineations off the coast of California (Vacquier et al 1961). A rather severe attitude is taken to opponents and the arguments against drift are called "ad hoc and far fetched". The possibility that continental drift is produced by thermal convection is discussed much on the lines of Holmes (1944). A distinction is made between continental edges, such as that of South America, where the ocean floor is carried under the continent and quiescent edges where "the continent could ... be transported as on a conveyor belt". It is odd that, though there is a reference to Dietz (1962), there is no indication of the developments that were about to take place which transformed "continental drift into sea floor spreading" and "plate tectonics". In spite of this deficiency the talk had, I think, some influence in moving opinion in England towards acceptance of continental movement. I was surprised by the amount of support shown in the subsequent discussion. E. B. Bailey said that he had been convinced by G. W. Lamplugh and had lectured on drift (to a nongeological audience at the Old Vic theater!) in 1910. Unfortunately neither published his views, though Lamplugh is mentioned by Holmes in the discussion following his paper (Holmes 1929), and Bailey (1929) went so far as to say, "Wegener may perhaps be telling us the truth".

SEA-FLOOR SPREADING

By the early 1960s there was strong evidence of continental movement, of the absence of sunken continents beneath the oceans, and of the youth of the oceans. The consequences of these ideas had not, however, been thoroughly explored and there existed no comprehensive description of the development of oceans or continents. The development of a coherent picture covering a great range of phenomena was the work of the 1960s, but the origin of some of the ideas goes back much further.

Thirty years before, Holmes (1929) had published a paper in which he discussed the geological effects of radioactive heating in the mantle. He suggested that there will be a rising convection current under a continent which will stretch and thin the continental crust to form a new ocean, a blob of continental material being left in the center to form the mid-ocean ridge (Figure 5a). The two halves of the continent are carried away, riding on the horizontal limb of the convection current and leaving a new ocean between them. On the far side of the two continental halves the moving material meets another convection current and both turn vertically downwards with a thickening of the continent and the formation of an ocean deep. Essentially the same account is given 15 years later in Holmes (1944) except that the ridge is now basalt and not a blob of sial (Figure 5b). Idealized forms for the motion were considered and the forces calculated by Pekeris (1935) and Hales (1936).

A number of writers adopted and modified the views of Holmes: among these were

Griggs (1939), Meinesz (1948, 1952), and Heezen (1960) (Figure 5c). The main attraction of such ideas was, and is, that they provide an adequate force to split the ocean floor and a mechanism for carrying the split continental pieces away from each other. They also account very naturally for the striking fact, established by Menard (1958), that the central valleys of most mid-ocean ridges are rather accurately halfway between the two continental edges. Some doubt was still felt about the possibility of the downturning of the currents and both Carey (1958) and Heezen (1960) avoided the difficulty by supposing that the earth is expanding to accommodate the extra width of the oceans.

The developing ideas on convection currents were brought together in a masterly paper by Hess (1962b) and in brief notes by Dietz (1961, 1962). The main difference

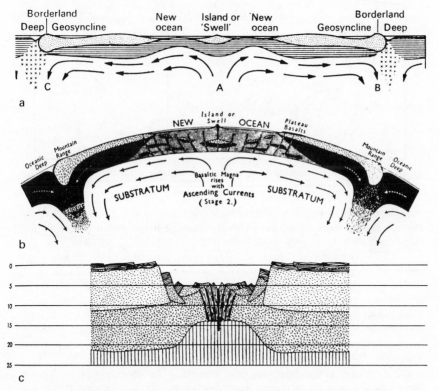

Figure 5 The development of ideas about sea-floor spreading. (*a*) Holmes (1929): note the blob of sial forming the mid-ocean ridge and the vertically descending currents at the leading edges of the continents. (*b*) Holmes (1944): as in 1929 but with a basaltic ridge (by permission from Thomas Nelson & Sons Ltd.) (*c*) Heezen (1960): an early stage in the formation of an ocean on an expanding earth with fracture and intrusion on the ridge axis but no convection currents. (Heezen also gives a diagram with convection currents but with no movement of the continents.) (By permission from *Scientific American*) (*d*) See next page.

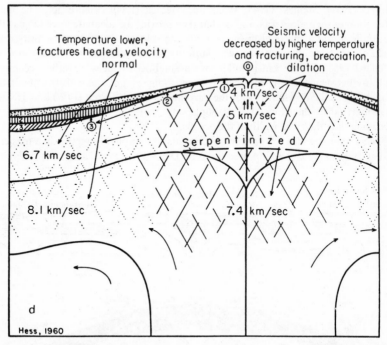

Temperature lower,
fractures healed, velocity
normal

Seismic velocity
decreased by higher temperature
and fracturing, brecciation,
dilation

① 4 km/sec
② 5 km/sec
③
S e r p e n t i n i z e d
6.7 km/sec
8.1 km/sec   7.4 km/sec
d
Hess, 1960

Figure 5d   Hess (1962b) showing a convection current rising to the ocean floor under the axis of the ridge and splitting the crust.

from the ideas of Holmes is that the oceanic crust is supposed to be broken to form new sea floor (Figure 5d) whereas Holmes shows it being stretched and thinned (Meyerhoff 1968, Dietz 1968, Hess 1968). The effect of these papers was great; the phrase "sea-floor spreading" (devised by Dietz) put the emphasis on the site of the creation of new crust and took it away from the motions of the continents. Both Holmes and Hess were professional petrologists (as well as many other things) and both of them provided petrological mechanisms to assist motion, in addition to the thermal buoyancy forces. For Holmes it is the basalt-eclogite transformation in the downgoing limb of the current; for Hess it is serpentization of periodotite in the up-going limb by water from the mantle. The importance of these processes is still in doubt.

The next stage in the development of ideas was the interpretation by Vine & Matthews (1963) of the magnetic lineations on the sea floor. Mason (1958) and Mason & Raff (1961) had described magnetic lineations on the sea floor off California. These lineations implied that the sea floor was magnetized in parallel stripes. The stripes were found to have no relation to the topography of the sea floor or to the form of the buried basement. They were crossed by faults showing displacements of over 1000 km (Vacquier et al 1961). These very striking phenomena defied explanation for some years; for example, Bullard & Mason (1963), in a paper written in the summer of 1960, had no useful interpretation. A little later Vine & Matthews (1963) suggested that Hess' idea that a strip of new ocean floor was continually being formed on the axis of the mid-ocean ridge would provide a double tape recording of the intensity and the reversals of the earth's magnetic field. Each magnetic stripe was magnetized when that piece of ocean floor was formed in the central valley on the ridge axis. Matthews and Vine came to this idea while analyzing magnetic surveys they had made in the Indian Ocean. The first clue was the discovery of reversely magnetized sea mounts on the ridge (Cann & Vine 1966). Suggestions closely parallel to those of Vine and Matthews were made in a paper written independently and at about the same time by L. W. Morley of J. T. Wilson's department in Toronto. By regrettable errors of editorial judgment both *Nature* and *Science* rejected this paper. Some extracts from it and a commentary were published much later by Lear (1967).

In 1965 Hess, McKenzie, Matthews, Vine, and Wilson were all in Cambridge and rapid progress was made. Wilson devised "transform faults" to explain the offsets of the ridge axis and the magnetic pattern. It had generally been supposed that the offset sections of ridge axis were moving apart and that the faults were analogous to the "transcurrent faults" seen on the continents. Wilson (1965) suggested that the sections of ridge axis were not moving apart and that the motions of the faults were those of two plates moving away from each other and away from the ridge axis (Figure 6); with this arrangement only the section of the fault joining the two ridge axes is active. On the older view it is unclear how the fault can end; the similar difficulty with transcurrent faults on land, such as the Great Glen Fault in Scotland or the San Andreas fault in California, had usually been met by saying that they "ran out to sea," i.e. passed beyond the ken of geologists.

In March 1964, while these discoveries were being made, the Royal Society held a discussion on continental drift, the proceedings of which were later published as a book (Blackett, Bullard & Runcorn 1965). A number of people from the USA attended and there were some complaints that the meeting had been packed with believers and that the opponents had not been given a proportionate representation. This was not a subtle plot; the fact was that almost everyone working in the field in England had become convinced a year or two before a comparable near unanimity was reached in the USA.

In 1965 Vine moved to Hess' department in Princeton where he produced two papers (Vine & Wilson 1965, Vine 1966) which left no doubt of the correctness of the explanation of the magnetic stripes. Soon after this, Sykes (1967) showed that the earthquakes on the ridge axis and on the transform faults had the expected types of

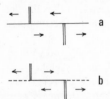

Figure 6   Faults on the axis of a mid-ocean ridge (adapted from Wilson 1965): (*a*) a transcurrent fault; (*b*) a transform fault (the dashed section is inactive).

focal mechanisms. These successes convinced the major groups working on marine geology and great effort was expended in analyzing the considerable stores of unpublished magnetic data and in obtaining new data. The result was several maps showing the age of the basement beneath the sediments for a large part of the oceans and an extension of the history of reversals back into the Jurassic. In this very substantial undertaking, the work at the Lamont Geological Observatory, especially that of Heirtzler et al (1968) and Le Pichon (1968), was particularly significant.

The turning point for opinion in America was marked by a conference in New York in November 1966, sponsored by NASA (Phinney 1968). At this the worldwide evidence from the magnetic lineations and the earthquake epicenters on the ocean ridges was presented. The effect was striking. As we assembled on the first day, Maurice Ewing came up to me and said, I thought with some anxiety, "You don't believe all this rubbish, do you, Teddy?" At the end of the meeting I was to sum up in favor of continental movement and Gordon Macdonald against; on the last day Macdonald was unable to attend and no one else volunteered to take his place. I attempted to say what I thought Macdonald would have said, but it was unconvincing and I left it out of the published account. In my summary I pointed out how far we had gone from the traditional geological interest in the continents and their mountain systems and recommended a return to these problems. I also stressed the importance of deep-sea drilling in the verification and extension of ideas about the sea floor.

PLATE TECTONICS

The transition from sea-floor spreading to plate tectonics is largely a change of emphasis. Sea-floor spreading is a view about the method of production of new ocean floor on the ridge axis. The magnetic lineations give the history of this production back into the late Mesozoic and illuminate the history of the now aseismic parts of the ocean floor. This naturally directed attention to the relation of the sea floor to the continents. There are two approaches: in the first, one looks back in time to earlier arrangements of the continents; in the second, one considers the current problem of the disposal of the rapidly growing sea floor.

At the same time as the ideas of sea-floor spreading were developing, there was a revival of interest in what had been Wegener's strongest argument in favor of drift: the geometric, structural, stratigraphic, and paleoclimatological fits which suggested the former existence of the giant continent of Pangea. The early work on the geometric fits was, as has already been said, sloppy, amateurish, and, in some degree, dishonest. Fits were shown in sketches which implied unrecognized, or at any rate unstated, distortions. Among the reasons for this were the expectation that the fits would not be close and the difficulty of making fits objectively in the precomputer age.

The first man to make careful geometric fits was Carey (1955, 1958) who worked with transparent caps which could be moved over the surface of a globe. The resulting fits around the Atlantic were much better than anyone expected; the impact of his 1958 paper was less than it might have been because the maps were published only on a small scale in a duplicated conference report which contained much other material of a more or less controversial character connected with Carey's theoretical views on the expansion of the earth.

Influenced by Carey's results, and stung by Jeffreys' oft repeated disbelief in the closeness of the fits, I began to consider how one could best make fits that were, in some defined sense, best fits using the data from maps, which are much more reliable and detailed than any globe. In this I was joined by Everett and Smith. We realized that the movement of a continent could be defined by three parameters and that the problem was to choose a criterion of goodness of fit and use it to determine the parameters. By analogy with least squares fitting it is convenient to express the criterion as the finding of the values of the three parameters that minimize a symmetric function of the coordinates of points on the two continental edges after one of them has been displaced. A vague memory of undergraduate mechanics led me to Routh's *Rigid Dynamics* and to Euler's (1776) theorem that any motion of a sphere over itself can be regarded as a rotation about some pole of rotation. The problem was then to find the position of this pole and the amount of the rotation (many other choices of parameters are possible, for example the "Euler Angles," but our choice seems to have established itself). The results for the fit around the Atlantic were spectacular, the root mean square gaps and overlaps being only about 50 km. The paper describing this work (Bullard, Everett & Smith 1966) was given at the Royal Society symposium in 1964. It had an impact much greater than we had expected; the reason was, I think that we had provided professionally drawn diagrams on a large scale (perhaps fortunately, computer plotters were not available to us) and had confined ourselves rigidly to the geometric problem. Our intention was to produce the facts about the geometric fits of the continental edges and to leave all other questions, such as the corresponding structural and stratigraphic relations, to be considered by the local experts who could use our map as a base for their investigations. We thus avoided arguing many controversial questions and did not allow the opponents of drift to concentrate on these and to ignore the irrefutable geometry. We stressed certain arbitrary features of our reconstruction, such as the abolition of Iceland, the retention of Rockall Bank as a continental fragment, and the rotation of Spain. All these have since been justified by other evidence and greatly strengthened our position. The subsequent geochronological and structural studies have also been favorable to our fit (e.g. Hurley & Rand 1968). The present summary provides an opportunity for me to say that it seems that Everett and Smith have had less than their share of credit for this work. The fit around the Atlantic has gradually, with repeated quotation and reproduction, become "Bullard's fit" (an undetected editorial change once led me to join in this injustice); in fact, the work was almost all done by Everett and Smith and I only provided the methodology and wrote the paper. In one sense our deliberate concentration on the geometric fits was an undesirable limitation which inhibited us from thinking of wider questions. We should at any rate have fitted the ridge axis and earthquake epicenters to the continental edges and thus connected with the work of Hess. We intended to do this but did not do so since we wished to get the paper finished in time for the report of the Royal Society conference (Blackett et al 1965). The corresponding fit of the continents around the Indian Ocean cannot be done by geometry alone and is a much harder task; it has been performed by Smith & Hallam (1970) and by McKenzie & Sclater (1971), following an earlier and unsatisfactory attempt by Wilson (1963). The Atlantic now needs re-examination in the light of new bathymetric and magnetic data, particularly those from the Arctic Ocean.

The work on continental fits is concerned with the movement of large portions of the earth's surface which may be thought of as rigid, aseismic, spherical caps or plates carried hither and thither by convection currents. Some of these plates are so large that it is not easy to believe that they behave as rigid bodies. That they do so was demonstrated by McKenzie & Parker (1967) and by Morgan (1968) (who introduced many of the most convincing arguments and coined the phrase "plate tectonics"). Numerous checks are possible from the focal mechanisms of earthquakes and from the geometry of magnetic lineations and transform faults (Le Pichon 1968).

The recognition of large, rapidly growing, aseismic plates raises an acute problem in their disposal. There are only two possibilities: either the earth is expanding rapidly or sea floor is being destroyed. There are many objections to rapid expansion, though it still has its supporters, and plate destruction is the generally accepted process.

If plates are being destroyed, the likely sites are along the broad earthquake belts associated with island arcs. It has long been known that these earthquakes are distributed on planes dipping beneath the arcs at about 45° and reaching depths of up to 700 km. These deep earthquakes were discovered by Turner (1922), but decisive evidence was only available much later in the work of Stoneley (1931) and Scrase (1931). The inclined plane of earthquakes was at first interpreted as a thrust fault, which in a sense it is (Visser 1936, Benioff 1954), and only much later as a downgoing plate diving under the continent, heating, and becoming incorporated in the mantle. It is frequently referred to as the Benioff zone.

Surprisingly this inclined plane of earthquakes is not mentioned by Dietz, Hess, or Heezen in the early papers on sea-floor spreading or in the report of the NASA conference (Phinney 1968). Most of these papers have the downgoing limb of the convection current going vertically down. I do not know who was the first to show the solid plate bending over and moving obliquely down beneath the continent. There is an almost insensible gradation from the rival theories of thrust fault and tectogene of the 1930s to the descending plate. The idea is clear in Oliver & Isacks (1967) and in Elsasser (1967). The real existence of the cold, descending plate was demonstrated when it was shown that it transmitted seismic waves with higher velocity and less attenuation than the surrounding mantle. This was first suggested by Katsumata (1960) in a paper published in Japanese. Isacks & Molnar (1971) have collected all the available information.

After 1967 the subject of plate tectonics blossomed into a major scientific enterprise with an enormous bulk of publications. It would be difficult to pursue the details and perhaps not very useful; there are many textbooks on the current state of the subject (e.g. Le Pichon et al 1973). The main lines of work have been: (a) the application of the ideas to particular areas (e.g. the Red Sea, the Mediterranean, the Bay of

Biscay, the Galapagos Islands); (b) the explanation of the topography and heat flow of the ridge as a result of cooling of the outward moving plate; (c) the vindication of the ideas of sea-floor spreading by the results of the JOIDES drilling; and (d) the development of the theory of convection in the mantle (this is a subject of extreme difficulty and still has a long way to go).

THE OBJECTIONS

In an earlier section the main objections raised in the 1920s and 1930s to Wegener's views are listed. It is interesting to see how these have been met in the theory of plate tectonics: (a) The magnetic lineations and the results of deep-sea drilling make it clear that the separation of Greenland from Europe occurred in the Cretaceous not in the Quaternary. (b) Nothing has been heard for a long time of Washington's (1923) claim that the rocks on the two sides of the Atlantic do not match and there is much evidence that they do match (e.g. Hurley 1972, Miller 1965, Stoneley 1966). (c) It seems likely that from the Devonian to the Permian there was a large continental mass bearing some resemblance to Wegener's Pangea but all recent reconstructions show a wide ocean between the south coast of Asia and the Australian-Indian block, which was attached to South Africa (Smith & Hallam 1970). It is probable also that Eurasia was split into several plates; only the most westerly, which included everything west of the Urals, are attached to North America and Africa. The large continent was a temporary thing; it is clear that North America and Europe were separate in the lower Paleozoic and it seems likely that the continents have a long history of splitting, movement, and collision. There is nothing exceptional about the formation of the Atlantic in the Mesozoic; Africa is splitting from Arabia today and the Ur-Atlantic closed 450 m.y. ago. (d) Thermal convection seems to provide a plausible mechanism for movement; this is discussed below. (e) The continents do not push through the ocean floor, they are carried on the horizontal limb of a convection current. The ocean floor achieves its motion relative to the leading edge of the continent by diving beneath it, the solid oceanic plate being bent over and broken in the process. As would be expected, the formation of mountains is not a a simple process. Some, such as the Andes, derive a large part of their material from the melting plate which supplies andesitic lavas. The Himalayas are the result of a collision of two continental blocks. Most mountains also contain material derived from the floor of the deep ocean (ophiolites), perhaps in some places scraped from the descending plate.

    In recent years the ideas of plate tectonics have met with very general acceptance but there are still a number of respected figures who are convinced of the fixity of the continents. This note is concerned with the history of ideas and it would be inappropriate to attempt a detailed refutation of the opponents of the ideas presented. It is, however, desirable to show the nature and origin of the objections and how they fit into the scheme we have presented. The objectors may be divided into three classes: (a) some paleontologists who believe that what is known of the distribution of animals and plants in the past is incompatible with the proposed motions of the continents; (b) Sir Harold Jeffreys who believes the motions to be incompatible with the known rheological properties of matter; and (*c*) Dr. V. V. Belousov who is concerned that the proposed scheme does not provide for the great vertical motions that are observed in the sedimentary basins of the continents.

The main proponents of fixed continents among the paleontologists are D. I. Axelrod and A. A. Meyerhoff. I am clearly incompetent to discuss the paleontological evidence and will merely say that the absence of a consensus of paleontological opinion after 60 years of discussion does not encourage me to look there for decisive criteria. References to the extensive literature will be found in Axelrod (1963) and Meyerhoff & Meyerhoff (1972). It is desirable that someone or some group should produce a review of the paleontological and related paleoclimatological evidence and arguments adduced by the objectors. The papers are long and indigestible and contain a number of statements which seem implausible, such as the claim that there was glaciation at the equator in the Pleistocene.

Meyerhoff's attack on the central evidence for movement (paleomagnetism, the magnetic lineations, the continental fits, and the earthquake epicenters and focal mechanisms) seems to me unconvincing. Like Wegener, he argues too hard. For example, he says, correctly, that published maps show bad correlation of magnetic anomalies between the south end of the Rykjanes ridge and the Azores, but he does not pause to consider whether this is really so or whether the published observations are not dense enough to show the real continuity in the presence of numerous transform faults. In fact, observations published subsequently to Meyerhoff's paper show satisfactory continuity in this area (Vogt & Avery 1974). There is a methodological difference between Meyerhoff and his opponents which is of general interest. There are places where there are excellent, well-separated linear magnetic anomalies stretching for hundreds or thousands of kilometers and where correlations and symmetry are irrefutable. But there are other areas where complicated things have happened and where the anomalies are close together, unclear, and are broken up by transform faults. The believers in movement attempt to extend the ideas derived from the simple areas into the more complicated ones and thereby understand complex events that are unclear from the local evidence. The unbelievers seize on the doubts about the difficult areas and use the difficulties to shed doubt on the work elsewhere, which is for most people perfectly clear. The difference is not peculiar to magnetic survey. Most subjects have, and need, their lumpers and their splitters; the people who like to see a simple pattern against which to judge the complexity of the detail and the people for whom nature is too complex for any theory and the details are their own reward. It is, I fear, inevitable that the first group reach the great generalizations (see e.g. Watson 1968).

The objections of Jeffreys are of a different kind. He is the deviser of much of the theoretical framework of geophysics and has convinced himself that motion is impossible on mechanical grounds. He says that he is not an expert in magnetism but, since motion cannot occur, the observations must have been misinterpreted and that it is not his business to find out what has gone wrong. His objections are of long standing. In the first edition of his celebrated book, *The Earth* (Jeffreys 1924), he shows that the forces proposed by Wegener are inadequate by many orders of magnitude to produce significant motions. This is repeated in all subsequent editions and is, beyond question, correct. The first edition also argues that the widespread fields of gravity anomalies of one sign show that strengths of the order of 10^8 dyn/cm^2 must exist at depths between 100 and 400 km; this would prohibit motion.

The second edition of 1929 adds the statement that the African and South American coastlines will not fit; this is correct and is repeated in all subsequent editions. However, it is irrelevant as Wegener's diagram, and all the later ones I have seen, do not fit the coastlines, they fit the continental edges at the outer margin of the shelf (it must be confessed, though, that Wegener's diagram is in places far from clear). For the continental edge the fit is excellent.

The third edition of 1952 and the fourth of 1959 were the first that had to cope with the ideas of convection in the mantle put forward by Holmes. Jeffreys considers that if such currents exist they could break the crust but believes that the broken blocks would pile up and stop the surface movement before it had gone far. He still thinks that the material to a depth of 600 km is likely to have a strength of around 3×10^8 dyn/cm^2, though he has lost his main earlier reason for believing this. The widespread gravity anomalies which previously needed strength at depth to support them could now be due to the density differences in the convecting system. Vertical motions of the surface, not accompanied by gravity anomalies, could also be related to motions in the mantle. Of this and related matters Jeffreys says only, "The suggestion of systematic convection currents complicates the question", which indeed it does.

In the 1970 edition of *The Earth* the previous argument is repeated including the odd statement that the main evidence for drift is paleontological. In a note on the last page it is admitted that the continents around the Atlantic do fit and that this might be due to the break-up of a large continent formed early in the earth's history (this cannot be true as the broken rocks are of all ages up to lower Mesozoic). The main objection to motion is still the belief that materials within the earth have a finite strength. This belief is based largely on the experiments of Lomnitz (1956). These experiments were conducted at atmospheric pressure and room temperature on time scales of up to a week. There seems no reason to suppose that they have any relevance to the properties of materials within the earth and the resulting "law" is in fact contrary to experiments with ceramics at high temperatures and pressures where viscous creep is observed. Theory provides a number of mechanisms which are important at high temperatures and some of which give zero strength. The whole question has been discussed by McKenzie (1968) who gives references to both experiments and theory.

Our last objector is Beloussov (1968). He makes the important observation that plate tectonics does not account for the large vertical motions that have taken place in and around the continents; for example, in the North Sea. This is true and needs to be said; the pendulum has swung so far that the traditional geological interest in vertical motions, rising mountains, and deepening sedimentary basins has been largely replaced by discussion of very large horizontal motions. It is certainly time to return to the older problems, but to do so it does not seem necessary to throw away what we have learned about horizontal motion. Belousov (1962) calls continental motion a "completely sterile idea" which is only a reasonable opinion for one who discounts all the evidence for large horizontal motions. It is only fair to say that Belousov (1962) is a translation of a book written in 1956; I do not know if he would be so vehement today. Belousov believes that continental crust can be converted to ocean floor; the process involves great petrological difficulties. He also says that the equality of the continental and oceanic heat flows excludes the possibility of horizontal movement; the position of plate tectonics in this matter is certainly unsatisfactory in that the equality is not easily explained and appears as a kind of accident. The difficulty is the more surprising as the variation in heat flow across an ocean is in good agreement with the expected cooling as the plates move away from the ridge.

CONCLUSION

The development of some of the ideas of plate tectonics goes back to the work of Wegener in 1912 and Du Toit and Holmes in the 1920s. Its development as a viable theory has involved fundamental changes in Wegener's ideas about the kinematics and mechanics of the motions, but it is still Wegener's theory in the sense that it is concerned with the movement of large plates over horizontal distances of thousands of kilometers. Its wide acceptance in the 1960s and 1970s is in large part the result of the study of paleomagnetism and the sea floor. The introduction of lines of investigation that were not part of the stock-in-trade of most students of the earth has raised great difficulties in communication, the more so because the new work was for some years largely carried out by people whose training had been in physics. Physicists are not the easiest people with whom to carry on a discussion; they have a high opinion of their own abilities and are apt to believe in Lord Rutherford's classification of the sciences: "There's physics and there's chemistry, which is a sort of physics, and there's stamp collecting". The surprising thing is that in the last 20 years common ground has been found among the very heterogeneous group that makes up the students of the earth; a common language and a degree of mutual tolerance have been developed. Nothing succeeds like success and the success has drawn some of the brightest young men to the study of the earth. The object of this essay is to give an account of the rather devious route by which the success was achieved.

Just as I was finishing, the news came of the death of Maurice Ewing. It was he, more than any other man, who provided the fuel for the revolution in earth science. I once asked him "Where do you keep your ship?". He replied, "I keep my ship at sea". Over nearly 40 years he and his colleagues kept their ships at sea and provided the major part of our new knowledge.

Literature Cited[1]

Axelrod, D. I. 1963. Fossil floras suggest stable not drifting continents. *J. Geophys. Res.* 68 : 3257–63

Bacon, F. 1620. *Instauratio Magna (Novum Organum)*. London: Billium. 360 pp.
Bailey, E. B. 1929. The Palaeozoic mountain

[1] This list of references is unsatisfactory in that, although it is long, it leaves out many important papers. I have concentrated on the earlier period since bibliographies and textbooks are easily available for the work of the last ten years. The list is also unsatis-

systems of Europe and America. *Brit. Assoc. Advan. Sci., 96th Meet.* 57–76. London: Brit. Assoc. Advan. Sci. 704 pp.

Behm, H. W. 1923. *Entwicklungsgeschichte des Weltalls, des Lebens und des Menschen.* Stuttgart: Franckische Verlagshandlung. 232 pp. 2nd Ed.

Beloussov, V. V. 1962. *Basic Problems in Geotectonics.* New York: McGraw. 809. pp.

Beloussov, V. V. 1968. Some problems of development of the earth's crust and upper mantle of oceans. In *The Crust and Upper Mantle of the Pacific Area,* ed. L. Knopoff et al, 449–59. Am. Geophys. Union Monogr. 12

Benioff, H. 1954. Orogenesis and deep crustal structure—additional evidence from seismology. *Geol. Soc. Am. Bull.* 65: 385–400

Blackett, P. M. S. 1956. *Lectures on Rock Magnetism.* Jerusalem: Weizmann Sci. Press. 131 pp.

Blackett, P. M. S., Bullard, E. C., Runcorn, S. K., Ed. 1965. *A Symposium on Continental Drift.* London: Royal Society. 323 pp. (also *Phil. Trans. Roy. Soc. London. A* 258: 1–323)

Blackett, P. M. S., Clegg, J. A., Stubbs, P. H. S. 1960. An analysis of rock magnetic data. *Proc. Roy. Soc. A* 256: 291–322

Bruhnes, B. 1906. Recherches sur le direction d'aimentation des roches volcaniques. *J. Phys. Radium* (4) 5: 705–24

Bullard, E. C. 1939. Submarine geology. *Sci. Progr. London* No. 134: 237–48

Bullard, E. C. 1940a. Geophysical study of submarine geology. *Nature* 145: 764–66

Bullard, E. C. 1940b. The geophysical study of submarine geology. *Proc. Roy. Inst. Gt. Brit.* 31: 139–47

Bullard, E. C. 1954. A comparison of oceans and continents. *Proc. Roy. Soc. A* 222: 403–7

Bullard, E. C. 1956. The floor of the ocean. *Mem. Proc. Manchester Lit. Phil. Soc.* 97: 47–58

Bullard, E. C. 1961. Forces and processes at work in ocean basins. In *Oceanography,* ed. M. Sears, 39–50. Washington DC: Am. Assoc. Advan. Sci. (Publ. No. 67)

Bullard, E. C. 1962. *Proc. Geol. Soc. London* No. 1602: 154–55

Bullard, E. C. 1964. Continental drift. *Quart.*

J. Geol. Soc. London 120: 1–33

Bullard, E. C. 1968. Reversals of the earth's magnetic field. *Phil. Trans. Roy. Soc. London A* 263: 481–524

Bullard, E. C. 1975. The effect of World War II on the physical sciences. *Proc. Roy. Soc. A.* In press

Bullard, E., Everett, J. E. Smith, A. G. 1965. The fit of the continents around the Atlantic. *Phil. Trans. Roy. Soc. London A* 258: 41–51

Bullard, E. C., Gaskell, T. F. 1938. Seismic methods in submarine geology. *Nature* 142: 916

Bullard, E. C., Gaskell, T. F. 1941. Submarine seismic investigations. *Proc. Roy. Soc. London A* 177: 476–99

Bullard, E. C., Mason, R. G. 1963. The magnetic field over the oceans. In *The Sea,* ed. M. N. Hill, 3: 175–217. New York: Interscience

Cann, J. R., Vine, F. J. 1966. An area of the crest of the Carlsberg Ridge: petrology and magnetic survey. *Phil. Trans. Roy. Soc. London A* 259: 198–217

Carey, S. W. 1955. Wegener's South America-Africa assembly, fit or misfit? *Geol. Mag.* 92: 196–200

Carey, S. W. 1958. The tectonic approach to continental drift. In *Continental Drift, A Symposium,* 177–355. Hobart, Aust.: Univ. Tasmania. 363 pp.

Chamberlain, R. T. 1928. In *Theory of Continental Drift, A Symposium,* ed. W. A. J. M. van Watershoot van der Gracht, 87. Tulsa, Okla.: Am. Assoc. Petrol. Geol. 240 pp.

Cox, A., Doell, R. R. 1960. Review of paleomagnetism. *Geol. Soc. Am. Bull.* 71: 645–768

Cox, A., Doell, R. R. 1961. Palaeomagnetic evidence relating to a change in the earth's radius. *Nature* 189: 45–47

Cox, A., Doell, R. R., Dalrymple, G. B. 1963a. Geomagnetic polarity epochs and Pleistocene geochronology. *Nature* 198: 1049–51

Cox, A., Doell, R. R., Dalrymple, G. B. 1963b. Geomagnetic polarity epochs: Sierra Nevada II. *Science* 142: 382–85

Daly, R. A. 1926. *Our Mobile Earth.* New York: Scribner. 342 pp.

Daly, R. A. 1942. *The Floor of the Ocean.* Chapel Hill: Univ. North Carolina Press. 177 pp.

factory in another way: it contains too many of my own papers. This superfluity is due, not to a superabundance of contributions to the subject, but to my having used the reviews that I have written since 1939 to describe the development of thought. I might have used other people's reviews, but I thought it more interesting and conducive to a coherent account if I followed my own successive steps to understanding.

Deacon, M. 1971. *Scientists and the Sea 1650–1900*. London: Academic. 445 pp.

Dietz, R. S. 1961. Continent and ocean basin evolution by spreading of the sea floor. *Nature* 190:854–57

Dietz, R. S. 1962. Ocean-basin evolution by sea-floor spreading. In *The Crust of the Pacific Basin*, 11–12. Am. Geophys. Union Monogr. No. 6

Dietz, R. S. 1968. Reply. *J. Geophys. Res.* 73:6567

Du Toit, A. L. 1927. *A Geological Comparison of South America with South Africa*. Washington DC: Carnegie Inst. (Publ. No. 381). 158 pp.

Du Toit, A. L. 1937. *Our Wandering Continents*. Edinburgh: Oliver & Boyd. 366 pp.

Elsasser, W. M. 1967. Convection and stress propagation in the upper mantle. In *The Application of Modern Physics to the Earth and Planetary Interiors*, ed. S. K. Runcorn, 223–45. New York: Interscience

Euler, L. 1776. Formulae generales pro translatione quacunque corporum rigidorum. *Novi Comment. Acad. Sci. Petropolitanae* 20:189–207. [*Opera Omnia Ser. 2*, 1968, ed. C. Blank, 9(2):84–98]

Ewing, M., Crary, A. P., Rutherford, H. M. 1937. *Geol. Soc. Am. Bull.* 48:753–802

Ewing, M., Heezen, B. C. 1956. Some problems of Antarctic submarine geology. In *Antarctica in the Geophysical Year*, 75–81. Am. Geophys. Union Monogr. No. 1

Ewing, M., Press, F. 1956. Surface waves and guided waves. In *Handb. Phys.* 47:119–39

Field, R. M. 1938. Geophysical exploration of ocean basins. *Quart. J. Geol. Soc. London* 94:IV–VII

Gelletich, H. 1937. Über magnetitführende eruptive Gänge und Gangsysteme in mittlenen Teil des südlichen Transvaals. *Beitr. Angew. Geophys.* 6:337–406

Gevers, T. W. 1950. Life and work of Dr. Alex L. Du Toit. *Trans. Geol. Soc. S. Afr.* 52 (Suppl.):1–109

Gilbert, W. 1600. *De Magnete*. London: Short. 240 pp.

Graham, J. W. 1949. The stability and significance of magnetism in sedimentary rocks. *J. Geophys. Res.* 54:131–67

Gregory, J. W. 1929. The geological history of the Atlantic Ocean. *Quart. J. Geol. Soc. London* 85:LXVIII–CXXII

Griggs, D. 1939. A theory of mountain building. *Am. J. Sci.* 237:611–50

Hales, A. L. 1936. Convection currents in the earth. *Mon. Notic. Roy. Astron. Soc. Geophys. Suppl.* 3:372–79

Hallam, A. 1973. *A Revolution in the Earth Sciences*. Oxford: Clarendon. 127 pp.

Harrison, C. G. A., Funnell, B. M. 1964. Relationship of palaeomagnetic reversals and micro-palaeontology in two late Caenozoic cores from the Pacific Ocean. *Nature* 204:566

Heezen, B. C. 1960. The rift in the ocean floor. *Sci. Am.* 203(4):99–110

Heirtzler, J. R., Dickson, G. O., Herron, E. M., Pitman, W. C., Le Pichon, X. 1968. Marine magnetic anomalies, geomagnetic fields reversals, and motions of the ocean floor and continents. *J. Geophys. Res.* 73:2119–36

Hess, H. H. 1954. Geological hypothesis and the earth's crust under the oceans. *Proc. Roy. Soc. A* 222:341–48

Hess, H. H. 1962a. Richard Montgomery Field. *Trans. Am. Geophys. Union* 43:1–3

Hess, H. H. 1962b. History of ocean basins. In *Petrologic Studies: A Volume to Honor A. F. Buddington*, 599–620. New York: Geol. Soc. Am.

Hess, H. H. 1968. Reply. *J. Geophys. Res.* 73:6569

Holmes, A. 1929. Radioactivity and earth movements. *Trans. Geol. Soc. Glasgow* 18:559–606 (discussion, 614–15)

Holmes, A. 1944. *Principles of Physical Geology*. London: Nelson. 532 pp.

Hospers, J. 1951. Remnant magnetism of rocks and the history of the geomagnetic field. *Nature* 168:1111–12

Hospers, J. 1953. Reversals of the main geomagnetic field, I, II, *Proc. Kon. Ned. Akad. Wetensch. B* 56:467–476, 477–491

Hospers, J. 1954. Reversals of the main geomagnetic field, III. *Proc. Kon. Ned. Akad. Wetensch. B* 57:112–21

Hurley, P. M. 1972. Can the subduction process of mountain building be extended to Pan-African and similar orogenic belts? *Earth Planet. Sci. Lett.* 15:305–14

Hurley, P. M., Rand, J. R. 1968. Review of age data in West Africa and South America relative to a test of continental drift. In *The History of the Earth's Crust*, (ed. R. A. Phinney), 153–60. Princeton NJ: Princeton Univ. Press. 244 pp.

Hutton, J. 1795. *Theory of the Earth*, Vol. 1. Edinburgh: Cadell, Junior & Davies

Irving, E. 1964. *Palaeomagnetism*. New York: Wiley. 399 pp.

Isacks, B., Molnar, P. 1971. Distribution of stresses in the descending lithosphere from a global survey of focal mechanism solutions of mantle earthquakes. *Rev. Geophys. Space Phys.* 9:103–74

Jeffreys, H. 1924. *The Earth*. Cambridge: Cambridge Univ. Press. 278 pp. (Later editions 1929, 1952, 1959, 1970)

Katsumata, M. 1960. The effect of seismic zones upon the transmission of seismic waves (in Japanese with English summary). *Kensin-Siho* 25(3): 1960

Koenigsberger, J. K. 1938. Natural residual magnetism of eruptive rocks. *Terr. Magn. Atmos. Elec.* 43: 119–30, 299–320

Lake, P. 1922. Wegener's displacement theory. *Geol. Mag.* 59: 338–46

Lear, J. 1967. Canada's unappreciated role as scientific innovator. *Sat. Rev.* 1967 (Sept. 2): 45–50

Le Pichon, X. 1968. Sea floor spreading and continental drift. *J. Geophys. Res.* 73: 3661–97

Le Pichon, X., Francheteau, J., Bonnin, J. 1973. *Plate Tectonics.* Amsterdam: Elsevier. 300 pp.

Lomnitz, C. 1956. Creep measurements in igneous rocks. *J. Geol.* 64: 473–79

McDougall, I., Tarling, D. H. 1963. Dating polarity zones in the Hawaiian Islands. *Nature* 200: 54–56

McKenzie, D. P. 1968. The geophysical importance of high temperature creep. In *The History of the Earth's Crust*, ed. R. A. Phinney, 28–44. Princeton, NJ: Princeton Univ. Press. 244 pp.

McKenzie, D. P., Parker, R. L. 1967. The North Pacific: an example of tectonics on a sphere. *Nature* 216: 1276–80

McKenzie, D. P., Sclater, J. G. 1971. The evolution of the Indian Ocean since the late Cretaceous. *Geophys. J. Roy. Astron. Soc.* 25: 437–528

Mason, R. G. 1958. A magnetic survey off the west coast of the United States between latitudes 32° and 36°N, longitudes 121° and 128°W. *Geophys. J. Roy. Astron. Soc.* 1: 320–29

Mason, R. G., Raff, A. D. 1961. A magnetic survey off the west coast of North America, 32°N to 42°N. *Geol. Soc. Am. Bull.* 72: 1259–65

Meinesz, F. A. V. 1948. Major tectonic phenomena and the hypothesis of convection currents in the earth. *Quart. J. Geol. Soc. London* 103: 191–207

Meinesz, F. A. V. 1952. Convection-currents in the earth and the origin of continents. *Proc. Kon. Ned. Akad. Wetensch. B* 55: 427–553

Menard, H. W. 1958. Development of median elevations in the ocean basins. *Geol. Soc. Am. Bull.* 69: 1179–86

Meyerhoff, A. A. 1968. Arthur Holmes: originator of spreading ocean floor hypothesis. *J. Geophys. Res.* 73: 6563–65

Meyerhoff, A. A., Meyerhoff, H. A. 1972. The new global tectonics: major inconsistencies. *Am. Assoc. Petrol. Geol. Bull.* 56: 269–336

Miller, J. A. 1965. Geochronology and continental drift. *Phil. Trans. Roy. Soc. Lon-*

don *A* 258: 180–91

Morgan, W. J. 1968. Rises, trenches, great faults and crustal blocks. *J. Geophys. Res.* 73: 1959–82

Nagata, T. 1951. Reverse thermo remenant magnetism. *Nature* 169: 704–5

Nagata, T. 1953. *Rock Magnetism.* Tokyo: Maruzen. 225 pp. (Revised ed. 1961)

Ninkovich, D., Opdyke, N., Heezen, B. C., Foster, J. H. 1966. Paleomagnetic stratigraphy, rates of deposition and tephrochronology in North Pacific deep-sea sediments. *Earth Planet. Sci. Lett.* 1: 476–92

Oliver, J., Isacks, B. 1967. Deep earthquake zones, anomalous structures, and the lithosphere. *J. Geophys. Res.* 72: 4259–75

Pekeris, C. L. 1935. Thermal convection in the interior of the earth. *Mon. Notic. Roy. Astron. Soc. Geophys. Suppl.* 3: 343–67

Pepper, J. H. 1861. *The Playbook of Metals.* London: Routledge. 504 pp.

Phinney, R. A., Ed. 1968. *The History of the Earth's Crust.* Princeton, NJ: Princeton Univ. Press. 244 pp.

Rothé, J. P. 1954. La zone seismique mediane Indo-Atlantique. *Proc. Roy. Soc. London* 222: 387–97

Rudwick, M. J. S. 1972. *The Meaning of Fossils.* London: Macdonald. 287 pp.

Runcorn, S. K. 1956. Paleomagnetic comparisons between Europe and North America. *Proc. Geol. Assoc. Can.* 8: 77–85

Runcorn, S. K. 1962. Paleomagnetic evidence for continental drift and its geophysical cause. In *Continental Drift*, ed. S. K. Runcorn, 1–40. New York: Academic

Scrase, F. J. 1931. Reflected waves from deep focus earthquakes. *Proc. Roy. Soc. A* 132: 213–35

Smith, A. G., Hallam, A. 1970. The fit of the southern continents. *Nature* 225: 139–44

Snider, A. 1858. *La création et ses mystères dévoilés.* Paris: A. Franck. 487 pp.

Stoneley, R. 1931. On deep-focus earthquakes. *Gerlands Beitr. Geophys.* 29: 417–35

Stoneley, R. 1966. The Niger delta region in the light of the theory of continental drift. *Geol. Mag.* 103: 385–97

Sykes, L. R. 1967. Mechanism of earthquakes and nature of faulting on mid-ocean ridges. *J. Geophys. Res.* 72: 2131–53

Taylor, F. B. 1910. Bearing of the Tertiary mountain belt on the origin of the earth's plan. *Geol. Soc. Am. Bull.* 21: 179–226

Termier, P. 1925. *Ann. Rep. Smithson. Inst.* 1924: 219–36

Turner, H. H. 1922. On the arrival of earthquake waves at the antipodes and on the measurement of the focal depth of an earthquake. *Mon. Notic. Roy. Astron. Soc. Geophys. Suppl.* 1 : 1–13

Vacquier, V., Raff, A. D., Warren, R. E. 1961. Horizontal displacements in the floor of the Pacific ocean. *Geol. Soc. Am. Bull.* 72 : 1251–58

van Watershoot van der Gracht, W. A. J. M. 1928. *Theory of Continental Drift, A Symposium.* Tulsa : Am. Assoc. Petrol. Geol. 240 pp.

Vine, F. J. 1966. Spreading of the ocean floor : new evidence. *Science* 154 : 1405–15

Vine, F. J., Matthews, D. H. 1963. Magnetic anomalies over ocean ridges. *Nature* 199 : 947–49

Vine, F. J., Wilson, J. T. 1965. Magnetic anomalies over a young ocean ridge off Vancouver island. *Science* 150 : 485–89

Visser, S. W. 1936. Some remarks on the deep-focus earthquakes in the international seismological summary. *Gerlands Beitr. Geophys.* 48 : 254–67

Vogt, P. R., Avery, O. E. 1974. Detailed magnetic surveys in the northeast Atlantic and Labrador Sea. *J. Geophys. Res.* 79 : 363–89

Washington, H. S. 1923. Comagmatic regions and the Wegener hypothesis. *J. Nat. Acad. Sci.* 13 : 339–47

Watson, J. D. 1968. *The Double Helix.* London : Weidenfeld & Nicholson. 226 pp.

Wegener, A. 1912a. Die Entstehung der Kontinente. *Petermanns Geogr. Mitt.* 58 : 185–95, 253–56, 305–8

Wegener, A. 1912b. Die Entstehung der Kontinente. *Geol. Rundsch.* 3 : 276–92

Wegener, A. 1915. *Die Entstehung der Kontinente und Ozeane.* Braunschweig : Vieweg. (Other editions 1920, 1922, 1924, 1929, 1936, 1962. Translation of 1922 edition into English by J. G. A. Skerl in 1924 as *The Origin of Continents and Oceans,* Methuen ; 1929 edition translated by J. Biram, published by Dover in 1966; there are also translations into French, Russian, Spanish, and Swedish.)

Willis, B. 1944. Continental drift, ein Märchen. *Am. J. Sci.* 242 : 509–13

Wilson, J. T. 1963. Continental drift. *Sci. Am.* 209 (April) : 86–100

Wilson, J. T. 1965. A new class of faults and their bearing on continental drift. *Nature* 207 : 343–47

PRECAMBRIAN TECTONIC ENVIRONMENTS

×10031

Carl R. Anhaeusser

Economic Geology Research Unit, University of the Witwatersrand,
Johannesburg, South Africa

INTRODUCTION

Since the concept of sea-floor spreading was first introduced by Dietz (1961) and Hess (1962) a voluminous literature has developed to support Mesozoic continental drift. Notable advances in paleomagnetism, polar wandering, and sea-floor spreading (Runcorn 1962, Vine & Matthews 1963, Vine 1966) were followed by formulation of the theory of plate tectonics (McKenzie & Parker 1967, Le Pichon 1968, Morgan 1968, Isacks et al 1968, McKenzie 1972). Emergence of the plate tectonic or "new global tectonic" theory has provided a unifying worldwide explanation for tectonic processes (Dewey & Bird 1970). New models for mountain building involve plate motions at divergent and convergent plate boundaries. Relegated to virtual obsolescence (Dickinson 1971) are the classical geosynclinal concepts widely held until the mid-1960s.

The concept of plate tectonics compelled geologists to reexamine sedimentation, igneous activity, deformation, and metamorphism on both regional and global scales. Initially, attention was devoted to plate tectonic theory as it affected contemporary geosynclines, island arcs, mountain ranges, and oceanic domains (Mitchell & Reading 1969, Bird & Dewey 1970, Dewey & Bird 1970, Coleman 1971, Dickinson 1971, Gilluly 1971, Oxburgh & Turcotte 1971), but the success of these efforts led to the application of the plate tectonic model to Paleozoic and late Precambrian orogenic terranes such as the Uralides and Appalachian/Caledonian Orogen (Hamilton 1970, Bird & Dewey 1970, McElhinny & Briden 1971). A sequel has been a growing tendency to apply the theory to the Precambrian record and to evaluate the possible role of the plate tectonic mechanisms during early continental evolution (Dewey & Horsfield 1970). Some believe that ancient mobile belts, and even the Archean greenstone belts, represent the impact scars of protocontinental fragments colliding with each other, with island arcs, or with oceanic crust (Gibb 1971, Gibb & Walcott 1971, White et al 1971, Condie 1972, Talbot 1973).

How justified is application of the plate tectonic hypothesis to earlier geologic time? Before making judgment it may be advisable to examine critically the

31

geological record. To attempt this, consideration will be given to an appraisal of the similarities and differences in geosynclinal and geotectonic processes through geologic time, especially in the Precambrian.

PHANEROZOIC GEOTECTONIC CHARACTERISTICS

Geosynclines and Orogeny

The concepts of "geosynclines" and "orogeny" have long been regarded by many as virtually synonomous in terms of sedimentation, mountain building, deep burial, regional metamorphism, granitization, igneous activity, and erosion. Lack of precise definition not only caused confusion but also impeded broader understanding of global geotectonic events. Aubouin (1965) tried to resolve the situation in his review of classical descriptive data and classification schemes of geosynclinal or orogenic cycles. He defined a basic model which he believed could be successfully applied to geosynclinal regions and which involved a foreland or craton and mio- and eu-geosynclinal furrows or troughs and mio- and eu-geanticlinal ridges or highs. Variations in type of volcanism, sedimentation, and plutonism were related to specific parts of the model with deformation migrating towards the craton with time to end the mobility of the geosyncline.

Classical descriptive models, like that of Aubouin, are based on much geological data and offer a valuable conceptual framework for classification of tectonic elements. Distinctions could be made between various orogenic episodes throughout the world based on characteristic lithologies, metamorphism, and deformational styles. Zwart (1969) found little difficulty comparing and contrasting the Carboniferous Hercynian orogenic belt of Europe with the Tertiary Alpine mountain chain. The Hercynian belt he found

> is characterized by low-pressure metamorphism, high geothermal gradients, wide extent, abundant granites, few initial magmatic rocks and a restricted amount of uplift and erosion. The Alpine foldbelt, in contrast, exhibits high-pressure meta-morphism, low geothermal gradients, narrowness of the chain, scarcity of granites, abundant ophiolites, and strong and rapid uplift.

Zwart furthermore contended that the Hercynian and Alpine belts represented two extremes of mountain building and that older orogenic belts, possessing intermediate characteristics, were represented by the Caledonian chain of Europe and the Grenville Province in Canada.

Significant differences were found to exist between the complex circum-Pacific orogenic areas and those of the Atlantic region (Crook 1969, Matsumoto 1967). Crook distinguished between "Pacific geosynclines" and "Atlantic geosynclines," the latter exemplified by the model of Aubouin (1965) and applied to the Alpine, Caledonian, and Appalachian geosynclines. It was concluded that although Atlantic and Pacific geosynclines are superficially similar, detailed examination reveals significant differences which appear to reflect their contrasting geotectonic positions. Pacific geosynclines seem newly formed on a largely simatic floor, marginal to sialic cratons; Atlantic geosynclines appear to occupy, in part, the sites of older geosynclines, on the trends of which they are discordantly overprinted (Hercynian overprinted by Alpine; Grenville overprinted by Appalachian/Caledonian). Furthermore, the Atlantic geosynclines appear to have lain between cratonic blocks, if conventional drift reconstructions are accepted.

These examples serve to remind us that plate tectonic theory has not been responsible for all the recent advances in the earth sciences and that we still owe much to the efforts and observations of geologists in the past. Many distinguishing features of orogenic belts, including aspects of volcanism, magmatism, sedimentation, metamorphism, and deformation, from a wide range of environments, were thus immediately available for reevaluation, reinterpretation, and translation into terms of the new global tectonics when the concept emerged.

Plate Tectonics and Orogeny

We know that the earth's surface is traversed by extensive linear tectonic features, represented on the continents and along some continental margins by orogenic belts of diverse age and in the oceans by mid-ocean ridges, trenches, and transform faults which, according to the theory of sea-floor spreading, cannot be older than 200–300 m.y. (Hess 1962). The youngest major tectonic episodes are represented by the Tethyan (Alpine/Himalayan) and circum-Pacific orogenic belts, both of which probably originated during the early Triassic. Plate tectonic theory requires that plates be generated at oceanic ridges and destroyed at continent-ocean boundaries and complex trench-transform systems (Isacks et al 1968), thus there is no difficulty in linking the young mountain belts (Andes) and modern island arcs (Japan) to a causative process in sea-floor spreading and plate tectonics. The theory also explains development of intracontinental Alpine/Himalayan-type orogenic belts by collision of continental crustal blocks along consuming plate margins (Dewey & Bird 1970, Mitchell & Reading 1969).

If tectonic features such as the Paleozoic Hercynian, Caledonian, Appalachian, Uralide, and Tasman orogenic belts are linked to the concept of plate tectonics (Hamilton 1970, Dewey & Bird 1970, Bird & Dewey 1970), then it might be concluded that the continents have drifted more than once in geologic time.

Dewey & Bird (1970) and Dewey & Horsfield (1970) maintain that plate tectonics is too powerful a mechanism in explaining contemporary relationships among continents, orogenic belts, and oceans to be disregarded in favor of ad hoc, nonactualistic models for the older orogenic belts on the continents. These and other authors advocate that sea-floor spreading has been responsible for growth and evolution of continents for at least 3.0 b.y.[1] Thus the controversy over the validity of continental drift has taken a dramatic new turn, and inquiry now does not ask whether drift has taken place but rather how often has it occurred throughout geologic time?

Following the advent of the plate tectonic concept little time was lost in reexamining previously studied orogenic belts. Several notable contributions demonstrated how orogenic belts record events controlled by plate divergence and

[1] b.y. = 10^9 years.

convergence (Mitchell & Reading 1969, McKenzie 1969, Hamilton 1970, Dewey & Bird 1970, Bird & Dewey 1970). Mid-oceanic rises were recognized as sites of major divergence while convergent boundaries mark sites of orogeny involving plate consumption, continental collision, stratal crumpling, thrusting, crustal thickening, and isostatic uplift, together with magmatic and metamorphic events (Dewey & Horsfield 1970). Thus the new synthesis, unlike the classical geosynclinal approach, does not require a regular sequence of orogenic events to be recognized in the same order in all mountain belts (Dickinson 1971).

Oceanic Ridges and Transform Faults

Plate tectonics postulates a number of rigid plates whose boundaries mark the seismic belts of the world (Barazangi & Dorman 1969, Isacks et al 1968). Le Pichon (1968) suggested that six plates cover the globe, these being generated at oceanic ridges where Vine & Matthews (1963) first correlated linear magnetic anomalies with sea-floor spreading.

Although the oceanic lithosphere (crust of V_p 6–6.5 km/sec layer extending to upper mantle with $V_p \sim 8$ km/sec layer) is approximately 6–8 km thick, McKenzie (1972) indicated that the base of plates is now considered to be at the top of the low-velocity zone (~ 80 km). He reported that plate motions are no longer believed to be related to mantle motions and that ridges and trenches do not appear to be confined to the rising and sinking limbs of convection cells.

The essentially linear configuration of oceanic ridges is disturbed intermittently by offsetting transform faults at which sites crustal surface appears to be neither created nor destroyed. Some fracture zones, like those in the eastern Pacific, are extensive. Morgan (1968) suggested that they lie on small circles centered about poles of relative motion of diverging plates.

Oceanic ridges are locations of active basaltic volcanism ranging in composition from high-alumina olivine tholeiites to low-alumina tholeiites (Miyashiro et al 1969). Other characteristics include low K_2O, TiO_2, and P_2O_5, rather high Al_2O_3, high CaO, and very high Na/K ratios (Cann 1971, Engel et al 1965). These subalkaline, generally saturated, oceanic tholeiites are the dominant basalts of abyssal ocean floors and mid-oceanic ridges and are believed to overlie partly residual peridotitic lithosphere. Because the axes of the ridges are characterized by high heat flow these associations, including heterogenous mixtures of gabbroic and peridotitic composition (ophiolite complexes), appear to derive from relatively shallow depths (~ 20–60 km) by high degrees of melting of upper mantle "pyrolite" (Green 1972a, Ringwood 1969). Silica-undersaturated alkali basalts, derived from the upper reaches of the low velocity zone (70–100 km), occur less commonly and are found in oceanic volcanic islands, seamounts, and island chains.

Oceanic and Continental Edge Arc-Trench Systems

The arc-trench regions are sites of subduction or plate consumption where oceanic crust, created at mid-oceanic ridges, returns to the mantle. Precisely how this is accomplished remains problematical. Sufficient is known of the complex arc-trench systems to conclude that their distinctive properties link them genetically to processes accompanying oceanic lithospheric plate destruction. The island chains and trenches are the most seismically active regions on earth, being the sites of deep-focus earthquakes originating at depths of 300–700 km. Epicenters define zones about 250 km wide (Benioff seismic zones) dipping beneath continents or island arcs to depths of 650–700 km. The sea-floor spreading concept suggests oceanic crust is thrust or dragged down the Benioff zone where it undergoes progressive metamorphism, phase variation, and fractional melting. Hydrated oceanic basalts undergo dehydration and progressive metamorphism through green-schist, amphibolite, granulite, and eclogite facies as the temperature of the initially cool slab rises in a changing pressure realm (Ringwood & Green 1966, Ringwood 1969, Green 1972a). Partial melts of the descending slab, derived from various pressure-temperature regimes, make their way to surface and contribute to continental crust development. Where the overriding plate margin is capped by oceanic crust the magmatic arc is an intra-oceanic island arc structure (Aleutians, Philippines) with crustal thicknesses ranging between 12–25 km (Engel 1970); where it is capped by continental crust, it is a volcano-plutonic complex near the continental margin (Andes). The distinctive calc-alkaline volcanic suite (basalt, andesite, dacite, rhyolite) and plutonic equivalents (gabbro, diorite, granodiorite, granite) make up the bulk of the magma introduced in these two situations with tholeiites and shoshonites occurring in subordinate quantities (Jakeš & White 1972).

Extrusive and intrusive phases of magmatism vary systematically across island arcs with tholeiitic magmas emplaced closest to the trenches and increasingly more alkaline magma-types (tholeiitic to calc-alkaline to high-K calc-alkaline to shoshonitic) occurring across the arcs towards the continents. This increase in potassium has been related genetically to the depth of the dipping Benioff zone beneath the arcs (Kuno 1959).

Sedimentation within arc-trench systems is also divergent, with volcano-clastic detritus being dispersed from volcanic centers and deposited in a variety of marine and terrestrial environments within and outside the confines of the arc. Arc-type eugeosynclinal sequences accumulate, and pelagic sediments, basalts, oceanic crust, and clastic turbidites all feed steadily into subduction zones until, ultimately, a characteristic regime of deformation leads to the chaotic tectonic style of the mélanges (Dickinson 1971).

The occurrence of alpine-type peridotites without high-temperature contact aureoles in orogenic regions along continental margins has led to the general consideration that these rocks have been emplaced tectonically, rather than by igneous activity. Evidence from New Caledonia and New Guinea suggests that at the onset of compressional impact between oceanic and continental plates, some obduction or overriding can occur where large oceanic-mantle crustal slabs are thrust over, or into, continental edges contemporaneously with blueschist metamor-phism (Coleman 1971). Thereafter, continued oceanic lithospheric motions convert to subduction with all the attendant seismic activity and andesitic volcanism and plutonism.

In 1961 Miyashiro drew attention to the existence of "paired metamorphic belts." He recognized that in the Japanese island arc and in other parts of the circum-

Pacific region there occur adjacent and parallel metamorphic belts of similar age and distinctly different mineral paragenesis. In any one pair, the inner belt (that closer to the continent) is characterized by the stability of andalusite at higher levels and of sillimanite in deeper zones. The outer belt, nearer the ocean, is characterized by the presence of jadeite-quartz and glaucophane. Different temperature-pressure conditions were necessary for the development of these two kinds of metamorphic belts: the inner required high temperature and low pressure whereas the outer required a combination of low temperature and high pressure. Oxburgh & Turcotte (1971) related these changing pressure-temperature relationships to processes accompanying descending lithospheric slabs beneath island arcs. Unconsolidated sediments accumulate in the trenches and are subjected to high pressures before they and the underlying oceanic lithosphere descend and melt along the slip zone. Magmas are generated and heat is transferred upward by magmatic convection. There is thus a direct link between the variations of magma-type, heat flow, and metamorphism across arc structures.

Continental Orogeny

The Alps, the Himalayas, and the Urals are cited as examples of orogenic belts produced as a result of continental blocks colliding during the final stages of ocean floor contraction or consumption (Mitchell & Reading 1969, Hamilton 1970, Dewey & Bird 1970). The mountain belts that form are asymmetrical, being characterized by assemblages thrust over and onto the consumed plates and by remobilized basement near the site of collision. Upthrust wedges of oceanic crust (ophiolites) occur in suture zones (Alpine Ivrea zone, Himalayan Indus suture) regarded as the principal join-lines of impacted continental fragments (Dewey & Horsfield 1970, Gilluly 1972). As collision-type orogenic belts result from compressional forces, magmatic activity within them is less marked, no paired metamorphic zonation is apparent, and the dominant metamorphic assemblages are of the high pressure blueschist facies (lawsonite-glaucophane-jadeite).

PRECAMBRIAN EARTH HISTORY

The Phanerozoic Era which accounts for but one eighth of earth history has clearly undergone an extremely varied and complex series of events. Equally diverse, if not more so, were events taking place during Precambrian times. Coupled with the fact that this part of the geological record is not everywhere well preserved, the task of unraveling early earth history remains problematical.

The Early Precambrian Crust

Two types of Archean terranes are known: those dominated by high-grade gneisses and migmatites, and those consisting of granites and greenstone belts (low metamorphic grade cratonic areas). High-grade gneiss terranes, such as those in West Greenland (McGregor 1973), expose the oldest continental rocks yet found (3.70–3.75 b.y., Moorbath et al 1972). In granite-greenstone terranes some of the oldest rocks dated (ca 3.4 b.y.) include granites from the Barberton-Swaziland areas in southern Africa. Reviewing the geological history of this region, Anhaeusser (1973b) pointed out, however, that age dating had still to be attempted on the thick pile (7 km) of chemically unique mafic and ultramafic rocks described from the base of the greenstone belt by Viljoen & Viljoen (1969a). These rocks are considered to be older than the surrounding granites and migmatities and may represent a primitive oceanic-type crust.

The nature and development of the earth's crust is of fundamental importance in understanding the evolution of continents and the changing geotectonic environments, and must be examined closely. In Archean terranes argument revolves around which existed first: high-grade gneisses, migmatites, and granites, or greenstone belts? On balance, the early existence of some sialic crust has been favored (Poldervaart 1955, Ramberg 1964). It has been postulated that even in greenstone terranes, a thin sialic crust may have preceded the volcanic sequences (Anhaeusser et al 1969, Viljoen & Viljoen 1969b), but further investigations in southern Africa and Western Australia suggest that a primordial crust consisting of stratiform oceanic-type mafic/ultramafic assemblages existed before the granite events in these areas (Anhaeusser 1973a,b, Glikson 1971, 1972). To explain the diversity of opinion relating to the nature of the crust upon which greenstone belts were developed, Windley (1973) suggested that greenstone belts may have formed diachronously, those in southern Africa forming before those in other shield areas. Should this concept prove correct, it might be necessary to interpret each Archean region separately.

Investigators in the high-grade gneiss areas (Windley 1973, Windley & Bridgwater 1971) contend that the granite-greenstone terranes, which occupy the cratonic areas of the shields, are underlain by granulite complexes that were in existence in Archean times. These granulites are thus regarded as the low crustal residue from which granites were derived, the latter migrating upwards to invade the borders of overlying greenstone belts.

Green & Ringwood (1967) subdivided the granulite facies into low-intermediate- and high-pressure regimes. Low-pressure (high-temperature) granulite facies rocks are difficult to distinguish from amphibolites with which they are characteristically intimately associated. These granulites could form by partial melting at the base of crust 8–9 km thick if Archean thermal gradients approached $100°C$ km^{-1} (Fyfe 1973), the latter caused by greater residual heat and higher radioactive element heat production. This presupposes that an overburden at least 8 km thick existed prior to granulite development. If this was of mafic/ultramafic composition it would account for the countless number of inclusions of this composition engulfed in high-grade migmatite-gneiss terranes like those in Greenland, India, Norway, Australia, and elsewhere.

Shield areas are continental blocks of crystalline rocks of Precambrian age containing ancient cratonic nuclei consisting of complex, low metamorphic grade, granite-greenstone terranes unaffected by any major tectono-thermal event for the last ~ 2.5 b.y. Medium- to high-grade metamorphic areas (amphibolite-granulite facies rocks) are widespread on the shields but occur essentially as linear mobile belts, many of which tend to encircle and rework the Archean granites and greenstones of the cratons (Anhaeusser et al 1969, Mason 1973). Medium- and high-pressure granulite facies rocks occur in the mobile belts but are virtually unknown on the cratons. Using modern geothermal gradients (20–30°C km^{-1}; Fyfe 1973), these rocks probably formed at the base of crust 30–50 km thick. Even substituting higher Archean gradients, depths of burial in these belts must have exceeded the 8 km lower limit mentioned earlier. If, as has been suggested, granulite facies rocks preceded or acted as basement to the greenstones, both in present-day cratons and in reworked mobile belts, then crustal stabilities inherent in this argument are irreconcilable with the deformational styles so characteristic of the granite-greenstone terranes. These tectonic aspects are discussed later.

Archean Greenstone Belts

Greenstone belts are defined as the distinctive low-grade metavolcanic and sedimentary assemblages ($\sim 2.5 \geq 3.5$ b.y. old) which occur as scattered remnants on the cratons, forming an essential part of the latter. Numerous accounts of their geological characteristics are available (Anhaeusser et al 1969, Anhaeusser 1971a,b, 1973b, Baragar & Goodwin 1969, Engel 1970, Goodwin 1971, Goodwin & Ridler 1970, Glikson 1972, Macgregor 1951, Viljoen & Viljoen 1969a,b, 1971, Wilson 1973). Despite differences that are apparent from one shield area to the next, Anhaeusser et al (1969) stressed the worldwide uniformity of the stratigraphy, structure, metamorphism, mineralization, associated granites, and geotectonic setting of the greenstone belts. Refinements to the model originally proposed center mainly around increased knowledge of the geochemistry of the volcanic, sedimentary, and granitic rocks in these regions (Anhaeusser 1973a, Baragar & Goodwin 1969, Brooks & Hart 1972, Hart et al 1970, Hallberg 1972, Viljoen & Viljoen 1969a,b, 1971).

Of particular significance has been the recognition in Canada, Western Australia, and southern Africa of a suite of distinctive mafic and ultramafic rocks invariably confined to lower stratigraphic divisions of greenstone belts and referred to as basaltic and peridotitic "komatiites" (Viljoen & Viljoen 1969a). The only justifiable modern analogs of these volcanic rocks occur in present-day abyssal or oceanic settings, but the unique geochemistry of the komatiites (high Mg, high Ca/Al ratio, low alkalis) still sets them apart (Anhaeusser 1973b, Viljoen & Viljoen 1969a).

Evidence of ultramafic liquids, extruded as fluid magmas in an aqueous environment, was provided by Viljoen & Viljoen (1969a), and experimental studies (Green 1972a) suggest that the temperature of extrusion of such rocks, which approach the "pyrolite" model compositions of the upper mantle, was at least 1600–1650°C. The experimental studies further suggest that the peridotitic komatiite could have derived from high degrees of melting (60–80%) of pyrolite whereas that of the basaltic komatiite would be lower ($\sim 40\%$). Conditions of melting in the Archean must therefore have been considerably more intense than those operating to yield modern mafic/ultramafic magmas. It has been estimated that geothermal gradients were two or three times those at present (Brooks & Hart 1972, Fyfe 1973) and may have been caused by rapid diapiric ascent of mantle source rock. The high degrees of mantle melting led Green (1972a,b) to suggest a more catastrophic and rapid magma genesis. He postulated that greenstone belts represent altered equivalents of lunar maria, formed as a result of major impacts triggering partial melting at depths of 150–300 km.

Regardless of origin, the unique geochemical properties of the mafic and ultra-mafic rocks cannot be overemphasized and have to be considered when discussing crustal evolution.

In average greenstone belts, rocks possessing tholeiitic and calc-alkaline chemical affinities occur stratigraphically higher than the mafic/ultramafic successions (Anhaeusser et al 1969, Anhaeusser 1971b, Baragar & Goodwin 1969, Viljoen & Viljoen 1971). Volcanic, pyroclastic, and chemical sedimentary assemblages of diverse types appear, often characterized by a cyclical mode of development (Anhaeusser 1971b, Goodwin 1971, Viljoen & Viljoen 1971). Chemically, these volcanic assemblages are similar to the modern volcanic complexes of island arcs, supporting the view that the greenstone belts evolved partly as island arc-like complexes in an oceanic-type geotectonic setting (Anhaeusser 1973b, Engel 1970, Engel & Kelm 1972, Folinsbee et al 1968, Goodwin & Ridler 1970, Hart et al 1970, White et al 1971).

Details of Archean volcanism have yet to be resolved as geochemical data are available only from selected areas in Canada, Western Australia, and southern Africa. Nevertheless, a remarkably similar pattern of development is emerging from these widely separated regions. Hallberg (1972) has demonstrated, however, that Canadian greenstone belts appear to have more abundant andesites. He holds the view that in some instances the proportion of intermediate volcanic material has been greatly overemphasized, and that the assessment of these belts in terms of calc-alkaline differentiation sequences, rather than tholeiitic piles, may be un-warranted.

The lower, predominantly volcanic, assemblages of greenstone belts are frequently overlain by a wide variety of sediments (Anhaeusser et al 1969), the latter sometimes associated with subordinate volcanism. The available chemistry of these lavas appears to reflect the changing conditions of the crust as the greenstone belts evolved (Anhaeusser 1971b). Near Kirkland Lake, Ontario, flows of alkali mafic trachytes and leucitic lavas and pyroclasts occur with Timiskaming sediments (Cooke & Moorhouse 1969), and in the Barberton area, South Africa, trachyandesites are associated with Fig Tree sediments. Alkalic basalts like these are rare in the Archean, and their position towards the top of greenstone belt piles suggests they may have been contaminated during their upward journey through a progressively thickening protocontinental crust.

The greenstone belt sedimentary sequences generally commence with immature detritus largely derived from the initial volcanic episode and consist mainly of graywacke-shale associations, banded iron formations, jaspilites, and cherts. These predominantly volcanogenic and chemical sediments are often succeeded sequentially by clastic sediments (conglomerates, quartzites, subgraywackes, shales) derived from heterogeneous provenance areas, including granitic-metamorphic terranes.

Petrologic and geochemical studies of Archean sedimentary successions have revealed secular increases of ensialic components with stratigraphic height. Such findings have been used (Condie et al 1970) to support arguments favoring the progressive unroofing of prevolcanic, granite-migmatite basements. In the Barberton greenstone belt in South Africa, however, the basal volcanic pile, which is approximately 15 km thick, contains no unequivocal clastic ensialic components (Anhaeusser 1973b). Instead, the earliest sediments interlayered with the volcanic members consist of chemical precipitates (banded iron formations, cherts, carbonates) and volcanogenic debris (tuffs, agglomerates). Anhaeusser maintained that incipient detrital sedimentation commenced only after island arc-like emergence of the volcanic pile. This sedimentation was supplemented by detritus derived from the erosion of granites, gneisses, and migmatites that evolved progressively as a consequence of the foundering and partial melting of primitive ensimatic lithosphere.

Archean greenstone belts or "orogens" have characteristics resembling Pacific/Alpine-type orogenic belts, and attempts have been made to compare and equate the ancient volcanic assemblages with the ophiolite suites of geosynclines or island arcs, while the sedimentary successions have been compared with geosynclinal flysch and molasse assemblages. Anhaeusser et al (1969) contended, however, that the evolutionary development of the granite-greenstone terranes was sufficiently distinctive to warrant consideration in terms other than those generally applied to younger orogenic belts. Support for this viewpoint is provided by Douglas & Price (1972) and Engel & Kelm (1972) who outlined some of the analogous and contrasting features of Archean and post-Archean orogenic belts and their changing tectonic styles.

PRECAMBRIAN GEOTECTONIC EVOLUTION

Australia, Canada, and southern Africa collectively possess a remarkably complete record of crustal evolution spanning more than 3.5 b.y. These areas therefore hold the

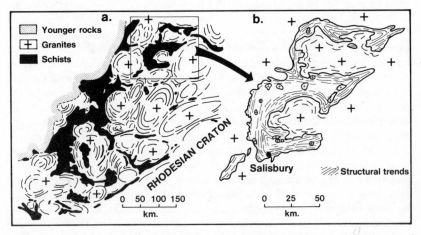

Figure 1   Archean structural style explained by (*a*) the "gregarious batholith" map of Rhodesia (after Macgregor 1951) and (*b*) the "granite-greenstone pattern" exemplified by the arcuation in the Salisbury–Mt. Darwin greenstone belt (after Anhaeusser et al 1969).

clues to many of the problems associated with the earth's early history, and it is from them that we must seek to confirm or reject proposals equating Phanerozoic tectonic styles and events with those now fossilized in Proterozoic and Archean terranes.

Archean Structural Style

Most maps depicting greenstone belts within ancient cratons reveal a distinctive pattern of relationships between the belts themselves and their surrounding granitic terrane. In marked contrast to the extensive, linear, Pacific/Alpine-type orogenic belts, with length-to-width ratios exceeding 100 to 1 (Engel & Kelm 1972), those of the Archean (rarely exceeding 5 to 1) display highly irregular patterns well illustrated by Macgregor's (1951) "gregarious batholith" map of the Rhodesian craton (Figure 1a). Referred to also as the "granite-greenstone pattern" by Anhaeusser et al (1969), the ancient greenstones, which at an earlier stage probably covered the entire Rhodesian and Transvaal cratons of southern Africa (Anhaeusser 1973b), were fragmented, partly assimilated and migmatized, and complexly folded in response to gravitational adjustments during the upwelling of granitic melts. Little doubt exists that the granites were responsible for most of the thermal and dynamic metamorphism of the greenstone belts into which they intruded. They were furthermore responsible for the typical arcuate structures illustrated in Figure 1b.

The distinctive deformational styles of the Archean complexes are unique. Broadly, their tectonic history is seen as having developed in two stages. During the early part of the first stage, an extensive, primitive, ensimatic lithosphere underwent localized partial melting, and calc-alkaline island arc-type volcanism developed in subsiding troughs while adjacent areas were invaded and migmatized by early sodic (trondhjemitic) melts (Anhaeusser 1973a,b). These melts may have been derived by processes involving large-scale gravitational instability of ensimatic lithosphere following upon which the lower part of the basaltic crust may have converted to eclogite ($\rho \simeq 3.5$) overlying mantle peridotite ($\rho \simeq 3.3$) as suggested by Ringwood & Green (1966). This gravity instability would initiate the foundering of lithospheric slabs, eventually causing eclogite masses to descend into the mantle where, at depths of 100–150 km, they would undergo fractional melting giving rise to the calc-alkaline volcanic and plutonic magma series (Green & Ringwood 1968). Liquids of this composition would, due to their lower densities ($\rho \simeq 2.8$), be extruded mainly as lava sequences, whereas the plutonic equivalents would rise diapirically, thereby initiating processes of protocontinental growth and accretion.

On this rapidly changing crust, surviving greenstone rafts would be situated in a thoroughly unstable geotectonic setting—one which would promote gravity slumping of the troughs of lavas and sediments during the latter phases of Stage 1 (Figure 2a). As the gravity-induced deformation proceeded in a manner imitated experimentally by Ramberg (1967), variably plunging isoclinal folds formed in preferentially developing synclinoria (Figure 2b), and steeply inclined longitudinal faults or slides were generated, the latter frequently eliminating intervening anticlinal folds. Deeply infolded greenstone belts may have had their root-zones affected by differential anatectic melting. These melts were probably responsible for the discrete diapiric tonalite/trondhjemite plutons frequently found emplaced around greenstone belt

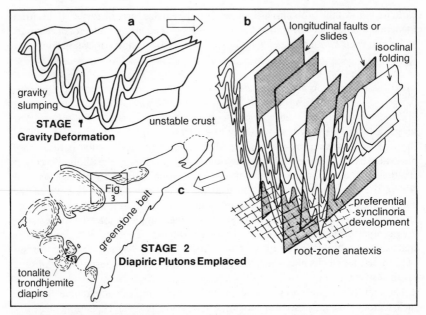

Figure 2   Schematic illustrations depicting the episodic Archean greenstone belt deformation. Stage 1: Gravity deformation on unstable crust; (*a*) passive slumping and warping, (*b*) intensified isoclinal folding, faulting, and root-zone anatexis. Stage 2:Emplacement of diapiric plutons responsible for the structures shown in Figure 3.

margins (Stage 2, Figure 2c) and, less commonly, as small stocks in the axial zones of some greenstone belts (e.g. the Salisbury–Mt. Darwin belt, Rhodesia; Macgregor 1951; Figure 1b). Support for root-zone anatexis follows from gravity studies carried out over some greenstone belts. Despite surface and underground observations of vertical to subvertically dipping lithological units, the gravity investigations suggest the presence of rounded or saucer-shaped keels not extending to great depths.[2]

Structures produced during Stage 1 were predominantly linear features with folds possessing wave lengths and amplitudes smaller than those found in Pacific-Alpine orogenic belts (Engel & Kelm 1972). By contrast, the deformation produced by the Stage 2 diapiric granite plutons was responsible for the intensification of Archean structural complexity. The pluton or, in some cases, batholith emplacement, either singly or collectively, produced many of the striking arcuate schist belt "tongues" found protruding into the surrounding granites. Structures typically resulting from granite diapirism have been described by Clifford (1972). In Figure 3, a further example is provided which schematically illustrates part of the Barberton greenstone belt in South Africa. Briefly, the deformational events appear to have taken place

[2] Average depth of 3–4 km, extending possibly to approximately 8 km (B. W. Darracott, personal communication, 1973 and manuscript in preparation).

sequentially in the following order: 1. The diapiric plutons prized off, stoped, and assimilated much of the greenstone belt margins, often becoming compositionally more basic and full of inclusions as the lavas were engulfed. 2. As the granites rose, concomitant downsagging of the adjacent greenstones occurred and the pluton margins developed a pronounced foliation and lineation. 3. Greenstone xenoliths close to the granite contacts were aligned in the foliation directions of the gneisses, parallel to the greenstone belt margins which also developed a strong schistosity parallel to the granite contacts. 4. Differential compression of the greenstone belt formations produced isoclinal folding, pebble and pillow flattening, and mineral reorientation. 5. A variety of fold styles developed because of the competency contrasts that existed between the various lithological units. 6. Early planes of weakness were reactivated, producing transcurrent faults, drag and disharmonic folds, and numerous attendant second and higher order faults, fractures, and joints. 7. Late-stage vertical adjustments produced superimposed small- and large-scale folds (conjugate, chevron, and kink-band folds).

In summary, therefore, the distinctive structural style of the Archean complexes, which initially involved deep, gravity-induced infolding, adds support for a thin, unstable early crust. That the initial deformations developed essentially under

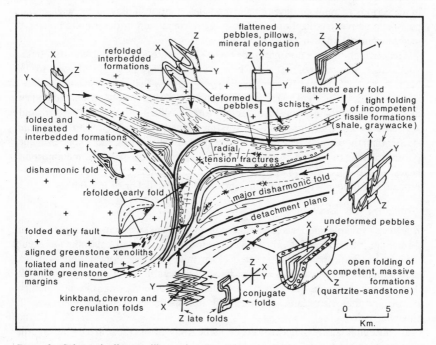

Figure 3   Schematic diagram illustrating the main tectonic elements and strain indicators in an Archean greenstone belt deformed by emplacement of marginal diapiric granite plutons. Region depicted forms part of the Barberton greenstone belt, South Africa.

gravitational influences (as opposed to compressional tectonics accompanying crustal shortening) is borne out by the following three features characteristic of most greenstone belts: 1. very low grades of regional (dynamic) metamorphism; 2. the irregular distribution or absence of all-pervasive, thoroughly penetrative structures (cleavage-schistosity); and 3. the preferred tendency for synclines to form and anticlines to be faulted out by high-angled slides.

Metamorphism and Plutonism

Archean metamorphism, like Archean tectonism, is distinctively a low-pressure/high-temperature (Abukuma) variety. Absent, except for an isolated occurrence described from India (reported in Shackleton 1973), are blueschist metamorphic assemblages like those found in Phanerozoic orogenic belts. Absent too, is the "paired metamorphism" so characteristic of the circum-Pacific island arcs. Thus on two counts this metamorphic disparity has to be accounted for by proponents of Archean sea-floor spreading and plate tectonics.

Other significant features of Archean metamorphism (Engel 1970) include the telescoped nature of the metamorphic aureoles at granite-greenstone contacts and the bimodal nature of the mineral facies (amphibolite-greenschist). The metamorphism and deformation are intimately related, the events usually comprising two or more periods of mineral development and deformation, followed by widespread,. post-tectonic recrystallization and late-stage retrogression in the waning phases of metamorphism (Anhaeusser et al 1969).

Areas of highest metamorphic grade occur adjacent to K-enriched, granitic phases (including stocks and pegmatites), whereas the metamorphic grade adjacent to the Na-rich plutons is generally low. This supports the view that at the time of emplacement the sodic diapirs had lost much of their heat and were semi-consolidated bodies, more competent than the greenstones they forced aside.

Archean geothermal gradients, it was mentioned earlier, appear to have been steeper than those today. The high heat flow presumably enabled metamorphism and partial melting to occur at shallow depths (hence the low-pressure metamorphic facies series). This heat was probably dissipated and absorbed in the destructive processes of lithospheric conversion (ensimatic to ensialic) prior to 3.0 b.y.

Archean Plate Tectonics

Evidence available supports the suggestion that greenstone belts developed in, or formed an integral part of, an oceanic-type (ensimatic) crustal environment. Some form of sea-floor spreading this far back in time does not appear unreasonable, although just how extensive any such oceanic regions might have been in the Archean is not known. Hess (1962) postulated that at an early stage the earth had no oceans and that the oceans, seawater, and atmosphere evolved progressively with time.

Because analogies exist between events in the Archean and those in the Phanerozoic orogenic belts and arc-trench systems (even gold mineralization is common to both environments), several attempts have been made to relate plate tectonic theory to the Archean (Dewey & Horsfield 1970, White et al 1971, Talbot 1973). Undoubtedly, most support for this hinges around attempts to explain the nature and diversity of Archean volcanism. Geochemical characteristics, discussed earlier, permit direct comparisons to be made with both abyssal and island arc volcanism and the inevitable consequence (whether justifiable or not) has led to the modern mechanism being related to the ancient setting.

Archean Tectonics in Greenland

Discussions of Archean tectonics and crustal development have drawn largely on evidence from the greenstone-granite terranes. Until recently, little was known of the Greenland Archean, which consists of high-grade metamorphic gneisses and amphibolites (McGregor 1973, Windley & Bridgwater 1971). Strongly contrasting tectonic events are envisaged for this region by Bridgwater et al (1974) who summarized the Greenland findings and suggested that a dominantly horizontal tectonic regime had been responsible for the thickening of the sialic crust in this area between 2.80–3.75 b.y. ago.

The following points listed by Bridgwater et al (1974) outline the main features of the Greenland Archean, all of which differ markedly from those of the greenstone-granite terranes: 1. Supracrustal rocks occur only as relatively thin tabular units intercalated with and folded together with rocks of granitic origin. 2. The metamorphic grade is high (amphibolite to granulite facies) both in the supracrustal rocks and in the granite-derived gneisses. 3. Original structures are only rarely preserved. 4. The dominant movements were subhorizontal and tangential (thrusts and isoclinal recumbent folds), and were succeeded by vertical and transcurrent movements only at a late tectonic stage. 5. Deformation is intense and penetrates the whole complex. 6. Granitic parents of the gneisses (the dominant element of the stratigraphy) were injected laterally as subhorizontal sheets. 7. Layered anorthosite-leucogabbro rocks are important in the stratigraphy.

These authors concluded that the differences between the high- and low-grade terranes were due to their different tectonic settings and that there appears to be a wider range of tectonic regimes within the Archean than has generally been recognized.

PROTEROZOIC GEOTECTONIC EVOLUTION

Today we may confidently state that the continents drifted apart from one or more macrocontinents (Pangaea, Laurasia, Gondwanaland) during Mesozoic times. Convincing plate tectonic arguments have also lent support to Paleozoic drift. The numerous similarities between Archean orogens and those of the modern Pacific/Alpine-type likewise make any suggestion of ancient sea-floor spreading most alluring. It is, however, the intervening timespan of the Proterozoic that provides "drifters" with their greatest challenge.

Cratonization and the Development of Continents

The Proterozoic has generally been used to define a period of earth history extending between approximately 0.7–2.5 b.y. Originally intended to express *time of first life*, the term "Proterozoic" has fallen into misuse and now connotes, to many, the period immediately following the stabilization of the granite-greenstone terranes and prior to the Paleozoic. As this stabilization did not occur simultaneously on all continents, the tectonic usage of the term Proterozoic should be avoided. In South Africa, for example, the crust was stable 3.0 b.y. ago and a number of interior cratonic basins developed prior to 2.5 b.y. (Pongola, Messina, Dominion Reef, and Witwatersrand sequences). Stages in the evolution of the crust of southern Africa were outlined by Anhaeusser (1973b), who attempted to demonstrate how the developing Archean granite suite had successively invaded, fragmented, and migmatized an originally ensimatic crust, thereby converting it into a ensialic cratonic nucleus that underwent progressive crustal thickening from below (granite underplating; Engel 1970).

The vast granite "flooding" in the Archean was unique in the earth's history, being unrelated to any currently accepted process involving metamorphism and orogenesis. These anorogenic granites, it has been suggested, were largely derived from a fractional process in the mantle (Roering 1967). The secular influx of granite resulted in protocontinental growth and the eventual development of vast shield areas which formed the continental nuclei.

It has long been debated whether or not the continental areas have grown and differentiated through geologic time. Engel (1963) argued that the tectonic structure of North America supported the concept of continental growth about an older core by marginal accretion. This, he believed, resulted from successive sheaths of granite and granitized sediment being welded to continent-ocean interfaces following reactions (orogenies) incorporating preexisting sial, oceanic crust, and mantle. Similar but unidirectional continental accretion appears possible across Australia and Antarctica, whereas in South Africa there is some support for east to west accretionary development dating back from 3.5 b.y. (Anhaeusser 1973b).

Whatever processes were involved, it is clear that the deeply infolded Archean greenstone depositories were eventually incorporated into the evolving cratons which became extensive stable platforms on which interior basins developed—the latter only mildly folded, faulted, and regionally metamorphosed (Anhaeusser 1973b, Douglas & Price 1972, Goodwin 1971). Absent in the Proterozoic, except for the apparently unique Coronation geosyncline, located on the northwest margin of the Canadian shield (Hoffman 1973), are geosynclines or orogenic belts of the Phanerozoic Pacific/Alpine type.

The degree of preservation and the vast expanses of the sequences that developed on the shields suggest that long-continued relative tectonic stability prevailed. Supporting this is the nature of the sediment found in the interior basins. In contrast to the thick, immature detritus, deposited rapidly in subsiding Archean greenstone depositories, the sediments in Proterozoic basins are comprised of mature orthoquartzites, dolomites, limestones, chert, and banded and granular iron formations. The volcanic associations likewise show change from essentially sequential, K-deficient, komatiitic, oceanic tholeiitic, and calc-alkaline magma types to extensive (flood basalt), thick, nonsequential, chemically diverse volcanism including continental tholeiitic basalts, andesites, trachytes, and rhyolites, showing evidence of K-enrichment.

Proterozoic Tectonics

Two structural regimes—the cratons and the mobile belts—predominated during the Proterozoic. The cratons, which are variably sized, consist of complex granitic terranes incorporating the Archean greenstone belts. They form composite crustal units unaffected by any major tectono-thermal events for the last 2.5 b.y. By contrast, the mobile belts are linear, or curvilinear, high-grade metamorphic belts, which tend to encircle the ancient cratonic nuclei of shield areas. They may incorporate one or more tectono-thermal event in a single belt (Churchill, Limpopo, Mozambique).

As mentioned earlier, the nature of the sediment deposited on the cratons suggests that relative tectonic stability prevailed over these platformal regions. Epeirogenic adjustments from time to time caused faulting, fissuring, and rifting. Block faults produced horsts and grabens, and basin edge uplift resulted in the development of yoked basins, arkosic wedges, and alluvial fans. Structures once initiated were frequently recurrent and self-perpetuating, and provided local controls on erosion, sediment-variety, and depositional styles.

The interior basins, in places, demonstrate a migratory shift of basin axes across cratons (e.g. Kaapvaal craton, South Africa; Anhaeusser 1973b). This phenomenon possibly results from epeirogenic tilting of the cratons. A broad depositional periodicity may also be found on the cratons with cycles of uplift, erosion, and deposition occupying time intervals ranging mainly between 200–300 m.y. Proterozoic magmatic activity is largely a response to crustal flexuring accompanied by fissure volcanism, emplacement of basic dykes and sills, and less frequently, the intrusion of massive layered igneous complexes.

While comparative calm prevailed on the cratons, the surrounding mobile belts were severely tectonized. The changes from one tectonic regime to the other occur over short distances (10–100 km), the boundaries between the structural provinces coinciding with the limits of tectonic overprinting of pre-existing continental basement or supracrustal rocks. These boundaries are not suture lines marking the sites of former subduction zones (Douglas & Price 1972), as the mobile belts lack calc-alkaline volcanic arc assemblages, ophiolite suites, and low-pressure metamorphic facies assemblages.

Precambrian Mobile Belts

Traversing the shields and surrounding the cratons are the mobile belts which form an integral part of the crystalline continental basement. Characterized by complex folding, faulting, high-grade metamorphism, and granitization, these areas are considered to represent reworked cratonic material, with or without infolded younger supracrustal rocks (Anhaeusser et al 1969, Douglas & Price 1972, Wynne-Edwards 1972, Mason 1973). The high-grade metamorphism, linear transcurrent dislocation, and high heat flow (Carte & Van Rooyen 1969) suggest that these belts are related to mantle activity.

Argument that the mobile belts represent suture zones resulting from continent-continent collision is not well supported by those familiar with the circum-cratonic mobile belts of central and southern Africa. Apart from the geometrical difficulties inherent in imposing collision-type reasoning to the areas rimming the old cratons, many of these mobile belts provide evidence suggesting that they were stable areas prior to ca 3.1 b.y. (Van Breemen & Dodson 1972). In the Limpopo belt of southern Africa, the Messina formation represents a cover sequence of about this age, onto which are superimposed a number of subsequent tectono-thermal events extending from 2.69 b.y. to the present (Mason 1973).

Experimental studies suggest 15–30 km depths of burial would be necessary to produce the distinctive high-pressure/high-temperature mineral assemblages (granulites) characteristic of the mobile belts (Wynne-Edwards 1972). Instead of collision zones, these high-grade gneiss belts appear rather to have been involved in vertical tectonic motions, with superimposed strain caused by transcurrent movements, some possibly caused by craton rotation (e.g. Rhodesia). High heat flow measurements and positive gravity anomalies recorded over mobile belts suggest a shallower mantle in these regions. Tectonic disturbances within or adjacent to these weak zones of the shields (areas of folding, dislocation, and brittle fracture) are responsible for igneous activity and give rise to anorthositic complexes like those in the Grenville Province, West Greenland, the Limpopo belt, and elsewhere (Wynne-Edwards 1971, Windley & Bridgwater 1971, Mason 1973). They may also have triggered the mafic and ultramafic magmatism found in some of these belts (Giles Complex) as well as on the cratons (Great Dyke, Bushveld, Sudbury).

Some of the mobile belts on the continents may be related to present-day transform faults in the oceans (De Loczy 1970). De Loczy maintained, for example, that the numerous indentations or offsets of the South American coastline are linear features traceable directly into mobile belts on the continents. Should their relationship to transform faults be proved, it may suggest that the continents have been sited over a mantle convection system that has remained stationary since initiation (possibly during the Archean), and over with Pangaea, Laurasia, or Gondwanaland were positioned before being rent apart by Mesozoic drift.

In summary, the mobile belts with their high-grade metamorphism, circumcratonic disposition, variable geochronological histories, and early development of platform-type cover sequences, suggest that mechanisms other than continental collision will have to be sought to explain their particular characteristics.

Proterozoic Plate Tectonics

Although much theoretical support exists for Proterozoic plate tectonics and seafloor spreading, few accounts are available where these ideas have been tested. Gibb (1971) and Gibb & Walcott (1971) proposed that the Slave and Superior Provinces of the Canadian shield were once contiguous and formed a single protocontinent that split apart 2.4–1.6 b.y. ago leaving the great arc of eastern Hudson Bay. The boundary between the Churchill and Superior structural provinces they regarded as a suture, 3200 km in length, characterized by distinctive rock types and geophysical anomalies. These authors suggested, furthermore, that rifting and oceanic crust developed between the Slave and Superior provinces enabling sea-floor spreading to take place. No explanation was given as to how the Churchill Province (minimum width 1000 km) had reconsolidated to an ensialic crust sufficiently stable to withstand shelf and interior basin deposition of extensive platform dolomites, limestones, blanket orthoquartzites, arkosic wedges, and "Superior-type" iron formations, of the type described by Davidson (1972).

Paleomagnetic data appear to favor continents, or fragments of continents, drifting about more than once in the geologic past (McElhinny & Briden 1971), but paleomagnetists concede that much has still to be done to provide convincing support or rejection of this possibility.

Cluster patterns of available radiometric age data on continental basement rocks throughout the world did not, according to Hurley & Rand (1969), support pre-Mesozoic drift. They concluded that two (or one) ancient nuclei had remained as coherent masses with growth patterns largely peripheral and concentric about the ancient nuclei in their predrift positions. Engel & Kelm (1972) likewise rejected pre-Permian global tectonics concluding that

> the ensialic nature of most Proterozoic orogenic belts and their lithic components as well as their interrelations to Archean terranes indicate that they evolved in large part on and between the closely spaced Archean protocontinental clusters. These underwent deformation, refractionation, and thickening as the Proterozoic orogens evolved, but without the major fragmentation, widespread dispersion, and recollision of continents typical of the post-Permian.

DISCUSSION

The evolution of the earth can be divided into three principal stages. During the Archean, which ended ca 2.5 b.y. ago, foundations were laid for the development of the continental masses. The continental nuclei, comprising the ancient greenstone-granite terranes, may have evolved diachronously during the earliest geologic times, but from about 3.0 b.y. ago settled down to a unified steady state of growth and consolidation. Initial events were complex and have been made obscure by subsequent crustal reworking.

Although the nature of the primitive crust is still debated, evidence from the oldest recognizable rock sequences—the ancient greenstone belts—points to an early ensimatic lithosphere which underwent progressive phases of fragmentation, migmatization, and granitization. The ancient greenstone belts appear as remnants of this primitive environment and are analogous to modern island arc systems. Despite the many similarities, however, the earliest Precambrian displays a unique sequence of events. Differences in the types and styles of volcanism, sedimentation, deformation, and metamorphism support this viewpoint. The primitive, unstable environment underwent adjustment by gravitational influences manifested in the "flooding by granite diapirism" of vast ensimatic lithospheric slabs. The step-wise development of anorogenic granites culminated in the welding together of protocontinental landmasses which became the shield nuclei (cratons).

Once established as unified shield areas, the continents continued to grow thicker and more widespread, and much of the Proterozoic was given over to consolidation and stabilization built largely upon Archean foundations. Intracontinental orogenesis or mobile belt tectonism developed as a response to mantle activity beneath the landmasses, and vertical tectonic movements exposed extensive areas of high-grade metamorphic terrane. In these, as well as in neighboring cratonic areas, igneous magmatic activity often accompanied the mobile belt readjustments.

The enlarged ensialic landmasses, continuously harassed by mantle processes, themselves became progressively unwieldy and, as a consequence, fragmented in a manner accounted for in terms of the new global tectonics. The continents did not split apart in any random manner but broke up in Phanerozoic times mainly along linear mobile belts that had undergone successive stages of tectono-thermal reactivation, in some cases traceable back in time into the early Proterozoic or Archean.

ACKNOWLEDGMENTS

Thanks are extended to my colleague, Dr. A. Button, for his very helpful criticism of the original manuscript. Useful additional comments and suggestions were offered by Prof. R. Hargraves, Princeton University; Prof. R. W. Hutchinson, University of Western Ontario; and Dr. R. Mason, J.C.I. Geological Research Department, Randfontein, Transvaal.

Literature Cited

Anhaeusser, C. R. 1971a. The Barberton Mountain Land, South Africa—a guide to the understanding of the Archaean geology of Western Australia. *Spec. Publ. Geol. Soc. Aust.* 3:103–19

Anhaeusser, C. R. 1971b. Cyclic volcanicity and sedimentation in the evolutionary development of Archaean greenstone belts of shield areas. *Spec. Publ. Geol. Soc. Aust.* 3:57–70

Anhaeusser, C. R. 1973a. The geology and geochemistry of the Archaean granites and gneisses of the Johannesburg-Pretoria dome. *Spec. Publ. Geol. Soc. S. Afr.* 3:361–85

Anhaeusser, C. R. 1973b. The evolution of the early Precambrian crust of southern Africa. *Phil. Trans. Roy. Soc. London. Ser. A* 273:359–88

Anhaeusser, C. R., Mason, R., Viljoen, M. J., Viljoen, R. P. 1969. A reappraisal of some aspects of Precambrian shield geology. *Bull. Geol. Soc. Am.* 80:2175–2200

Aubouin, J. 1965. *Geosynclines.* Amsterdam: Elsevier. 335 pp.

Baragar, W. R. A., Goodwin, A. M. 1969. Andesites and Archean volcanism of the Canadian shield. *Oregon. Dep. Geol. Miner. Ind. Bull.* 65:121–42

Barazangi, M., Dorman, J. 1969. World seismicity map of E.S.S.A. coast and geodetic survey epicenter data for 1961–1967. *Bull. Seismol. Soc. Am.* 59:369–80

Bird, J. M., Dewey, J. F. 1970. Lithosphere plate-continental margin tectonics and the evolution of the Appalachian orogen. *Bull. Geol. Soc. Am.* 81:1031–60

Bridgwater, D., McGregor, V. R., Myers, J. S. 1974. A horizontal tectonic regime in the Archaean of Greenland and its implications for early crustal thickening. *Precambrian Res.* 1(2):In press

Brooks, C., Hart, S. 1972. An extrusive basaltic komatiite from a Canadian Archean metavolcanic belt. *Can. J. Earth Sci.* 9:1250–53

Cann, J. R. 1971. Major element variations in ocean floor basalts. *Phil. Trans. Roy. Soc. London. Ser. A* 268:495–505

Carte, A. E., Van Rooyen, A. I. M. 1969. Further measurements of heat flow in South Africa. *Spec. Publ. Geol. Soc. S. Afr.* 2:445–48

Clifford, P. M. 1972. Behavior of an Archean granitic batholith. *Can. J. Earth Sci.* 9:71–77

Coleman, R. G. 1971. Plate tectonic emplacement of upper mantle peridotites along continental edges. *J. Geophys. Res.* 76:1212–22

Condie, K. C. 1972. A plate tectonics evolutionary model of the South Pass Archean greenstone belt, southwestern Wyoming. *Int. Geol. Congr., 24th, Montreal, Canada* 1:104–12

Condie, K. C., Macke, J. E., Reimer, T. O. 1970. Petrology and geochemistry of early Precambrian graywackes from the Fig

Tree Group, South Africa. *Bull. Geol. Soc. Am.* 81 : 2759–76

Cooke, D. L., Moorhouse, W. W. 1969. Timiskaming volcanism in the Kirkland Lake area, Ontario, Canada. *Can. J. Earth Sci.* 6 : 117–32

Crook, K. A. W. 1969. Contrasts between Atlantic and Pacific geosynclines. *Earth Planet. Sci. Lett.* 5 : 429–38

Davidson, A. 1972. The Churchill province. In *Variations in Tectonic Styles in Canada,* ed. R. A. Price, R. J. W. Douglas, 381–433. Toronto: Geol. Assoc. Can. Spec. Pap. 11. 688 pp.

De Loczy, L. 1970. Role of transcurrent faulting in South American tectonic framework. *Am. Assoc. Petrol. Geol. Bull.* 54 : 2111–19

Dewey, J. F., Bird, J. M. 1970. Mountain belts and the new global tectonics. *J. Geophys. Res.* 75 : 2625–47

Dewey, J. F., Horsfield, B. 1970. Plate tectonics, orogeny and continental growth. *Nature* 255 : 521–25

Dickinson, W. R. 1971. Plate tectonic models of geosynclines. *Earth Planet. Sci. Lett.* 10 : 165–74

Dietz, R. S. 1961. Continent and ocean basin evolution by spreading of the sea floor. *Nature* 190 : 854–57

Douglas, R. J. W., Price, R. A. 1972. Nature and significance of variations in tectonic styles in Canada. In *Variations in Tectonic Styles in Canada,* ed. R. A. Price, R. J. W. Douglas, 625–88. Toronto: Geol. Assoc. Can. Spec. Pap. 11. 688 pp.

Engel, A. E. J. 1963. Geologic evolution of North America. *Science* 140 : 143–52

Engel, A. E. J. 1970. The Barberton Mountain Land: clues to the differentiation of the earth. In *Adventures in Earth History,* ed. P. E. Cloud, 431–45. San Francisco: Freeman. 992 pp.

Engel, A. E. J., Engel, C. G., Havens, R. G. 1965. Chemical characteristics of oceanic basalts and the upper mantle. *Bull. Geol. Soc. Am.* 76 : 719–34

Engel, A. E. J., Kelm, D. L. 1972. Pre-Permian global tectonics: A Tectonic test. *Bull. Geol. Soc. Am.* 83 : 2325–40

Folinsbee, R. E., Baadsgaard, H., Cumming, G. L., Green, D. C. 1968. A very ancient island arc. In *The Crust and Upper Mantle of the Pacific Area,* ed. L. Knopoff, C. L. Drake, P. J. Hart, Geophys. Monogr. 12 : 441–48. Washington, DC: Am. Geophys. Union

Fyfe, W. S. 1973. The granulite facies, partial melting and the Archaean crust. *Phil. Trans. Roy. Soc. London. Ser. A* 273 : 457–61

Gibb, R. A. 1971. Origin of the great arc of eastern Hudson Bay: A Precambrian continental drift reconstruction. *Earth Planet. Sci. Lett.* 10 : 365–71

Gibb, R. A., Walcott, R. I. 1971. A Precambrian suture in the Canadian shield. *Earth Planet. Sci. Lett.* 10 : 417–22

Gilluly, J. 1971. Plate tectonics and magmatic evolution. *Bull. Geol. Soc. Am.* 82 : 2383–96

Gilluly, J. 1972. Tectonics involved in the evolution of mountain ranges. In *The Nature of the Solid Earth,* ed. E. C. Robertson, 406–39. New York: McGraw-Hill. 677 pp.

Glikson, A. Y. 1971. Primitive Archaean element distribution patterns: chemical evidence and geotectonic significance. *Earth Planet. Sci. Lett.* 12 : 309–20

Glikson, A. Y. 1972. Early Precambrian evidence of a primitive ocean crust and island nuclei of sodic granite. *Bull. Geol. Soc. Am.* 83 : 3323–44

Goodwin, A. M. 1971. Metallogenic patterns and evolution of the Canadian shield. *Spec. Publ. Geol. Soc. Aust.* 3 : 157–74

Goodwin, A. M., Ridler, R. H. 1970. The Abitibi orogenic belt. In *Symposium on Basins and Geosynclines of the Canadian Shield,* ed. A. J. Baer, 1–30. Ottawa: Geol. Surv. Can. Pap. 70–40. 265 pp.

Green, D. H. 1972a. Magmatic activity as the major process in the chemical evolution of the earth's crust and mantle. *Tectonophysics* 13 : 47–71

Green, D. H. 1972b. Archaean greenstone belts may include terrestrial equivalents of lunar maria? *Earth Planet. Sci. Lett.* 15 : 263–70

Green, D. H., Ringwood, A. E. 1967. An experimental investigation of the gabbro to eclogite transformation and its petrological applications. *Geochim. Cosmochim. Acta* 31 : 767–833

Green, T. H., Ringwood, A. E. 1968. Genesis of the calc-alkaline igneous rock suite. *Contrib. Mineral. Petrol.* 18 : 105–62

Hallberg, J. A. 1972. Geochemistry of Archaean volcanic belts in the Eastern Goldfields region of western Australia. *J. Petrol.* 13 : 45–56

Hamilton, W. 1970. The Uralides and the motion of the Russian and Siberian platforms. *Bull. Geol. Soc. Am.* 81 : 2553–76

Hart, S. R., Brooks, C., Krogh, T. E., Davis, G. L., Nava, D. 1970. Ancient and modern volcanic rocks: a trace element model. *Earth Planet. Sci. Lett.* 10 : 17–28

Hess, H. H. 1962. History of ocean basins. In *Petrologic Studies: A Volume to Honor A. F. Buddington,* ed. A. E. J. Engel, H. L.

James, B. F. Leonard, 599–620. Boulder: Geol. Soc. Am. 660 pp.

Hoffman, P. 1973. Evolution of an early Proterozoic continental margin: the Coronation geosyncline and associated aulacogens of the northwestern Canadian shield. *Phil. Trans. Roy. Soc. London. Ser. A* 273:547–81

Hurley, P. M., Rand, J. R. 1969. Pre-drift continental nuclei. *Science* 164:1229–42

Isacks, B., Oliver, J., Sykes, L. R. 1968. Seismology and the new global tectonics. *J. Geophys. Res.* 73:5855–99

Jakeš, P., White, A. J. R. 1972. Major and trace element abundances in volcanic rocks of orogenic areas. *Bull. Geol. Soc. Am.* 83:29–40

Kuno, H. 1959. Origin of Cenozoic petrographic provinces of Japan and surrounding area. *Bull. Volcanol.* 20:37–76

Le Pichon, X. 1968. Sea-floor spreading and continental drift. *J. Geophys. Res.* 73:3661–97

Macgregor, A. M. 1951. Some milestones in the Precambrian of Southern Rhodesia. *Proc. Trans. Geol. Soc. S. Afr.* 54:xxvii–lxxi

Mason, R. 1973. The Limpopo mobile belt—southern Africa. *Phil. Trans. Roy. Soc. London. Ser. A* 273:463–85

Matsumoto, T. 1967. Fundamental problems in the circum Pacific orogenesis. *Tectonophysics* 4:595–613

McElhinny, M. W., Briden, J. C. 1971. Continental drift during the Palaeozoic. *Earth Planet. Sci. Lett.* 10:407–16

McGregor, V. R. 1973. The early Precambrian gneisses of the Godthåb District, West Greenland. *Phil. Trans. Roy. Soc. London. Ser. A* 273:343–58

McKenzie, D. P. 1969. Speculations on the consequences and causes of plate motions. *Geophys. J.* 18:1–32

McKenzie, D. P. 1972. Plate tectonics. In *The Nature of the Solid Earth,* ed. E. C. Robertson, 323–60. New York: McGraw-Hill. 677 pp.

McKenzie, D. P., Parker, R. L. 1967. The North Pacific: an example of tectonics on a sphere. *Nature* 216:1276–80

Mitchell, A. H., Reading, H. G. 1969. Continental margins, geosynclines, and ocean floor spreading. *J. Geol.* 77:629–46

Miyashiro, A. 1961. Evolution of metamorphic belts. *J. Petrol.* 2:277–311

Miyashiro, A., Shido, F., Ewing, M. 1969. Diversity and origin of abyssal tholeiite from the Mid-Atlantic Ridge near 24° and 30° north latitude. *Contrib. Mineral. Petrol.* 23:38–52

Moorbath, S., O'Nions, R. K., Pankhurst,

R. J., Gale, N. H., McGregor, V. R. 1972. Further rubidium-strontium age determinations on the very early Precambrian rocks of the Godthaab District, West Greenland. *Nature Phys. Sci.* 240:78–82

Morgan, W. J. 1968. Rises, trenches, great faults, and crustal blocks. *J. Geophys. Res.* 73:1959–82

Oxburgh, E. R., Turcotte, D. L. 1971. Origin of paired metamorphic belts and crustal dilation in island arc regions. *J. Geophys. Res.* 76:1315–27

Poldervaart, A. 1955. Chemistry of the earth's crust. In *Crust of the Earth,* ed. A. Poldervaart, 119–44. Boulder: Geol. Soc. Am. Spec. Pap. 62. 762 pp.

Ramberg, H. 1964. A model for the evolution of continents, oceans and orogens. *Tectonophysics* 2:159–74

Ramberg, H. 1967. *Gravity Deformation and the Earth's Crust as Studied by Centrifuged Models.* London: Academic. 214 pp.

Ringwood, A. E. 1969. Composition and evolution of the Upper mantle. In *The Earth's Crust and Upper Mantle,* ed. P. J. Hart, Geophys. Mongr. 13:1–17. Washington DC: Am. Geophys. Union

Ringwood, A. E., Green, D. H. 1966. An experimental investigation of the gabbro-eclogite transformation and some geophysical implications. *Tectonophysics* 3:383–427

Roering, C. 1967. Non-orogenic granites in the Archaean geosyncline of the Barberton Mountain Land. *Inform. Circ. 35, Econ. Geol. Res. Unit, Univ. Witwatersrand, Johannesburg.* 13 pp.

Runcorn, S. K. 1962. Palaeomagnetic evidence for continental drift and its geophysical cause. In *Continental Drift,* ed. S. K. Runcorn, 1–40. New York: Academic. 338 pp.

Shackleton, R. M. 1973. Problems of the evolution of the continental crust. *Phil. Trans. Roy. Soc. London. Ser. A* 273:317–20

Talbot, C. J. 1973. A plate tectonic model for the Archaean crust. *Phil. Trans. Roy. Soc. London. Ser. A* 273:413–27

Van Breemen, O., Dodson, M. H. 1972. Metamorphic chronology of the Limpopo belt, southern Africa. *Bull. Geol. Soc. Am.* 83:2005–18

Viljoen, M. J., Viljoen, R. P. 1969a. The geology and geochemistry of the lower ultramafic unit of the Onverwacht Group and a proposed new class of igneous rocks. *Spec. Publ. Geol. Soc. S. Afr.* 2:55–85

Viljoen, M. J., Viljoen, R. P. 1969b. The geochemical evolution of the granitic rocks of the Barberton region. *Spec. Publ.*

Geol. Soc. S. Afr. 2:189–218

Viljoen, R. P., Viljoen, M. J. 1971. The geological and geochemical evolution of the Onverwacht volcanic group of the Barberton Mountain Land, South Africa. *Spec. Publ. Geol. Soc. Aust.* 3:133–49

Vine, F. J. 1966. Spreading of the ocean floor —new evidence. *Science* 154:1405–15

Vine, F. J., Matthews, D. H. 1963. Magnetic anomalies over oceanic ridges. *Nature* 199:947–49

White, A. J. R., Jakeš, P., Christie, D. M. 1971. Composition of greenstones and the hypothesis of sea-floor spreading in the Archaean. *Spec. Publ. Geol. Soc. Aust.* 3:47–56

Wilson, J. F. 1973. The Rhodesian Archaean craton—an essay in cratonic evolution. *Phil. Trans. Roy. Soc. London* 273:389–411

Windley, B. F. 1973. Crustal development in the Precambrian. *Phil. Trans. Roy. Soc. London. Ser. A* 273:321–41

Windley, B. F., Bridgwater, D. 1971. The evolution of Archaean low- and high-grade terrains. *Spec. Publ. Geol. Soc. Aust.* 3: 33–46

Wynne-Edwards, H. R. 1972. The Grenville province. In *Variations in Tectonic Styles in Canada,* ed. R. A. Price, R. J. W. Douglas, 263–334. Toronto: Geol. Assoc. Can. Spec. Pap. 11. 688 pp.

Zwart, H. J. 1969. Metamorphic facies series in the European orogenic belts and their bearing on the causes of orogeny. In *Age Relations in High-Grade Metamorphic Terrains,* ed. H. R. Wynne-Edwards, 7–16. Toronto: Geol. Assoc. Can. Spec. Pap. 5. 228 pp.

THE ORIGIN OF BIRDS   ×10032

John H. Ostrom

Department of Geology and Geophysics and the Peabody Museum of Natural History, Yale University, New Haven, Connecticut 06520

The theory of the origin of birds from reptiles has been generally accepted for more than a century. Even before Darwin, some naturalists had recognized that birds and living reptiles possess several anatomical characters in common that are not known in either amphibians or mammals. Following the publication of *Origin of Species,* both Gegenbaur (1863) and W. K. Parker (1864) commented on the affinities between reptiles and birds, but Haeckel (1866) apparently was the first to suggest a direct evolutionary relationship. That relationship was firmly established by T. H. Huxley (1867) in his classic paper on the classification of birds, despite the fact that he did not explicitly suggest derivation of birds from reptiles or propose a specific reptilian ancestor. He did, however, propose a major new taxonomic category, which he termed the Sauropsida, for the inclusion of both reptiles and birds, and he cited 14 anatomical features that occur in birds and reptiles (but not in mammals) as supporting evidence for classifying them together.

Although general agreement about reptilian-avian affinities has prevailed ever since Huxley's classification paper, the specific reptilian ancestor of birds has remained a matter of debate. Over the years, several different reptilian groups have been suggested, but for the past fifty years or more the general consensus has placed the source of birds among a group of primitive archosaurian reptiles of Triassic age—the Thecodontia. Thecodonts are also judged to have given rise to crocodilians, pterosaurs (flying reptiles), and all dinosaurs, as well.

Two separate events hold special significance for the question of bird origins: publication of *Origin of Species* (Darwin 1859) and the discovery in 1861 of the first (recognized) specimen of *Archaeopteryx.* Without the first, the question of origins could not have been raised; and without the second, no answer could possibly satisfy everyone. Indeed, we can well reason that until the first event had occurred, the significance of the *Archaeopteryx* specimens could not be understood. But even today, with our present understanding of the evolutionary process and the existence of five specimens of *Archaeopteryx,* unanimity on the immediate ancestry of birds is highly unlikely.

SOME EARLY IDEAS ON BIRD ORIGINS

Discovery of the single feather impression (von Meyer 1861a) and the first skeletal remains of *Archaeopteryx* (von Meyer 1861b) in slabs of Solnhofen lithographic

55

A

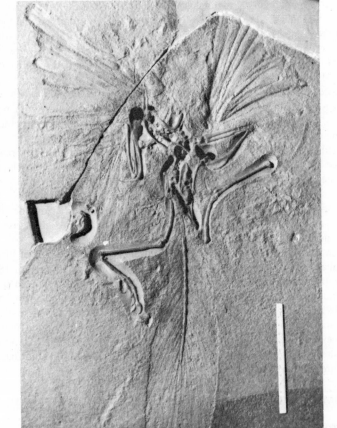

B

limestone (Tithonian age) generated great interest and controversy (see Figure 1). Initially there was some doubt as to the authenticity of these first specimens, chiefly because birds of such great antiquity had not been anticipated. The debate quickly shifted, however, to the question of whether these remains were those of a true bird or just a feathered reptile. Most authorities, incuding Sir Richard Owen and T. H. Huxley, labeled them as unquestionably avian, but von Meyer (1862) was less certain: "The fossil feather of Solenhofen therefore, even if agreeing perfectly with those of our [modern] birds need not necessarily be derived from a bird." (Translated by W. S. Dallas in *Ann. Mag. Natur. Hist. London* (3) 9:369.) Wagner (1861), on the other hand, regarded the remains as reptilian and proposed to name them *Griphosaurus* (enigmatic reptile). From that emphatic conclusion, Wagner went on to make the first published statement on the evolutionary significance of "*Archaeopteryx*" [Translated by W. S. Dallas in *Ann. Mag. Natur. Hist. London* (3) 9:266]:

> In conclusion, I must add a few words to ward off Darwinian misinterpretation of our new Saurian. At the first glance of the *Griphosaurus* we might certainly form a notion that we had before us an intermediate creature, engaged in the transition from the Saurian to the bird. Darwin and his adherents will probably employ the new discovery as an exceedingly welcome occurrence for the justification of their strange views upon the transformation of animals. But in this they will be wrong.

As we know, Wagner's first prediction came true: the specimens of *Archaeopteryx* repeatedly have been cited as evidence for (even proof of) evolution. Six years later, Huxley (1868a, 1868b), then apparently concerned more with defending Darwin's theory of evolution than with establishing bird origins, cited the specimen of *Archaeopteryx* and the very similar small dinosaur *Compsognathus* (also from the Solnhofen limestones). Recognizing that as contemporaneous species one could not be descended from the other, he reasoned that as a reptile-like bird and a bird-like reptile, respectively, those two specimens closed the gap between the reptilian and avian classes, thus reinforcing his "Sauropsida."

Huxley's reputation and legendary eloquence persuaded some, but not all, and for the next several decades various scholars presented their views on the affinities of *Archaeopteryx*. One theory, held by Vogt (1879), Wiedersheim (1884, 1885), and most recently by Petronievics (1921, 1927, 1950), derived *Archaeopteryx* from a lizard ancestry apparently because of its long, lizard-like tail. An opposing school led by Owen (1875), Seeley (1881), and (on occasion) Wiedersheim (1883, 1886) postulated a pterosaur-like ancestor. Following the original example of Huxley, a dinosaurian link was advocated by Marsh (1877), Gegenbaur (1878), Williston (1879), T. J. Parker (1882), Baur (1883, 1884, 1885, 1886), and in recent years by Boas (1930), Lowe (1935, 1944), and Holmgren (1955). Space does not permit a detailed review of each theory or of the respective evidence. It is sufficient to note

←

Figure 1 (opposite)   A: The solitary feather impression reported by von Meyer (1861a), the first indication of birds as ancient as Jurassic. Scale in mm. B: The London specimen of *Archaeopteryx lithographica* discovered less than two months after the feather impression. Scale equals 10 cm.

here that neither the lacertilian nor the pterosaurian ancestry won any lasting support.

The dinosaurian theory, however, was well received for a number of years, although it was not without critics. Paradoxically, it appears that early acceptance of this theory was in large part due to incomplete evidence and faulty understanding of dinosaurian relationships. The roster of dinosaurians known during the last half of the nineteenth century was much smaller than that known today. Also, all dinosaur taxa were grouped together in a single category (Dinosauria) and treated as closely related animals, contrary to the present scheme where they are placed in two separate orders (Ornithischia and Saurischia) that are usually considered as only remotely related, although derived from a common ancestral group—the Thecodontia.

The evidence for dinosaur-bird affinities was multiple. *Iguanodon* and *Hypsilophodon* (ornithischians) possessed distinctly bird-like pelves in which the pubic bone was directed down and backward as in modern birds. *Compsognathus* (a saurischian) had a remarkably bird-like foot and ankle joint. In addition, Huxley (1870) had noted the bird-like design of the ilium in *Megalosaurus* (saurischian) and also had pointed out that the ischium of several other dinosaurs was very bird-like. Some critics, however, such as Dollo (1882, 1883), Dames (1884), W. K. Parker (1887), and Furbringer (1888), argued that such anatomical similarities were adaptive only and had no evolutionary significance. Mudge (1879) pointed to the large number of nonavian features present in various dinosaurian kinds and made the (then) very astute observation that "The dinosaurs vary so much from each other that it is difficult to give a single trait that runs through the whole. But no single genus or set of genera, have many features in common with birds, or a single persistent, typical element or structure which is found in both." (Mudge 1879) Mivart (1881) then made the extreme suggestion that carinate (flying) birds arose from pterosaurs, and ratites (nonflying) birds arose from dinosaurs.

Gradually, as new discoveries came to light, it became apparent that the "Dinosauria" were much more diverse and less closely related than Owen and Huxley had originally thought. Relationships between different kinds became doubtful and the matter of a dinosaurian ancestry of birds became less certain. Although the actual situation was much more complex, it can be reduced to this question: Which is more important, the bird-like pelvis of certain ornithischian dinosaurs or the bird-like foot and tarsus of theropod saurischians? The question was unanswerable then and the theory of dinosaurian ancestry gradually lost favor as it became evident that all varieties of dinosaurs known then were too specialized in one feature or another.

Furbringer (1888), in his classic monograph on birds, concluded that direct descent of birds from any known type of dinosaur was not possible; birds were monophyletic; the resemblances between dinosaurs and birds are all "convergent analogies" and "parallels"; and the stem of birds lies in a common sauropsid ancestor lying between the Dinosauria, the Crocodilia, and the Lacertilia. This was the beginning of the compromise explanation—the common ancestor hypothesis—subsequently advocated by Osborn (1900), Broom (1908, 1913), and Heilmann (1926).

THE COMMON ANCESTOR THEORY

At the time of Furbringer's monograph, the ancestry of the dinosaurs was unknown, hence Furbringer's hypothetical ancestor intermediate between dinosaurs, crocodilians, and lizards. Osborn (1900) also postulated a hypothetical ancestor among proganosaurs (primitive "rhynchocephalian" reptiles) of Permian age as the source of all archosaurs including birds, and argued that the avian line most probably separated off the "bipedal" dinosaurian stem rather than originating independantly from among "quadrupedal" proganosaurs. Again, the critical stage is an unknown, *hypothetical* reptile. Broom (1908) similarly wrote that "probably both dinosaurs and birds were derived" from (Triassic) bipedal "Rhynchocephaloid" reptiles.

Until that time, almost nothing was known about primitive "archosaurs" (which we now group together in the Order Thecodontia). That was especially true of the more progressive thecodonts, the pseudosuchians. Although several specimens that we now recognize as pseudosuchians had been discovered from Late-Triassic rocks at Elgin (i.e. *Ornithosuchus*) prior to the turn of the century, it was not until Broom (1913) recognized the distinctive nature of *Euparkeria* remains from the Lower Triassic of South Africa that a particular reptilian group stood out as the probable ancestral stock of dinosaurs, birds, and all other archosaurs. Broom (1913) was the first to recognize this:

> There cannot, I think, be the slightest doubt that the Pseudosuchia have close affinities with the Dinosaurs, or at least with the Theropoda. This has been recognized by Marsh, v. Huene and others. In fact there seems to me little doubt that the ancestral Dinosaur was a Pseudosuchian. . . . There is still another group to which some Pseudosuchian has probably been ancestral, namely the Birds. For a time one or other of the Dinosaurs was regarded as near the avian ancestor. . . . Seven years ago . . . I argued that the bird had come from a group immediately ancestral to the Theropodous Dinosaurs. The Pseudosuchia, now that it is better known, proves to be just such a group as is required. In those points where we find the Dinosaur too specialized we see the Pseudosuchian still primitive enough.

Thus the stage was set for Heilmann's (1926) classic treatise on *The Origin of Birds*.

The impact of Heilmann's book cannot be exaggerated: the time was exactly right. On the question of bird origins, its impact has been second only to the original discovery of *Archaeopteryz*. So thorough and extensive is the work that few have challenged any part of it and almost no one has raised the question of bird origins since its publication half a century ago. In fact, it has been widely accepted as the last word on the subject and virtually all subsequent workers have simply reiterated Heilmann's conclusions.

The Origin of Birds first reviewed the evidence of relevant fossil birds, namely *Archaeopteryx* (and *Archaeornis*, the Berlin specimen), *Ichthyornis*, and *Hesperornis*. Next, Heilmann compared certain aspects of embryologic stages in living reptiles and birds. This was followed by a discussion of a variety of anatomical topics, chiefly of modern birds, ranging from secondary sexual characters to brain morphology, sense organs, digital claws, and retardation of primary feathers. These three sections were intended to demonstrate the close relationship between reptiles and birds. The final chapter, entitled "The Proavian," consists of a detailed comparison of the skeletal anatomy of *Archaeopteryx* with all of the avian ancestral candidates that had been espoused by previous investigators, namely pterosaurs, predentates (ornithopod ornithischian dinosaurs), coelurosaurs (small theropod saurischian dinosaurs), and pseudosuchians. Heilmann found that among these varied groups, *Archaeopteryx* most closely resembled the coelurosaurian dinosaurs in a number of anatomical features. Yet, he concluded (Heilmann 1926):

> From this it would seem a rather obvious conclusion that it is amongst the coelurosaurs that we are to look for the bird-ancestor. And yet, this would be too rash, for the very fact that the clavicles are wanting would in itself be sufficient to prove that these saurians could not possibly be the ancestors of the birds. . . . We have therefore reasons to hope that in a group of reptiles closely akin to the Coelurosaurs we shall be able to find an animal wholly without the shortcomings here indicated for bird ancestors. Such a group is possibly the Pseudosuchians . . . .

Considering the state of our knowledge then, that conclusion was about the only one possible; ". . . all our requirements of a bird ancestor are met in the Pseudosuchians, and nothing in their structure militates against the view that one of them might have been the ancestor of the birds" (Heilmann 1926).

In the half century since *The Origin of Birds*, a few divergent views have been expressed, but Heilmann's conclusions have held fast. Boas, in 1930, suggested that birds descended from the "Compsognathides," arguing that the "postpubis" (= the backwardly directed pubis of birds) had been overlooked in *Compsognathus*. A much better-known view, which has received considerable criticism, is that of Lowe (1935, 1944), who assessed the osteology of *Archaeopteryx* and *Archaeornis* as nonavian and almost completely reptilian. On those grounds, he concluded that these creatures were not birds but feathered *dinosaurs!*—and too specialized to have given rise to true birds. As I will try to show later, Lowe was essentially correct in his assessment of the skeletal evidence, *but* he failed to recognize how ideally intermediate many of those structures were between reptilian and avian conditions. Most unfortunate, however, was his extreme conclusion that *Archaeopteryx* was a dinosaur.

As Simpson (1946) correctly pointed out: "*Archaeopteryx* and *Archaeornis* are intermediate between reptiles and birds in structure and their bearing on the origin of birds is unchanged by the purely verbal question of whether to call them birds or reptiles" (Simpson 1946). He labeled Lowe's dinosaur designation as "nothing short of fantastic." Any persisting thoughts that dinosaurs might be related to birds would seem to have been permanently laid to rest by Simpson (1946):

> Almost all the special resemblances of some saurischians to birds, so long noted and so much stressed in the literature, are demonstrably parallelisms and convergences. These cursorial forms developed strikingly bird-like characters here and there in the skeleton and in one genus or another. They never showed a general approach to avian structure (as do *Archaeopteryx* and *Archaeornis*). . . . It is not a matter for

argument but a simple fact of observation (if one accepts the published data of Heilmann and other authorities not questioned by Lowe) that *Archaeopteryx* is intermediate between the pseudosuchian *Euparkeria* and *Columba* [pigeon] in every one of
these basic characters.

DeBeer (1954), in his monograph on *Archaeopteryx*, repeated many of Simpson's
conclusions, dismissing the pterosaurs as avian ancestors because of the quite
different design of the flight apparatus, and rejecting dinosaurs because of "spur-
ious resemblances" (the forelimb and backwardly directed pubis) and lost structures
(such as clavicles) which are retained in birds. Like Heilmann and Simpson, deBeer
also concluded that birds are descendant from pseudosuchians.

The very next year, a final attempt to derive birds from coelurosaurian dinosaurs
was made by Holmgren (1955), apparently without knowledge of the emphatic
contrary pronouncements by Simpson and deBeer. Holmgren's effort was unsuccessful and his paper remains little-known today.

And so the matter stands. Over the last two decades most authors (Swinton
1958, 1960, 1964, Heller 1959, Van Tyne & Berger 1959, Welty 1962, deBeer 1964,
Galton 1970, Romer 1966, Pettingill 1970, Brodkorb 1971) have adhered to the
pseudosuchian ancestry of birds and accepted the "convergent or parallel" explanation for the bird-like features of some theropod dinosaurs. The avian identification
of *Archaeopteryx* is accepted by all, although some, like Swinton (1960, 1964),
doubt that it lies on the main line of descent to modern birds. Others, like
Simpson (1946), do favor a main line position and, improbable though that position
may be, there are no features in any of the *Archaeopteryx* specimens that preclude
such a central position. In any case, most authorities today would agree that the
question of bird origins and the origin of *Archaeopteryx* are one and the same
problem. In seeking the identity of bird ancestors, it may be useful to examine
modern birds, but the answer is *not* to be found there—some 150 million or more
years removed from avian beginnings. The place to look is at *Archaeopteryx*.

THE DINOSAURIAN NATURE OF *ARCHAEOPTERYX*

It has repeatedly been observed that the *Archaeopteryx* specimens are very bird-
like, but also possess a number of reptilian features (teeth, long tail, abdominal
ribs or gastralia, etc). The actual fact is that these specimens are not particularly
like *modern* birds at all. If feather impressions had not been preserved in the
London and Berlin specimens, they never would have been identified as birds.
Instead, they would unquestionably have been labeled as coelurosaurian dinosaurs.
Notice that the last three specimens to be recognized were all misidentified at first,
and the Eichstätt specimen for 20 years was thought to be a small specimen of the
dinosaur *Compsognathus*.

In reviewing the transition from reptile to bird, *Archaeopteryx* must occupy a
central position. In this review, *Archaeopteryx* is accepted as a true (albeit very
primitive) bird. I personally believe *Archaeopteryx* lies very close to bird origins
and probably is directly ancestral to all later birds. However, even if it should be an
aberrant form removed from the main lineage of avian evolution, as the oldest known

bird it constitutes the *only* hard anatomical evidence available that pertains to bird origins and the early stages of bird evolution. Of necessity, the following discussion must concentrate on the *nonavian* characters of *Archaeopteryx* and not on the avian features. The latter merely preview what was to follow after *Archaeopteryx*, whereas our objective here is to recognize what came before. Accordingly, the emphasis here will be on the dinosaur-like nature of *Archaeopteryx*. This is not to be confused with the so-called "bird-like" features of dinosaurs so frequently alluded to in the past. For example, notice that Simpson's (1946) remarks explained the "special resemblances of some *saurischians* to birds" as parallelisms and convergences. That does not account for the nonavian, dinosaur-like features of *Archaeopteryx*, which was a contemporary of several very similar theropod dinosaurs!

By now the reader must suspect that I am attempting to resurrect the old dinosaurian theory of bird origins that was so emphatically rejected by Heilmann and others. But it will be recalled that Heilmann rejected any evolutionary relationship between birds and theropod dinosaurs because of the putative absence of clavicles (the presumed antecedants of the avian furcula or wish-bone) in all known theropods. First of all, that absence is not true. Clavicles have been reported in at least two theropods: in *Oviraptor* by Osborn (1924) and in *Segisaurus* by Camp (1936). They also appear to be present in *Velociraptor mongoliensis*, in a specimen just reported by Kielan-Jaworowska & Barsbold (1972). It must also be pointed out that the absence of clavicles in any particular specimen is negative evidence only and thus inconclusive. Unless the clavicle were preserved in natural articulation (as in the three specimens just mentioned), by the nature of its shape it very likely would be misidentified as a rib fragment. But aside from these facts, the absence of clavicles is unimportant anyway because this particular bone is membranous in origin rather than endochondral, and thus may have been present in some theropods as a membranous element—unossified, but not lost—and thus not preservable in a fossil specimen.

It would thus appear that the *only evidence* that Heilmann or anyone else has offered for rejection of a direct evolutionary relationship between dinosaurs and *Archaeopteryx does not in fact exist*. So let us examine the evidence that does exist.

At the present time there are five known specimens of *Archaeopteryx*, aside from the single feather impression (von Meyer 1861a). The original specimen (the so-called London specimen) reported by von Meyer (1861b) is a nearly complete skeleton lacking only one foot, most of both hands and most of the skull, but preserving excellent impressions of the plumage (Figure 1B). That specimen (including the counterpart slab) is in the British Museum (Natural History), London. The second specimen (Figure 2), the Berlin specimen (previously designated *Archaeornis siemensi*), was found in 1877 (Giebel 1877). It is virtually 100% complete and is by far the most spectacular of all the specimens. Currently, it resides in the Museum für Naturkunde in East Berlin. The third specimen (Figure 3A), unrecognized until several years after its discovery, was described by Heller (1959). It consists of a partial skeleton that is largely disarticulated and rather poorly preserved. At the present time, it is displayed in the private industrial museum of Solnhofen Actien-Verein near the village of Solnhofen, Germany. The fourth specimen (Figure 3B),

Figure 2 The Berlin specimen of *Archaeopteryx* once referred to a distinct genus, *Archaeornis siemensi*, but now considered as probably conspecific with the London specimen. Scale equals 10 cm.

Figure 3 A: The third specimen of *Archaeopteryx* cf. *lithographica*, usually referred to as the Maxberg specimen, was unrecognized at first because the feather impressions are obscure. Scale equals 10 cm. B: The Teyler *Archaeopteryx* specimen that was misidentified since 1855 because of its fragmentary nature. Scale equals 10 cm.

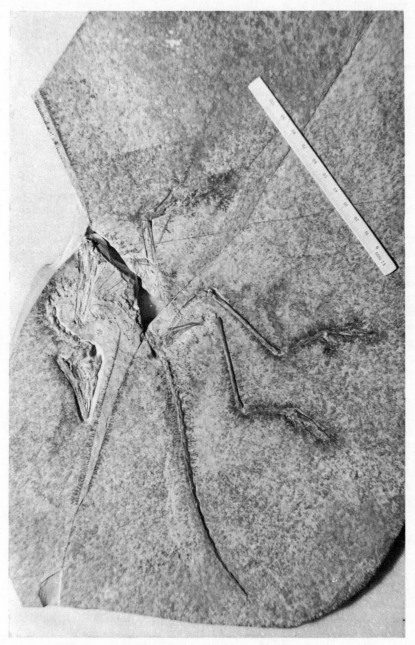

Figure 4   The remarkable Eichstätt specimen of *Archaeopteryx* sp. reported in 1973. For twenty years it was believed to be a small theropod dinosaur, *Compsognathus*. Scale equals 10 cm.

on display in the Teyler Museum, Haarlem, Netherlands, was originally discovered in 1855, but because of its fragmentary nature and the very faint feather impressions, it was not recognized until 1970 (Ostrom 1970, 1972). Originally, this specimen was thought to be a pterosaur (von Meyer 1857, 1860). F. X. Mayr (1973) announced recognition of yet another specimen that had lain misidentified for more than two decades (labeled *Compsognathus*). This fifth specimen (Figure 4) is a small but entirely complete skeleton with only the faintest impressions of feathers. After initial studies are completed (now underway by P. Wellnhofer of Munich), this latest specimen will be exhibited in the Willibaldsburg Castle in Eichstätt, Germany.

Skull and Jaws

Heilmann (1926, Figure 5) presented a detailed reconstruction of the skull of *Archaeopteryx* based on the Berlin specimen. That reconstruction has been widely utilized by many subsequent authors, many of whom have failed to note that it is a restoration and that the actual specimen is so badly crushed that it does not permit such detailed conclusions with any high degree of confidence. The bone is so badly fractured in most places that few of the cranial or mandibular sutures can be recognized. That being the case, detailed comparisons with the skull and jaws of a pseudosuchian or theropod are not very meaningful. Fortunately, however, this situation may be changed with Wellnhofer's studies of the new Eichstätt specimen, which does preserve many details of the skull and jaws. Until that study is available, though, the only cranial and mandibular features of *Archaeopteryx* that are beyond dispute are the following: 1. the skull profile is triangular with a deep expanded temporal region and a sharply tapered snout; 2. the orbit is very large, circular, and contains a sclerotic ring; 3. there is a moderately large triangular ant-orbital fenestra; 4. the nares are narrow and elliptical; 5. upper and lower jaws bear short, moderately sharp isodont teeth in sockets; 6. the mandible is long and very shallow and bears a long retroarticular process; and 7. the quadrate appears to have sloped forward (descends anteriorly). Unfortunately, the temporal region is so poorly preserved that the presence of either temporal fenestra is indeterminate. Most of these features are characteristic of both pseudosuchians and small theropods, but three that are not presently known in any pseudosuchian are the very shallow condition of the lower jaw, the forwardly inclined quadrate, and the elongated retroarticular process. All three of these conditions are typical of theropods.

Vertebral Column

All presacral vertebrae of *Archaeopteryx* appear to have been narrow-waisted and amphicoelous, conditions that are found in both theropods and pseudosuchians. X rays seem to show that none of the vertebral centra have saddle-shaped articular surfaces as in all modern birds. Both the London and Berlin specimens preserve small lateral pleurocoels in the posterior dorsal vertebrae, a condition that is typical of theropods, but is not known in pseudosuchians. The vertebral formula of *Archaeopteryx* (10 cervicals, 12 to 15(?) dorsals, 5 or 6 sacrals, and 20 caudals)

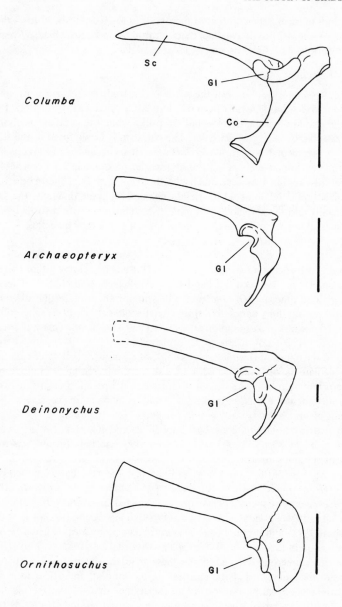

Figure 5 The scapula-coracoid (in lateral view) of the common pigeon (*Columba*) compared with that of *Archaeopteryx,* a theropod (*Deinonychus*) and a Late Triassic thecodont (*Ornithosuchus*). Scapulae are all drawn to unit length for easy comparison. The vertical lines are all equal to 2 cm. Abbreviations: Co, coracoid; Gl, glenoid; Sc, scapula.

corresponds more closely with that of theropods (9 to 10 cervicals, 13 to 14 dorsals, 4 or 5 sacrals, and 20 to 40 caudals) than with that of pseudosuchians (7 cervicals, 13 to 18 dorsals, 2 or 3 sacrals, and 30 to 40 caudals).

Pectoral Arch

The striking similarities between the scapulo-coracoid of *Archaeopteryx* and that of most theropods has generally been overlooked by previous investigators. In both forms the scapula is long, narrow and strap-like while the coracoid is short and subrectangular, or nearly semicircular. The coracoid is firmly fused to the scapula, has a concave medial border, a convex lateral margin, and bears a conspicuous biceps tubercle immediately below the supracoracoid foramen. That morphology is quite unbird-like, but it is very much like that of a variety of theropods. In contrast, the scapula and coracoid of pseudosuchians are quite different; the scapula flares distally and the coracoid is a broad, triangular plate with a strongly convex antero-medial border and little or no development of a biceps tubercle.

Forelimb

Since early in this century (Osborn 1903, 1917) it has been known that there were striking similarities between the hand of *Archaeopteryx* and that of at least one small theropod dinosaur, *Ornitholestes*. [It is important to note that discovery of this taxon was not fully announced until after Broom's (1908, 1913) suggestion that birds and dinosaurs had a common ancestry.] These similarities were well presented by Heilmann (1926), but dismissed. Curiously, Lowe (1935, 1944) and others have referred to the hands of *Ornitholestes* as "bird-like," and while it can be considered as intermediate between the specialized fused, three-fingered hand of modern birds and the primitive five-fingered hand of (for example) pseudosuchians, it is clearly inaccurate to describe it as "bird-like." It would be much more accurate to describe the hands of *Archaeopteryx* as "*Ornitholestes*-like" (see Figure 6).

The discovery of *Ornitholestes* was important, and it was not overlooked, but coming as it did (Osborn 1903, 1917) about the time that Broom was arguing against direct dinosaur-bird relationships, it was not enough to reverse the trend, especially in view of the "convergent and parallel" theories that were then coming to the forefront. In recent years, the same basic hand and wrist structure exhibited in the *Archaeopteryx* specimens has been recognized in several other small theropods, such as *Deinonychus* (Ostrom 1969a, 1969b), *Velociraptor* (Kielan-Jaworowska & Barsbold 1972), and probably *Coelurus* and *Stenonychosaurus* (Russell 1969).

Comparison of the hand of *Archaeopteryx* with that of *Ornitholestes* or *Deinonychus* reveals numerous detailed similarities: in each, the relative lengths of the fingers are the same, with I the shortest, II the longest, and III of intermediate length; the phalangeal proportions of all fingers are similar; and the metacarpals are alike with metacarpal I being short and metacarpals II and III of greater, subequal lengths. The similarities between *Archaeopteryx* and these theropods extend also to the wrist where the dominant element in all is a large lunate distal carpal bone that articulates tightly with the first and second metacarpals but not with the third. This same arrangement is found in a variety of theropods (*Deinonychus*,

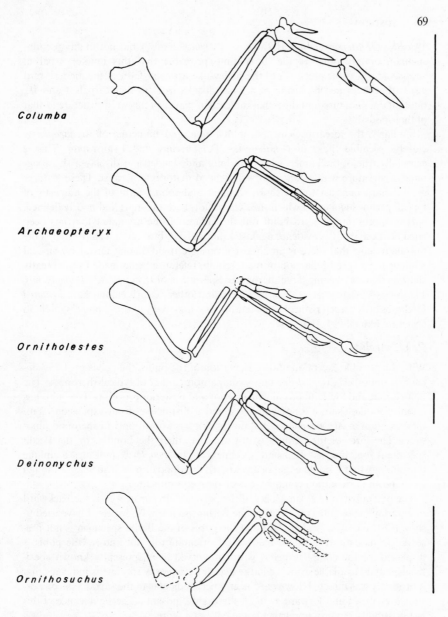

Figure 6 Outline drawings of the forelimbs of a pigeon and *Archaeopteryx* compared with those of two theropods (*Ornitholestes* and *Deinonychus*) and a thecodont (*Ornithosuchus*). *Archaeopteryx* and *Ornitholestes* are of Late Jurassic age, *Deinonychus* is Early Cretaceous, and *Ornithosuchus* is Late Triassic. The humeri of all are drawn to unit length so that the relative proportions of all limb components of all examples can be readily compared. The scales are indicated by the vertical lines, all of which equal 5 cm.

Velociraptor, Stenonychosaurus, Coelurus = *Ornitholestes*?), but not in any pseudo-suchian, or any other reptile. It is quite probable that this lunate carpal in *Archaeopteryx* is the precursor of the pulley-like articular facet of the modern bird wrist (that is formed by fusion of a large distal carpal to metacarpals I and II), but that does not discount the remarkable resemblance of this wrist structure to that of theropods.

Similarly, the forearm elements (radius, ulna, and humerus) of *Archaeopteryx* closely resemble those of *Ornitholestes, Deinonychus,* and *Velociraptor.* This is especially true of the humeri which are long and slender and display a slight sigmoidal curvature with a long and well-defined deltopectoral crest. These features are in sharp contrast to the short, straight, and robust form of the humerus of typical pseudosuchians in which the deltopectoral crest is short and poorly defined.

It is appropriate here to point out the fallacy of one argument that has been cited (Tucker 1938) as evidence against a dinosaurian-bird relationship—that is the mistaken belief that there is an almost invariable trend among terrestrial bipedal animals for marked reduction of the forelimb, reference being made to the greatly shortened arms of some large theropods. Specimens of *Ornitholestes, Deinonychus, Velociraptor, Deinocheirus, Struthiomimus,* and others clearly show that *elongated* forelimbs are characteristic of some theropods—and many of those possess forelimb structure like that of *Archaeopteryx.*

Pelvic Arch

With the possible exception of the orientation of the pubis, the pelvis of *Archaeopteryx* is not especially bird-like but is comparable to that of typical theropods. The ilium is long and low with prominent anterior and posterior processes. The pubis has a long rod-like shaft, a long symphysis, and a distinctive distal expansion. With the exception of *Archaeopteryx,* that morphology is known only in theropod dinosaurs. The form of the ischium is not clear in either the London or the Berlin specimens, but the new Eichstätt specimen of *Archaeopteryx* indicates a unique morphology that is not like that of any known theropod or pseudosuchian, although its shape and orientation could have been derived from either.

The one apparently avian aspect of the pelvis of *Archaeopteryx* is the backward orientation of the pubis preserved in the Berlin specimen. Elsewhere, I have tried to show (Ostrom 1972, 1973, 1974b) that the pubis of the Berlin specimen is not preserved in its natural position. The correct orientation is not known (the pubis is displaced, or the complete pelvis is not preserved in all presently known specimens), but all available evidence in the five *Archaeopteryx* specimens indicates that it probably was directed downward nearly perpendicular to the sacral axis, or possibly even down and forward as in all theropod dinosaurs. The significance of this last possibility requires no further comment (see Figure 7).

Hindlimb

The entire hindlimb of *Archaeopteryx* is remarkably bird-like in all major features, except that the fibula is complete, reaching all the way to the tarsus and is not fused to the tibia, and the metatarsals are not entirely fused. Except for being somewhat more slender (a consequence of its smaller size), the entire hindlimb is also very much like that of nearly all theropods and quite unlike that of pseudo-suchians. For example, the following features of *Archaeopteryx* are also typical of nearly all theropods, as well as most modern birds: 1. a nearly straight femur with a slight anterior-posterior curvature, a distinct head sharply offset from the shaft, and well-defined "greater and lesser trochanters"; 2. a straight tibia with a distinct cnemial crest; 3. a thin splint-like fibula that is closely applied to the tibia;

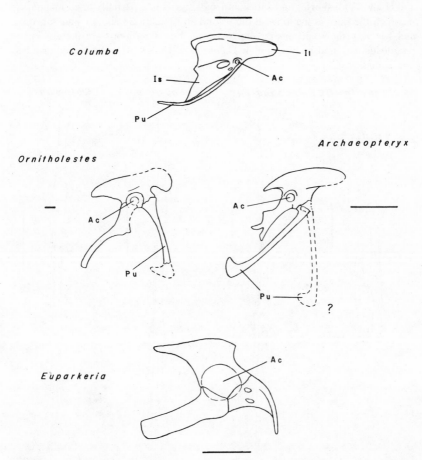

Figure 7   Pelves (in lateral view) of *Archaeopteryx*, pigeon, a Late Jurassic theropod (*Ornitholestes*), and an Early Triassic thecodont (*Euparkeria*). The ilia are all drawn to unit length and horizontal lines all equal 2 cm. The pelvis of *Archaeopteryx* is drawn as pre-served in the Berlin specimen. The dashed outline of the questionable pubis indicates my best estimate of the original position of the pubis in life, intermediate between that of typical theropods and modern birds. Abbreviations: Ac, acetabulum; Il, ilium; Is, ischium; Pu, pubis.

4. a well-defined mesotarsal ankle joint with the proximal tarsals (astragalus and calcaneum) firmly articulated or fused to the distal ends of the tibia and fibula, a well-developed ascending process of the astragalus, and at least two distal tarsals in compact articulation with the three principal metatarsals; 5. a four-toed foot in which the outer (V) toe is lost, the inner toe or hallux (I) is shortened and lies at the rear of the foot where it can oppose the other toes, and the principal toes (II, III, and IV) are symmetrical with III the longest (and main axis of the foot) and II and IV of shorter and subequal lengths; and 6. a partially fused metatarsus in which the third is the longest, the second and fourth are shorter and subequal, and I is very short and incomplete. None of these conditions are present in the pseudosuchian hind limb, although *Lagosuchus* (Romer 1971, 1972) may have had a mesotarsal joint.

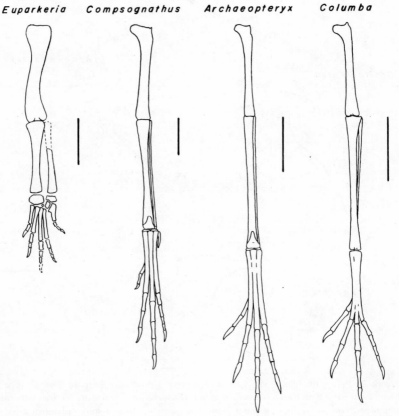

Euparkeria   *Compsognathus*   *Archaeopteryx*   *Columba*

Figure 8 Hind limbs (in anterior view) of *Archaeopteryx* and a pigeon compared with those of a Late Jurassic theropod (*Compsognathus*) and an Early Triassic thecodont (*Euparkeria*). Femora are all drawn to unit length for convenient comparison. Vertical lines equal 3 cm.

THE AVIAN CHARACTERS OF *ARCHAEOPTERYX*

Aside from the feather impressions, the most significant avian feature of *Archaeopteryx* is the furcula or wish bone that is preserved in the London and Maxberg specimens, and perhaps in the Berlin specimen as well. This bone is unique to birds and is thought to be derived from paired clavicles. As noted earlier, Heilmann concluded that since clavicles were not known in any theropod dinosaur (not true), that group could not be ancestral to birds. The important point, however, is not that bit of negative evidence, but the positive evidence of a furcula in *Archaeopteryx*. Its presence, together with feathers, confirms the avian identification of *Archaeopteryx*.

The presence of a furcula seems paradoxical together with the apparent absence of a sternum in all five *Archaeopteryx* specimens. (DeBeer's identification of a sternum in the London specimen has not been verified by subsequent studies.) The function of the furcula is not known, but it is suspected to have something to do with powered flight because it is present in almost all flying birds but is reduced in most flightless birds. However, the apparent absence of the sternum, together with the very short, theropod-like coracoids, indicates that *Archaeopteryx* was not a powered flier, and may not have had any gliding or parachuting abilities (Ostrom 1974a). Accordingly, it is quite possible that the original function of the furcula may not have been related to flight at all, because (despite the plumage) *Archaeopteryx* seems to have lacked all other avian skeletal conditions that are known to be important for flight.

THE ORIGIN OF *ARCHAEOPTERYX*

In a preceding section, bird origins were equated with the origin of *Archaeopteryx*. Not everyone will agree with that equation, but the burden of proof must rest with the dissenters. The general consensus (past and present) holds that *Archaeopteryx* was a true, even though extremely primitive, bird. Thus, regardless of whether it is accepted as a main line form ancestral to any later bird, those specimens represent a stage very close to the beginnings of the class Aves. Accordingly, the origin of *Archaeopteryx* is the key question.

Although *Archaeopteryx* has often been described as "bird-like," it should be clear from the preceding sections that those specimens are more dinosaurian (theropodous) in their osteology than they are avian (or pseudosuchian). The only osteological feature of *Archaeopteryx* that is exclusively avian is the furcula. The presumed bird-like orientation of the pubis in the Berlin specimen is probably not correct, but due to post-mortem displacement. The bird-like feet and hind legs are equally theropodous and all of the other so called bird-like features (hands, arms, pelvis, and skull) are actually more like those of theropod dinosaurs than they are bird-like.

In 1888, Furbringer explained the resemblances between *dinosaurs* and birds (not *Archaeopteryx* and dinosaurs!) as convergent analogies and parallels. Simpson (1946) offered the same explanation nearly 60 years later and most subsequent workers have apparently accepted his assessment. Implicit in that explanation is the common (thecodont) ancestry of dinosaurs and birds. The critical question, however, is: Is it more probable that *Archaeopteryx* acquired the large number of derived "theropod" characters by convergence or in parallel at the same time that these same features were being acquired by some coelurosaurian theropods—presumably from a common ancestor? Or is it more likely that these many derived characters are common to some small theropods and *Archaeopteryx* because *Archaeopteryx* evolved directly from such a theropod? There is absolutely no question in my mind that the last explanation is far more probable.

The fact that so many traits occurred almost simultaneously in both groups is strong prima facie evidence of a direct evolutionary relationship between theropods

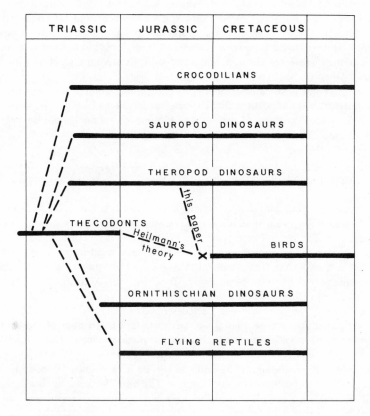

Figure 9 Generalized phylogeny of thecodont descendants to contrast the generally accepted "common ancestry" hypothesis of bird and dinosaur origins (the Heilmann theory) and the theory of bird origins proposed here. X marks the position of *Archaeopteryx*.

and *Archaeopteryx*. That conclusion is further reinforced by the fact that the only contrary evidence to date (Heilmann 1926) was not only negative, but is now known to be incorrect.

CONCLUSIONS

The data presented above lead me to the conclusion that the dinosaurian theory of bird origins is correct after all and that *Archaeopteryx* and all other birds are descendant from some small, Middle or Late Jurassic theropod dinosaur. Contrary to some previous conclusions, there is no positive evidence against a thecodont → theropod → *Archaeopteryx* → modern birds, evolutionary sequence.

ACKNOWLEDGMENTS

I am indebted to A. J. Charig [British Museum (Natural History), London], H. Jaeger (Humboldt Museum für Naturkunde, East Berlin), T. Kress (Museum Solnhofen Aktienverein, Maxberg, Germany), C. O. van Regteren Altena (Teyler Stichting, Haarlem, Netherlands), and P. Wellnhofer (Bayerische Staatssammlung, Munich) for their gracious assistance and permission to study the *Archaeopteryx* specimens in their care. I am also indebted to my colleagues and students at Yale who provided a most critical sounding board for many of the ideas expressed here. Figure 4 is by permission of F. X. Mayr and P. Wellnhofer.

Literature Cited

Baur, G. 1883. Der Tarsus der Vögel und Dinosaurier. *Morphol. Jahrb.* 8:417–56
Baur, G. 1884. Dinosaurier und Vögel. *Morphol. Jahrb.* 10:446–54
Baur, G. 1885. Bemerkungen über das Becken der Vögel und Dinosaurier. *Morphol. Jahrb.* 10:613–16
Baur, G. 1886. Zur Vögel—Dinosaurier—Frage. *Zool. Anz.* 8:441–43
Boas, J. E. V. 1930. Über das Verhältnis der Dinosaurier zu den Vögeln. *Morphol. Jahrb.* 64:223–47
Brodkorb, P. 1971. Origin and evolution of birds. *Avian Biology,* ed. D. S. Farner, J. R. King, Vol. I:19–55. London: Academic. 586 pp.
Broom, R. 1908. On the early development of the appendicular skeleton of the ostrich with remarks on the origin of birds. *Trans. S. Afr. Phil. Soc.* 16:355–68
Broom, R. 1913. On the South-African pseudosuchian *Euparkeria* and allied genera. *Proc. Zool. Soc. London* 1913: 619–33
Camp, C. L. 1936. A new type of small bipedal dinosaur from the Navajo Sandstone of Arizona. *Univ. Calif. Publ. Dep. Geol. Sci. Bull.* 24:39–56

Dames, W. 1884. Ueber *Archaeopteryx*. *Palaeontol. Abh.* Bd. 2, 3:119–98
Darwin, C. 1859. *On the Origin of Species by means of Natural Selection, or the Preservation of Favoured Races in the Struggle for Life.* London: John Murray. 502 pp. 2nd ed., 1869
deBeer, G. R. 1954. *Archaeopteryx lithographica.* London: Brit. Mus. (Natur. Hist.). 68 pp.
deBeer, G. R. 1964. *Archaeopteryx. A New Dictionary of Birds,* ed. A. L. Thomson, 58–62. London: Nelson. 928 pp.
Dollo, L. 1882. Première note sur les dinosauriens de Bernissart. *Bull. Mus. Roy. Hist. Natur. Belgique.* I:161–80
Dollo, L. 1883. Note sur la présence chez les oiseaux du "troisième trochanter" des dinosauriens et sur la fonction de celui-ci. *Bull. Mus. Roy. Hist. Natur. Belgique* II:13–20
Furbringer, M. 1888. *Untersuchungen zur Morphologie und Systematik der Vögel.* Amsterdam: Holkema. 1751 pp.
Galton, P. M. 1970. Ornithischian dinosaurs and the origin of birds. *Evolution* 24:448–62
Gegenbaur, C. 1863. Vergleichend-anatom-

ische Bemerkungen über das Fusskelet der Vögel. *Arch. Anat. Phys. Med.* 450–72

Gegenbaur, C. 1878. *Grundriss der vergleichenden Anatomie.* Leipzig: W. Engelmann. 655 pp.

Giebel, C. 1877. Neueste Entdeckung einer zweiten *Archaeopteryx lithographica. Z. Gesamte Naturwiss.* 49 : 326–27

Haeckel, E. H. P. A. 1866. *Generelle Morphologie der Organismen.* Berlin: G. Reimer. 574 pp.

Heilmann, G. 1926. *The Origin of Birds.* London: Witherby. 208 pp.

Heller, F. 1959. Ein dritter *Archaeopteryx*-Fund aus den Solnhofener Plattenkalken von Langenaltheim/Mfr. *Erlanger Geol. Abh.* 31 : 1–25

Holmgren, N. 1955. Studies on the phylogeny of birds. *Acta Zool.* 36 : 243–328

Huxley, T. H. 1867. On the classification of birds and on the taxonomic value of the modifications of certain of the cranial bones observable in that class. *Proc. Zool. Soc. London* 1867 : 415–72

Huxley, T. H. 1868a. Remarks upon *Archaeopteryx lithographica. Proc. Roy. Soc. London* 16 : 243–48

Huxley, T. H. 1868b. On the animals which are most nearly intermediate between the birds and reptiles. *Ann. Mag. Natur. Hist. London* (4) 2 : 66–75

Huxley, T. H. 1870. Further evidence of the affinity between the dinosaurian reptiles and birds. *Quart. J. Geol. Soc. London* 26 : 12–31

Kielan-Jaworowska, Z., Barsbold, R. 1972. Results of the Polish-Mongolian Palaeontological Expeditions — Part IV *Palaeontol. Pol.* 27 : 5–13

Lowe, P. R. 1935. On the relationships of the Struthiones to the dinosaurs and to the rest of the avian class, with special reference to the position of *Archaeopteryx. Ibis* 13 : 398–432

Lowe, P. R. 1944. An analysis of the characters of *Archaeopteryx* and *Archaeornis.* Were they reptiles or birds? *Ibis* 86 : 517–43

Marsh, O. C. 1877. Introduction and succession of vertebrate life in America. *Proc. Am. Assoc. Advan. Sci.* 1877 : 211–58

Mayr, F. X. 1973. Ein neuer *Archaeopteryx* Fund. *Paläontol. Z.* 47 : 17–24

Mivart, Saint G. 1881. A popular account of chamaeleons. *Nature* 24 : 309–353

Mudge, B. F. 1879. Are birds derived from dinosaurs? *Kansas City Rev. Sci.* 3 : 224–26

Osborn, H. F. 1900. Reconsideration of the evidence for a common dinosaur—avian stem in the Permian. *Am. Natur.* 34 : 777–99

Osborn, H. F. 1903. *Ornitholestes hermanni,* a new compsognathoid dinosaur from the Upper Jurassic. *Bull. Am. Mus. Natur. Hist.* 19 : 459–64

Osborn, H. F. 1917. Skeletal adaptations of *Ornitholestes, Struthiomimus, Tyrannosaurus. Bull. Am. Mus. Natur. Hist.* 35 : 733–71

Osborn, H. F. 1924. Three new Theropoda, *Protoceratops* zone, central Mongolia. *Am. Mus. Nov.* 144 : 1–12

Ostrom, J. H. 1969a. A new theropod dinosaur from the Lower Cretaceous of Montana. *Yale Peabody Mus. Natur. Hist. Postilla* 128 : 1–17

Ostrom, J. H. 1969b. Osteology of *Deinonychus antirrhopus,* an unusual theropod from the Lower Cretaceous of Montana. *Bull. Yale Peabody Mus. Natur. Hist.* 30 : 1–165

Ostrom, J. H. 1970. *Archaeopteryx:* Notice of a "new" specimen. *Science* 170 : 537–38

Ostrom, J. H. 1972. Description of the *Archaeopteryx* specimen in the Teyler Museum, Haarlem. *Proc. Kon. Ned. Akad. Wetensch. Ser. B* 75 : 289–305

Ostrom, J. H. 1973. The ancestry of birds. *Nature* 242 : 136

Ostrom, J. H. 1974a. *Archaeopteryx* and the origin of flight. *Quart. Rev. Biol.* 49 : 27–47

Ostrom, J. H. 1974b. On the origin of *Archaeopteryx* and the ancestry of birds. *Proc. Centre Nat. Rech. Sci.* In press

Owen, R. 1875. Monograph of the fossil reptiles of the Liassic formations II. Pterosauria. *Palaeontol. Soc. Monogr.,* 41–81

Parker, T. J. 1882. On the skeleton of *Notornis mantelli. Trans. Proc. N. Z. Inst.* 14 : 245–58

Parker, W. K. 1864. Remarks on the skeleton of *Archaeopteryx;* and on the relations of the bird to the reptile. *Geol. Mag.* (I) 1 : 55–57

Parker, W. K. 1887. On the morphology of birds. *Proc. Roy. Soc. London* 42 : 52–58

Petronievics, B. 1921. *Über das Becken, den Schultergürtel und einige andere Teile der Londoner Archaeopteryx.* Genf. Buchhandl. Georg. 31 pp.

Petronievics, B. 1927. Nouvelles recherches sur l'osteologie des Archaeornithes. *Ann. Paléontol.* 16 : 39–55

Petronievics, B. 1950. Les deux oiseaux fossiles les plus anciens (*Archaeopteryx et Archaeornis*). Ann. Geol. Pen. Balkan 18 : 89–127

Pettingill, O. S. 1970. *Ornithology in Laboratory and Field.* Minneapolis: Burgess. 524 pp.

Romer, A. S. 1966. *Vertebrate Paleontology.*

Chicago: Univ. Chicago Press. 368 pp.
Romer, A. S. 1971. The Chañares (Argentina) Triassic reptile fauna. X. Two new but incompletely known long-limbed pseudosuchians. *Harvard Univ. Mus. Comp. Zool. Breviora* 378 : 1–10
Romer, A. S. 1972. The Chañares (Argentina) Triassic reptile fauna. XV. Further remains of the thecodonts *Lagerpeton* and *Lagosuchus*. *Harvard Univ. Mus. Comp. Zool. Breviora* 394 : 1–7
Russell, D. A. 1969. A new specimen of *Stenonychosaurus* from the Oldman Formation (Cretaceous) of Alberta. *Can. J. Earth Sci.* 6 : 595–612
Seeley, H. G. 1881. Prof. Carl Vogt on the *Archaeopteryx*. *Geol. Mag.* (2) 8 : 300–9
Simpson, G. G. 1946. Fossil Penguins. *Bull. Am. Mus. Natur. Hist.* 87 : 1–95
Swinton, W. E. 1958. *Fossil Birds*. London: British Mus. (Natur. Hist.). 63 pp.
Swinton, W. E. 1960. The origin of birds. *Biology and Comparative Physiology of Birds*, ed. A. J. Marshall, Vol. I : 1–14. New York: Academic. 518 pp.
Swinton, W. E. 1964. Origin of birds. *A New Dictionary of Birds*, ed. A. L. Thomson, 559–62. London: Nelson. 928 pp.
Tucker, B. W. 1938. Functional evolutionary morphology: The origin of birds. *Evolution*, ed. G. R. de Beer, 321–36. Oxford: Clarendon. 350 pp.
Van Tyne, J., Berger, A. J. 1959. *Fundamentals of Ornithology*. New York: Wiley. 624 pp.
Vogt, C. 1879. *Archaeopteryx*, ein Zwischenglied zwischen den Vögeln und Reptilien. *Naturforscher* 42 : 401–4
von Meyer, H. 1857. Beiträge zur näheren Kenntniss fossiler Reptilien. *Neues Jahrb. Mineral. Geol. Palaeontol.* 1857 : 437

von Meyer, H. 1860. *Zur Fauna der Vorwelt. Reptilien aus dem lithographischen Schiefer des Jura in Deutschland und Frankreich*, 64–66. Frankfurt am Main: H. Keller. 142 pp.
von Meyer, H. 1861a. Vögel-Federn und *Palpipes priscus* von Solnhofen. *Neues Jahrb. Mineral. Geol. Palaeontol.* 1861 : 561
von Meyer, H. 1861b. *Archaeopteryx lithographica* (Vogel-Feder) und *Pterodactylus* von Solnhofen. *Neues Jahrb. Mineral. Geol. Palaeontol.* 1861 : 678–79
von Meyer, H. 1862. *Archaeopteryx lithographica* aus dem lithographischen Schiefer von Solenhofen. *Palaeontographica* 10 : 53–56
Wagner, J. A. 1861. Ueber ein neues, augenblich mit Vögelfedern versehenes Reptil aus dem Solenhofener lithographischen Schiefer. *Sitzungsber. Bayer. Akad. Wiss.* 2 : 146–54
Welty, J. C. 1962. *The Life of Birds*. Philadelphia: Saunders. 546 pp.
Wiedersheim, R. E. E. 1883. *Lehrbuch der vergleichenden Anatomie der Wirbelthiere auf Grundlage der Entwicklungsgeschichte*. Jena: Fischer. 905 pp.
Wiedersheim, R. E. E. 1884. Die Stammesentwicklung der Vögel. *Biol. Zentralbl.* 3 : 654–95
Wiedersheim, R. E. E. 1885. Über die Vorfahren der heutigen Vögel. *Humboldt* 4 : 212–24
Wiedersheim, R. E. E. 1886. *Lehrbuch der vergleichenden Anatomie der Wirbelthiere auf Grundlage der Entwicklungsgeschichte*. Jena: Fischer. 890 pp.
Williston, S. W. 1879. "Are birds derived from dinosaurs?" *Kansas City Rev. Sci.* 3 : 457–60

EARLY PALEOZOIC ECHINODERMS ✖10033

Georges Ubaghs
Laboratoire de Paléontologie Animale, University of Liège, B-4000 Liège, Belgium

INTRODUCTION

The oldest known remains of echinoderms are Early Cambrian in age. They consist of isolated plates, possibly eocrinoids, that occur in the Montenegro Member of the Campito Formation in eastern California, close to the base of the Cambrian (Durham 1971). In the same region and in western Nevada, helicoplacoids, eocrinoids, and edrioasteroids are associated in the Poleta Formation, in a horizon which is about middle Early Cambrian in age (Durham, 1964, 1967, 1971; Durham & Caster 1963, 1966). Several specimens of helicoplacoids have also been found in Lower Cambrian rocks of northwestern Alberta (Durham 1964). All these occurrences are within the lower part of the *Olenellus* biozone and are therefore significantly older than the edrioasteroid (Schuchert 1919), lepidocystoid (Durham 1968b, Foerste 1938, Sprinkle 1973), and camptostromatoid echinoderms (Durham 1966, 1968a) discovered near the top of the upper part of the Lower Cambrian in eastern North America.

The taxonomic diversity of echinoderms increased during Middle Cambrian times, with the appearance of representatives of Homostelea, Stylophora, Ctenocystoidea, Cyclocystoidea, and possibly Holothuroidea (Durham 1971) and Crinoidea (Sprinkle 1973). To these six classes were added the enigmatic genera *Cymbionites* and *Peridionites,* from early Middle Cambrian rocks of Queensland, Australia, considered by some authors (Durham 1971, Whitehouse 1941) to represent two distinct classes: the Cycloidea and Cyamoidea. Near the top of the Cambrian the Homoiostelea appeared, and by the end of the Middle Ordovician all of the 20 classes of echinoderms currently recognized, except Blastoidea (Silurian-Permian), were represented.

Such an early appearance of the major groups of echinoderms, coupled with the astonishing diversity and high degree of specialization shown by the oldest known representatives of the phylum, indicates that the common ancestor and the initial phases in the history of these animals certainly must be traced back to Precambrian times. Consequently, at present paleontology cannot furnish direct evidence bearing on the problem of the origins and relationships of echinoderms, nor on the manner in which their essential features were acquired.

In this paper only the main echinoderm classes represented in Lower and Middle Cambrian rocks are considered. They may be divided into two main groups: the nonradiate echinoderms that show no trace of radial symmetry, and the radiate echinoderms in which radial symmetry has affected a more or less important portion of the body structures.

79

NONRADIATE ECHINODERMS

In Recent echinoderms, the bilateral symmetry of the larvae in invariably disturbed by an asymmetrical development of the left and right sides of the body. The radial symmetry, typically pentamerous, appears at a relatively late stage in development. Translated into phylogenetic terms, this means that the radially organized types might have been preceded by asymmetrical nonradiate forms.

Several classes, whose members are not radially organized, are among the earliest known echinoderms. Apparently these classes separated from a common ancestral stock before radial symmetry became a distinctive feature of the phylum, and their very existence suggests the possibility of an early radiation of nonradiate groups.

Five classes may belong to this first radiation: the Helicoplacoidea, Homostelea, Stylophora, Ctenocystoidea, and Homoiostelea. The latter, being unknown from rocks older than late Cambrian, is not considered here.

Helicoplacoidea

This class comprises three genera and six species, all from the Lower Cambrian *Olenellus* zone of western North America (Durham 1964, 1967, 1971; Durham & Caster 1963, 1966). Its members are distinguished primarily by 1. lack of any symmetry and 2. helicoid arrangement of the plates forming the test (Figure 1). It is true that a somewhat similar torsion of parts may occur in members of some other classes, but being associated with various types of organization, this feature must have evolved separately in each class.

The morphology of helicoplacoids is imperfectly known. For instance, the anus has not been observed, and the structure of the narrow band of platelets that is identified as an ambulacrum does not appear to be fully elucidated (Durham 1967, Nichols 1967). However, the existence of tube feet and therefore of a water-vascular system seems highly probable. Besides, evidence favors interpretation of the main water vessel as internal rather than external (Figure 1D). But such a position in early Cambrian helicoplacoids cannot imply (Durham 1967) that it is the result of the primitive condition of the system, for in contemporaneous eocrinoids and edrioasteroids the radial water vessels probably were external (see below). On the other hand, even if the principal function of the water-vascular system in helicoplacoids was respiration, it does not follow that all presumed respiratory organs (including hydrospires, pore-rhombs, diplopores, and epispires) of extinct echinoderms were ramifications of this system. They could have quite another origin, and, like the papulae of asteroids, be diverticula of the main body cavity (not of the hydrocoel).

The single ambulacrum divides the body into halves that define a sort of bilateral symmetry. But this is only apparent. for the water-vascular system can be developed only from a single hydrocoel. The lack of another hydrocoel indicates that the right and left coeloms were unequally developed in these early echinoderms, and that the organization, consequently, was asymmetrical.

The helicoplacoids were not attached during life, but their real mode of existence is uncertain. Many hypotheses have been formulated (Durham & Caster 1963, 1966, Fell 1965, Fell & Moore 1966, Nichols 1967, 1969, Termier & Termier 1969). Some specimens seem to have had their apical pole imbedded in the substrate at the time of burial (Durham 1971), a position that suggests they had adopted an upright living posture.

Relationships of the helicoplacoids with other echinoderms have been variously interpreted (Durham & Caster 1963, 1966, Fell 1965, Fell & Moore 1966, Nichols 1969, Ubaghs 1971). Possible affinities with Echinozoa are indicated by 1. the globoid shape of the test, 2. the lack of arms or similar outspread extensions, 3. the probable origin of new plates from about the apical pole, and 4. the lack of any organ or structure that could have served for fixation. But the helicoplacoids are nonradiate echinoderms, while all classes usually included in the subphylum Echinozoa have a radial symmetry, and no intermediate between these conditions is known. An increase in the efficiency of food gathering and other physiological systems in helicoplacoids seems to have been effected by torsion of structures, rather than by fivefold repetition of body sections as in all echinozoans. In addition, the method of expanding the test in this group is unlike that known in any other

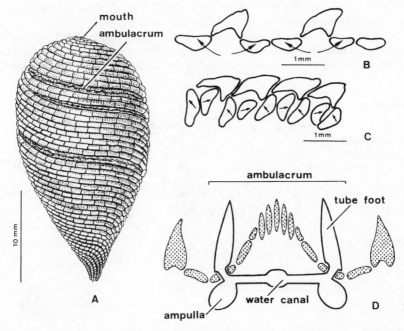

Figure 1 A: *Helicoplacus gilberti* Durham & Caster (1963), Lower Cambrian. B and C: sections across expanded (B) and contracted (C) test of *Helicoplacus*; arrows indicate direction of movement in expanding test (Durham 1964). D: Schematic cross section of ambulacrum of *Waucobella nelsoni* Durham, Lower Cambrian, as interpreted by Durham (1967).

echinoderms. The helicoplacoids have brought solutions of their own to particular problems, a fact that suggests distinctiveness in origin and evolution (Ubaghs 1971). In spite of their great antiquity and generally primitive organization, the helicoplacoids are highly specialized in several respects, probably in response to some particular mode of life. Their connection with other echinoderms, including Echinozoa, seems doubtful.

Homostelea

The classes Homostelea, Ctenocystoidea, Stylophora, and Homoiostelea are commonly referred (Caster 1968, Ubaghs 1968c) to the subphylum Homalozoa, which is essentially equivalent to the class Carpoidea of Jaekel (1918). Whether these four groups are related phylogenetically is uncertain. Their only common characteristics are: 1. a typical echinoderm skeleton, 2. a depressed body shape, and 3. an asymmetrical test, without any trace of radial symmetry or helicoid arrangement. All of them appear to have had a predominantly recumbent position in life.

The Homostelea comprise the single order Cincta which includes a few genera found in the Middle Cambrian (*Paradoxides* biozone) in Bohemia, Germany, France, Spain, and Morocco (Cabibel et al 1958, Jaekel 1918, Nichols 1967, 1969, Schroeder 1973, Termier & Termier 1973, Ubaghs 1968c, 1971). Figure 2 illustrates their main features. Only three controversial points are discussed here.

1. That the Homostelea reposed in life on one of their sides has never been seriously doubted. But that the face here designated as lower (Figure 2A) was the surface that rested on the substrate has recently been questioned (Termier & Termier

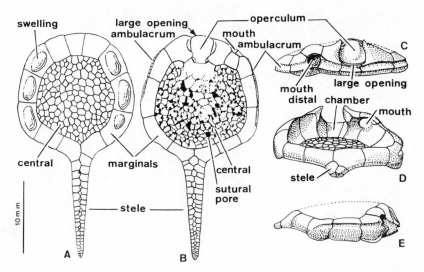

Figure 2   *Trochocystites bohemicus* Barrande, Middle Cambrian. A : lower face. B : upper face. C : distal view. D : proximal view; upper integuments and most part of stele lacking. E : right lateral view, most part of stele lacking (Ubaghs 1968c).

1973). The correctness of the generally accepted orientation is, however, strongly suggested by (*a*) the lower flattening and upper convexity of both theca and stele; (*b*) the presence of prominent swellings on marginals of the lower face, swellings that probably helped to anchor the animal in the mud; and (*c*) the concentration of sutural pores having presumed respiratory function on the upper face (Figure 2B).

2. The two main openings of the body have been variously interpreted. Detailed study (Ubaghs 1968c, 1971) of the grooves leading to the smaller of the two has revealed that they probably contained ambulacral structures and consequently that the smaller opening was the mouth (Figure 2C). On the other hand, there is no clear indication as to the precise nature of the larger opening, which looks as if it could have been opened and closed by a hinged opercular plate on its upper side. The infundibular shape of the distal chamber (Figure 2D) below this plate and the protruding lower lip of its outer aperture suggest the existence of a protruding organ designed for jet expulsion of excreta and possibly genital products (Ubaghs 1968c).

3. The single appendage of the body has been recently interpreted as a modified arm (Caster 1968) or a feeding organ (Termier & Termier 1969, 1973) capable of wriggling movements like the arms of an ophiuroid (Termier & Termier 1973). Such interpretations are opposed by the very structure of the appendage itself, which is clearly a stem or peduncle: it contains a narrow axial canal, has no external opening, and carries no groove that could have housed an ambulacrum. It seems to have been fairly rigid, as shown by its aspect and lack of flexible union with the theca or between component ossicles (Ubaghs 1968c, 1971). Its depressed form and the complete absence of any root-like device proves it was not used as a vertical support, but was resting on the sea-bed or, as suggested by some specimens, thrust obliquely into the sediment.

The Homostelea are not really primitive echinoderms. The occurrence of a well-organized stalk, the marked asymmetry of the body, the closeness of mouth and anus (indicating a probable twisting or bending of the gut), and the high degree of differentiation of the opercular region are judged to be specialized features. Detailed comparison of Homostelea with other echinoderms reveals no real resemblance and no certain homology. The Homostelea appear as an isolated group of unknown origin, affinities, and descent.

Ctenocystoidea

The class Ctenocystoidea has been proposed for a single species, *Ctenocystis utahensis,* found in Early Middle Cambrian rocks (Spence Shale) of northern Utah (Robinson & Sprinkle 1969). This form, and an associated Stylophora, are the oldest known carpoid echinoderms.

Ctenocystis has no peduncle and no arm (Figure 3). Its depressed and slightly asymmetrical body has a.double-layered frame of marginal plates that encloses minute central plates. Mouth and anus are located at opposite ends of the body. Anus is identified as a small pyramid protruding between marginals. The region around the mouth includes a complex structure, probably serving to collect food. Occurrence of a third orifice (? hydropore, ? gonopore) is suggested by a sulcus in inner surface of the right anterolateral supermarginal (Figure 3A).

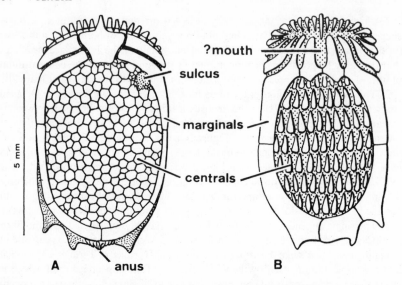

Figure 3   *Ctenocystis utahensis* Robinson & Sprinkle (1969), Middle Cambrian, reconstruction. A: upper face. B: lower face.

Of all primitive echinoderms, *Ctenocystis* most nearly approaches bilateral symmetry. Asymmetry of organization is, however, clearly indicated by: 1. difference in size, shape, and number of posterolateral infermarginals on each side of anteroposterior axis (Figure 3B); 2. unequal development of spines borne by these infermarginals; and 3. occurrence of a single opening (or sulcus) on the right side of the upper face without a similar aperture on the left side.

Complete absence of a peduncular and brachial appendages, as well as the unique nature of the feeding apparatus, are features that easily distinguish *Ctenocystis* from all other carpoid echinoderms. It shares, however, some characters with the Stylophora, such as the location of mouth and anus at opposite ends of the body, the occurrence of a marginal framework, and the probable association of the right anterolateral supermarginal with an opening. Besides, in Stylophora, a varying number of marginals, like those of *Ctenocystis*, are composed of a lower and an upper plate. Such similarities may suggest a common ancestor. Yet *Ctenocystis* differs so markedly from any known stylophorans that an attribution to that class would not be advisable.

Stylophora

The class Stylophora (Middle Cambrian–Middle Devonian) is the most diversified of the homalozoan classes. It comprises two orders, the Cornuta (Middle Cambrian–Upper Ordovician) and the Mitrata (Lower Ordovician–Middle Devonian), which are obviously related (Caster 1952, Jefferies 1967, 1968a,b, 1969, 1973, Jefferies & Prokop 1972, Ubaghs 1961, 1967, 1968b, 1970, 1971).

The oldest known stylophoran has been discovered, along with *Ctenocystis*

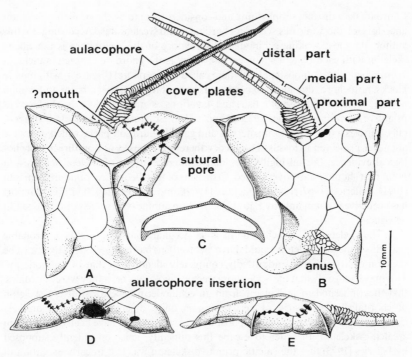

Figure 4 Ceratocystis perneri Jaekel, Middle Cambrian, reconstruction. A: upper face. B: lower face. C: cross section of theca. D: anterior side, aulacophore lacking. E: right lateral side (Ubaghs 1971).

utahensis, in early Middle Cambrian rocks (Spence Shale) of northern Utah (Robinson & Sprinkle 1969). It is a small undescribed cornute, with a boot-shaped theca, and a well-defferentiated marginal frame (R. A. Robinson, personal communication). Very different is the late Middle Cambrian *Ceratocystis perneri* Jaekel, found in the Jince beds of Bohemia (Figure 4). Thus, two well-marked types of cornutes were already in existence by Middle Cambrian time, indicating a long previous evolution. No mitrate has been discovered in deposits older than Lower Ordovician. The recent discovery of a Lower Ordovician cornute with distinct mitrate features suggests possible derivation of the mitrates from a cornute ancestor (Jefferies & Prokop 1972).

The stylophorans are traditionally regarded as echinoderms. Jefferies, however, has advocated the rather revolutionary theory that they are better seen as primitive chordates with echinoderm affinities (Jefferies 1967, 1968a,b, 1969, 1973, Jefferies & Prokop 1972).[1] Consequently he has called them Calcichordata, placed them in

[1] Possible relationships of carpoid echinoderms with chordates have been suggested previously by Matsumoto (1929) and Gislén (1930). The latter, however, has expressly written that it was not his intention to assume certain primitive Paleozoic echinoderms to be the actual ancestors of the higher Deuterostomia.

Chordata (as a distinct subphylum), and considers them to be the probable common ancestors of the tunicates, cephalochordates, and vertebrates. According to this author, the most striking echinoderm-like feature of the Stylophora is the calcite skeleton with each plate a single crystal, whereas the chordate characters include: 1. branchial slits; 2. a postanal tail (here called aulacophore) (Figure 4) with muscle blocks, notochord, dorsal nerve cord, and segmental ganglia; 3. a brain and cranial nervous system like those of a fish; and 4. various asymmetries like those of Recent primitive chordates (Jefferies 1973). Almost all these presumed chordate features— gills, muscles, notochord, brain, nerves, and ganglia—are soft parts, which of course are never preserved in fossils. The theory therefore is based not on direct morpho- logical evidence, but on highly speculative interpretations of skeletal features—a point that should not be overlooked. Criticisms of this theory are found in Denisson (1971), Jefferies (1967, 1968a), Nichols (1969), and Ubaghs (1970, 1971). Support has been given to it mainly by Eaton (1970). We examine only some points judged to be critical.

1. The skeleton of an echinoderm is so peculiar that it is highly improbable that this type of skeleton could have evolved in more than one phylum. The skeleton of Stylophora is admittedly completely identical to that of echinoderms. This may mean only that the Stylophora were echinoderms or, if they were chordates, that they inherited their skeleton from an echinoderm ancestor. But the skeleton of chordates, particularly of vertebrates, is quite different, and it is unlikely (Jefferies 1967, 1968a, 1973) that a phosphatic skeleton could have been derived from a calcitic skeleton. In order to escape this difficulty, it is conveniently supposed by Jefferies (1973) that the calcitic primitive skeleton was lost in early evolution of vertebrates, and replaced by an entirely new one. Perhaps it would be safer (and certainly more in accordance with the principle of economy of hypothesis) to accept evidence at its face value: the Stylophora had an echinoderm-like skeleton simply because they were echinoderms.

2. The skeleton of the Stylophora, with its calcitic constitution, histologic struc- ture, and crystallographic properties, is not only echinoderm-like, but also indicates obvious echinoderm affinities in features of its component elements: marginal plates similar to those of Homostelea or Ctenocystoidea; plated covering, like the plating of many echinoderms; movable spines similar to echinoid spines; an anal pyramid resembling that of many fossil and living echinoderms,[2] sutural pores like those present in many primitive echinoderms; and an articulated appendage (aulacophore) resembling the food-collecting arms of some extant ophiuroids (Nichols 1967, 1969, 1972, Ubaghs 1961, 1968b, 1970, 1971). Various markings (apophyses, grooves, ridges) on the inner side of the theca may also be interpreted in echinoderm terms, whereas none unequivocally indicate that they housed chordate structures such as a brain, a notochord, or an axial nerve cord.

3. The articulated process (aulacophore) of the Stylophora has been interpreted either as an ambulacrum-bearing or armlike process (Ubaghs 1961, 1967, 1968b,

[2] This structure is interpreted by Jefferies as being the mouth. It is, however, so obviously the vent that such an experienced specialist of echinoderms as Bather (1913) took it as the starting point of his interpretation of the stylophoran genus *Cothurnocystis*.

1970, 1971) or as a tail provided with notochord, dorsal nerve cord, segmented ganglia, and muscle blocks (Jefferies 1967, 1968a,b, 1969, 1973). The first interpretation implies that the thin plates covering the presumed ambulacral tract could open in life, whereas the second requires that they be permanently closed, attached to and protecting underlying soft structures. According to Jefferies, in at least some species, the way these plates are articulated indicates that they could not open outwards as cover plates. The present author can only say that he has never observed any structure that could prevent such movements. But the manner in which these plates are preserved is perhaps a more certain indication of their ability to move apart then the disputable nature of very small articulations, which are difficult to observe in fossils usually preserved as natural molds. In some species, the majority of specimens have these plates in a wide open position, forming a regular row on each side of the aulacophore (Figure 5I). They have kept their original order and arrangement with adjacent ossicles, and still imbricate in a distal direction as they surely did in life. If they had been mesodermal plates firmly attached to underlying soft structures, they would have merely collapsed after death, but certainly not separated into two opposed rows while preserving their original arrangement.

4. The aulacophore of the Mitrata approximates the aulacophore of the Cornuta very closely. Both appendages are composed of the same three parts and comprise the same component elements. Clearly, these parts and elements are homologous (Figure 5). Jefferies, however, in order to remain consistant with his interpretation of the theca and the orientation of mitrates that he advocates, is forced to give the mitrate aulacophore an orientation opposite to that of the cornute appendage, and accordingly to consider the massive ossicles of the medial and distal regions to be "ventral" in cornutes and "dorsal" in mitrates. Jefferies & Prokop (1972) try to solve this difficulty by supposing that in giving rise to the mitrates, their presumed cornute ancestors have lost the medial and distal regions of their aulacophore, and evolved entirely new, yet similar, structures from the remaining proximal region. It will probably be agreed that the explanation is difficult to believe, when it would be so much more satisfactory to take things as they are and reorient mitrates without regard to any theory or preconceived interpretation. But, if this thecal orientation problem is corrected, then many of the alleged homologies with chordates fall apart.

There is certainly no sufficient reason for doubting that the Stylophora were real echinoderms. Of course, they are fairly distinct from other members of the phylum, but probably no more so than the Helicoplacoidea, the Homostelea, or the Ctenocystoidea. Despite their general primitive nature, they exhibit many morphological and functional specializations that prevent us from regarding them as possible ancestors of other echinoderms. Apparently they became extinct by Middle Devonian time and left no descendants.

RADIATE ECHINODERMS

Radiate echinoderms appear in the geological column near the very base of the Cambrian. Their earliest known representatives are members of two quite different

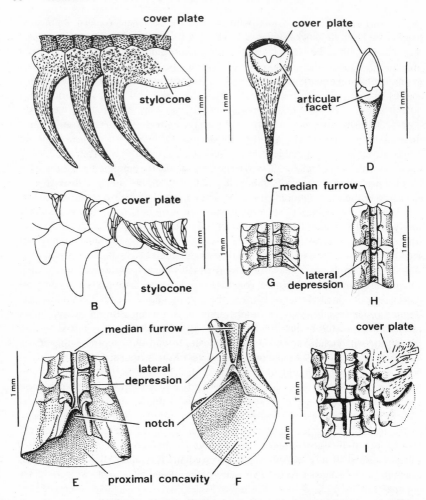

Figure 5 Comparison of cornute and mitrate aulacophores. A and B: lateral aspect of mid- and small portion of distal part of aulacophore in (A) the cornute *Reticulocarpos hanusi* Jefferies & Prokop (1972), Lower Ordovician, and (B) the mitrate *Chinianocarpos thorali* Ubaghs (1970), Lower Ordovician. C and D: distal aspect of stylocone in same species (Jefferies & Prokop 1972, Ubaghs 1970). E and F: adoral aspect of stylocone in (E) the cornute *Phyllocystis crassimarginata* Thoral, Lower Ordovician, and (F) the mitrate *Peltocystis cornuta* Thoral, Lower Ordovician (Ubaghs 1970). G and H: adoral aspect of two distal ossicles in (G) the cornute *Phyllocystis crassimarginata* Thoral, Lower Ordovician (Ubaghs 1970) and (H) the mitrate *Mitrocystites mitra* Barrande, Lower Ordovician (Ubaghs 1968b). I: adoral aspect of two distal ossicles, with cover plates preserved in wide open position (not drawn on left side), in the cornute *Cothurnocystis fellinensis* Ubaghs (1970), Lower Ordovician.

classes, the eocrinoids and the edrioasteroids, to which in late Lower Cambrian or early Middle Cambrian were added the camptostromatoids and the cyclocystoids. The other classes with radiate symmetry are all post-Cambrian, except the crinoids and the holothuroids (Durham 1971), possibly represented in the Burgess Shale fauna (Middle Cambrian) of British Columbia.

Eocrinoidea

The eocrinoids are the earliest known pelmatozoan echinoderms. They appear close to, or a little above, the base of the Cambrian (Durham 1971), where they are associated with helicoplacoids and earliest edrioasteroids (Durham 1964). They are represented by 3 genera in the upper part of the Lower Cambrian and by 6 genera and 17 species in the Middle Cambrian (Sprinkle 1973). This is the most important record for these echinoderms, which apparently became extinct by Ordovician time.

The oldest known eocrinoids were already so diversified—a fact that implies a long previous history and considerable adaptative divergence—that one may question whether they form a homogeneous group. Thus the Lower Cambrian genus *Lepidocystis* (Figure 6A) has been considered either as an eocrinoid (Foerste 1938, Regnéll 1945, Termier & Termier 1969) or as the type of a discrete class (Durham 1968b). Nevertheless, compared to later pelmatozoans, the structure of some of these early forms still appears remarkably primitive. It suggests, among other things, that 1. the stem or stalk developed from the aboral end of the calyx as an irregularly plated outgrowth with a large central lumen continuous with the thecal cavity (Figure 6A, B); 2. originally the plating of the calyx was irregular and composed of a large and indefinite number of elements (Figure 6A, B, D); and 3. the radiate symmetry did not affect the whole body, but was apparent only in the arrangement of the ambulacra and finger-like food-gathering appendages (brachioles) (Figure 6A–D). However, as early as late Middle Cambrian time, eocrinoids occurred that were provided with a five fold body organization (Figure 6C), or with a long stalk composed of a very short ringlike ossicle and a definite root-structure (Figure 6D). How these transformations were accomplished is unknown.

The brachioles, even if they served the function of arms in crinoids, were very different in origin, constitution, and mode of development. They certainly did not contain outgrowths from the general coelomic cavity or reproductive organs, nor were they provided with entoneural nerve cords. They did have a ciliated food-groove, sensory and motor nerves, and, in the opinion of the present writer (Ubaghs 1970), tube feet and radial water vessels—a view recently opposed by Sprinkle (1973), who believes that in eocrinoids and other related groups such as rhombifers and blastoids, the water-vascular system was reduced and lacked tube feet. However, Breimer & Macurda (1972) have furnished considerable evidence that a well-developed crinoid-like water-vascular system, with tentaculated extensions into the brachioles, was present in blastoids. And if so in blastoids, why not in eocrinoids? It is true that no trace of an ambulacral system is found in their ambulacra and brachioles, but the same condition occurs. in crinoids, because the ambulacral system in their arms lies away from the supporting ossicles and leaves no markings on the skeleton.

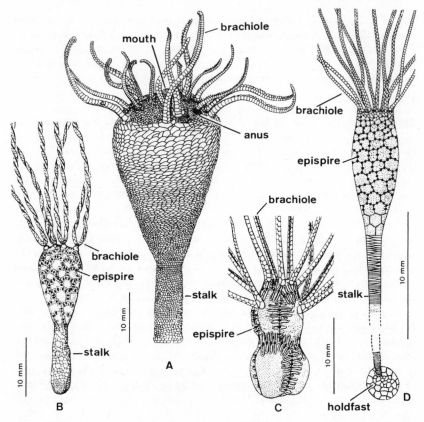

Figure 6   Reconstructions of various Lower and Middle Cambrian eocrinoids. A: *Lepido-cystis wanneri* Foerste, Lower Cambrian (adapted from Durham 1968b and Sprinkle 1973). B: *Gogia spiralis* Robinson (1965), Middle Cambrian. C: *Lichenoides priscus* Barrande, Middle Cambrian (Ubaghs 1968a). D: *Akadocrinus jani* Prokop (1962), Middle Cambrian.

On the other hand, it has been repeatedly assumed (Chauvel 1966, Durham 1967, 1971, Sprinkle 1973) that the presumed respiratory structures, such as epispires, hydospires, diplopores, etc, present in most noncrinoid pelmatozoans were the terminal organs of an internal water-vascular system. This interpretation would imply an organization of the system different from that of Recent echinoderms, which, in fact, offer the only direct evidence of the structure of the water-vascular system, and thus furnish the correct basis for interpretation of the system in extinct forms (Nichols 1972). In all living classes the structure of the system is similar—a fact that strongly suggests that it was inherited from a common ancestor. On the other hand, there is no valid reason for thinking that the extinct radiate forms arose from another source (Durham 1967). It appears probable that the extinct pelmatozoans probably had a water-vascular system very similar to that of living crinoids, with radial canals and tube feet intimately associated with the food-grooves. As to the presumed respiratory thecal structures, they are better seen as outfolds or infolds of the body wall, rather than parts of a hypothetical internal plexus of hydrocoelian origin.

The phyletic relationships of the Eocrinoidea are obscure. Some authors (Fell 1963b, Jaekel 1918) have taken them as possible ancestors of all echinoderms. But the eocrinoids are definitely too pelmatozoan in structure and habit to have given rise to nonradiate forms or to the Echinozoa or Asterozoa. They do not even seem to be the source of the Crinoidea, from which they differ mainly in lack of true arms (Sprinkle 1973, Ubaghs 1968a). They may represent the basic stock from which the other brachiole-bearing groups (Blastoidea, Parablastoidea, Rhombifera, and possibly Diploporita) were derived, for they share the same fundamental organization (Sprinkle 1973).

Crinoidea

The origin of the crinoids has long been puzzling. The earliest known representatives, discovered in Lower Ordovician deposits of England (Bates 1968) and France (Ubaghs 1969), already had all the diagnostic characters of the class and belong to two separate orders. Some of their features—especially the nature of their arms—are quite distinct from those shown by any eocrinoids, which, in spite of their name and earlier appearance, are probably not ancestral (Ubaghs 1968a).

Of considerable interest, therefore, is the discovery of a new pelmatozoan with armlike appendages in the Middle Cambrian (Burgess Shale) of western Canada (Sprinkle 1973). This form, *Echmatocrinus brachiatus* Sprinkle (Figure 7A), with its high conical calyx made up of numerous irregularly arranged plates and its irregularly multiplated stalk, resembles such archaic eocrinoids as *Gogia* (Figure 6B). But it has large uniserial arms with soft appendages (probably tube feet), and these arms seem to be very similar to those of crinoids. The existence of such a creature had been predicted to some extent (Ubaghs 1969). *Aethocrinus* (one of the oldest known crinoids) shows a stem composed of irregularly interlocking plates which grades into the calyx and passes distally into a mass of irregular tiny plates, while the large lumen merges into the cavity of the calyx. The plates of the calyx are relatively numerous and their arrangement in circlets is not quite regular (Figure 7B). These features are not strictly intermediate between those of *Echmatocrinus* and those of any typical crinoid, and a wide gap still separates the two forms. Yet, compared to later representatives, *Aethocrinus* allows recognition of the probable existence of a more primitive stage, which would not be unlike *Echmatocrinus*.

Neither *Echmatocrinus* nor the earliest known crinoids bear any resemblance to the oldest recorded asterozoans. They certainly do not support Fell's hypothesis (1963a) that the asterozoan echinoderms descended from a crinoid ancestor.

Edrioasteroidea

The edrioasteroids are among the earliest known echinoderms. They already occur in the same beds as those containing helicoplacoid and eocrinoid remains (Durham

1964). Two species have been found in late Upper Cambrian beds of Newfoundland (Schuchert 1919), and about ten others have been described from Middle Cambrian deposits of central and western Europe and British Columbia (Termier & Termier 1969). They flourished mainly in Ordovician times and apparently became extinct during the Pennsylvanian period.

Cambrian edrioasteroids are poorly known (Termier & Termier 1969). It is clear, however, that they did not differ markedly from later members of the class.

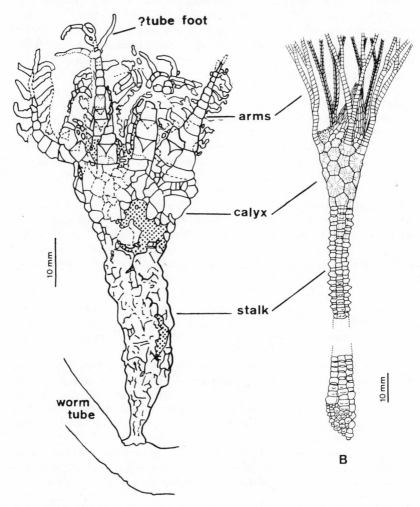

Figure 7   Comparison of *Echmatocrinus* with one of the two earliest known crinoids. A: *Echmatocrinus brachiatus* Sprinkle, Middle Cambrian (adapted from Sprinkle 1973). B: *Aethocrinus moorei* Ubaghs (1969), Lower Ordovician.

Their ambulacral grooves (at least in *Cambraster*) are surprisingly similar to those of asteroids (Ubaghs 1971), and probably contained a main water vessel running externally along the midline of each one, and tube feet provided with internal ampullae (Bather 1915, Nichols 1972, Ubaghs 1971). Such similitude does not imply a close relationship with asterozoan echinoderms, as is shown by the structure of the earliest known members of that group (Fell 1963a). The rather questionable resemblance offered by the pharyngeal skeleton of some Recent dendrochirote holothuroids with the ambulacral plates of edrioasteroid affinities also is not particularly significant (Fell 1965, Fell & Moore 1966). That radiate edrioasteroids could have given rise to nonradiate echinoderms (Termier & Termier 1969), is no more than a hypothesis. Present knowledge suggests that the edrioasteroids are best seen as a separate class, of unknown origin, affinity, and descent.

Cyclocystoidea

The recent discovery of a typical cyclocystoid in early Middle Cambrian rocks in Queensland (Australia) (Henderson & Shergold 1971) strengthens the viewpoint that these echinoderms comprise a distinct line of evolution. Since their earliest known appearance, they show no close similarity with any other echinoderm classes. Their origin is unknown, and their organization, relationships, and mode of life are uncertain. Various and sometimes opposing interpretations of their structure have been proposed (Kesling 1963, 1966, Nichols 1967, 1969, 1972).

It has been suggested that they could be derived from a diploporite cystoid ancestor (Kesling 1966), but the earliest known diploporites are Lower Ordovician (Tremadocian) in age. In radial symmetry, flattened form, presence of a marginal ring of small plates, and submarginal circlet of large plates, they offer a superficial likeness to the Middle Cambrian edrioasteroid *Cambraster*, but a detailed comparison fails to detect any compelling resemblance. The cyclocystoids probably are the most enigmatic group of echinoderms and are in need of a complete revision.

Camptostromatoidea

The genus *Camptostroma*, originally described as a floating hydrozoan and subsequently included in the Scyphozoa, has proved to be provided with an echinoderm-like skeleton (Durham 1966, 1968a). A single species from the Lower Cambrian Kinzers Formation of Pennsylvania, is associated with lepidocystoid eocrinoids. This strange and rather poorly known creature has been supposed to have been pelagic or bathypelagic (Durham 1966, 1968a), but it seems best to regard it as benthonic (in view of its heavy skeleton). It may have been free rather than sessile (Nichols 1969), for no attachment organ has been observed. It has been compared to some holothurians, but the peripheral position and plated structure of its armlike appendages, the lack of recognizable quinqueradiate and bilateral symmetries, and the apparent shortness of its oral-aboral axis differentiate it from holothuroids (Durham 1966, 1968a). It has some likeness to the ophiocistioids in general body shape, location of mouth and anus on opposite faces, and appearance of the free appendages, but such similarities may be superficial. In any case, they are not sufficient to prove relationship. *Camptostroma* has been placed in a separate class

(Durham 1966), but such a procedure serves more to conceal our ignorance than to express the real state of our present knowledge.

Cymbionites and Peridionites

These two genera have been proposed on the basis of fossils found in lowermost Middle Cambrian strata of Queensland, Australia (Whitehouse 1941). That they are echinoderms can hardly be doubted, for their plates are composed of single calcite crystals and they exhibit the honeycomb microstructure characteristic of the skeleton of the phylum. However, nothing more is certain about these remains. It is not known whether they are complete or partial skeletons, the nature of associated soft parts is enigmatic. Without information on the organization of the animals represented, it seems imoper to assign them to any particular group of echinoderms. Nevertheless, they have been regarded as representatives of two new classes forming a distinct subphylum (Durham 1971, Whitehouse 1941), or as representatives of extinct phyla (Cuénot & Tetry 1949), or as eocrinoids with reduced thecae (Schmidt 1951). In addition *Cymbionites* has been viewed as a cystoid, and *Peridionites* as possibly a benthonic ctenophore (Gislén 1947). All these interpretations are purely speculative and lack any firm morphological foundation (Ubaghs 1968d).

CONCLUSIONS

1. Remains of echinoderms showing all the essential characteristics of the phylum are known from the base of the Cambrian upward. In its present state, therefore, paleontology throws no light on the ultimate affinity or origin of these animals, nor on the manner in which their organization developed.

2. Forms with radial symmetry, forms with no trace of radial symmetry, forms that were attached during life (pelmatozoic), and forms that were free-living (eleutherozoic) all appear at about the same geological level. Thus, on strictly stratigraphical (chronological) grounds, it cannot be decided which of these conditions is the most primitive.

3. The differentiation of the classes took place before or possibly during Lower Palaeozoic time, and certainly was accomplished before the end of the Silurian period.

4. The origin of most classes is unknown, for their earliest representatives already show the diagnostic class features.

5. No phylogenetically intermediate forms between two classes have ever been found. No class is known with certainty to have given rise to another. It is assumed, but not known, that the eocrinoids are the stock from which the other classes of brachiole-bearing pelmatozoans evolved. Yet the ascendancy of none of these classes has ever been established with certainty.

6. The lack of radial symmetry in helicoplacoids and carpoid or homalozoan echinoderms (i.e. Homostelea, Ctenocystoidea, Stylophora, Homoiostelea) is judged to be primitive. These forms are interpreted as (*a*) having differentiated before the appearance of radial symmetry, (*b*) having followed an evolutionary path different

from that of radiate echinoderms (Bather 1929), and (c) belonging to an early pre-radiate radiation of various forms testing the adaptive possibilities of asymmetry.

7. The nonradiate echinoderms, however, share at least three important features with the radiate echinoderms: (a) they are asymmetrical; (b) their skeleton has the same histological, mineralogical, and crystallographic characteristics; and (c) they probably were provided with a water-vascular system. This may mean that the nonradiate echinoderms separated from the common ancestral stock of the phylum only after they had acquired these three fundamental features.

8. Three major steps or grades in the general evolution of echinoderms may be recognized (Figure 8): (a) The first grade, which is hypothetical, corresponds to the pre-echinodermal bilateral stage (*Echinodermata bilateralia*), classically represented by the Dipleuraea in phylogeny and the Dipleurula larva in ontogeny. (b) The second grade, characterized by an asymmetrical structure (*Echinodermata asymmetrica*), comprises the nonradiate echinoderms (Helicoplacoidea and Homalozoa)

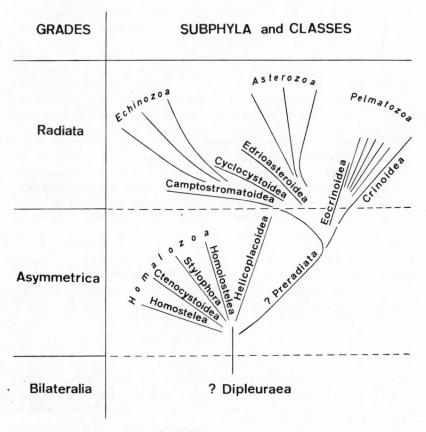

Figure 8   Phyletic diagram for early echinoderms.

and probably the unknown ancestors of the radiate echinoderms. It may correspond in ontogeny of Recent forms to the asymmetric phase of development. (c) The third grade includes all the other echinoderms in which radial symmetry has been secondarily imposed on a pre-existing asymmetry (*Echinodermata radiata*), as it is in ontogeny of living classes.

9. At present, paleontology cannot prove or disprove the phylogenetic theories that have been proposed to explain the organization of the phylum, because these theories need to include events that must have occurred in Precambrian time and because larvae are totally lacking from the fossil record. For instance, we are not certain that the Precambrian ancestors of the free-living Cambrian echinoderms were not attached. We may only observe that the presumed organization of some of them, particularly the Helicoplacoidea, Stylophora, and Ctenocystoidea, does not suggest the existence of attached ancestors. The evidence points instead to a common ancestral stock of free-living and asymmetrical organisms, with opposite mouth and anus, an elongate body divided into two or three coelomic regions, a functional hydrocoel (or lophophore), and possibly a calcitic endoskeleton— that is to say, to a stock having at least some of the features found in all larvae of the Recent echinoderms before metamorphosis.

10. The hypothesis according to which the eocrinoids and other brachiole-bearing echinoderms were characterized by a reduced water-vascular system, without tube feet, and the assumption that the presumed respiratory calycinal structures, like epispires, diplopores, or hydrospires, were extensions from the hydrocoel are judged to be unproved, contrary to evidence furnished by blastoids and generally inconsistent with the structure of the water-vascular system in living classes.

11. Interpretation of stylophoran carpoids as chordates rather than echinoderms is rejected, for they possess a skeleton typical for echinoderms and many features strictly comparable to structures commonly occurring among echinoderms. On the other hand, they have not a single skeletal feature that demonstrates chordate nature.

12. The theory according to which the asterozoan echinoderms evolved from a crinoid or a crinoid-like ancestor finds no support in the structure of any known Cambrian Pelmatozoa, including *Echmatocrinus* and the earliest known crinoids.

ACKNOWLEDGMENTS

With the kind permission of the Cambridge Philosophical Society, small parts of the text, the main conclusions, and some figures have been borrowed from an article published by the author in 1971. The section on Stylophora has been critically reviewed by James. Sprinkle.

Literature Cited

Bates, D. E. 1968. On 'Dendrocrinus' cambriensis Hicks, the earliest known crinoid. *Palaeontology* 11:406–9
Bather, F. A. 1913. Caradocian Cystidea from Girvan. *Trans. Roy. Soc. Edinburgh* 49:359–529

Bather, F. A. 1915. *Studies in Edrioasteroidea*. Wimbledon, England: Fabo
Bather, F. A. 1929. Une classe d'Echinodermes sans trace de symétrie rayonnée. *Assoc. Fr. Avan. Sci.*, 435–38
Breimer, A., Macurda, D. B. Jr. 1972. The

phylogeny of the fissiculate blastoids. *Verh. Kon. Ned. Akad. Wetensch. Afd. Natuurkunde* 26(3):390

Cabibel, J., Termier, H., Termier, G. 1958. Les Echinodermes mésocambriens de la Montagne Noire. *Ann. Paléontol.* 44:281–94

Caster, K. E. 1952. Concerning *Enoploura* of the Upper Ordovician and its relation to other carpoid Echinodermata. *Bull. Am. Paleontol.* 34:1–56

Caster, K. E. 1968. Homoiostelea. *Treatise on Invertebrate Paleontology,* ed. R. C. Moore, Pt. S, 581–627. Geol. Soc. Am. & Univ. Kansas

Chauvel, L. 1966. Echinodermes de l'Ordovicien du Maroc. *Cah. Paléontol.* 120 pp.

Chauvel, L. 1971. Les Echinodermes Carpoïdes du Paléozoïque inférieur marocain. *Notes Serv. Géol. Maroc* 31:49–60

Cuénot, L., Tetry, A. 1949. Deux prétendus Echinodermes du Cambrien d'Australie. *13th Congr. Int. Zool.,* 568

Denisson, R. H. 1971. The origin of the vertebrates: a critical evaluation of current theories. *Proc. N. Am. Paleontol. Conv.* 2:1132–46

Durham, J. W. 1964. The Helicoplacoidea and some possible implications. *Yale Sci. Mag.* 39:24–25, 28

Durham, J. W. 1966. *Camptostroma,* an Early Cambrian supposed scyphozoan, referable to Echinodermata. *J. Paleontol.* 40:1216–20

Durham, J. W. 1967. Notes on the Helicoplacoidea and early echinoderms. *J. Paleontol.* 41:97–102

Durham, J. W. 1968a. Camptostromatoids. See Ref. 8, pp. 627–31

Durham, J. W. 1968b. Lepidocystoids. See Ref. 8, pp. 631–34

Durham, J. W. 1971. The fossil record and the origin of the Deuterostomata. *Proc. N. Am. Paleontol. Conv.* 2:1104–32

Durham, J. W., Caster, K. E. 1963. Helicoplacoidea: a new class of echinoderms. *Science* 140:820–22

Durham, J. W., Caster, K. E. 1966. Helicoplacoidea. See Ref. 8, Pt. U, pp. 131–36

Eaton, T. H. 1970. The stem-tail problem and the ancestry of chordates. *J. Paleontol.* 44:969–79

Fell, H. B. 1963a. The phylogeny of seastars. *Phil. Trans. Roy. Soc. London Ser. B* 246:381–435

Fell, H. B. 1963b. The evolution of the echinoderms. *Smithson. Rep.* 1962:457–90

Fell, H. B. 1965. The early evolution of the Echinozoa. *Breviora* 219:1–17

Fell, H. B., Moore, R. C. 1966. General features and relationships of echinozoans. See Ref. 8, Pt. U, pp. 108–18

Foerste, A. E. 1938. Echinodermata. In Resser, C. E., Howell, B. F.: Lower Cambrian *Olenellus* zone of the Appalachians. *Bull. Geol. Soc. Am.* 49:212–13

Gislén, T. 1930. Affinities between the Echinodermata, Enteropneusta, and Chordonia. *Zool. Bidr. Uppsala* 12:199–304

Gislén, T. 1947. On the Haplozoa and the interpretation of *Peridionites. Zool. Bidr. Uppsala* 25:402–8

Henderson, R. A., Shergold, J. H. 1971. *Cyclocystoides* from Early Middle Cambrian rocks of northwestern Queensland, Australia. *Palaeontology* 14:704–10

Jaekel, O. 1918. Phylogenie und System der Pelmatozoen. *Paläontol. Z.* 3:1–128

Jefferies, R. P. S. 1967. Some fossil chordates with echinoderm affinities. *Symp. Zool. Soc. London* 20:163–208

Jefferies, R. P. S. 1968a. Fossil chordates with echinoderm affinities. *Proc. Geol. Soc. London* 1649:128–35

Jefferies, R. P. S. 1968b. The subphylum Calcichordata (Jefferies, 1967); primitive fossil chordates with echinoderm affinities. *Bull. Brit. Mus. Natur. Hist. Geol.* 16:243–339

Jefferies, R. P. S. 1969. *Ceratocystis perneri* Jaekel—A Middle Cambrian chordate with echinoderm affinities. *Palaeontology* 12:494–535

Jefferies, R. P. S. 1973. The Ordovician fossil *Lagynocystis pyramidalis* (Barrande) and the ancestry of Amphioxus. *Phil. Trans. Roy. Soc. London Ser. B* 265:409–69

Jefferies, R. P. S., Prokop, R. J. 1972. A new calcichordate from the Ordovician of Bohemia and its anatomy, adaptations and relationships. *Biol. J. Linn. Soc.* 4:69–115

Kesling, R. V. 1963. Morphology and relationships of Cyclocystoides. *Contrib. Univ. Mich. Mus. Paleontol.* 18:157–76

Kesling, R. V. 1966. Cyclocystoids. See Ref. 8, Pt. U, 188–210

Matsumoto, H. 1929. Outline of a classification of Echinodermata. *Sci. Rep. Tôhoku Imp. Univ. Ser. 2* 13:27–32

Nichols, D. 1967. The origin of echinoderms. *Symp. Zool. Soc. London* 20:209–29

Nichols, D. 1969. *Echinoderms.* London: Hutchinson. 192 pp.

Nichols, D. 1972. The water-vascular system in living and fossil echinoderms. *Palaeontology* 15:519–38

Prokop, R. 1962. *Akadocrinus* nov. gen., a new crinoid from the Cambrian of the Jnce area (Eocrinoidea). *Sb. Ustred. Ustavu Geol. Oddil Geologicky* 27:31–39

Regnéll, G. 1945. Non-crinoid Pelmatozoa from the Paleozoic of Sweden. A taxonomic study. *Medd. Lunds Geol. Mineral. Inst.* 108:255 pp.

Regnéll, G. 1960. "Intermediate" forms in Early Palaeozoic echinoderms. *21st Int. Geol. Congr.* 22:71–80

Robinson, R. A. 1965. Middle Cambrian eocrinoids from western North America. *J. Paleontol.* 39:355–64

Robinson, R. A., Sprinkle, J. 1969. Ctenocystoidea: new class of primitive echinoderms. *Science* 166:1512–14

Schmidt, H. 1951. Whitehouse's Ur-Echinodermen aus dem Cambrium Australiens. *Paläontol. Z.* 24:142–45

Schroeder, R. 1973. Carpoideen aus dem Mittelkambrium Nordspaniens. *Palaeontographica A* 141:119–42

Schuchert, Ch. 1919. A Lower Cambrian edrioasteroid, Stromatocystites walcotti. *Smithson. Misc. Coll.* 70(1):1–8

Sieverts-Doreck, H. 1951. Uber Cyclocystoides Salter & Billings und eine neue Art aus dem belgischen und rheinischen Devon. *Senckenbergiana* 32:9–30

Sprinkle, J. 1973. Morphology and evolution of blastozoan echinoderms. *Spec. Publ. Mus. Comp. Zool. Harvard Univ.* 248 pp.

Termier, H., Termier, G. 1969. Les Stromatocystoïdes et leur descendance. Essai sur l'évolution des premiers Echinodermes. *Geobios* 2:131–56

Termier, H., Termier, G. 1973. Les Echinodermes Cincta du Cambrien de la Montagne Noire (France). *Geobios* 6:243–65

Ubaghs, G. 1961. Sur la nature de l'organe appelé tige ou pédoncule chez les Carpoïdes Cornuta et Mitrata. *C. R. Acad. Sci.* 253:2738–40

Ubaghs, G. 1967. Le genre Ceratocystis Jaekel (Echinodermata, Stylophora). *Univ. Kans. Paleontol. Contrib.* 22:16 pp.

Ubaghs, G. 1968a. The eocrinoids. See Ref. 8, pp. 455–94

Ubaghs, G. 1968b. Stylophora. See Ref. 8, pp. 495–564

Ubaghs, G. 1968c. Homostelea. See Ref. 8, pp. 565–80

Ubaghs, G. 1968d. Cymbionites and Peridionites. Unclassified Middle Cambrian echinoderms. See Ref. 8, pp. 634–37

Ubaghs, G. 1969. Aethocrinus moorei Ubaghs, n. gen. n. n., le plus ancien Crinoïde dicyclique connu. *Univ. Kans. Paleontol. Contrib.* 38:25 pp.

Ubaghs, G. 1970. Les Echinodermes carpoïdes de l'Ordovicien inférieur de la Montagne Noire (France). *Cah. Paléontol.* 122 pp.

Ubaghs, G. 1971. Diversité et spécialisation des plus anciens echinodermes que l'on connaisse. *Biol. Rev.* 46:157–200

Whitehouse, F. W. 1941. Early Cambrian echinoderms similar to larval stages of Recent forms. *Mem. Queensl. Mus.* 12:1–28

INTERACTION OF ENERGETIC   ✖10034
NUCLEAR PARTICLES IN SPACE WITH
THE LUNAR SURFACE

Robert M. Walker

Laboratory for Space Physics, Washington University, St. Louis, Missouri 63130

This paper gives a brief review of the interaction of energetic particles in space with the lunar surface. Being an almost atmosphereless body of great antiquity, the moon can be considered as a giant space probe whose surface contains a virtually complete record of the energetic particle environment back to almost the beginning of the solar system. Deciphering this record is, however, far from simple. Tied intimately to the problem of the history of the radiations is the concomitant problem of temporal changes in the lunar surface. Progress in the one problem leads to progress in the other; for example, if it can be shown that the average flux and energies of galactic cosmic rays have not changed in time, then cosmic ray exposure ages can be used to date the time of formation of specific features such as lunar craters.

In the spirit of the instructions given by the editors, I have not tried to be encyclopedic; rather, I have attempted to summarize the major results obtained and to outline major current areas of research.

Three classes of energetic particles strike the present moon: solar wind particles, solar flare particles, and galactic cosmic rays. The properties of these different kinds of particles, as well as the physically measurable effects that they produce, are summarized first. Following this, a discussion is given of the new information gained about these radiations from the Apollo missions. The last part of the review briefly treats the problem of lunar surface dynamics and summarizes the conclusions that can be drawn concerning various rate processes characteristic of the lunar surface.

The above inventory of energetic particles impinging on the moon is not complete. Considering only purely external particles, there is growing evidence for a transient "suprathermal" component of particles in interplanetary space (Borg et al 1972). Whether these should be considered as mini-solar flare particles or whether they represent a distinct component is not clear. There is also an internal-external component; atoms thermally released from the lunar surface can be accelerated by interactions with the interplanetary medium and then restrike the lunar surface at much higher energies (Manka & Michel 1971). Only a brief mention of this fascinating phenomenon is given here.

ENERGETIC PARTICLES IN SPACE AND THEIR INTERACTION WITH LUNAR SURFACE MATERIALS

Solar Wind

The solar wind plasma consists of a variety of different elements whose abundances are at least approximately those of the sun itself. At the position of the moon the energy of the particles is ~ 1 keV/amu (unlike an ordinary gas, the solar wind is an isovelocity gas) and the flux is $\sim 3 \times 10^8$ p/cm^2/sec. Although the composition, average energy, and flux all change somewhat depending on solar activity, the solar wind is a constant component of the interplanetary medium. For a review of solar wind properties, we refer the reader to the book by Hundhausen (1972).

Two principal effects are produced by the solar wind on lunar materials. The first and most studied effect is the direct implantation of the solar wind ions into the surfaces of lunar materials directly exposed to the sun. The direct implantation depth is small (~ 500 Å) but the depth distribution may, of course, be modified by subsequent diffusion. The relative shallowness of the lunar rubble layer (regolith), typically about 10 m in thickness, means that large numbers of directly implanted ions can be present. Approximately 5×10^{25} atoms are present in a column 1 cm^2 in area and 10 m deep. If the solar wind were captured with unit efficiency (this is, of course, an unrealistic assumption) for a period of 4×10^9 yr, an almost equal number of atoms from the sun would be added to the surface! Thus a pinch of moon dust is also, in a very real sense, a pinch of stardust. Since the atoms that are added (mostly hydrogen and helium) are not normally abundant in the lunar target materials, directly implanted solar wind ions produce large and easily measurable effects.

A second striking effect, first demonstrated by Maurette and his collaborators at the University of Paris, Orsay (Borg et al 1970, Dran et al 1970, Bibring & Maurette 1972, Bibring et al 1972a,b, Bibring et al 1973), is the radiation damage produced by the bombardment. These workers showed that angular crystalline grains from the interior of lunar rocks have their outer crystalline regions converted to amorphous layers when subjected to laboratory beams of 1 keV/amu atoms. The grains also become noticeably rounded. As we shall see later, both these effects are seen in typical lunar dust grains.

The presence of large numbers of implanted ions of solar composition also leads to the formation of simple molecules in the implanted regions (see, for example, Pillinger et al 1973). This fascinating subject is beyond the scope of this article and is not further treated.

Solar wind ions also cause atomic sputtering of exposed surfaces. However, as originally shown by Wehner et al (1963), the rates are very small; this process is probably not important for most lunar samples.

Solar Flares

Energetic particles are injected into the interplanetary medium at sporadic intervals during solar flare events. (For an early review of solar energetic particle events, the reader is referred to Fichtel & McDonald 1967). The frequency of solar flares is generally higher at the maximum of the sunspot cycle, but they can occur at any time. Different flares vary considerably in the total numbers and energy spectra of the emitted nuclear particles. Several larger flares generally contribute the bulk of the high energy particles during a total solar cycle. For example, in the period 1966 to 1972 some 70% of the total solar flare protons ≥ 10 MeV in energy were emitted in a single large flare that occurred early in August 1972.[1] Although the precise value of the high energy crossover is still not certain, solar flare particles dominate the average flux of energetic particles in the energy range from 100 keV/amu to ~ 100 MeV/amu.

In contrast to solar wind particles, which have a more or less constant energy, solar flare particles have a differential energy spectrum that drops steeply with increasing energy. It is conventional to express the differential spectrum either as a power law in kinetic energy, $dN/dE = AE^{-\gamma}$ (with γ typically between 2 and 4), or an exponential in particle rigidity, $dJ/dR = J_0 \exp(-R/R_0)$, where J is the particle flux and R is the rigidity (with R_0 typically between 40 and 200 MV). Published data on individual flares shows that some flares are better described by a power law and others by an exponential function; in any event, over a limited energy range, either representation is equally satisfactory.

The sharply decreasing energy spectrum means that solar flare effects are confined to the near surface regions of lunar materials. For proton-induced effects, solar flares will dominate in the first ~ 5 g/cm^2 of material, while for solar flare iron particles, which have a much higher rate of ionization loss, the flare particles will dominate only to a depth of $\lesssim 1$ g/cm^2 of material. Since typical ablation distances for meteorites are much larger than these characteristic depths (Fleischer et al 1967), the study of the long-term history of solar flares only became possible when non-ablated samples were returned from the moon.

Three principal effects produced by solar flares have so far been studied: (a) The production of radioisotopes in nuclear collisions (for particles with energies > 10 MeV/amu), (b) the production of trapped electrons due to ionization, and (c) the production of nuclear particle tracks. The first two effects are due mostly to solar flare protons; in contrast, nuclear particle tracks can be revealed in lunar minerals only when the atomic number of the particles is $\gtrsim 20$.

Reedy & Arnold (1972) give theoretical curves for the production of different radioisotopes as a function of depth for a particular lunar rock. To obtain such curves it is necessary to know the excitation functions for different targets, the chemistry of the rock, and the incident spectrum of solar particles. The probability that a given proton will suffer a nuclear collision before it slows down and comes to rest by ionization loss does not reach unity until a proton energy of $\gtrsim 300$ MeV. For this reason, reactions by secondary nuclear particles generated in nuclear collisions do not play an important role in the production of radioisotopes by solar flare particles.

Although trapped electrons can be studied in principle by any number of solid

[1] J. H. King. 1973. Goddard Spaceflight Center, Preprint Number X-601-73-324.

state methods, only one technique, thermoluminescence, has so far proven useful in lunar samples. When free charge carriers are produced in an insulator, they can become trapped at localized defect sites in the crystal at energy levels that lie in the normally forbidden band. Continued irradiation at low temperatures continues to fill the electron-hole traps. If the samples are subsequently heated, the trapped charge carriers will be thermally liberated; in certain solids, containing appropriate recombination centers, the charge carriers recombine with the emission of visible photons.

As discussed in detail by Hoyt et al (1973), the amount of thermoluminescence stored in a lunar sample represents a dynamic equilibrium between the addition of trapped charge carriers due to ionization and the loss of charge carriers due to thermal excitation at ambient lunar temperatures. Thermoluminescence measurements are thus analogous to radioisotope measurements in that the effects produced have a characteristic lifetime.

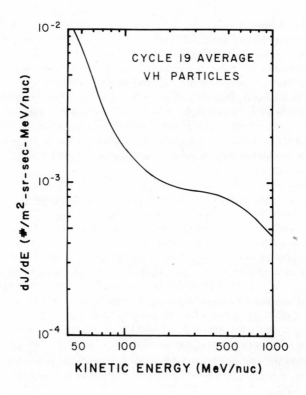

Figure 1   Solar cycle averaged VH differential energy spectrum for solar cycle 19. The low energy part (< 100 MeV/nuc) is taken from the Surveyor III spectrum (see text) and does not represent solar cycle 19 values (from Yuhas & Walker 1973).

The radiation damage produced by a slow, heavy particle may become sufficiently intense to disrupt the atomic structure and produce a latent track in the material. Only heavy particles [atomic number $\gtrsim 20$ near the end of their range (typically the last 10 to 20 μm for iron nuclei)] produce enough damage to give visible tracks in silicate minerals. Such tracks can be either seen directly by transmission electron microscopy or, as is more generally the case, chemically etched to produce a track (hole) that is visible by optical microscopy. The subject of track etching has been reviewed extensively in recent years (Fleischer et al 1965, Price & Fleischer 1971, Fleischer et al 1974b). Because of the drop in elemental abundances beyond iron, most cosmic ray or solar flare tracks are produced by particles with $20 \leq Z \leq 28$ (VH ions).

Solar flare particles also produce stable isotopes and directly implanted ions; to date, however, little work has been reported on these effects in lunar samples.

Galactic Cosmic Rays

Although occasional solar flares may produce many particles in the GeV range, the time *average* flux of nuclear particles > 100 MeV in energy is dominated by galactic cosmic ray nuclei that come from outside the solar system (for a recent review of cosmic ray physics, see Meyer 1969). At high energies ($\gtrsim 100$ GeV/amu), galactic cosmic rays are invariant in time with a decreasing power law energy spectrum given by $dN/dE = BE^{-2.5}$. At lower energies, however, the cosmic rays are greatly modulated by the interplanetary plasma, changing in number density and energy depending on the degree of solar activity. The net effect is to give a time-averaged energy spectrum that is quite flat at low energies, as shown in Figure 1. This figure also shows the sharp rise in particle flux at energies < 100 MeV/amu due to solar flares.

The principal effects due to cosmic rays are the production of stable isotopes and radioisotopes by the proton and helium components and the production of tracks by ions with atomic numbers > 20. Some tracks are also produced by the heavy recoil nuclei from high energy spallation reactions (Fleischer et al 1974b). Such tracks are very short and have not played an important role in lunar sample studies. Almost all the high energy nuclei suffer inelastic nuclear collisions before coming to rest; direct implantation effects are thus negligible. Trapped electrons produced by galactic particles have been studied in lunar core stems where it has been suggested that they could be used to give independent measures of lunar heat flow (Hoyt et al 1971); however, this possibility has still to be realized in practice.

Figure 2, taken from Reedy & Arnold (1972), shows the calculated rates of production for various radioisotopes as a function of depth in a lunar rock. Because of the production of many secondary particles, the calculation of the depth dependence of nuclide production is more difficult than in the case of solar flares. If the product nucleus is much lighter than the target nucleus (high energy spallation product), the production tends to drop off immediately with a characteristic scale length determined by the nuclear mean free path. However, the rate of production of a product nucleus close in mass to a target nucleus (low energy spallation product) first increases with depth, reaches a maximum at ~50 g/cm², and then decreases with a characteristic mean free path of about ~150 g/cm².

The radioactive nuclei can be used to measure or set limits on the exposure ages of different samples and to examine the constancy of cosmic ray protons in time.

Except for very young samples, however, spallation exposure ages are usually determined by measuring the total accumulated amount of stable spallation isotopes. For many reasons, including their normally low initial abundance in the samples prior to irradiation, rare gas isotopes are usually used in this work.

Production rates for different rare gas isotopes, including ^{3}He, ^{21}Ne, ^{38}Ar, ^{83}Kr, ^{124}Xe, and ^{126}Xe can be calculated for rocks of different chemistry using empirical rates for different target elements, as, for example, those given by Bogard et al (1971). A better method for determining exposure ages that automatically takes into account the rock chemistry as well as any peculiarities of shielding during irradiation is the

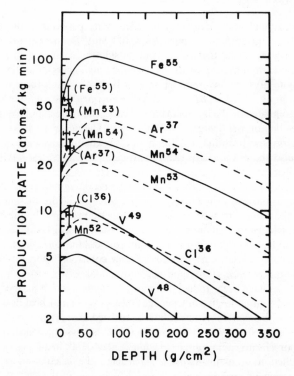

Figure 2   The calculated production rates of various radionuclides from galactic cosmic ray particles as a function of depth in the moon assuming the chemical composition of rock 12002 (from Reedy & Arnold 1972). The points for the radionuclides whose symbols are enclosed in parentheses are the experimental results of Finkel et al (1971) and D'Amico et al (1971) for this rock.

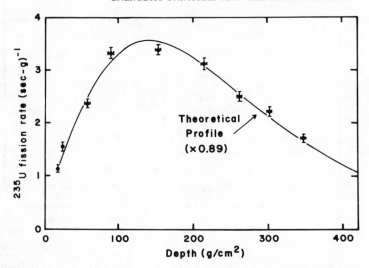

Figure 3 Measured variation in the rate of neutron-induced uranium fission as a function of depth below the lunar surface. This curve is from Woolum & Burnett (1974), who were responsible for the neutron probe experiment flown on the Apollo 17 mission.

Kr-Kr method first proposed by Marti (1967). In this technique, the rate of production of radioactive ^{81}Kr $(T_{1/2} = 2.1 \times 10^5$ yr) is coupled with measures of the accumulated stable spallation isotope ^{83}Kr. For a rock that has received all its irradiation in a single geometry, with no temporal changes in shielding, this age strictly measures the total time since the rock was first exposed.

However, as discussed in detail by Drozd et al (1974), great care must be exercised in attributing a physical significance to an "exposure age" determined by standard techniques. If changes in shielding have occurred during the time a sample was accumulating a significant fraction of its spallation isotopes, then the spallation age is not a measure of a well-defined time of exposure to cosmic rays. In general, a combination of techniques, such as those using tracks or neutron effects as discussed below, must be employed to specify the exposure history of a given specimen.

The secondary nuclear cascade contains a large number of high energy neutrons. These, in turn, are moderated by the lunar material, producing a slowing down spectrum. Specific neutron reactions such as an epithermal neutron resonance or 1/V-type reaction with a large cross section will thus have a characteristic depth dependence different from that of reactions produced by the total secondary nuclear cascade. The low energy neutron flux as a function of depth in the lunar surface was measured directly on the Apollo 17 mission by Woolum & Burnett (1974). Figure 3, taken from their paper, shows the rate of ^{235}U fission as a function of depth as measured by nuclear track detectors. The experimental values are close to the theoretical estimates of Lingenfelter et al (1972). However, more recent measurements of other neutron detectors included in the experiment (Burnett & Woolum

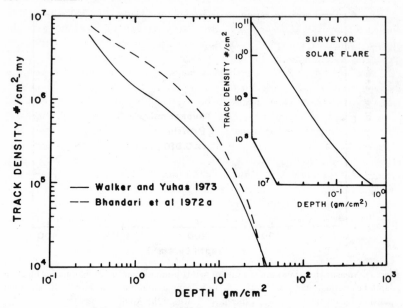

Figure 4   Minimum track density vs depth below an infinite plane surface. This calculation was done for a plane of observation perpendicular to the rock surface and an assumed track length of 10 μm. Shown for comparison is the track density profile given by Bhandari et al (1972a) for the *average* track density. Also shown in insert at right is the minimum track density vs depth calculated from the Surveyor III spectrum (see text). In the case of a rock, the actual track densities at shallow depths depend critically on the rates of erosion.

1974) indicates that the theory underestimates the fraction of epithermal neutrons at depth.

The rate of track production vs depth falls off much more rapidly than the rate of fall off of proton spallation effects. At low energies where nuclear interactions are unimportant, this is due to the very large ionization losses of the heavy, track-producing ions. At higher energies the fall off reflects both the drop in the number of incident particles capable of reaching a given depth and the rapid loss of heavy particles because of their larger nuclear interaction cross sections. In Figure 4 we show the best current estimate taken from Walker & Yuhas (1973) for the rate of track production vs depth. The difference in the characteristic scale length for the drop in production rates of heavy-ion tracks vs spallation products (~ 5 g/cm^2 vs ~ 150 g/cm^2) gives a very valuable method of deciphering the radiation history of lunar samples.

NEW INFORMATION ON ENERGETIC PARTICLES IN SPACE FROM THE APOLLO MISSIONS

Solar Wind

The intensity of the solar wind is sufficiently high and the techniques for detection sufficiently refined that it was possible to gain important new information about the solar wind by direct exposure of materials brought to the moon from earth and then returned for analysis. The most extensive measurements of this kind were the "window shade" experiments of the group at the University of Bern. In these experiments a section of metallic foil ~4000 cm^2 in area was unrolled and exposed to the solar wind during the time the astronauts were on the moon. The foil was subsequently returned to earth and analyzed by mass spectrometric methods for rare gases. Variations of this experiment were flown on all the Apollo missions; however, the total exposed area in the Apollo 17 mission was restricted to only 28 cm^2 (included as part of the Lunar Surface Cosmic Ray Experiment) and has yet to be analyzed.

The results of this series of experiments are summarized in the *Apollo 16 Preliminary Science Report* (Geiss et al 1972). The ratios between He and Ne isotopes were measured; preliminary results are given also for Ar. Whereas the ^{4}He flux during the different missions varied by a factor of four, relative elemental and isotope abundances are the same within the error limits except for the ^{3}He/^{4}He and ^{4}He/^{20}Ne ratios. These two ratios show anticorrelation (see Figure 5) which implies larger

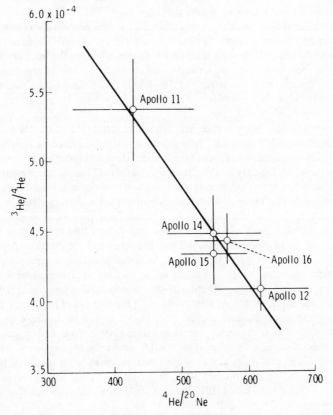

Figure 5   Correlation between the ^{3}He/^{4}He and ^{4}He/^{20}Ne solar wind abundance ratios as determined for the five Apollo foil exposure times (from Geiss et al 1972).

fluctuations in the ^{4}He abundances than in the abundances of the other isotopes. This was expected on theoretical grounds (Geiss 1972). Another important result was to establish a correlation between the ^{4}He/^{3}He ratio and the geomagnetic index K_p. This dependence, which is also seen for the ^{36}Ar/^{20}Ne ratio, is in accordance with satellite observations of He/H enhancements during geomagnetic storms.

A quite different experiment designed to measure heavy solar wind particles was flown on the Apollo 17 mission. The experiment, known as the Lunar Surface Cosmic Ray Experiment 17 (LSCRE 17), consisted of a box containing mica, glass, plastic, and metal detectors. Once on the moon, the box was opened and one set of detectors was exposed in the sun and another in the shade. By analyzing small etch pits produced in the mica, Zinner et al (1974) were able to show that the abundance of iron atoms relative to protons was 4.1×10^{-5}. This is within a factor of two of the most recent estimates of the solar abundance of 2.6×10^{-5} (Cameron 1973a). Furthermore, it was shown that the abundances of the heavier elements extending to the end of the periodic table could not be enriched relative to protons by more than a factor of four. As will be described shortly, this is a quite different situation than has been found for low energy solar flare particles.

As might be anticipated, the lunar samples themselves contain abundant amounts of implanted solar wind atoms. For example, a typical lunar soil sample has a helium content of 0.3 cm^3 STP/g. This is far higher than is found in the interior of a lunar crystalline rock. Both grain size separates and etching experiments have been used to show that the gas resides principally in the outer layers of the dust grains (Eberhardt et al 1970, 1972). In particular, the etching studies have been used to infer an implantation depth of ~ 0.2 μm.

The rare gas data in themselves are difficult to interpret because there is a clear depletion of the light rare gases. The implantation depth inferred by Eberhardt et al (1970, 1972) is also much larger than the range of directly implanted solar wind ions. Also, as shown by Ducati et al (1972) and Frick et al (1973a), the temperature release of laboratory-implanted ions occurs at much lower temperatures than those found for lunar samples. The latter group showed further that radiation damage prior to implantation increased the temperature release. They propose a model in which particles of higher energy (solar flares) produce defects that act as traps that can retain solar wind diffusing gases. Thus the observed penetration depths and concentrations of rare gases may reflect the distribution of flare-produced defect traps rather than the energy and fluxes of the directly implanted solar wind ions. This interpretation is also consistent with the observations of Leich et al (1974) who measured the depth profiles of solar protons on the surfaces of lunar rock samples.

From these observations it appears clear that it would be desirable to measure elements with a high degree of retention whose distributions and concentrations would reflect the original implantation of solar wind ions. Measurements of carbon, nitrogen, silicon, and other elements have been made in lunar materials by Smith et al (1973), Taylor & Epstein (1973), and Müller (1974), and apparent surface correlations have been found. It is, however, important to point out that all these measurements have been made on bulk samples or grain-size separates and that no implantation profiles have been made on individual grains.

Moreover, it is this author's belief that great care must be taken in the interpretation of surface-correlated effects at this time. Even after prolonged cleaning of surfaces in an ultrasonic bath, a number of small particles 1 to 10 μm in size can be seen adhering to the surfaces of lunar dust grains. There is also evidence that very tiny particles $\sim$ 1000 Å in size may be attached to the surfaces, though this is less certain. Blanford et al (1974) have also shown that the surfaces of many dust grains are coated with small accretionary glass splashes. Much more work needs to be done to measure the concentration and depth distribution of implanted ions in lunar soil grains.

There is experimental evidence of an increased solar wind flux in the past. Geiss (1973) and Eberhardt (1974) concluded from the total amounts of trapped xenon in Apollo 15 (Bogard & Nyquist 1972, Hubner et al 1973), Apollo 16 (Bogard & Nyquist 1973), and Apollo 17 (Eberhardt et al 1974) deep drill cores that the average solar wind flux during the last several billion years was larger by about a factor of three than it is today. If one takes the observed concentrations of carbon and nitrogen (Smith et al 1973), which are even more retained than xenon, this factor might be as large as ten.

Two considerations make it important to establish changes in the solar wind (and/or solar flares) in the past. To explain the unexpectedly low observed flux of solar neutrinos, Fowler (1972) suggested that mixing of the solar interior could lead to a reduced neutrino flux accompanied by a reduced solar luminosity. On this model one would expect periodic changes in the solar luminosity on a time scale of about 10 m.y. (see e.g. Cameron 1973b) and it is important to look for correlated solar wind and solar flare effects in lunar cores. Second, models of the evolution of solar-type stars point toward a very active early sun. In this view the early sun was rapidly rotating and was decelerated by the magnetically coupled solar wind (Kraft 1967). For the early convective phase of the sun (Hayashi phase), Sonett (1974) arrives at a primordial solar wind flux seven or eight orders of magnitude higher than present fluxes; this high initial flux then decayed exponentially with a relatively short time constant.

A totally different approach to the study of solar wind effects has been taken by Maurette and his collaborators at the University of Paris, Orsay (Borg et al 1970, Dran et al 1970, Bibring & Maurette 1972, Bibring et al 1972a, b, Bibring et al 1973). Figure 6 shows a high voltage electron microscope photograph of a lunar dust grain. It can be seen that there is a thin coating surrounding the grain. This coating, which has been shown by Maurette and his co-workers to be amorphous, is identical in appearance to amorphous layers produced by laboratory ion implantation experiments. Further, the coating thickness has been shown to give a measure of the energy of the implanted ions. By measuring coating thicknesses in a number of grains taken from different locations in the Apollo 15 deep drill stem, Maurette and his co-workers find that the average energy of solar wind-implanted ions has not changed in the period represented by the deep drill stem (as discussed later, these samples probably cover an interval of 5×10^8 to 10^9 yr). Although excursions of both low and high solar wind energies are seen, no systematic pattern has emerged so far from this work.

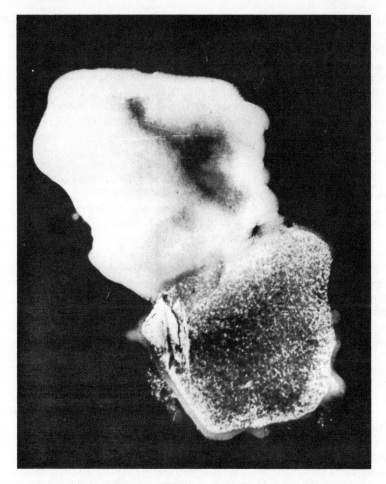

Figure 6   Ultra-high resolution transmission electron microscope photographs of micron-sized lunar dust grains. Grains of this size typically have internal track densities of 10^{10} to $10^{12}/cm^2$, indicating a high abundance of very low energy particles in space. Many grains are also coated with an amorphous layer probably produced by solar wind induced radiation damage. Rounding, probably due to solar wind erosion, is another characteristic feature. Sticking of the grains due to interaction of the amorphous layers can also be seen. (Photo courtesy of M. Maurette.)

A striking anomaly in the rare gas patterns, noted by a number of investigators at the first lunar science conference, is the presence of an amount of surface-correlated ^{40}Ar far in excess of what would be expected from either solar-wind implantation or from the in situ decay of ^{40}K. This component, dubbed "little orphan argon" by some of our colleagues (Frick et al 1973b), is generally attributed to a reimplanted component of thermally released ^{40}A diffusing from the interior of the moon. The mechanism envisaged (Manka & Michel 1971) is the photoionization of the emitted atoms followed by their acceleration due to the $\mathbf{v} \times \mathbf{B}$ field of the impinging solar wind.

This process has also been invoked to explain ^{244}Pu fission xenon gases found in Apollo 14 gas-rich breccias (Behrmann et al 1973a), although here the surface correlation of the fission gas has yet to be demonstrated.

Recent studies of the lunar atmosphere (Hodges et al 1973, Hoffman et al 1973) have confirmed the basic mechanism proposed and have further shown that there exists a large ^{40}Ar component corresponding to a degassing of about 0.4% of the total rate of production of ^{40}Ar in the moon (Hodges et al 1974).

Solar Flares

Radioisotope data indicate that the average flux and energy spectrum of solar flare particles has not changed appreciably in the last several million years. Figure 7, taken from Finkel et al (1971), shows the variation of ^{53}Mn $(T_{1/2} = 3.7 \times 10^6$ yr) and ^{26}Al $(T_{1/2} = 7.4 \times 10^5$ yr). The spectral form for the ^{26}Al is identical, within errors, to data for ^{22}Na $(T_{1/2} = 2.6$ yr, not shown) being represented by a rigidity spectrum of the form $dJ/dR = J_0 \exp(-R/80 \text{ MV})$. The data for ^{53}Mn is somewhat flatter, indicating either a change in past solar flare activity or, more probably, the effects of rock erosion.

Preliminary work by Begemann et al (1972) and Boeckl (1972) indicated that in the time span represented by the half-life of ^{14}C $(T_{1/2} = 5.7 \times 10^3$ yr) solar flares might have been considerably more intense than at present. However, analyses of the depth dependence of thermoluminescence by Hoyt et al (1973) indicate that this is not the case. In the energy region from 20 to 100 MeV they find values consistent within 20% with those derived by Wahlen et al (1972) from their ^{26}Al data. The characteristic decay time of the thermoluminescence was estimated to be $\sim 5 \times 10^3$ yr.

Although solar flare tracks were seen in the very first samples returned from the moon, the interpretation of the results in terms of the long-term behavior of heavy solar flare ions was limited by the lack of information of the properties of contemporary heavy solar flare particles. This situation was remedied, at least in part, by the return of a glass filter from the Surveyor III spacecraft on the Apollo 12 mission. Analysis of the particle tracks in the glass, which was exposed for a period of approximately three years on the moon, showed that the energy spectrum of solar flare particles could be represented by energy spectrum of the form $dN/dE = AE^{-3}$ (see Fleischer et al 1974b for a more complete discussion).

However, observations of solar flare tracks in most lunar rocks indicated a much flatter depth dependence than that expected from the Surveyor spectrum. This difference could be reconciled with the Surveyor spectrum, provided it was assumed that erosion of the rock surfaces occurred at a rate of 3 to 8 Å/yr (see Fleischer et al 1974b for a fuller discussion). Although the erosion hypothesis was plausible, it has only recently been verified. Two split rocks were returned from the Apollo 17 mission in which the exposures of the bottom portions of the split surfaces occurred with a very small solid angle. It was thus possible to accumulate a long exposure time

Depth variation of cosmogenic nuclides in a lunar surface rock and lunar soil

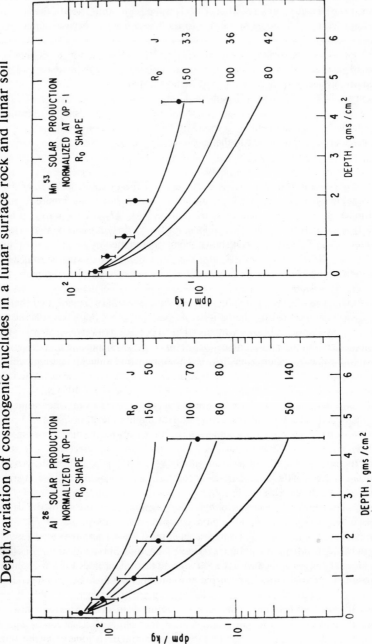

Figure 7   Observed and calculated solar production for long-lived cosmogenic nuclides. Solid lines are theoretical curves which have been normalized to the experimental point at OP-1. J is the 4π integral flux above 10 MeV in units of protons/cm^2 sec. R_0 is the mean rigidity in units of MV (from Finkel et al 1971).

($\sim 10^6$ yr) without a concomitant amount of erosion. The spectra found in these rocks are similar to those found in the Surveyor glass (Crozaz et al 1974b, Hutcheon et al 1974). The erosion rates given above refer to processes in which material is removed in steps ranging from atomic dimensions to the order of 10 μm. They should not be confused with macroerosion rates where material is removed in larger sized steps. The magnitude of the microerosion rate is larger than can be accounted for by solar wind sputtering, and its origin is still unknown.

The characteristic steep Surveyor-type spectrum has also been found in other samples including crystals that were incorporated in the interior of gas-rich meteoritic breccias and lunar breccias (Lal & Rajan 1969, Pellas et al 1969, Price et al 1973). These crystals were probably incorporated in the breccias at an early time in the history of the solar system and indicate that the solar flares have had similar energy spectra throughout most of the lifetime of the sun. The absolute flux of heavy solar flare particles is not well established, even for contemporary solar flares, and this remains an important problem area.

One of the most important results of the lunar solar flare studies was the demonstration by Price et al (1971) that heavy particles are enriched relative to lighter particles at energies < 10 MeV/nucleon. The original demonstration of this effect is shown in Figure 8. For many years prior to this observation it had been believed the relative abundances of particles in solar flares were representative of the solar abundances. This may be true of the more energetic particles but is certainly not true for the low energy ones. The reasons for the progressive enrichments of heavy particles at low energies are not fully understood but probably are related to the fact that heavy particles are only partially ionized in the source regions (see Fleischer et al 1974b for a fuller discussion). Recent measurements of fossil tracks in lunar samples by Bhandari et al (1973a) indicate that the low energy, heavy particle enrichments are a characteristic of ancient, as well as modern, solar flares.

New information on solar flares was also obtained in the Apollo 16 cosmic ray experiment. This experiment had been designed to study particles in the interplanetary medium during quiet sun conditions. A small solar flare that occurred during the mission obviated this goal but made it possible to extend measurements of solar flare heavy particles down to much lower energies than had previously been possible (Burnett et al 1972, Fleischer & Hart 1972, Price et al 1972, Braddy et al 1973, Fleischer & Hart 1973a). The enrichment of heavy particles relative to light particles was confirmed by this experiment.

Lunar soil samples vary considerably in their state of irradiation (see Figure 9). Occasional soil samples are found (e.g. the Conelet Crater sample 14141) that have had simple irradiation histories. However, most soils show a tremendous spread in the track densities of individual grains, indicating complex irradiation histories. Some 5 to 10% of grains in the 50 to 150 μm range are found to have large track gradients, indicating that they have been exposed directly on the surface of the moon.

Poupeau et al (1973) have found positive evidence for an early active sun in measurements of the track densities in samples returned on the Soviet Luna 16 mission. They found that feldspar crystals, identified as anorthosites and therefore likely of highland origin (and hence old), had much higher track densities on the average than feldspars that were characteristic of mare basalts (and hence younger). Unfortunately, attempts to reproduce these observations in typical mature soil samples from mare regions have been unsuccessful (Crozaz et al 1974a).

Short-lived radioactivities induced by solar flares vary considerably from mission

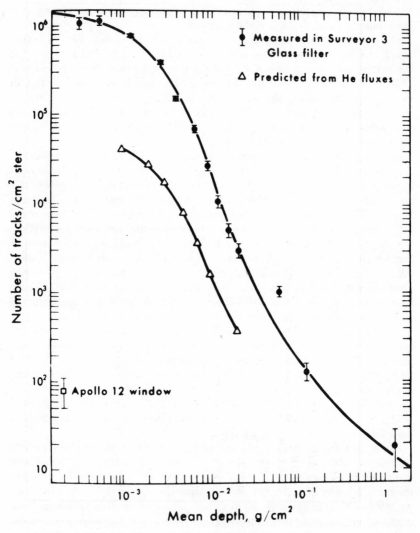

Figure 8 Observed densities of Fe tracks penetrating to a given depth of the Surveyor III glass and Apollo 12 glass window compared with densities predicted assuming Fe/He solar particle ratio is the same as the photospheric ratio. This illustrates the progressive enrichment of Fe nuclei in solar flares at low energies (from Price et al 1971).

to mission depending on the solar flare activity prior to flight. On the Apollo 17 mission Rancitelli et al (1974) have shown that the activities induced by bombardments in different directions are noticeably different. Their results indicate an anisotropy in both the number and the energy distribution of solar flare particles.

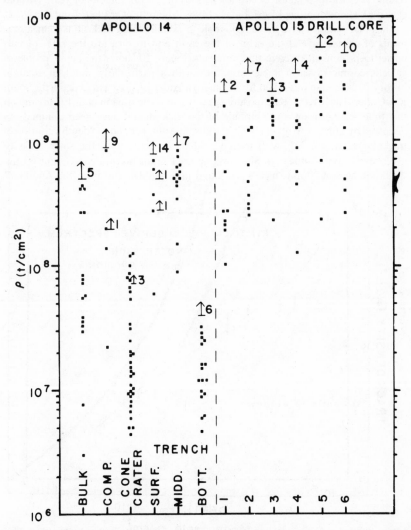

Figure 9   Representative distribution of track densities measured by the Washington University group in several Apollo 14 and 15 samples (from Crozaz et al 1972). The crystals are all feldspars and are greater than 100 mesh in size. The samples labeled "Cone Crater" (14141) and "trench bottom" are clearly very much less irradiated than the other (more typical) samples, many of which have no measured crystals with $< 10^8$ tracks/cm^2.

Galactic Cosmic Rays

Meteoritic radioisotope and track studies performed prior to the Apollo missions had established the constancy of the proton component and iron components of cosmic rays within a factor of two for periods of several millions of years (Arnold et al 1961, Cantelaube et al 1967, Lal et al 1969, Maurette et al 1969, Tamhane 1972).

The lunar samples have made it possible to extend these conclusions to measurements of cosmic rays at the orbit of the earth and to measure the ratio of iron nuclei to protons at discrete time intervals. Figure 10 shows the depth profile of galactic cosmic ray iron nuclei in a rock with a particularly simple irradiation history. This sample was taken from a sharp angular boulder associated with South Ray Crater. The ^{81}Kr–^{83}Kr exposure age was 2 m.y. (Behrmann et al 1973b) and on this time scale erosion and chipping of the rock should have been minimal. As discussed by Yuhas & Walker (1973), the absolute number and the depth variation of tracks agree within 30% with what would be expected from the solar cycle 19 average of heavy cosmic rays. Similar steep track gradients and concordant proton spallation ages of 20 m.y. have been found in a boulder chip from the Apollo 17

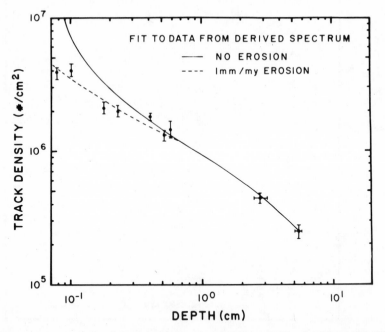

Figure 10   Track density vs depth in lunar rock 68815 (from Walker & Yuhas 1973). This rock has a particularly young Kr-Kr exposure age of 2×10^6 yr and, unlike most lunar rocks, appears to have had a single one-stage exposure history. The solid line shows the calculated track density using the energy spectrum for galactic cosmic rays coupled with the average solar flare contribution. The dotted line below 0.6 cm includes the effects of an assumed microerosion rate of 10^{-7} cm/yr.

mission (Crozaz et al 1974b) and in boulder samples from North Ray Crater dated at 50 m.y. (although in the latter case rock erosion had to be invoked to bring the proton and iron nuclei data into agreement). Thus averaged over periods of 2, 20, and 50 m.y., there is good evidence that the energy spectrum of iron particles has not changed appreciably and further that the iron to proton ratio has remained constant within 20%.

It must be noted, however, that most lunar rocks do not have the depth dependence shown in Figure 10. Very flat depth dependences tend to be the rule rather than the exception. The conventional interpretation of such flat spectra is that the rocks have had complicated irradiation histories and that a large fraction of the tracks measured were accumulated while the rock had considerably more shielding than in its final state. Although this interpretation is quite reasonable, it remains to be established with certainty. In particular, there is one rock, 14310, that presents an enigma at the present writing. It has a very flat cosmic ray track gradient and it had previously been inferred from this that the rock could not have been on the surface of the moon in its present geometry for more than about 3 m.y. (Crozaz et al 1972). This would imply, in turn, that the ^{53}Mn induced by solar flares on the surface of this rock should be below the saturation level. However, recent measurements of the ^{53}Mn show that it is saturated (J. R. Arnold, private communication). These data suggest that either galactic cosmic rays or perhaps the solar flares have changed in the interval from 5 to 15 m.y. The general conclusion that both solar flares and cosmic rays have been constant in time should thus be treated with a reasonable amount of caution.

The abundance of elements much heavier than iron in nature is very low. The first measurements of the abundance of such ions, which in the jargon of the trade are called VVH particles ($Z \geq 30$), were made by studying the frequency of very long tracks in meteorite samples (Fleischer et al 1967). Recently Bhandari et al (1974) have reported measurements on a variety of meteoritic and lunar samples, indicating that the VVH to VH abundances have been constant for at least 10^9 yr.

In contrast, Plieninger et al (1973) have given track results in lunar rock 10049 that they interpret as showing a change in the chemical composition in atomic number group from $Z = 20$ to 28. Specifically they find a much higher ratio of the lighter charges with respect to iron than measured in the contemporary radiation. These results are extremely interesting but cannot yet be considered definitive. Much more work on the chemical abundances of all charge groups remains to be done.

Several unexpected results emerged from the analysis by the Berkeley group[2] of the plastic detectors included in the Apollo 17 Lunar Surface Cosmic Ray Experiment. At energies of 1 to 10 MeV/amu where the contribution of solar particles was small, the abundance patterns were found to be anomalous, matching neither the solar abundances nor those of high energy cosmic rays. The oxygen spectrum showed a plateau in this region while the spectra of Mg + Si and of $Z \geq 20$ particles steeply decreased in the same energy interval. The Li Be B/CNO ratio was found to be <0.06, indicating that the anomalous oxygen particles, while of probable galactic origin, cannot have passed through nearly as much matter as have galactic cosmic rays.

[2] J. H. Chan, P. B. Price. Anomalies in the composition of interplanetary heavy ions with $0.01 < E < 40$ MeV/amu. Submitted for publication to *Astrophys. J. Lett.*

LUNAR SURFACE DYNAMICS

Age of Formation of Specific Lunar Features

The dating of specific lunar events such as the formation of a particular crater or the emplacement of a large boulder is interesting from a number of points of view. There are essentially two ways to be certain that an exposure age as measured by either track or rare gas techniques represents the time of formation of a particular feature. If a number of samples taken in the vicinity of a particular crater give essentially the same spallation age as, for example, in the case of North and South Ray Craters (Marti et al 1973, Drozd et al 1974), then one can be reasonably sure that the exposure ages refer to a specific event in time. Another and even better way is to obtain concordance between proton spallation ages and fossil nuclear track ages. This is because the two types of phenomena have very different characteristic attenuation lengths. Concordancy assures that the exposure history has been simple and thus that the age is meaningful. There are at least four features on the moon whose ages we now know with certainty. They include Cone Crater (25 m.y., Apollo 14 mission) (Turner et al 1971, Crozaz et al 1972, Lugmair & Marti 1972), North Ray Crater (50 m.y., Apollo 16 mission), South Ray Crater (2 m.y., Apollo 16 mission), and the time of the Station 6 boulder emplacement of Apollo 17, 21 m.y. (Crozaz et al 1974b). Many other ages have been reported and associated with other lunar features. However, not all of these ages satisfy the two criteria given above and additional work is necessary to establish their validity.

Surface Exposure Ages of Rocks

In this section and in those below we are concerned with the measurement of rate processes that are affected by the flux of micrometeoroids striking the moon. One such measurement is the survival lifetime of small lunar rocks. Proton spallation ages are generally much higher than the track exposure ages of lunar rocks. The differences are attributable to the deeper penetration distances of the protons and the track ages are generally taken to be more representative of the "true surface exposures". In two recent publications we have given a complete summary of all track age determinations (Crozaz et al 1974b, Fleischer et al 1974b); only the highlights are presented here.

Most track age measurements rely on a single-point track density determination at a known position in the rock. To obtain an age it is assumed that the rock has always been in the same position. However, as previously noted, detailed measurements of the depth profile of tracks indicate that this assumption is generally not correct. Thus single-point determinations give *maximum* surface exposure ages. A review of the data for 24 selected rocks that fall in the 1 to 10 kg range gives an average surface exposure age of 13 m.y. This is somewhat higher than the calculated value of 2 to 6 m.y. originally calculated by Gault et al (1972) for the time that rocks in the 1 to 10 kg range would survive on the lunar surface before being catastrophically ruptured by micrometeoroid impacts. However, if it is kept in mind that the track determinations for the most part give maximum surface exposure ages, the disagreement does not seem serious.

Erosion Rates

A related problem is the measurement of erosion rates caused by the chipping of lunar rocks by small micrometeoroids. The calculated values (Hörz et al 1974) based on present-day micrometeoroid fluxes range from 0.5 to 2 mm per m.y. These numbers are in good agreement with mass-wastage erosion rates estimated from comparison of track data with proton spallation data (see Fleischer et al 1974b for a complete discussion) or from the depth profiles of radioisotopes in lunar rocks (Finkel et al 1971, Wahlen et al 1972). Erosion measurements and predictions based on current micrometeoroid flux data thus appear to be in good agreement.

The Absolute Production Rate of Impact Pits

The absolute production rate of impact pits is still another parameter that is related to the micrometeoroid flux (for a recent review of the present status of lunar microcrater measurements, see Hörz et al 1974). In Table 1 we show measurements of crater densities for craters with a central pit diameter > 500 μm in size for surfaces whose ages have been estimated either by track data or by radioisotope data. The results indicate an experimental rate of impact pit production of 1–3 craters/cm^2 m.y. This value is considerably smaller than the approximately 8 pits/cm^2 estimated by Hörz et al (1974). However, it should be noted that most of the rock surfaces in Table 1 are older than 1 m.y. For such surfaces, erosional effects on 500 μm size pits should be important. The surfaces may not be in a pure production state, as was assumed. Although there appears to be a real difference between calculated and predicted rates, some or all of this may be due to the effects of erosion.

Table 1   Selected surface exposure ages and crater counts for craters >0.05 cm in diameter

Rock	No. of craters counted	$\rho_c (\text{cm}^{-2})$	Surface exposure age (m.y.)	Cratering rates cm^{-2} m.y.$^{-1}$
12017	12[a]	2.3	$\lesssim 0.7$[f]	$\gtrsim 3.3$
12038	30[a]	3.6	$\lesssim 1.3$[g]	$\gtrsim 2.8$
12054	4[b]	0.4	0.05 to 0.5[h]	$\lesssim 0.8$ to 8
14301	54[c]	2.5	> 1.5[i]	$\lesssim 1.7$
14303	10[d]	2.3	$\lesssim 2.5$[j]	$\lesssim 0.9$
14310	30[b]	2.0	1.5 to 3[k]	$\gtrsim 0.7$ to 1.3
60315	14[e]	3.4	> 1.5[l]	$\gtrsim 2.3$
61175	196[e]	13.2	> 1.5[m]	$\lesssim 8.8$
62295	34[e]	4.0	$\lesssim 2.7$[n]	$\lesssim 1.5$
68415	57[e]	2.3	> 1.5[o]	$\lesssim 1.5$

[a] Hörz et al 1971.
[b] Hartung et al 1972.
[c] Morrison et al 1972.
[d] Hartung et al 1973.
[e] Neukum et al 1973.
[f] Fleischer et al 1971.

[g] Bhandari et al 1971.
[h] Schönfeld (Personal communication).
[i] Keith et al 1972.
[j] Bhandari et al 1972a.

[k] Crozaz et al 1972.
[l] Clark & Keith 1973.
[m] Eldridge et al 1973.
[n] Bhandari et al 1973b.
[o] Rancitelli et al 1973.

Perhaps the best evidence for past fluctuations in the rate of micropitting is the recent work by Hartung et al (1974). Using solar flare tracks to date individual microcraters, these authors found many more young pits than old ones. This interesting line of investigation certainly deserves to be pursued. However, the particles measured by Hartung et al were typically much smaller than those responsible for most of the dynamic effects discussed here. Thus even if their observations were confirmed, it might turn out to be a separate problem from the one discussed above.

Rates of Turnover and Accretion of the Lunar Surface

Although considerable data of several types exists on a number of lunar soils and soil columns, the lack of appropriate conceptual models limits the conclusions that currently can be drawn. In what follows below we summarize certain aspects of this complex problem.

If an assemblage of crystals containing no particle tracks is thrown onto the lunar surface as a distinct layer several centimeters thick and then only slightly stirred, the track data permit a simple interpretation. Most of the particles will receive tracks from galactic cosmic rays and the distribution of track densities will reflect the depth dependence of track production. This appears to be the case with the Conelet Crater sample (14141) shown in Figure 9. Occasional crystals that have been exposed to solar flares and hence contain very large track densities will be seen, but these will be rare. In Table 2 we summarize the results for several lunar soils that have been irradiated in the simple way described above.

The existence of layers several centimeters thick that have lain apparently undisturbed for several million years immediately sets certain boundary conditions

Table 2   Lightly and (rather) simply irradiated soil samples

Sample	Depth	Model age
12028[a] Coarse-grained layer in core	4 cm thick at 18 cm depth	$\lesssim 15$ m.y. (slightly stirred)
14141[b,c,d,e] Conelet Crater	Scoop sample (~ 4 cm)	8 to 15 m.y.
15401[f] Green glass	Scoop sample (~ 4 cm)	0.5 m.y. (stirred at least once)
73241[g] Trench sample	0 to 5 cm	~ 20 m.y.
74220[f] Orange glass	5 to 7.5 cm	4 to 7 m.y. (slightly stirred)

[a] Crozaz et al 1971.
[b] Bhandari et al 1972a.
[c] Crozaz et al 1972.
[d] Hart et al 1972.

[e] Phakey et al 1972.
[f] Fleischer & Hart 1973c.
[g] Fleischer et al 1974a.

on the rate of turnover of the lunar surface. Gault et al (1972) originally estimated that the first 6 cm of the lunar regolith would be turned over, on the average, once in a million years. These predictions were in striking disagreement with the measurements shown in Table 2. However, the original estimates of turnover rates have been considerably modified. The revised estimates of Gault et al (1974) reported at the Fifth Lunar Science Conference now permit soil layers to exist in an essentially undisturbed condition for very long times. In fact, the current theoretical estimates, particularly for deep layers, may now be too low.

Several different techniques have been applied to the deep drill stems (~ 3 m deep) taken on the Apollo 15, 16, and 17 missions in order to develop a chronology of deposition. Each of these stems contains distinct layers of material and appears to represent regions of the moon where accretional processes have dominated over stirring processes in recent times. A chronology of core deposition can be built up from track data in two ways. Lal and his colleagues (Arrhenius et al 1971) originally showed a correlation between median track densities and the visual layering observed in cores (see Figure 11). They estimate the age of a given layer by assuming that the

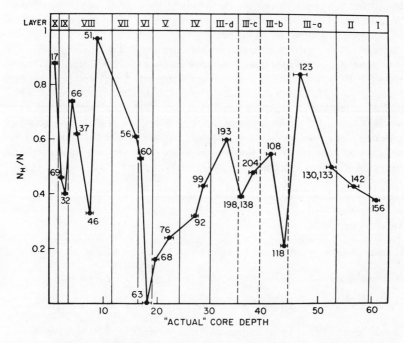

Figure 11   Track data for different visible layers denoted by Roman numerals. The fraction (number of grains with $\rho > 10^8$ plus grains with gradients)/(all grains counted), or N_H/N, is plotted against location of sample in the double core of Apollo 12 (from Arrhenius et al 1971). N_H can also be thought of as the number of grains irradiated without much shielding. The numbers refer to the number of grains counted at each position.

particles making up the layer had a common time of deposition. From the median track densities, the measured layer thicknesses, and the rate of production of tracks as a function of depth, they have inferred the accretional chronology of several cores as shown in Table 3. A somewhat different method has been used by Fleischer and his colleagues (Fleischer et al 1974a) to analyze cores. They search through a number of crystals, looking for those with the lowest track densities. They then argue that this crystal, if not the others, could not have existed at its position in the layer for more than a certain minimum time. Results based on the minimum track density method are also shown in Table 3. It should be noted that neither of these methods is above criticism. They are both model age determinations and contain certain implicit assumptions that may or may not be correct. However, it is interesting to note that the general trend of the data seems to agree moderately well with that determined by totally independent methods.

The stratigraphic chronology of lunar cores has also been investigated by neutron-induced changes in Sm and Gd isotopes (Russ et al 1972) and most recently by spallation isotopes (Pepin et al 1974). In the Apollo 15 deep drill both effects show a characteristic rise with depth followed by a decrease; because of the different depth dependence of the neutron effects, the details of these depth effects are different. The existence of a well-defined maximum shows clearly, however, that the material has not been well stirred for a long time.

Several different models are consistent with the observed data. The simplest is to assume that a pre-irradiated layer, encompassing most of the drill stem, was thrown out $\sim 5 \times 10^8$ yr ago and has remained quiescent, with the exception of having a capping layer added, since that time. However, the data are also consistent with a gradual accretionary buildup for a period of $\sim 4 \times 10^8$ yr followed by a quiescent period of $\sim 5 \times 10^8$ yr. All layers of the deep drill show crystals with very high track densities indicating that they were once exposed to solar flares on the surface of the moon. The crystals at the bottom of the stem are thus measuring solar properties at least as long ago as 5×10^8 to 10^9 yr.

Table 3 Typical layer track model ages in lunar cores

Modeling method	Sample	
Minimum density	Apollo 12 double core[a]	Layers 2 to 4 cm have typical ages of 6 to 30 m.y.
Minimum density	Apollo 15 deep drill[b]	Layers 0.5 to 2 cm have ages of 0.5 to 5 m.y.
Median density	Apollo 12 double core[c]	Layers 2 to 10 cm have ages of 5 to 40 m.y.
Median density	Apollo 15 deep drill[d]	Layers 1.5 to 5 cm have ages of 1 to 40 m.y.

[a] Fleischer & Hart 1973b.   [c] Arrhenius et al 1971.
[b] Fleischer & Hart 1973c.   [d] Bhandari et al 1972b.

The continued measurement of the chronology of lunar cores, coupled with measurements of solar wind and solar flare effects at different times in the past, represents one of the unfinished tasks of the Apollo sample analysis program.

SUMMARY

The Apollo landing missions have contributed important new information on all types of energetic particles in space, including those found in the solar wind, solar flares, and galactic cosmic rays. Part of this information has come from experiments expressly flown to the moon for this purpose, some of it has come from man-made materials returned from the moon for other purposes, and a large part of it has come from the study of the returned lunar samples.

The Apollo sample collection represents a treasurehouse of information, much of it still to be extracted. In principle, these samples contain a virtually complete record of the energetic particle environment at the orbit of earth back to a time close to the beginning of the solar system. The first-order look at the samples indicates surprisingly little change in the energetic particle environment in the past. However, at this writing, certain suggestive results have been obtained that indicate that measurable changes may be found, particularly in samples exposed at early stages of the solar system.

Much more work needs to be done on all types of samples; the continued study of core samples, many of which have not yet ever been opened, appears particularly important. The field is ripe for new experimental approaches to supplement those already in use. This is particularly important for the study of the surface-correlated properties associated with solar wind implantation effects. The development of better conceptual models for the development of the regolith is also badly needed if we are to extract the maximum information from the samples.

The effects produced by energetic particles impinging on the lunar surface provide the basic tools for measuring the dynamic processes that continue to change and shape the lunar surface. Although much has been learned, there is a great need to further exploit the splendid Apollo sample collection and to correlate this work with surface and orbital photography to better establish the chronology of the detailed development of the lunar surface.

ACKNOWLEDGMENTS

It is a great privilege to have participated in the scientific adventure of Apollo, and I wish to acknowledge my gratitude to all those who made this possible. A significant portion of the material reviewed was gathered in the course of writing a book on nuclear tracks in solids with R. L. Fleischer and P. B. Price. My thanks are due to my colleagues at Washington University who participated in many of the original researches and who gave me permission to reproduce unpublished material. Preparation of the manuscript would have been impossible without the dedicated service of H. Ketterer. This work was supported under NASA Grant NGL 26-008-065.

Literature Cited

Arnold, J. R., Honda, M., Lal, D. 1961. Record of cosmic ray intensity in the meteorites. *J. Geophys. Res.* 66:3519–31

Arrhenius, G. et al 1971. The exposure history of the Apollo 12 regolith. *Proc. 2nd Lunar Sci. Conf., Geochim. Cosmochim. Acta Suppl.* 2 3:2583–98

Begemann, F., Born, W., Palme, H., Vilcsek, E., Wänke, H. 1972. Cosmic-ray produced radioisotopes in Apollo 12 and Apollo 14 samples. *Proc. 3rd Lunar Sci. Conf., Geochim. Cosmochim. Acta Suppl.* 3 2:1693–1702

Behrmann, C. J., Drozd, R. J., Hohenberg, C. M. 1973a. Extinct lunar radioactivities: Xenon from ^{244}Pu and ^{129}I in Apollo 14 breccias. *Earth Planet. Sci. Lett.* 17:446–55

Behrmann, C. et al 1973b. Cosmic-ray exposure history of North Ray and South Ray material. *Proc. 4th Lunar Sci. Conf., Geochim. Cosmochim. Acta Suppl.* 4 2:1957–74

Bhandari, N. et al 1971. High resolution time averaged (millions of years) energy spectrum and chemical composition of iron-group cosmic ray nuclei at 1 A.U. based on fossil tracks in Apollo samples. *Proc. 2nd Lunar Sci. Conf., Geochim. Cosmochim. Acta Suppl.* 2 3:2611–19

Bhandari, N. et al 1972a. Collision controlled radiation history of the lunar regolith. *Proc. 3rd Lunar Sci. Conf., Geochim. Cosmochim. Acta Suppl.* 3 3:2811–29

Bhandari, N., Goswami, J. N., Lal, D. 1972b. Apollo 15 regolith: A predominantly accretion or mixing model? *The Apollo 15 Lunar Samples,* 336–41. Houston: Lunar Sci. Inst.

Bhandari, N., Goswami, J., Lal, D., Tamhane, A. 1973a. Time averaged flux of very very heavy nuclei in solar and galactic cosmic rays. *Lunar Science IV,* 69–71. Houston: Lunar Sci. Inst.

Bhandari, N., Goswami, J., Lal, D. 1973b. Surface irradiation and evolution of the lunar regolith. *Proc. 4th Lunar Sci. Conf., Geochim. Cosmochim. Acta Suppl.* 4 3:2275–90

Bhandari, N., Padia, J. T., Lal, D. 1974. On the constancy of cosmic ray composition in the past. *Lunar Science V,* 54–56. Houston: Lunar Sci. Inst.

Bibring, J. P. et al 1972a. Solar wind implantation effects in the lunar regolith. *Lunar Science III,* 71–73. Houston: Lunar Sci. Inst.

Bibring, J. P. et al 1972b. Ultrathin amorphous coatings on lunar dust grains. *Science* 175:753–55

Bibring, J. P. et al 1973. Solar wind and lunar wind microscopic effects in the lunar regolith. *Lunar Science IV,* 72–74. Houston: Lunar Sci. Inst.

Bibring, J. P., Maurette, M. 1972. Stellar wind radiation damage in cosmic dust grains: Implications for the history of early accretion in the solar nebula. *On the Origin of the Solar System,* ed. H. Reeves, 284–92. Paris: Centre Nat. Rech. Sci. 383 pp.

Blanford, G., Fruland, R. M., McKay, D. S., Morrison, D. A. 1974. Lunar surface phenomena: Solar flare track gradients, microcraters, and accretionary particles. *Proc. 5th Lunar Sci. Conf., Geochim. Cosmochim. Acta Suppl.* 5 3:2501–26

Boeckl, R. S. 1972. A depth profile of ^{14}C in the lunar rock 12002. *Earth Planet. Sci. Lett.* 16:269–72

Bogard, D. D., Funkhouser, J. G., Schaeffer, O. A., Zähringer, J. 1971. Noble gas abundances in lunar material—Cosmic-ray spallation products and radiation ages from the Sea of Tranquility and the Ocean of Storms. *J. Geophys. Res.* 76:2757–79

Bogard, D. D., Nyquist, L. E. 1972. Noble gases in the Apollo 15 drill cores. *The Apollo 15 Lunar Samples,* 342–46. Houston: Lunar Sci. Inst.

Bogard, D. D., Nyquist, L. E. 1973. ^{40}Ar/^{36}Ar variations in Apollo 15 and 16 regolith. *Proc. 4th Lunar Sci. Conf., Geochim. Cosmochim. Acta Suppl.* 4 2:1975–85

Borg, J., Dran, J. C., Durrieu, L., Jouret, C., Maurette, M. 1970. High voltage electron microscope studies of fossil nuclear particle tracks in extraterrestrial matter. *Earth Planet. Sci. Lett.* 8:379–86

Borg, J. et al 1972. Search for low energy $(10 \lesssim E \lesssim 300 \text{ keV/amu})$ nuclei in space: Evidence from track and electron diffraction studies in lunar dust grains and in Surveyor III material. *Lunar Science III,* 92–94. Houston: Lunar Sci. Inst.

Braddy, D., Chan, J., Price, P. B. 1973. Charge states and energy-dependent composition of solar-flare particles. *Phys. Rev. Lett.* 30:669–71

Burnett, D. et al 1972. Solar cosmic ray, solar wind, solar flare, and neutron albedo measurements. *Apollo 16 Preliminary Science Report, NASA SP-315,* 15–19–32

Burnett, D. S., Woolum, D. S. 1974. Lunar neutron capture as a tracer for regolith dynamics. *Proc. 5th Lunar Sci. Conf., Geochim. Cosmochim. Acta Suppl.* 5 2:2061–74

Cameron, A. G. W. 1973a. Abundances of the

elements in the solar system. *Space Sci. Rev.* 15:121–46

Cameron, A. G. W. 1973b. Major variations in solar luminosity? *Rev. Geophys. Space Sci.* 11:505–10

Cantelaube, Y., Maurette, M., Pellas, P. 1967. Traces d'ions lourds dans les mineraux de la chondrite de Saint Severin. *Radioactive Dating and Methods of Low-Level Counting*, 215–229. Vienna: IAEA. 744 pp.

Clark, R. S., Keith, J. E. 1973. Determination of natural and cosmic ray induced radionuclides in Apollo 16 lunar samples. *Proc. 4th Lunar Sci. Conf., Geochim. Cosmochim. Acta Suppl. 4* 2:2105–13

Crozaz, G. et al 1972. Solar flare and galactic cosmic ray studies of Apollo 14 and 15 samples. *Proc. 3rd Lunar Sci. Conf., Geochim. Cosmochim. Acta Suppl. 3* 3: 2917–31

Crozaz, G., Taylor, G. J., Walker, R. M. 1974a. Early active sun?: Radiation history of distinct components in fines. *Proc. 5th Lunar Sci. Conf., Geochim. Cosmochim. Acta Suppl. 5* 3:2591–96

Crozaz, G. et al 1974b. Lunar surface dynamics: Some general conclusions and new results from Apollo 16 and 17. *Proc. 5th Lunar Sci. Conf., Geochim. Cosmochim. Acta Suppl. 5* 3:2475–99

Crozaz, G., Walker, R., Woolum, D. 1971. Nuclear track studies of dynamic surface processes on the moon and the constancy of solar activity. *Proc. 2nd Lunar Sci. Conf., Geochim. Cosmochim. Acta Suppl. 2* 3: 2543–58

D'Amico, J. D., DeFelice, J., Fireman, E. L., Jones, C., Spannagel, G. 1971. Tritium and argon radioactivities and their depth variations in Apollo 12 samples. *Proc. 2nd Lunar Sci. Conf., Geochim. Cosmochim. Acta Suppl. 2* 2:1825–39

Dran, J. C., Durrieu, L., Jouret, C., Maurette, M. 1970. Habit and texture studies of lunar and meteoritic materials with a 1 MeV electron microscope. *Earth Planet. Sci. Lett.* 9:391–400

Drozd, R. J., Hohenberg, C. M., Morgan, C. J., Ralston, C. E. 1974. Cosmic-ray exposure history at the Apollo 16 and other lunar sites: Lunar surface dynamics. *Geochim. Cosmochim. Acta* 38:625–42

Ducati, H., Kalbitzer, S., Kiko, J., Kirsten, T. Müller, H. W. *Rare gas diffusion studies in individual lunar soil particles and in artificially implanted glasses. MP1 H—1972—V36, Max-Planck-Institut für Kernphysik, Heidelberg*. Unpublished

Eberhardt, P. 1974. The solar wind as deduced from lunar samples. *Proc. 3rd Solar Wind Conf., Asilomar, Pacific Grove, Calif.* In press

Eberhardt, P. et al 1970. Trapped solar wind noble gases, exposure age and K/Ar-age in Apollo 11 lunar fine material. *Proc. Apollo 11 Lunar Sci. Conf., Geochim. Cosmochim. Acta Suppl. 1* 2:1037–70

Eberhardt, P. et al 1972. Trapped solar wind noble gases in Apollo 12 lunar fines 12001 and Apollo 11 breccia 10046. *Proc. 3rd Lunar Sci. Conf., Geochim. Cosmochim. Acta Suppl. 3* 2:1821–56

Eberhardt, P. et al 1974. Solar wind and cosmic radiation history of Taurus Littrow regolith. *Lunar Science V*, 197–99. Houston: Lunar Sci. Inst.

Eldridge, J. S., O'Kelley, G. D., Northcutt, K. J. 1973. Radionuclide concentrations in Apollo 16 lunar samples by non-destructive gamma-ray spectrometry. *Proc. 4th Lunar Sci. Conf., Geochim. Cosmochim. Acta Suppl. 4* 2:2115–22

Fichtel, C. E., McDonald, F. B. 1967. Energetic particles from the sun. *Ann. Rev. Astron. Astrophys.* 5:351–98

Finkel, R. C. et al 1971. Depth variations of cosmogenic nuclides in a lunar surface rock and lunar soil. *Proc. 2nd Lunar Sci. Conf., Geochim. Cosmochim. Acta Suppl. 2* 2:1773–89

Fleischer, R. L., Hart, H. R. Jr. 1972. Composition and energy spectra of solar cosmic ray nuclei. *Apollo 16 Preliminary Science Report, NASA SP-315,* 15–2–11

Fleischer, R. L., Hart, H. R. Jr. 1973a. Enrichment of heavy nuclei in the 17 April 1972 solar flare. *Phys. Rev. Lett.* 30:31–34

Fleischer, R. L., Hart, H. R. Jr. 1973b. Mechanical erasure of particle tracks, a tool for lunar microstratigraphic chronology. *J. Geophys. Res.* 78:4841–51

Fleischer, R. L., Hart, H. R. Jr. 1973c. Particle track record in Apollo 15 deep core from 54 to 80 CM depths. *Earth Planet. Sci. Lett.* 18:420–26

Fleischer, R. L., Hart, H. R. Jr., Comstock, G. M., Evwaraye, A. O. 1971. The particle track record of the Ocean of Storms. *Proc. 2nd Lunar Sci. Conf., Geochim. Cosmochim. Acta Suppl. 2* 3:2559–68

Fleischer, R. L., Hart, H. R. Jr., Giard, W. R. 1974a. Surface history of lunar soil and soil columns. *Geochim. Cosmochim. Acta* 38:365–80

Fleischer, R. L., Price, P. B., Walker, R. M. 1965. Solid-state track detectors: Applications to nuclear science and geophysics. *Ann. Rev. Nucl. Sci.* 15:1–28

Fleischer, R. L., Price, P. B., Walker, R. M. 1974b. *Nuclear Tracks in Solids.* Berkeley: Univ. Calif. Press. In press

126   WALKER

Fleischer, R. L., Price, P. B., Walker, R. M., Maurette, M., Morgan, G. 1967. Tracks of heavy primary cosmic rays in meteorites. *J. Geophys. Res.* 72:355–66

Fowler, W. A. 1972. What cooks with solar neutrons? *Nature* 238:24–26

Frick, U. et al 1973a. Diffusion properties of light noble gases in lunar fines. *Proc. 4th Lunar Sci. Conf., Geochim. Cosmochim. Acta Suppl.* 4 2:1987–2002

Frick, U. et al 1973b. Diffusion of argon in lunar fines, or little orphan argon retrapped. *Lunar Science IV*, 266–68. Houston: Lunar Sci. Inst.

Gault, D. E., Hörz, F., Brownlee, D. E., Hartung, J. B. 1974. Mixing of the lunar regolith. *Lunar Science V*, 260–62. Houston: Lunar Sci. Inst.

Gault, D. E., Hörz, F., Hartung, J. B. 1972. Effects of microcratering on the lunar surface. *Proc. 3rd Lunar Sci. Conf., Geochim. Cosmochim. Acta Suppl.* 3 3:2713–34

Geiss, J., Buehler, F., Cerutti, H., Eberhardt, P., Filleux, C. 1972. Solar wind composition experiment. *Apollo 16 Preliminary Science Report, NASA SP-315*, 14–1–10

Geiss, J. 1972. Elemental and isotopic abundances in the solar wind. In *Solar Wind*, ed. C. P. Sonett, P. J. Coleman Jr., J. M. Wilcox, 559–81. NASA SP-308. 717 pp.

Geiss, J. 1973. Solar wind composition and implications about the history of the solar system. *13th International Cosmic Ray Conference, Conference Papers* 5:3375–98. Denver: Univ. Denver Press

Hart, H. R. Jr., Comstock, G. M., Fleischer, R. L. 1972. The particle track record of Fra Mauro. *Proc. 3rd Lunar Sci. Conf., Geochim. Cosmochim. Acta Suppl.* 3 3:2831–44

Hartung, J. B., Hörz, F., Aitken, F. K., Gault, D. E., Brownlee, D. E. 1973. The development of microcrater populations on lunar rocks. *Proc. 4th Lunar Sci. Conf., Geochim. Cosmochim. Acta Suppl.* 4 3:3213–34

Hartung, J. B., Hörz, F., Gault, D. E. 1972. Lunar microcraters and interplanetary dust. *Proc. 3rd Lunar Sci. Conf., Geochim. Cosmochim. Acta Suppl.* 3 3:2735–53

Hartung, J. B., Storzer, D., Hörz, F. 1974. *Lunar Science V*, 307–9. Houston: Lunar Sci. Inst.

Hodges, R. R. Jr., Hoffman, J. H., Johnson, F. S. 1974. The lunar atmosphere. *Lunar Science V*, 343–45. Houston: Lunar Sci. Inst.

Hodges, R. R. Jr., Hoffman, J. H., Johnson, F. S., Evans, D. E. 1973. Composition and dynamics of lunar atmosphere. *Proc. 4th Lunar Sci. Conf., Geochim. Cosmochim. Acta Suppl.* 4 3:2855–64

Hoffman, J. H., Hodges, R. R. Jr., Johnson, F. S., Evans, D. E. 1973. Lunar atmospheric composition results from Apollo 17. *Proc. 4th Lunar Sci. Conf., Geochim. Cosmochim. Acta Suppl.* 4 3:2865–75

Hörz, F., Hartung, J. B., Gault, D. E. 1971. Micrometeorite craters on lunar rocks. *J. Geophys. Res.* 76:5770–98

Hörz, F. et al 1974. Lunar microcraters: Implications for the micrometeoroid complex. *Planet. Space Sci.* In press. (This paper is a summary of papers presented at the COSPAR meeting in 1973, Konstanz, Germany.)

Hoyt, H. P. Jr. et al 1971. Radiation dose rates and thermal gradients in the lunar regolith: Thermoluminescence and DTA of Apollo 12 samples. *Proc. 2nd Lunar Sci. Conf., Geochim. Cosmochim. Acta Suppl.* 2 3:2245–63

Hoyt, H. P. Jr., Walker, R. M., Zimmerman, D. W. 1973. Solar flare proton spectrum averaged over the last 5×10^3 years. *Proc. 4th Lunar Sci. Conf., Geochim. Cosmochim. Acta Suppl.* 4 3:2489–2502

Hubner, W., Heymann, D., Kirsten, T. 1973. Inert gas stratigraphy of Apollo 15 drill core sections 15001 and 15003. *Proc. 4th Lunar Sci. Conf., Geochim. Cosmochim. Acta Suppl.* 4 2:2021–36

Hundhausen, A. J. 1972. *Coronal Expansion and Solar Wind.* Berlin: Springer. 238 pp.

Hutcheon, I. D. et al 1974. Rock 72315: A new lunar standard for solar flare and micrometeorite exposure. *Lunar Science V*, 378–80. Houston: Lunar Sci. Inst.

Keith, J. E., Clark, R. S., Richardson, K. A. 1972. Gamma-ray measurements of Apollo 12, 14, and 15 lunar samples. *Proc. 3rd Lunar Sci. Conf., Geochim. Cosmochim. Acta Suppl.* 3 2:1671–80

Kraft, R. P. 1967. Studies of stellar rotation V. The dependence of rotation on age among solar-type stars. *Astrophys. J.* 150:551–70

Lal, D., Lorin, J. C., Pellas, P., Rajan, R. S., Tamhane, A. S. 1969. On the energy spectrum of iron-group nuclei as deduced from fossil-track studies in meteoritic minerals. *Meteorite Research*, ed. P. Millman, 275–85. Dordrecht, Holland: Reidel. 940 pp.

Lal, D., Rajan, R. S. 1969. Observations on space irradiation of individual crystals of gas-rich meteorites. *Nature* 223:269–71

Leich, D. A., Tombrello, T. A., Burnett, D. S. 1974. Trapped solar hydrogen in lunar samples. *Lunar Science V*, 444–46. Houston: Lunar Sci. Inst.

Lingenfelter, R. E., Canfield, E. H., Hampel,

V. E. 1972. The lunar neutron flux revisited. *Earth Planet. Sci. Lett.* 16 : 355–69

Lugmair, G. W., Marti, K. 1972. Neutron and spallation effects in Fra Mauro regolith. *Lunar Science III,* 495–97. Houston : Lunar Sci. Inst.

Manka, R. H., Michel, F. C. 1971. Lunar atmosphere as a source of lunar surface elements. *Proc. 2nd Lunar Sci. Conf., Geochim. Cosmochim. Acta Suppl. 2* 2: 1717–28

Marti, K. 1967. Mass-spectrometric detection of cosmic-ray produced Kr^{81} in meteorites and the possibility of Kr-Kr dating. *Phys. Rev. Lett.* 18 : 264–66

Marti, K., Lightner, B. D., Osborn, T. W. 1973. Krypton and xenon in some lunar samples and the age of North Ray Crater. *Proc. 4th Lunar Sci. Conf., Geochim. Cosmochim. Acta Suppl. 4* 2 : 2037–48

Maurette, M., Thro, P., Walker, R., Webbink, R. 1969. Fossil tracks in meteorites and the chemical abundance and energy spectrum of extremely heavy cosmic rays. *Meteorite Research,* ed. P. Millman, 286–315. Dordrecht, Holland : Reidel. 940 pp.

Meyer, P. 1969. Cosmic rays in the galaxy. *Ann. Rev. Astron. Ap.* 7 : 1–38

Morrison, D. A., McKay, D. S., Heiken, G. H., Moore, H. J. 1972. Microcraters on lunar rocks. *Proc. 3rd Lunar Sci. Conf., Geochim. Cosmochim. Acta Suppl. 3* 3: 2767–91

Müller, O. 1974. Solar wind and indigenous nitrogen in Apollo 17 lunar samples. *Lunar Science V,* 534–36. Houston : Lunar Sci. Inst.

Neukum, G., Hörz, F., Morrison, D. A., Hartung, J. B. 1973. Crater populations on lunar rocks. *Proc. 4th Lunar Sci. Conf., Geochim. Cosmochim. Acta Suppl. 4* 3: 3255–76

Pellas, P., Poupeau, G., Lorin, J. C., Reeves, H., Audouze, J. 1969. Primitive low-energy particle irradiation of meteoritic crystals. *Nature* 223 : 272–74

Pepin, R. O., Basford, J. R., Dragon, J. C., Coscio, M. R. Jr., Murthy, V. R. 1974. *Lunar Science V,* 593–95. Houston : Lunar Sci. Inst.

Phakey, P. P., Hutcheon, I. D., Rajan, R. S., Price, P. B. 1972. Radiation effects in soils from five lunar missions. *Proc. 3rd Lunar Sci. Conf., Geochim. Cosmochim. Acta Suppl. 3* 3 : 2905–15

Pillinger, C. T. et al 1973. Formation of lunar carbide from lunar iron silicates. *Nature Phys. Sci.* 245 : 3–5

Plieninger, T., Krätschmer, W., Gentner, W. 1973. Indications for time variations in the

galactic cosmic ray composition derived from track studies on lunar samples. *Proc. 4th Lunar Sci. Conf., Geochim. Cosmochim. Acta Suppl.* 4 3 : 2337–46

Poupeau, G., Chetrit, G. C., Berdot, J. L., Pellas, P. 1973. Etude par la méthode des traces nucléaires du sol de la mer de la Fécondité (Luna 16). *Geochim. Cosmochim. Acta* 37 : 2005–16

Price, P. B., Braddy, D., O'Sullivan, D., Sullivan, J. D. 1972. Composition of interplanetary particles at energies from 0.1 to 150 MeV/nucleon. *Apollo 16 Preliminary Science Report, NASA SP-315,* 15–11–19

Price, P. B., Fleischer, R. L. 1971. Identification of energetic heavy nuclei with solid dielectric track detectors : Applications to astrophysical and planetary studies. *Ann. Rev. Nucl. Sci.* 21 : 295–334

Price, P. B., Hutcheon, I., Cowsik, R., Barber, D. J. 1971. Enhanced emission of iron nuclei in solar flares, *Phys. Rev. Lett.* 26 : 916–19

Price, P. B., Rajan, R. S., Hutcheon, I. D., MacDougall, D., Shirk, E. K. 1973. Solar flares, past and present. *Lunar Science IV,* 600–602. Houston : Lunar Sci. Inst.

Rancitelli, L. A., Perkins, R. W., Felix, W. D., Wogman, N. A. 1973. Lunar surface and solar process analyses from cosmogenic radionuclide measurements at the Apollo 16 site. *Lunar Science IV,* 609–11. Houston : Lunar Sci. Inst.

Rancitelli, L. A., Perkins, R. W., Felix, W. D., Wogman, N. A. 1974. Anisotropy of the August 4–7, 1972 solar flares at the Apollo 17 site. *Lunar Science V,* 618–20. Houston : Lunar Sci. Inst.

Reedy, R. C., Arnold, J. R. 1972. Interaction of solar and galactic cosmic-ray particles with the moon. *J. Geophys. Res.* 77 : 537–55

Russ, G. P. III, Burnett, D. S., Wasserburg, G. J. 1972. Lunar neutron stratigraphy. *Earth Planet. Sci. Lett.* 15 : 172–86

Smith, J. W., Kaplan, I. R., Petrowski, C. 1973. Carbon, nitrogen, sulfur, helium, hydrogen, and metallic iron in Apollo 15 drill stem fines, *Proc. 4th Lunar Sci. Conf., Geochim. Cosmochim. Acta Suppl. 4* 2: 1651–56

Sonett, C. P. 1974. The primordial solar wind. *Proc. 3rd Solar Wind Conf., Asilomar, Pacific Grove, Calif.* In press

Tamhane, A. S. 1972. *Abundance of heavy cosmic ray nuclei from the induced micrometamorphism in meteoritic minerals.* PhD thesis. Univ. Bombay, Bombay, India

Taylor, H. P. Jr., Epstein, S. 1973. O^{18}/O^{16} and Si^{30}/Si^{28} studies of some Apollo 15, 16, and 17 samples. *Proc. 4th Lunar Sci.*

128   WALKER

Conf., Geochim. Cosmochim. Acta Suppl. 4 2:1657–79

Turner, G., Huneke, J. C., Podosek, F. A., Wasserburg, G. J. 1971. ^{40}Ar–^{39}Ar ages and cosmic ray exposure ages of Apollo 14 samples. Earth Planet. Sci. Lett. 12:19–35

Wahlen, M. et al 1972. Cosmogenic nuclides in football-sized rocks. Proc. 3rd Lunar Sci. Conf., Geochim. Cosmochim. Acta Suppl. 3 2:1719–32

Walker, R., Yuhas, D. 1973. Cosmic ray track production rates in lunar materials. Proc. 4th Lunar Sci. Conf., Geochim. Cosmochim. Acta Suppl. 4 3:2379–89

Wehner, G. K., Kenknight, C., Rosenberg, D. L. 1963. Sputtering rates under solar wind bombardment. Planet. Space Sci. 11:885–95

Woolum, D. S., Burnett, D. S. 1974. In-situ measurement of the rate of ^{235}U fission induced by lunar neutrons. Earth Planet. Sci. Lett. 21:153–63

Yuhas, D., Walker, R. 1973. Long term behavior of VH cosmic rays as observed in lunar rocks. 13th International Cosmic Ray Conference, Conference Papers 2: 1116–21. Denver: Univ. Denver Press

Zinner, E., Walker, R. M., Borg, J., Maurette, M. 1974. Apollo 17 lunar surface cosmic ray experiment—measurement of heavy solar wind particles. Proc. 5th Lunar Sci. Conf., Geochim. Cosmochim. Acta Suppl. 5. 3:2975–89

MICROSTRUCTURES OF MINERALS   ✱10035
AS PETROGENETIC INDICATORS

J. D. C. McConnell
Department of Mineralogy and Petrology, Cambridge University,
Cambridge CB2 3EW, Great Britain

INTRODUCTION

The nature and scale of the microstructure in minerals have traditionally been used in petrology to define, in a qualitative way, the likely cooling history of the parent rock. Thus the existence of a low-temperature mineral polymorph formed in a sluggish transformation, or the existence of a coarse exsolution texture, may be cited as evidence of an extended time scale in the equilibration of the mineral concerned. In certain cases it is possible to be more specific and to obtain quantitative data on the cooling history of the mineral through the application of transformation process theory. This theory deals not only with ideal subsolidus behavior in a given mineral system but also with the likely behavior of the system as a function of both time and temperature. It is usual to describe the progress of a given transformation at a specified temperature as a function of time. This information is normally displayed on a diagram, described as a TTT diagram, in which the progress of the transformation is plotted in isothermal runs on a log time scale. In almost all cases the process thus described occurs under thermodynamically irreversible conditions, and certain characteristic features of the plot can be associated with specific aspects of the mechanism and kinetics of the process studied.

The application of time-temperature-transformation studies in mineralogy and petrology is long overdue, and it is fair to say that, in general, petrologists have been too preoccupied with ideal equilibrium behavior. In certain cases it is possible to reclaim a great deal of information from the microstructure of a mineral where adequate TTT data exist on the transformation concerned. The application of the theory is also extremely important in attempting to understand why certain transformations in minerals are particularly sluggish. Both of these aspects are of direct interest to the petrologist and are dealt with in this brief review. Transformation process theory relevant to the subsolidus behavior of minerals can conveniently be considered in three parts.

In the first instance it is desirable to know as much as possible about the ideal behavior of the mineral system as determined by direct experiment or thermodynamic investigation. A great deal is now known about the ideal thermodynamic behavior

129

of most of the important mineral systems and this aspect of the application of transformation process theory is already well advanced. However, there are some notable examples where metastable behavior appears to be the rule rather than the exception, and in these circumstances the existence of metastable behavior can only really be understood in terms of an analysis of the factors of mechanism and kinetics which render the ideal equilibrium state inaccessible. Low temperature behavior in the plagioclase feldspars is a particularly important example of a situation of this kind.

The second important aspect of the application of transformation process theory to subsolidus behavior involves the study of the mechanisms of processes. The study of mechanisms, as the term is defined here, is concerned with athermal aspects of the process, the spatial dimensions, distribution, and crystallographic relationships of parent and product crystalline phases. A considerable amount of information is already available on mineral microstructures in this context and the current application of electron microscopy in this field has been particularly fruitful. In the immediate future a great deal more data of this kind are likely to accrue, and the future development of this aspect of the subject is likely to be particularly healthy.

Unfortunately, the same comment cannot be made about the present state of development of the third and final aspect of transformation process theory as applied to minerals. This deals with thermal and kinetic aspects of mineral transformations. In this sphere, with a few notable exceptions, the necessary fundamental data on self-diffusion coefficients and reaction rates are virtually nonexistent. This is particularly unfortunate because it is precisely in this field in which data are of most potential use in the application of transformation process theory in petrogenesis. The application of rate data, particularly in the case of transformations which involve volume diffusion processes, is likely to provide the petrologist with an extremely powerful tool for the detailed investigation of the thermal history of minerals and rocks. It is hoped that, even if the present review achieves little else, it will stimulate some careful experimental work in this field. One important point arises in this context. Certain processes in particular mineral groups, such as the feldspars, are inordinately affected by the presence or absence of water vapor. Thus, in determining effective diffusion rates and their temperature dependence, it will often be necessary to establish whether or not these are affected by the presence of water vapor.

A comprehensive review of all the relevant aspects of transformation behavior in minerals is certainly outside the scope of this review. The primary objective here is to review the general situation with regard to the interpretation of mineral microstructures in petrogenetic terms, and to provide an analysis of certain systems and situations where it is already possible to obtain quantitative information on time-temperature and other aspects of the reaction behavior from mineral microstructures.

At the outset it is only fair to admit that not all subsolidus reactions and transformations in minerals necessarily lend themselves to quantitative description in time-temperature terms and certain transformations are unlikely to be of use to the petrologist in this context. Martensitic transformations (which involve very localized atomic displacements and occur by cooperative shear) belong in this class. On the other hand, transformations where the rates are determined by reaction or diffusion, kinetics are likely to be particularly useful in this context because the time factor is implicit in the description of such processes. Where transformations are primarily governed by rates associated with volume diffusion, the situation is particularly favorable for analysis in time-temperature terms. Exsolution, eutectoidal and peritectoidal reactions belong to this class, which also includes polymorphic transformations of the order-disorder type as a special case.

It is appropriate to make one general comment at this point. Metastable transformation behavior in minerals cannot be discussed in the same terms as equilibrium behavior and, in consequence, a general feeling exists that such behavior, because it cannot be rigorously characterized on an equilibrium phase diagram, is in some sense unpredictable. The situation is quite otherwise in relation to the application of transformation process theory, since in many cases metastable behavior can be defined more rigorously in time-temperature terms than alternative equilibrium behavior. In such circumstances it is possible to use TTT data for a highly irreversible and metastable equilibration process in defining the time-temperature aspects of petrogenesis of the mineral and rock concerned. Indeed, in this context it would seem that the usefulness of transformation behavior in a given mineral system is likely to increase in direct proportion to its complexity, particularly where reproducible but markedly contrasted behavior is observed over a wide range of possible cooling histories. In this instance the TTT plot for the system will comprise a number of well-defined behavior patterns spread over a large area of time-temperature space, and thus be appropriate to the study of a very wide range of possible cooling rates.

At this point the objectives of the application of transformation process theory to mineral microstructures should be clear. The subject deals with precise physical description of the actual behavior of minerals in time-temperature terms, and is equally applicable to all transformation processes whether or not they occur under ideal conditions of thermodynamic reversibility.

Transformation process theory, as discussed in this review, has been used for some considerable time in metallurgy. The TTT diagram originated in this field and was introduced to help explain the results of isothermal annealing experiments in certain alloys where a very wide range of possible behavior was observed. In applying the theory to mineral microstructures the petrologist is in fact attempting to reverse the usual sequence of analysis and thus deduce the annealing conditions from the microstructure produced in a natural equilibration cycle. In spite of the fact that a very large number of alloy systems have been studied by metallurgists, some problems of transformation behavior appear to be unique to mineral systems. This applies particularly to the existence of extremely sluggish reactions in minerals which may or may not occur even in natural equilibration processes which have operated on a geological time scale. Although the extended time scale available to the petrologist is particularly useful, it actually introduces no fundamentally new features into the general theory of transformation processes as defined in short-term experiments in the laboratory.

The main body of this review is divided into two parts. In the first part transformation process theory is explained and used to define two broad classes of transformation behavior of importance in mineral systems. Particular attention is devoted to those aspects of the theory which are of most direct concern to the petrologist in the elucidation of mineral microstructures in petrogenetic terms. The second part of the review deals with the analysis of several examples of transformation behavior in natural minerals where adequate data on all three aspects of the theory, as defined above, are available. In each of the chosen examples definite conclusions can be reached on time-temperature and other conditions under which these minerals are formed.

A definitive account of the theory of phase transformations in alloy systems is available (Christian 1965), and a review of the relevant data for exsolution processes in mineral systems has been presented recently by Yund & McCallister (1970). Nucleation in solids has been treated specifically by Russell (1970) and a similarly detailed account of spinodal decomposition in alloy systems is available (Hilliard 1970).

In the immediate future, application of transformation process theory is likely to be particularly important in the study of petrogenetic aspects of the behavior of the feldspars and the pyroxenes, inasmuch as microstructures in both of these systems have already been studied in detail by electron microscopy. In this review attention is concentrated on some of the problems associated with behavior in these very important mineral systems. An excellent review of the microstructures observed in the feldspars by electron microscopy is available (McLaren 1974).

TRANSFORMATION PROCESS THEORY

Classification of Transformations in Mineral Systems: Continuous and Discontinuous Processes

For most, if not all, of the transformations of interest here, the closest possible approach to equilibrium behavior equates with a process of nucleation and growth. By its very nature this process is necessarily discontinuous in the sense that it corresponds to the appearance of a new crystalline phase, and some of the most important features of the time-temperature description of such transformations are associated with this discontinuity. In the study of transformation behavior in certain of the most important mineral systems it is frequently observed that nucleation does not occur readily, and in its absence alternative behavior exists. This behavior has the special characteristic that it is functionally continuous in respect to time, involves no equivalent of the nucleation incident, and is governed by constraints associated with the structure and symmetry of the parent crystalline phase. The microstructure associated with continuous transformation behavior of this kind, and the definition of the operative constraints, are discussed in a recent paper (McConnell 1971). In this review it is convenient to refer to such transformation behavior as continuous, implying that there is no discontinuity in the observed TTT behavior of the system.

Two aspects of the study of continuous transformation behavior are particularly relevant. In the first instance, an analysis of such behavior often helps directly in explaining the inherent difficulties associated with a true nucleation and growth mechanism, particularly where the discontinuous transformation is impeded by symmetry constraints. An excellent example of this type of situation is provided by the monoclinic to triclinic transformation in potassium feldspar. It is possible to understand the factors which often impede true nucleation and growth of the low temperature phase microcline in this case only by studying the alternative behavior of continuous type exhibited by orthoclase and adularia.

Continuous-type behavior is normally rate controlled by thermodynamics and the relevant diffusion constants alone. In general it is possible to be much more specific about the precise behavior of a system which transforms in this way in time-temperature terms. Thus the results of a study of such behavior can be applied directly in a petrogenetic sense.

By contrast, transformations that involve a nucleation incident, whether this occurs by a homogeneous or inhomogeneous mechanism, inevitably involve a discontinuity associated with the appearance of a new crystalline phase in the nucleation event. The nature of this discontinuity, and in particular the mechanism of the nucleation process, introduce an additional complexity into the description of the transformation process in time-temperature terms. Thus it will often be more difficult to draw up a definitive TTT diagram for a system which transforms in a functionally discontinuous manner since this will depend on the precise mechanism of nucleation which operates in any given circumstances.

The above definitions of transformation processes in terms of functionally continuous and discontinuous behavior are quite general in the sense that they may be applied to a very wide range of behavior in mineral systems. The definitions relate specifically to behavior and carry no direct implications about the ideal thermodynamic behavior of the systems to which they are applied. In general, however, the existence of continuous behavior will usually imply that a discontinuous transformation process is implicated whether or not any direct evidence for such a discontinuity is actually observed.

It would seem at the present stage in the application of transformation process theory to minerals that a great deal of careful observation, experiment, and analysis remains to be done. It is already obvious that the results of this type of study will be particularly important in not only defining the time-temperature behavior of minerals as actually observed, but also in defining the probable ideal behavior in certain systems where this is never evident.

Time-Temperature Description of Discontinuous and Continuous Transformation Behavior in Mineral Systems

Having defined two very broad classes of transformation behavior in minerals it is now appropriate to examine the details of their behavior in time-temperature terms.

Ideally a discontinuous transformation in a mineral or mineral system should commence as soon as the temperature differs from the critical value defined for the transformation. In practice, such prompt behavior is not observed except in circumstances where the structures of parent and product phases are very closely related and the process is not thermally activated. The high-low quartz transformation is of this kind. In transformations of interest here, the transforming system is likely to become appreciably supercooled due to the inherent difficulties associated with the appearance of a new crystalline phase by a nucleation mechanism. A detailed appreciation of the process of nucleation is important not only in relation to the mechanism of discontinuous transformation processes, but also in attempting to explain why, in certain circumstances, the probability of nucleation is extremely low and alternative metastable behavior results.

Where a transformation occurs at temperature T' and the high temperature phase is supercooled, this phase is thermodynamically unstable and the corresponding value of ΔG associated with the transformation at this temperature operates as the driving force for the transformation. An approximate relationship between the extent of supercooling ΔT and this free-energy term ΔG is given by $\Delta G = \Delta H \Delta T / T'$, implying that the driving force for the transformation increases approximately linearly with the extent of supercooling. Thus, in the absence of other factors, the development of a small region of the new, low temperature structure will be favored increasingly as the extent of supercooling increases. Since the new phase differs in either structure or chemical composition from the parent phase it cannot appear spontaneously but must grow within the existing structure. At the very earliest stages in the development of such a region it may be called an embryo. It is probably best to describe the embryo in terms of local fluctuations in the structure or composition of the parent phase. The actual transition between this state and that of a discrete nucleus usually involves a further complication. Differences in structure and cell dimensions of embryo and host crystal introduce a free-energy term associated with the strain energy involved in coexistence. Under conditions of slight supercooling this adverse strain-energy term will entirely outweigh the reduction in free energy associated with the initial growth of an embryo of the stable phase and the probability that a local fluctuation will develop into a viable nucleus will be inordinately small. With increased supercooling the development of a viable nucleus will be favored by the increased value of ΔG, and the nucleation incident, although still improbable, will eventually provide a mechanism for the transformation.

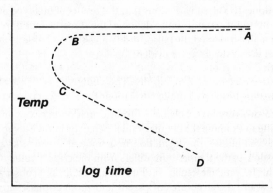

Figure 1 TTT diagram for the initiation of a transformation process by a nucleation event. Rates close to the transformation temperature are governed by ΔG, at low temperature by the activation energy for diffusion.

Since nucleation under conditions of slight supercooling is an extremely improbable event in any case, the nucleation observed in any system for a defined degree of supercooling will be a function of the time scale. Thus at slower rates of cooling the initial nucleation incident is likely to occur at smaller values of ΔT. This is likely to be the case whether the nucleation occurs through a homogeneous or inhomogeneous mechanism, and for very slow rates of cooling the nucleation is likely to differentiate critically between different possible mechanisms and operate exclusively on the basis of the lowest possible free-energy barrier for the process. In fact it would seem, from a study of the characteristics of the microstructure of minerals where exsolution has occurred at very slow rates of cooling, that the nucleation incident must frequently have occurred under quasi-equilibrium conditions in which a very delicate balance was maintained between the development of the nuclei and the controls exercised by diffusion, the extent of undercooling, and the time scale. In these circumstances the scale of the microstructure, as determined in the very first instance by the nucleation process, may provide a valuable indication of the original rate of cooling.

Figure 1, which illustrates the time-temperature characteristics of a typical discontinuous transformation, may be examined at this point. The temperature of the ideal transformation has been inserted as a full line on this diagram and the heavy dashed line ($A-B$) which approaches this idealized transformation temperature at very large values of t represents the nucleation incident. From the previous discussion it should be clear that behavior close to the transformation temperature, while not perhaps entirely predictable, is dominated by the value of ΔG, the overall change in free energy in the transformation. Thus the nucleation rate must necessarily go to zero at the transformation temperature regardless of the time involved in the equilibration process.

It should be noted at this point that where the mechanism of the nucleation process is inhomogeneous, and determined by the presence of dislocations or subgrain boundaries, the precise form of the nucleation curve at small ΔT will be affected.

At larger values of ΔT nucleation rates must increase rapidly due to the increase in the value of ΔG, the overall change in free energy in the transformation process. In Figure 1 this is illustrated by the part of the nucleation curve labeled as $B-C$. Where the growth of the nuclei is thermally activated due to local changes in structure, or where growth is rate controlled by diffusion, the rate of nucleation usually decreases with further decrease in the temperature. This situation is illustrated in Figure 1 by the approximately linear part of the nucleation curve lying between points C and D.

This type of situation, in which the rates at temperature close to the true transformation temperature are dominated by the thermodynamics of the transformation and at low temperatures by the activation energy associated with the process, is typical of most of the solid state transformations of interest here, whether they involve a change in local chemical composition or not. In considering further aspects of the processes of transformation in such systems it is usual, however, to distinguish between transformations in which there is no change in local chemical composition during the process, and transformations which involve volume diffusion in the growth of the new phase or phases.

This distinction is likely to be of particular importance in relation to the interpretation of microstructure in petrogenetic terms, since the case in which volume diffusion
operates is likely to provide much more data on the time-temperature aspects of the
processes involved. The petrogenetic implications of behavior of both types will be
discussed later.

So far this discussion has been concerned exclusively with the general principles
of time-temperature behavior of discontinuous transformation processes. A similar
analysis can be made for continuous behavior. As already noted, functionally
continuous behavior is usually observed in circumstances where for some special
reason a nucleation and growth mechanism is not observed or occurs only under
extremely slow cooling conditions. Continuous behavior is apparently quite common
in a number of mineral systems and it is now apparent that this class of behavior in
mineral systems is closely analogous to certain specific behavior patterns observed in
alloy systems described as spinodal decomposition. In this context the earliest stages
of spinodal behavior in alloys corresponds to a situation where a nucleation and
growth process is not observed and behavior of continuous character occurs in lieu
of true exsolution. Thus a considerable body of theoretical data and the results of
practical study of spinodal behavior in alloys are directly applicable to a number of
related but diverse problems in mineralogy. In what follows, the standard analysis
of spinodal behavior will be presented.

In circumstances appropriate to continuous transformation behavior the existence
of a discrete nucleation incident associated with a discontinuous transformation of
exsolution is either prohibited or highly improbable, and the phase involved may be
supercooled through a temperature interval ΔT where no change occurs. At a critical
value of ΔT the system produces local homogeneous fluctuations in chemical
composition. The necessary condition that such local fluctuations in the single crystal
should be stable is simply that a reduction in the free energy of the system exists which
more than balances the strain energy of coexistence of the regions of different
chemical composition.

Current theoretical treatment of spinodal decomposition (Cahn 1968, Hilliard
1970) indicates that there are three separate factors involved in defining the character
of the fluctuation wave behavior observed, in addition to the time factor associated
with their development. The first relates to the reduction in free energy associated
with the development of compositional fluctuations. In a spinodal system this may be
related to the magnitude of the second derivative of the free energy defined as a
function of the chemical composition. Where fluctuation behavior occurs this must
be negative. The second factor relates to the magnitude of the strain-energy term
associated with the coexistence of regions of different chemical composition and
lattice constants. Both the elastic constants and the lattice constants are involved
in defining the strain-energy term as a function of the amplitude of the fluctuations
in local chemical composition. The strain-energy term in this case is dependent upon
a fourth order tensor implying that, in a crystalline solid, favored fluctuations are
likely to have a specific habit and orientation. Where the system remains as a single
crystal the strain is necessarily a volume effect and, for fluctuations with prescribed
wave vector, the strain energy is independent of wavelength. The final factor governing

the development of fluctuations in the single crystal within the spinodal region is described as the gradient energy, which is proportional to the product of the heat of mixing and the square of the nearest neighbor distance in the structure. In effect the gradient-energy term defines a minimum distance over which an appreciable change in chemical composition is acceptable in energy terms. Thus, in the development of spinodal microstructures, fluctuation waves with wavelength shorter than a certain critical value are prone to disappear due to the gradient-energy effect.

The influence of all three of the factors defined above combine to dictate the behavior of the system as a function of the extent of undercooling (ΔT) and the time. At the highest temperature where spinodal fluctuations are stable (the spinodal temperature), only extremely long wavelength and small amplitude fluctuations are possible. These are formed very slowly due to the long diffusion distances involved. At lower temperatures the possibility of a substantial reduction in free energy in the system favors the development of fluctuations of shorter wavelength which grow more rapidly. At much lower temperatures the growth of even short wavelength fluctuations becomes slow due to the influence of falling temperature on the mobility of the diffusing species. The general characteristics of the TTT diagram for spinodal behavior are indicated in Figure 2. In form it is clearly comparable to the generalized TTT diagram for a discontinuous transformation as shown in Figure 1. It is important to note at this stage, and particularly from the point of view of the application of this behavior in a petrogenetic context, that spinodal behavior is highly reproducible and almost totally unaffected by factors which, in normal discontinuous transformations, are prone to make a substantial difference to transformation rates. Behavior in this general class is of particular interest to the petrologist. Spinodal behavior corresponds to a rather special type of transformation process clock which records thermal history in terms of the wavelength and amplitude of the fluctuations in chemical composition produced.

The concepts applied in the study of spinodal behavior in alloy systems have recently been used successfully in the analysis of the early stage evolution of micro-

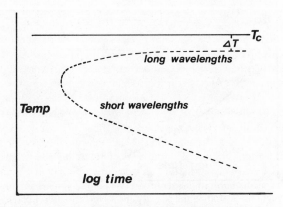

Figure 2   TTT diagram for spinodal behavior.

structure in the alkali feldspars. This particular example of the application of the theory will be discussed in some detail later.

Effects of Pressure Change on TTT Behavior

Pressure is likely to have both primary and secondary effects on the characteristics of the TTT diagram for a specified mineral transformation. Primary effects relate to the movement of the TTT diagram due to change in the equilibrium temperature for the transformation as a function of pressure. But, in moving the TTT diagram in this way, it is also necessary to consider certain secondary effects since the change in temperature is likely to move the whole TTT diagram parallel to the time axis due to the temperature dependence of the diffusion constants operative in the process. One may also anticipate that the diffusion constants and their temperature dependence may themselves be a function of pressure. Thus, in the general case, the TTT diagram may change shape and move with respect to both the temperature and time axes under conditions of changing total pressure. The primary and secondary factors which are likely to influence the TTT diagram as a function of changing total pressure have been illustrated schematically in Figure 3.

PETROGENETIC IMPLICATIONS

In this section the time-temperature aspects of the development of mineral micro-structures as defined by transformation process theory will be examined particularly from the point of view of defining situations in which the observed microstructure may be used to provide data on aspects of petrogenesis.

Microstructures Produced in Discontinuous Transformations Without Volume Diffusion

In this class of discontinuous transformation, which includes reconstructive poly-morphic transformations, the primary step in the process involves nucleation of the

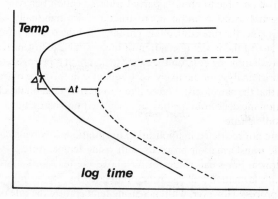

Figure 3   TTT diagrams illustrating the primary and secondary effects of pressure in terms of ΔT and Δt, respectively.

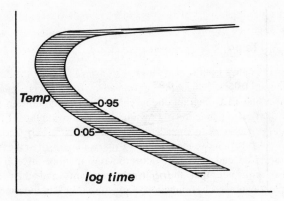

Figure 4   Typical TTT diagram for a transformation of discontinuous type which does not involve change in chemical composition.

new phase. The general factors already discussed will control this initial stage of the process. In most cases the nucleation step will dominate the TTT behavior of the system since, once the new phase has been nucleated, the reaction will proceed rapidly by a process of growth in which a reaction interface moves through the crystal.

The characteristic form of the TTT diagram for a reaction of this type, where the initial nucleation incident dominates the behavior of the transforming system, is shown systematically in Figure 4. Notice that on this diagram the TTT behavior is restricted to a relatively small time interval under isothermal conditions. The petrogenetic implications of the diagram are relatively obvious. For a wide range of possible cooling rates the final state of the system will correspond either to no observed transformation or to complete reaction. Only in a very narrow range of cooling rates is it possible to observe partial transformation behavior. This implies that where partial transformation microstructures are habitually observed in a reaction of this type, the rate-controlling parameter in the overall process is almost certain to be due to the initial nucleation of the new phase, in which case there is little possibility of utilizing data on the microstructure in a petrogenetic context.

From this brief analysis of the likely TTT behavior in this type of transformation it would seem that the petrologist is more likely to recover data on the characteristics of the nucleation incident from a study of associated microstructures, than specific data on the cooling rate.

TTT data are not restricted in application to situations involving cooling. Where a polymorphic transformation proceeds with rising temperature, a TTT diagram can also be drawn. For small values of ΔT above the transformation temperature the rates of transformation will again be dominated by the overall change in free energy in the process. However, with increasing temperature the rate of transformation will generally increase due to the temperature dependence of the activation energy for the process of reaction at the interface. The general characteristics of the

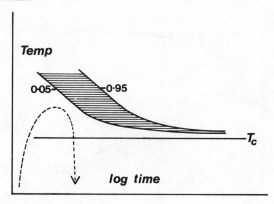

Figure 5 TTT diagram for a transformation of polymorphic type associated with rising temperature. Note that in the heating and cooling cycle marked in with a dotted line the temperature exceeds the transformation temperature but the transformation does not occur.

TTT diagram associated with rising temperature have been illustrated in Figure 5. The implications of this diagram are of general interest to the field of regional and thermal metamorphism. Where a heating cycle of relatively short period occurs, as indicated by the dotted curve in Figure 5, no evidence of transformation will be observed. Only in a long-term heating and cooling cycle is it likely that the transformation will take place under near-equilibrium conditions. The application of transformation process theory to the behavior of mineral systems in prograde thermal and regional metamorphism is likely ultimately to provide information on thermal history but, as far as I am aware, no exhaustive studies in such terms have yet been made.

Microstructures Formed in Discontinuous Transformations With Volume Diffusion

Transformations of this type, particularly where at low temperature the rates of transformation are governed by the kinetics of volume diffusion, are likely in the future to provide the petrologist with a great deal of information on time-temperature and other aspects of petrogenesis. Examples are fairly common in minerals and include exsolution as the most important case.

In applying transformation process theory to reactions in this group it is convenient to consider separately the initial nucleation incident, and the subsequent development of the microstructure under conditions where the dominant rate-controlling factor is defined by volume diffusion. These two aspects of the overall transformation process correspond, respectively, to a quasi-equilibrium stage during nucleation, in which the overall free-energy change in the reaction is important, and a strongly irreversible stage, in which the diffusion process is dominant in defining the rates of transformation.

Although a detailed analysis of nucleation phenomena at very slow rates of cooling, and thus under quasi-equilibrium conditions, does not appear to have been made, at least in relation to mineral microstructures, the main factors operating under such conditions can readily be evaluated. As anticipated in the previous general discussion of nucleation phenomena, the probability of a nucleus appearing under conditions of slight supercooling is likely to be controlled primarily by the time scale of equilibration. This condition is likely to obtain whether or not the actual process takes place by a homogeneous nucleation mechanism, providing the degree of super-cooling is not excessive. Once a nucleus is formed under such quasi-equilibrium conditions its growth is necessarily dependent on the overall change in free energy in the process and the operative diffusion constants which define the rate of supply of material to the growing nucleus. For very slow growth rates under quasi-equilibrium conditions, the free-energy term rather than the diffusion constants is likely to dominate the observed behavior and there will be ample time for the development of a local zone of depletion around the developing nucleus. In these circumstances the number of nuclei actually formed will be a function of the cooling rate and the operative diffusion constant. The spatial density of the nuclei will be governed by a relationship of the form $x \approx (Dt)^{1/2}$ where x is the separation of the nuclei, D is the effective diffusion constant at the temperature concerned, and t a measure of the time scale involved in the cooling cycle.

Where nucleation operates in this way under quasi-equilibrium conditions it is possible to use the scale of the initial nucleation process itself as a measure of the rate of cooling provided that the operative diffusion constants are known. Examination of exsolution microstructures in minerals endorses this conclusion. The situation is particularly well illustrated by exsolution in the pyroxenes where, for extremely slow cooling rates, relatively few, and widely spaced, exsolution lamellae are present. Although it is normal for petrologist to use such evidence in a qualitative way in the interpretation of cooling rates, it is probably not generally appreciated that the actual scale of the exsolution microstructure is almost certainly related quantitatively to D and the rate of cooling. In circumstances which are otherwise the same, a change in one order of magnitude of the scale of the exsolution micro-structure is equivalent to a change in cooling rate by two orders of magnitude, in terms of the simple relationship $x \approx (Dt)^{1/2}$. Unfortunately, although a great deal of information exists on the mechanism of exsolution in the pyroxenes, no definitive data on the relevant diffusion processes are available, and a more detailed analysis of the situation cannot yet be made.

Where the initial nucleation incident occurs under quasi-equilibrium conditions, the spatial characteristics of the microstructure as a whole are already defined at the nucleation stage. Thus the subsequent behavior of the system will be determined within the constraints exercised by this primary nucleation process. Within these constraints the later behavior of the system will now be determined by the details of the equilibrium phase diagram, the temperature dependence of the relevant diffusion coefficients, and the rate of cooling. Of these three factors the temperature dependence of the diffusion coefficients is likely to have the most important effect, particularly where the activation energy is large. In these circumstances and at constant cooling rate, the range of diffusion, x, as set initially in the nucleation stage, is likely to decrease exponentially with falling temperature according to the simple relationship $x \propto \exp -E_a/2RT$. Where the activation energy for the diffusion process, E_a, is very large, volume diffusion will rapidly become the rate-controlling factor in the growth of the exsolution microstructure. This is likely to lead to stranded diffusion profiles when the scale factor $(Dt)^{1/2}$ becomes appreciably smaller than the initial separation of the exsolved lamellae. Stranded profiles produced in this way due to the temperature dependence of the diffusion coefficients have been observed in the pyroxenes by Lorimer & Champness (1973). Where this effect obtains at a relatively early stage in the overall evolution of the microstructure it is possible that a second generation of exsolution lamellae may appear, and in this case their separation may again define the relevant scale factor appropriate to the lower temperature. In the pyroxenes as studied by Champness & Lorimer (1973), late-stage nucleation of this kind is observed at subgrain boundaries and dislocations. The final episode in the evolution of the pyroxene microstructure which they have observed within the stranded profile was late-stage coherent, Guinier-Preston-type zones which represent local segregation of Ca ions in the orthopyroxene host. At this stage in the evolution of microstructure a true nucleation and growth mechanism could not operate. The rate law for the growth of lamellae by diffusion, where the diffusion profiles for adjacent lamellae do not impinge, is given by the relationship $x = kt^{1/2}$, where x is the width of the lamellae, t the time, and k an isothermal rate constant. The rate of growth of the lamellae is therefore $dx/dt = \frac{1}{2}kt^{-1/2}$ for growth under isothermal conditions. The influence of the temperature dependence of the diffusion process decreases the growth rate very rapidly with falling temperature where the activation energy for diffusion is large, and most of the growth will take place in a small temperature interval close to the original temperature of nucleation. For example, where the activation energy is 60 kcal mole^{-1}, the scale factor for growth, $(Dt)^{1/2}$, will decrease by one order of magnitude in a temperature interval of 200°C below a nucleation temperature of 1000°C.

It has been noted above that the primary scale of the microstructure, as defined by the original nucleation incident under quasi-equilibrium conditions, can be used in conjunction with the relevant diffusion coefficient to define the probable rate of cooling. It is also possible to obtain data on the cooling rate by analyzing the stranded diffusion profile developed in the later stages of the evolution of the microstructure. Both techniques have been illustrated in the determination of the cooling rates for meteorites recently, and this particular application of the technique is described in the final section of this review.

This brief analysis of the main factors governing the evolution of microstructure in transformations such as exsolution, where diffusion is likely to operate as the rate-controlling parameter, indicates that a great deal of information may be obtained from a careful study of microstructure, providing critical data on the diffusion coefficients and their temperature dependence exist. Unfortunately, little data of this kind are available for any of the important mineral systems. It is to be hoped that this situation may be rectified in the near future. The potential advantage of the application of these principles to petrogenesis can hardly be overestimated.

Continuous Transformation Processes

The definition of continuous transformation behavior given earlier implies that in certain circumstances where the nucleation and growth mechanism associated with a discontinuous transformation does not operate, or is extremely sluggish, transformation behavior of continuous type is likely to be present. Because there are a large number of very sluggish discontinuous transformations known in mineral systems, the study of alternative continuous behavior is particularly important in the present context on at least two counts: 1. the factors that control these extremely sluggish nucleation and growth mechanisms can only reasonably be understood in terms of the study of the alternative behavior; and 2. where the nucleation mechanism renders the true equilibrium state totally inaccessible, or at least highly improbable, it is necessary to use TTT data for the alternative continuous behavior in a petrogenetic context. The characteristics of continuous behavior are such that the transformation process is governed by thermodynamics and the relevant diffusion mechanism. Hence, TTT data can be defined precisely and used with confidence in studying time-temperature aspects of the origin of the microstructures observed.

The time scale for the development of continuous transformation microstructure is likely to be related to that associated with the corresponding true nucleation and growth mechanism. It is true that spinodal behavior as observed in alloys is a particularly short-term effect, but this need not be true in general. The time scale for spinodal and other continuous behavior is determined, in the limit, by the magnitude of the diffusion constants involved. Where diffusion constants are very small the whole continuous behavior pattern will be shifted along the log time axis of the TTT diagram by a corresponding number of orders of magnitude. In general, the existence of extremely sluggish discontinuous transformations in natural mineral systems is likely to be paralleled by the existence of relatively slow continuous transformation processes. This point illustrates one of the important differences between mineralogical and metallurgical systems.

The second point at which these two types of systems diverge relates to differences in both structure and behavior, which have particularly important implications for the characteristics of continuous transformation behavior in the two cases. True spinodal behavior in alloy systems is extremely transitory since the boundaries between regions of different composition and lattice constants readily become semicoherent due to the introduction of dislocations. Thus spinodal behavior in alloy systems is normally the prelude to true nucleation and growth of separate crystalline phases. For this reason such behavior, when observed in alloy systems, is always referred to as spinodal decomposition. The situation is otherwise in the corresponding behavior in silicate systems and particularly in the feldspars, which have a fully three-dimensional framework structure of linked $(Si, Al)O_4$ tetrahedra. Similar loss of coherency here is extremely unlikely since the energy which would be associated with the introduction of edge dislocations in this context is prohibitively large (Brown & Willaime 1974). Constraints of this nature, which require that the structure should remain fully coherent in the presence of fluctuations of either order-parameter or composition, and hence remain as a true single crystal, are particularly important and may be explained further as follows.

In true spinodal behavior where the lattice remains fully coherent, the mismatch in dimensions between regions of different chemical composition involves strain energy. In a fully coherent system, this is a function of the volume fractions of the regions of divergent chemical composition and is independent of the wavelength of the microstructure, at least in the case of a simple lamellar system. However, the volume strain-energy term increases as the square of the amplitude of the chemical fluctuation, and hence builds up with the growth and development of the spinodal microstructure. In alloy systems, as the fluctuation wavelength increases gradually with time under isothermal conditions, a stage is reached at which it is more economical in energy terms for the system to lose coherence through the introduction of edge dislocations at the interface between regions of different composition and lattice constants, as illustrated in Figure 6. This situation is unlikely to occur in the feldspars until the wavelength of the fluctuations is very large indeed. In consequence, in these structures true spinodal or other fully coherent behavior may be observed in relatively coarse microstructures developed over a very long time interval. Energetic constraints which require that feldspar and similar structures remain fully coherent are an important feature of a number of continuous transformations in mineral systems for which there are not necessarily analogues in alloy systems. Indeed, while the basic theory of such processes is common to both mineral and metal systems, there is not necessarily a close relationship between the actual behavior patterns observed in the two cases.

Behavioral constraints related to coherency are probably less important in the nonframework silicate structures. Available evidence suggests that, in the case of the pyroxenes, the possibility of semicoherent boundaries associated with dislocations

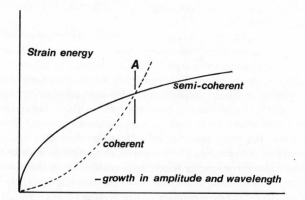

Figure 6   Schematic relationship between coherent strain energy and energy associated with semicoherent boundaries in the development of spinodal structure. At a certain stage in the development of the spinodal microstructure it is economical in energy terms to switch from a fully coherent condition to one involving boundary dislocations.

leads to true nucleation and growth phenomena at an early stage in continuous transformation processes in these minerals. The existence of boundary dislocation networks has been noted by Champness & Lorimer (1973), and evidence of spinodal behavior in the pigeonite from a lunar basalt has been presented by Champness & Lorimer (1971).

Thus it would seem that the detailed aspects of continuous behavior in different mineral systems is likely to be a function of the structure and the nature of possible defects. It should also be clear at this point that the factors governing this behavior are likely to be rather closely related to those involved in the corresponding processes of true nucleation and growth. For example, where a very strong coherency condition can be demonstrated from the study of continuous behavior, it is likely that the same factor will operate to limit the possibilities of nucleation in the corresponding discontinuous transformation process. In this context the constraints exercised by coherency are by no means the only important factors involved. Symmetry constraints are likely to be equally important. An account of the role of symmetry in this context has been given by McConnell (1971). Where constraints due to lattice coherency and symmetry are both involved, an extremely sluggish discontinuous transformation process can be anticipated. This situation is rather well illustrated by the monoclinic to triclinic transformation in potassium feldspar. The analysis provides a good example of the application of transformation process theory and goes some way towards explaining not only why this discontinuous transformation is extremely sluggish, but also how the alternative continuous behavior observed in adularia and orthoclase may be linked directly, in TTT terms, with normal spinodal behavior through the application of the general theory of continuous processes. The analysis also illustrates the sequence of steps which must be taken in assessing the potential of observed behavior and related microstructure in a petrogenetic context.

From crystallographic investigation the transformation in potassium feldspar is known to involve the ordering of Si and Al and, in consequence, the process must ultimately be controlled by a diffusive mechanism. The symmetry change from monoclinic to triclinic cannot take place by a truly continuous transformation process since the selection of ordered sites for Si and Al in the low temperature triclinic structure of microcline involves a choice of two alternative, and symmetry related, ordering schemes. This implies that a process of nucleation and growth of regions with degenerate symmetry must operate if the transformation proceeds under conditions close to true equilibrium. The question of whether or not the transformation in ideal circumstances is actually of first or second order is academic, since in practice the transformation must behave as though it were of first order. Thus the transformation in real terms is likely to involve a discontinuity associated with the development of regions which necessarily coexist and are ordered on either of the two possible symmetry related ordering schemes. The discontinuous transformation behavior therefore involves nucleation and growth of these alternatives. At this point an account of the study of the microstructures of adularia and orthoclase by electron microscopy is relevant (McConnell 1965, Nissen 1967). In both phases, while the overall symmetry remains monoclinic, it is possible to show that a distortion pattern exists in the crystal which is compatible with the intimate coexistence of the

146   McCONNELL

two ordering alternatives defined above. In adularia, the scale of the resulting microstructure for coexistence of these two types of ordered regions has a wavelength of approximately 100 Å. Thus in adularia what may certainly be described as a continuous transformation process results in the coexistence, on a very fine scale, of the two alternative types of embryo which, if they were given the opportunity, would ultimately develop as nuclei of the triclinic low temperature structure. In practice, however, a number of factors act to prevent the development of true nuclei of the low temperature phase. Ordering on the two alternative schemes is necessarily symmetry related and involves lattice distortions which are opposite in sense. Thus the growth of the embryo regions interfere with one another, and a balance is achieved in terms of the scale of the microstructure between the size of the embryos and the related volume strain and the extent of local order within them. A very strong condition of lattice coherency implies that the whole pattern of behavior must remain that of a true single crystal, and hence a metastable equilibrium state results in which the embryo regions can grow no larger (McConnell 1965). It is possible at this point to show how this continuous transformation behavior in metastable potassium feldspar structures fits in with the general TTT analysis given above.

The general aspects of spinodal theory as applied to fluctuations in chemical composition may readily be applied to continuous transformation behavior involving ordering, as follows. In the ordering case the fluctuations may be defined in terms of a local order parameter, and the free-energy reduction can be associated with ordering. The strain-energy term in conventional spinodal theory is here associated with the coexistence of ordered regions which distort in different senses. The gradient energy term rises also in exactly the same manner as in the normal spinodal treatment. Thus a very close analogy exists between continuous transformation behavior whether it depends on exsolution or ordering. One point of divergence should be noted, however. In dealing with the strain energy of coexistence of the symmetry related ordering alternatives in adularia, it is evident that at certain temperatures the development of the continuous microstructure may lead to a true free-energy minimum characterized by a very definite value for the wavelength of the fluctuations in order-parameter. In short, the development of a coarse transformation microstructure, as observed in microcline, still ultimately involves a discontinuous nucleation and growth process. In general, the presence of symmetry constraints associated with ordering provide a more precise control on the metastable equilibrium state than is the case in spinodal behavior involving chemical fluctuations.

Finally it remains to consider the probable form of the TTT diagram for discontinuous and continuous transformation behavior associated with the monoclinic-triclinic transformation in potassium feldspar.

Symmetry aspects of the problem and the related low diffusivity of Al and Si imply that the process of nucleation and growth in this transformation is possible only on an inordinately extended time scale at temperatures close to the true transformation temperature. On the other hand, continuous behavior is predictable. Relatively long wavelength fluctuations should be observed relatively close to the transformation temperature, but these are likely to develop very slowly. With larger values of undercooling, as in true spinodal behavior, there is a tendency to move both towards shorter wavelengths and faster times for the development of the microstructure.

This analysis of the transformation behavior in potassium feldspar has certain immediate petrogenetic implications. Where potassium feldspar crystallizes at high temperature and is slowly cooled through the transformation temperature, the appropriate path on the TTT diagram involves a period of ordering under monoclinic symmetry prior to reaching the monoclinic-triclinic transformation temperature. Subsequently, on crossing the inversion temperature, the monoclinic phase traverses a region of TTT space where a very wide range of possible behavior is demanded in terms of both wavelengths and amplitudes for the continuous transformation microstructure. The response of the potassium feldspar crystal to this wide variety of influences is likely to mean that no very specific wavelength is likely to be favored, and indeed the actual development of the low temperature order fluctuations appropriate to continuous type behavior may be inhibited. The resulting phase is orthoclase.

The situation is quite otherwise for potassium feldspar which forms under isothermal conditions at low temperature. In this case isothermal treatment is likely to favor one particular wavelength and a characteristic microstructure is likely to develop. This situation accords with the genesis of adularia which is known to have formed at relatively low temperatures and shows a very well-developed fluctuation microstructure.

This brief analysis of the factors governing both nucleation and alternative behavior in mineral systems where the process of transformation is sluggish indicates that in these circumstances information of direct petrogenetic significance can be deduced from the microstructure.

Another example of the application of this general theory relates to behavior in the plagioclase feldspars where the possibility of alternative behavior can be specified in terms of a combination of both ordering and compositional fluctuations. The theory has been used recently to explain both the development of antiphase domain structures in the plagioclases and the existence of quasi-exsolution effects in the peristerites, the labradorites, and the bytownites (McConnell 1974). Here it is sufficient to illustrate the effects observed in the bytownites since the same principles can be applied elsewhere in the plagioclase system.

It has long been known that in slowly cooled plagioclase in the composition range around 75% An, two types of behavior are possible (Bown & Gay 1958): 1. plagioclases of this composition may develop a true superlattice with b-type maxima indicating a doubling of the simple albite cell in the c direction; 2. plagioclases of the same composition may develop an antiphase structure in which the characteristic additional diffraction effects comprise pairs of maxima symmetrically disposed about the b-superlattice positions. Thus it is necessary to decide which state represents stable, and which metastable, behavior. The evidence (McConnell 1974) implies that the antiphase ordering scheme represents metastable behavior that occurs under conditions of supercooling. In the present context it can be identified with continuous behavior caused by the difficulty involved in the nucleation and growth of large regions of the ordered structure in single phase.

Because the true single phase state is not always observed in materials of this composition, and the ease with which the same state develops is clearly a function of chemical composition, it is possible to visualize behavior in this region by using a TTT diagram which illustrates the effect of composition on the rate of development of the single phase state. The rates of single phase ordering as a function of chemical composition have been shown schematically in Figure 7 for a given time section of the complete TTTX situation.

Clearly, where for a composition such as *A* on the diagram the transformation behavior misses the nose of the ordering region, no effects will be observed. For composition *B*, on the other hand, ordering on the basis of the true superlattice will be observed. An interesting possibility exists for compositions which in cooling pass close to the nose of the ordering region on an appropriate time scale of cooling. In this case fluctuations in chemical composition will allow ordering in the An-rich regions, and such fluctuations should develop spontaneously due to the associated reduction in free energy on ordering. In principle in a situation of this kind a spinodal effect will be observed due to the development of order-dependent compositional fluctuations.

This analysis of the situation accords with the evidence observed in certain plagioclases of composition 75% An from the Stillwater complex where a spinodal-type microstructure is present. Here the fluctuations have a wavelength of approximately 200 Å and show intimately associated regions that order on either the single lattice or an antiphase basis (McConnell 1974, McLaren 1974, Nord, Heuer & Lally 1974). Although order-dependent chemical fluctuations of spinodal type in such material can readily be explained in the above terms, the detailed analysis of the effects depends on an appreciation of the role of water vapor in defining the effective rate of Al-Si ordering. This aspect is discussed in greater detail later.

It is now obvious that a detailed analysis of transformation behavior of continuous type is of fundamental significance in studying the mechanism of sluggish transformations in minerals. In summary, the petrogenetic implications of continuous transformation processes are as follows:

(*a*) An appreciation of the mechanisms and kinetics of such behavior helps substantially with the understanding of the corresponding discontinuous transforma-

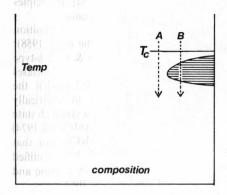

Figure 7 Effects of composition on ordering rate. The diagram represents a time section of a complete TTTX situation. For the rate of cooling specified in this section, specimen *A* shows no development of the ordered structure. Specimen *B* passes through a temperature-composition region where ordering is possible on this time scale.

tions which they replace, since in many cases the characteristics of the microstructure of such stranded states provide an insight into the reasons for extremely sluggish nucleation and growth processes. The petrologist should be conversant with these principles in discussing time-temperature aspects of such processes in petrogenesis.

(b) In many cases the experimental study of continuous behavior is possible where similar study of nucleation and growth phenomena are entirely out of the question from the point of view of the laboratory time scale. Where, for example, one is able to study spinodal behavior which occurs as an alternative to true exsolution, it is possible to use the data to define the likely maximum degree of undercooling associated with an observed nucleation and growth mechanism in nature. Knowledge of the strain-energy term, which defines the depression of the coherent spinodal, also permits one to define the position of the true solvus where this is inaccessible in direct experiments involving a nucleation and growth mechanism.

(c) Where TTT data are available on the development of the spinodal micro-structure these will normally define the rate-controlling diffusion process operative in the transformation. These rate data may then be applied directly to the analysis of the rate of growth of nuclei, once formed, in the corresponding discontinuous process, i.e. to extrapolate for the growth process in true exsolution to time values which are inaccessible in laboratory experiments. The pressure dependence of the transformation and of its diffusive mechanism may also be studied.

(d) Finally, microstructure observed in a natural continuous transformation process may be used to define time-temperature and other aspects of petrogenesis directly.

APPLICATION OF TRANSFORMATION PROCESS THEORY

The final section of this review deals with several specific examples of the application of transformation process theory to the petrogenetic aspects of the origin of mineral microstructures.

TTT Behavior in Fe-Ni Meteorites

Probably the most complete study of time-temperature and other aspects of the development of microstructure in minerals is provided by the recent study by Goldstein & Short (1967) on the microstructure of meteorites as indicators of the rate of cooling of the parent extraterrestrial body.

In the system Fe-Ni the low temperature behavior is defined by an equilibrium phase diagram characterized by the existence of an inversion interval which relates to the γ to α transformation in pure iron. The nature of the transformation behavior in this system has been known for a long time and leads to the characteristic Widmannstätten structures associated with the exsolution of the α phase, kamacite, on $\{111\}$ planes of the parent γ phase, taenite. The development of plates of kamacite in the taenite host in slowly cooled meteorites involves the diffusion of both Ni and Fe. Values for the compositionally dependent diffusion constants for Ni and Fe in this system, together with their temperature dependence, were determined by Goldstein, Hanneman & Ogilvie (1965).

In their study of the development of the microstructure in Fe-Ni meteorites, Goldstein & Short (1967) utilized the known diffusion coefficients and their temperature and composition dependence to compute the stranded chemical diffusion profiles associated with different extraterrestrial cooling rates. These were compared with the actual stranded diffusion profiles obtained by direct electron probe analysis of selected meteorites. In this way Goldstein and Short succeeded in showing that the cooling rates in the meteorites examined by them varied between 0.4 and 40°C/10^6 yr. Apart from deducing a value of approximately 100°C for under-cooling prior to nucleation, Goldstein and Short did not analyze the nucleation event itself. The calculations they carried out therefore assumed that a kamacite lamella grew in the host taenite without interference from neighboring lamellae. In a later paper Short & Goldstein (1967) pointed out that the cooling rate in Fe-Ni meteorites could be directly determined from the final kamacite plate thickness which they related directly to rate of cooling over a wide range of meteorite compositions. The analysis of meteorite microstructures in terms of cooling rates by Goldstein and Short provides an invaluable introduction to this field of application of transformation process theory, and their papers should be studied carefully by anyone who is even marginally interested in the petrogenetic implications of mineral microstructures.

TTT Behavior in the Alkali Feldspars

The second example of the application of transformation process theory to the development of mineral microstructures deals with the observation of spinodal behavior in short-term experiments on the alkali feldspars. In these experiments, the spinodal microstructures were introduced in isothermal runs and were examined

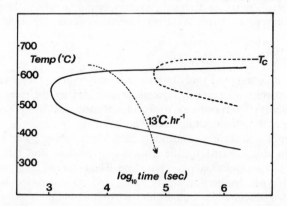

Figure 8 Determined TTT diagram for the development of spinodal and true exsolution microstructures in an alkali feldspar of composition 37 wt% KAlSi$_3$O$_8$. Spinodal micro-structures were observed within the full curve. Evidence of exsolution by a nucleation and growth mechanism was observed within the dashed curve. The lightly dashed line represents a cooling rate of 13°C hr^{-1}.

with the electron microscope by diffraction and direct resolution techniques (Owen & McConnell 1974).

The alkali feldspar used for these experiments had composition 37 wt% $KAlSi_3O_8$ and hence lay close in composition to the crest of the solvus for the alkali feldspar system. The development of spinodal microstructure in homogenized material was observed in the temperature interval between 350 and 630°C. At the highest temperatures the wavelength of the chemical fluctuations was large (300 Å) and the structure developed relatively slowly in accordance with the spinodal theory already outlined. In isothermal experiments at lower temperature the spinodal microstructure developed more rapidly and was characterized by a shorter wavelength. At the lowest temperatures at which the spinodal microstructure was observed the process was very slow and hence primarily rate controlled by the diffusion of the alkali ions. A TTT plot of the data for the development of spinodal microstructure in this alkali feldspar is provided in Figure 8. Evidence for spinodal behavior was observed within the full line on this diagram.

Diffraction study of the alkali feldspars in which spinodal behavior was detected invariably showed that the spinodal microstructure was associated with a single crystal diffraction pattern in which additional intensity was convoluted with the primary reciprocal lattice array. In certain experiments where the spinodal microstructure was allowed to continue to develop over long periods of time in the temperature range around 500°C this single lattice condition was maintained, implying that full lattice coherence is an important aspect of continuous transformation behavior in the feldspars. Thus the spinodal behavior observed in the alkali feldspars differs appreciably from that occurring in alloy systems where such behavior leads rapidly to true decomposition into two discrete phases.

At the highest temperatures used in isothermal treatment of this alkali feldspar, a second type of behavior was also observed. This was characterized by the presence of exsolved albite lamellae which showed characteristic multiple albite twinning, and yielded a second suite of diffraction maxima in the diffraction pattern. This true nucleation and growth mechanism for the exsolution of albite was also observed alone above the limiting spinodal temperature in the interval 630–650°C. The relevant data for this true nucleation and growth transformation have been inserted on the TTT diagram of Figure 8. Within the region outlined by the thick dashed line, evidence of true nucleation was observed in transmission studies in the electron microscope and in the corresponding electron diffraction pattern.

The TTT diagram derived experimentally for this alkali feldspar can be used directly to define the conditions associated with the development of similar spinodal microstructure in natural specimens. Electron-optical study of the feldspar specimen used in these experiments in its natural state (McConnell 1969) indicates that it had a well-developed spinodal microstructure with a scale of approximately 100 Å. Evidence of true nucleation and growth was not observed. These observations are consistent with a rate of cooling corresponding to the range of cooling rates over which the spinodal behavior can develop in the absence of a true nucleation and growth mechanism. From the evidence provided by the scale of the spinodal microstructure, the rate of cooling of the natural specimen was estimated as

$13° \pm 5°C \ hr^{-1}$ over the interval 600–400°C. Figure 8 also indicates that a cooling rate of $3°C \ hr^{-1}$ would have been sufficiently slow to produce ample evidence of true nucleation and growth. This particular alkali feldspar specimen was separated from a hand specimen of a hyalo-pantellerite from Pantelleria. Unfortunately there is no record of the thickness of the original lava flow from which it was collected.

Apart from their use in determining cooling rates in the development of spinodal microstructure in natural alkali feldspars, the data obtained from the experimental study of these microstructures can be used in a more general way to provide relevant information on longer-term equilibration in this system.

The lower limb of the TTT diagram of Figure 8 may be used to determine the activation energy for the rate-controlling diffusion of alkali ions during the development of spinodal microstructure at low temperatures by plotting the reciprocal of the time necessary for the appearance of the spinodal microstructure as a function of $T(K)^{-1}$. The observed rates must be adjusted by a factor $T/(T_s - T)$ where T_s is the spinodal temperature. This term arises because the free-energy drive for the process increases with increased supercooling (Hilliard 1970). On applying this correction to the rate data available on the alkali feldspar spinodal, an activation energy of 61 kcal $mole^{-1}$ was obtained. An approximate value for D, the effective rate-controlling diffusion constant for the development of the spinodal microstructure, may also be derived as follows. At 420°C a strong spinodal fluctuation with a wavelength of 90 Å was produced in a time interval of the order of four days. Using the simple relationship $x \approx (Dt)^{1/2}$, the value of D calculates at approximately $10^{-18} \ cm^2 \ sec^{-1}$.

Recently data on the self-diffusion of both Na and K in the alkali feldspar structure have been presented by Lin & Yund (1972). Their data indicate that the activation energy for self-diffusion of Na and K are, respectively, 19 and 70 kcal $mole^{-1}$. Extrapolation of their data to 420°C also indicates that the self-diffusion coefficient for Na is $\sim 10^{-11} \ cm^2 \ sec^{-1}$ and of K, $10^{-20} \ cm^2 \ sec^{-1}$ at this temperature. Comparison of these data with those actually derived above for the spinodal structure in the alkali feldspars indicates that the migration of K constitutes the rate-controlling step in the development of the microstructure. If the data on self-diffusion of Na were relevant alone in this context, the spinodal microstructure would develop at the same temperature in time of the order of seconds.

The overall rate-controlling factor in the development of spinodal microstructure, and presumably exsolution by nucleation and growth, in the alkali feldspars is apparently directly dependent on the self-diffusion constant for the potassium ion alone. In the corresponding situation in alloys (Hilliard 1970) the effective diffusion constant, D, depends on the diffusion constants for both migrating species since diffusion in metals operates through a vacancy mechanism and a flux of vacancies exists to balance the unequal migration of the two primary diffusing species. Whereas the ultimate mechanism of local diffusion of the cations in a silicate structure may involve vacancies, the necessity for local charge balance does not permit a vacancy flux when both species have the same valency.

Order-Dependent Spinodal Behavior in Certain Plagioclase Feldspars

The final example chosen here to illustrate the application of transformation process theory to mineral microstructure has been selected primarily to demonstrate that factors other than time and temperature can be important in defining mineral behavior. Where the time and temperature conditions are known, the characteristics of the microstructure may be used to define other aspects of the environment in which the reaction has taken place.

In the first instance it is necessary to appreciate that the scale of the microstructure observed in a diffusion-controlled reaction is governed by the operative diffusion constants for the process. Thus an extremely fine-scale microstructure may result even in long-term equilibration at high temperatures where the diffusion process is extremely slow.

A fine-scale spinodal microstructure with these characteristics is observed in plagioclase feldspars of composition 75% An from the Upper Gabbro of the Stillwater complex (McConnell 1974; see also McLaren 1974; Nord, Heuer & Lally 1974), and electron-optical study of the microstructure has shown that fluctuations in local chemical composition are related to fluctuations in the characteristics of the Al-Si ordering pattern. Regions characterized by the presence of true superlattice maxima (*b* maxima) occur in intimate association with regions showing antiphase behavior on a scale of approximately 200 Å. It is quite reasonable to assume that in this case the microstructure developed at a temperature of the order of 1000°C over an extended period of time. Theory appropriate to this type of order-dependent spinodal behavior has been presented earlier. Here the petrogenetic implications of the existence of this microstructure are discussed.

Unlike the situation in the alkali feldspars, processes of exsolution or quasi-exsolution in the plagioclase feldspars involve the migration of Si and Al over appropriate distances in the crystal. Such processes are known to be extremely sluggish under dry conditions in the laboratory but are found to be much faster under hydrothermal conditions.

The first step in the analysis of the situation observed in the plagioclase from the Stillwater complex involves an order of magnitude calculation of the operative diffusion constant for the development of the order-dependent spinodal microstructure associated with Al-Si diffusion. Using the observed wavelength of 200 Å, and a conservative estimate of 1000 yr for the time of cooling associated with the development of the microstructure, the effective diffusion constant D calculates as 10^{-23} cm^2 sec^{-1} using the simple relationship $x \approx (Dt)^{1/2}$. Hydrothermal treatment of this plagioclase from the Stillwater complex for periods of the order of weeks at temperatures above 900°C results in the homogenization of the sample and the development of a true superlattice throughout the specimen (McConnell 1974). From the hydrothermal experiments it may be deduced that the original spinodal behavior must have developed under quite different physical conditions during the history of the Stillwater complex, and that the most important contributory factor in the formation of the spinodal microstructure was probably the very low rates of diffusion of Al and Si under rather dry conditions.

To justify this conclusion, some additional order-of-magnitude data on the diffusion of Si and Al in feldspars under both wet and dry conditions are required. Data with adequate accuracy may be obtained from the results of ordering and disordering experiments on albite. Assuming next nearest neighbor exchange of Al and Si as the mechanism of the disordering process in albite at high temperatures, data of McKie & McConnell (1963) may be used to define an approximate value of the diffusion constant for the dry disordering process at 1000°C where the time appropriate to complete disorder is approximately 600 hr. Using a nearest neighbor exchange distance of 3 Å the effective diffusion constant calculates as 10^{-22} cm^2 sec^{-1} which is of the same order of magnitude as the diffusion constant determined above for the development of the order-dependent spinodal microstructure.

It is also possible to derive an appropriate order of magnitude value for the effective diffusion constant for Al-Si exchange under wet conditions by utilizing the data of MacKenzie (1957) for the ordering of low albite. In this case the magnitude of the appropriate diffusion constant is likely to be underestimated because the ordering transformation is likely to have a low or negative value for the entropy of activation. MacKenzie's experiments indicate that the maximum rate of ordering at 500°C corresponds to the attainment of complete order in approximately 200 hr under hydrothermal conditions. The corresponding value of the effective diffusion constant, which must be an underestimate as noted above, calculates as 10^{-22} cm^2 sec^{-1}. Using the value of 60 kcal mole^{-1} for activation energy for the process, the effective diffusion constant for a similar process at 1000°C would be 10^{-15} cm^2 sec^{-1} or approximately seven orders of magnitude greater than the same process taking place under dry conditions.

In spite of the fact that the calculations presented are likely to be inaccurate, it is certainly possible to say that dry and wet rates for processes involving Al-Si diffusion differ enormously. Thus, in conclusion it would seem that the only logical solution to the problem of the origin of the order-dependent spinodal behavior in the Stillwater plagioclase is to assume that the microstructure developed under extremely dry conditions.

This final example of the application of transformation process theory indicates that where aspects of petrogenesis other than time-temperature are important, these factors can be deduced by application of the general theory.

CONCLUSION

From this brief and rather selective review of the implications of transformation process theory in the study of mineral microstructures, it is clear that a great deal of theoretical and practical work remains to be done, particularly in the realm of study of nonequilibrium behavior. The results already available, however, show quite clearly that in combination, theory and experiment can provide a great deal of information on time-temperature aspects of petrogenesis, particularly in mineral systems where nonequilibrium behavior is the rule rather than the exception. Currently two main aspects of the subject merit substantial research effort. In the first place, it is important to place the theory of nonequilibrium behavior on a sound theoretical footing, and in this context the exhaustive analysis of behavior in systems such as the plagioclase feldspars constitutes a real challenge. Second, in the practical application of such theoretical studies in petrogenesis there is an immediate need for much more abundant data on the rates of processes, particularly for data on diffusion rates.

In the long term such studies will not only give the petrologist an invaluable insight into the nature of mineral processes, but will also provide quantitative data on time-temperature and other aspects of the history of rocks and minerals from a wide range of igneous and metamorphic environments.

Literature Cited

Bown, M. G., Gay, P. 1958. The reciprocal lattice geometry of the plagioclase feldspar structures. Z. Kristallogr. 111:1–14
Brown, W. L., Willaime, C. 1974. An explanation of exsolution orientations and residual strain in cryptoperthites. The Feldspars, 440–59. Manchester: Manchester Univ. Press
Cahn, J. W. 1968. Spinodal decomposition. Trans. AIME 242:166–80
Champness, P. E., Lorimer, G. W. 1971. An electron microscopic study of a lunar pyroxene. Contrib. Mineral. Petrol. 33:171–83
Champness, P. E., Lorimer, G. W. 1973. Precipitation (exsolution) in an orthopyroxene. J. Mater. Sci. 8:467–74
Christian, J. W. 1965. The Theory of Transformations in Metals and Alloys. London: Pergamon. 973 pp.
Goldstein, J. I., Hanneman, R. E., Ogilvie, R. E. 1965. Diffusion in the Fe-Ni system at 1 Atm and 40 Kbar pressure. Trans. AIME 233:812–20
Goldstein, J. I., Short, J. M. 1967. Cooling rates of 27 iron and stony-iron meteorites. Geochim. Cosmochim. Acta 31:1001–23
Hilliard, J. E. 1970. Spinodal decomposition. Phase Transformations, 497–560. Metals Park, Ohio: Am. Soc. for Metals.
Lin, T. H., Yund, R. A. 1972. Potassium and sodium self diffusion in alkali feldspar. Contrib. Mineral. Petrol. 34:177–84
Lorimer, G. W., Champness, P. E. 1973. Combined electron microscopy and analysis of an orthopyroxene. Am. Mineral. 58:243–48
MacKenzie, W. S. 1957. The crystalline modifications of NaAlSi$_3$O$_8$. Am. J. Sci. 255:481–516
McConnell, J. D. C. 1965. Electron optical study of effects associated with partial inversion in a silicate phase. Phil. Mag. 11:1289–1301
McConnell, J. D. C. 1969. Electron optical study of incipient exsolution and inversion phenomena in the system NaAlSi$_3$O$_8$-KAlSi$_3$O$_8$. Phil. Mag. 19:221–29
McConnell, J. D. C. 1971. Electron-optical study of phase transformations. Mineral. Mag. 38:1–20
McConnell, J. D. C. 1974. Analysis of the time-temperature-transformation behaviour of the plagioclase feldspars. The Feldspars, 460–77. Manchester: Manchester Univ. Press
McKie, D., McConnell, J. D. C. 1963. The kinetics of the low-high transformation in albite. I. Amelia albite under dry conditions. Mineral. Mag. 33:581–88
McLaren, A. C. 1974. Transmission electron microscopy of the feldspars. The Feldspars, 378–423. Manchester: Manchester Univ. Press
Nissen, H.-U. 1967. Direct electron-microscope proof of domain texture in orthoclase. Contrib. Mineral. Petrol. 16:354–60
Nord, G. L., Heuer, A. H., Lally, J. S. 1974. Transmission electron microscopy of substructures in Stillwater bytownites. The Feldspars, 522–35. Manchester: Manchester Univ. Press
Owen, D. C., McConnell, J. D. C. 1974. Spinodal unmixing in an alkali feldspar. The Feldspars. 424–39. Manchester: Manchester Univ. Press
Russell, K. C. 1970. Nucleation in solids. Phase Transformations, 219–68. Metals Park, Ohio: Am. Soc. for Metals
Short, J. M., Goldstein, J. L. 1967. Rapid methods of determining cooling rates of iron and stony iron meteorites. Science 156:59–61
Yund, R. A., McCallister, R. H. 1970. Kinetics and mechanisms of exsolution. Chem. Geol. 6:5–30

ARRAY SEISMOLOGY   ×10036

John Filson
Lincoln Laboratory, Massachusetts Institute of Technology,
Lexington, Massachusetts 02173

INTRODUCTION

During the summer of 1958 a "Conference of Experts" was convened in Geneva "to study the possibility of detecting violations of a possible agreement on the suspension of nuclear tests." The report of this conference (Department of State 1960) concluded that seismic control posts should be established to include "approximately ten short-period vertical seismographs dispersed over a distance of about 1.5–3 km and connected to a recording system by lines of cable." This is one of the early references to seismic arrays in the context of the problem of identification of underground nuclear explosions, and it marks the beginning of the development of these arrays for specific application to that problem.

For the purposes of this paper, a seismic array is defined as a suite of seismometers with similar characteristics recording at a central station for the comparison of arrival times, amplitudes, and wave forms of seismic phases propagating within the earth. The term seismic network usually implies a larger areal distribution of instruments and thus lacks the central recording feature which facilitates the direct comparison of the data from individual recording sites. Through the use of array data and simple processing techniques, the relative size of most seismic signals, with respect to the ambient seismic noise within the earth, may be increased. This is the primary reason for array application to the nuclear identification problem. At present one or more seismic arrays exist in at least ten countries of the world; the approximate locations of these arrays are shown in Figure 1. Not all of these are maintained for nuclear explosion monitoring and identification research; however, the governmental support of many is directed toward that end. The names, locations, and configurations of the arrays discussed in this paper and the references are given in Table 1.

Regardless of the political incentives for the construction and maintenance of arrays, they have offered seismologists dense spatial samples of the seismic field at the various sites on the earth's surface. This data has led to the refinement of velocity models of the earth's interior, the first observations of phases predicted theoretically, detailed observational analysis of wave propagation phenomena, and controversies concerning certain anomalous features brought to light by the arrays. Like the effect of the widespread use of powerful telescopes on modern astronomy,

157

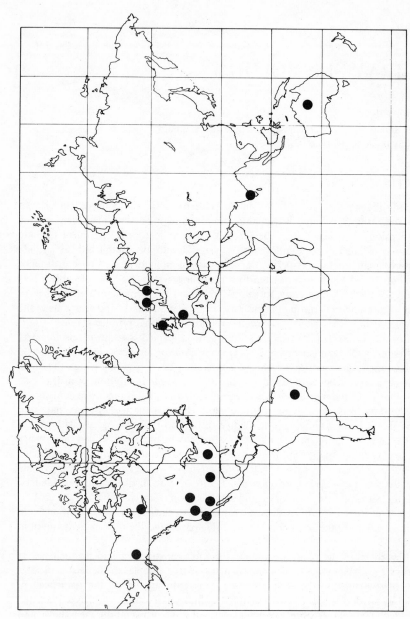

Figure 1   Locations of some of the arrays discussed in the text and listed in Table 1.

Table 1 The major seismic arrays[a]

Name	Location	Aperture and Instrumentation
Large Aperture Seismic Array LASA	46.7°N 106.2°W Eastern Montana	200 km, short and long period
Norwegian Seismic Array NORSAR	60.8°N 10.8°E Southern Norway	120 km, short and long period
Alaskan Long Period Array ALPA	65.2°N 147.7°W Central Alaska	80 km, long period
Blue Mountains Array BMO	44.8°N 117.3°W Western Oregon	4 km, short period
Cumberland Plateau Array CPO	35.6°N 85.6°W Eastern Tennessee	4 km, short period
Tonto Forest Array TFO	34.3°N 111.3°W Central Arizona	10 km, short period aperture extended temporarily in 1965 to 200 km
Uinta Basin Array UBO	40.3°N 109.6°W Northern Utah	4 km, short period
Variable Aperture Seismic Array VASA	53.1°N 113.3°W Central Alberta	160 km, short period Kanasewich et al (1973)
Gauribidanur Array GBA	13.6°N 77.4°E Southern India	20 km, "L"-shaped, short period, UK type
Yellowknife Array YKA	62.5°N 114.6°W Northwest Canada	UK type
Warramunga Array WRA	19.9°S 134.3°W Northern Australia	UK type
Eskdalemuir Array EKA	55.3°N 3.2°W Scotland	UK type
Brasilia Array BDF	15.7°S 47.9°W Central Brazil	UK type
Hagfors Array HFS	60.1°N 13.9°E Sweden	40 km, tripartite, short and long period
French Telemetered Array	Continental	350 km, short and long period, telemetered to Laboratorie de Geophysique, Paris
University of California Array	Western, central California	400 km, short period, telemetered to Berkeley
California Institute of Technology Array	Southern California	200 km, short period, telemetered to Pasadena
National Center for Earthquake Research Array (USGS)	Western California	400 km, short period, telemetered to Menlo Park
Wakayama Microearthquake Observatory	34.2°N 135.2°E Southern Honshu	170 km, short period, local recording. Used as an array by Kanamori (1967)

[a] The last five installations might be more accurately described as networks, but they can be or have been used as arrays.

the seismic arrays may have uncovered more questions than they have resolved. In this paper we hope only to demonstrate the ways in which arrays have been used as research tools in seismology and to point to some of these unresolved questions. Other recent reviews of array processing techniques (Capon 1973) and array applications (Davies 1973) are to be found in the literature.

THEORY

The Ray Parameter

As seismic waves propagate through the body of the earth, they eventually, because of the velocity structure within the earth, impinge upon the surface of the earth at some angle with respect to the vertical. The very simple law that predicts the passage of a seismic ray through the earth is given in terms of the ray parameter p where

$$p = R(\sin i)/v \qquad\qquad 1.$$

Here i is the angle the ray makes with the vertical and v the ray velocity at the radial distance R. The term $(\sin i)/v$ represents the apparent slowness (velocity^{-1}) of wave front across an imaginary geocentric spherical surface within the earth, and this slowness must appear the same to observers both directly above and below this surface (Snell's law). If this slowness is measured in central angle radians then distances along such imaginary surfaces must be normalized by the radial distance R. Thus p is constant along the entire ray path.

At the deepest point or bottom of the ray (R_b), where it turns back toward the surface,

$$p = R_b/v_b \qquad\qquad 2.$$

At the surface of the earth (R_0)

$$p = R_0 s \qquad\qquad 3.$$

where s is the apparent slowness of the ray measured along the surface R_0 by an array. Thus a measure of the slowness, s, is directly relatable to the velocity at the deepest point of ray penetration through the ratio R_b/R_0. A transformation of the expression for the depth R_b from an integral along the ray path to an integral along the surface (the Herglotz-Wiechert method; Bullen 1963, Båth 1968) can yield a seismic velocity distribution as a function of depth for the earth. The method is applicable only at depths where

$$(dv/dR) < (v/R) \qquad\qquad 4.$$

Of course the apparent slowness at the surface is just the slope of the time-distance curve, $dT/d\Delta$, for a particular seismic phase and versions of these curves were developed long before modern arrays were established. As Johnson (1967) points out: these curves are usually based solely on first arrivals; they are subject to error in source origin, time, and location; and they usually represent some smooth version of the original data. Thus seismic arrays have represented a new tool for making direct and detailed measurements of the ray parameter p measurements which may be inverted to yield velocity-depth distributions for the earth.

Array Processing

The vertical motion of a two-dimensional plane, represented as $v(t, x, y)$, may be decomposed into its frequency (ω) and wavenumber ($\mathbf{k}$) components by use of the three dimensional Fourier transform

$$V(\omega, k_x, k_y) = \int\int\int v(t, x, y)\exp\left[-i(\omega t - k_x x - k_y y)\right]\,dt\,dx\,dy \qquad 5.$$

where, in terms of the azimuthal angle θ, $k_x = |\mathbf{k}|\sin\theta$ and $k_y = |\mathbf{k}|\cos\theta$. If this motion is a monochromatic wave of unit amplitude propagating across the surface with a slowness $\mathbf{s}$ (now a vector), i.e.

$$v(t, x, y) = \exp\left[2\pi i(\omega_0 t - k_{0_x} x - k_{0_y} y)\right] \qquad 6.$$

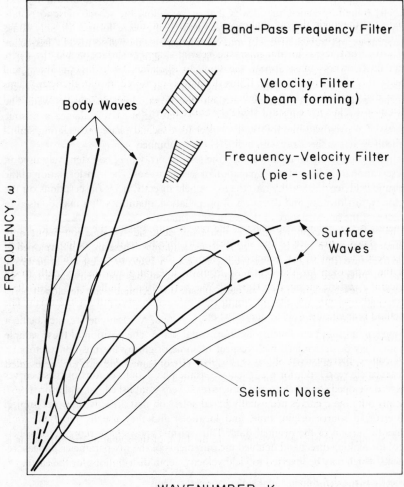

Figure 2 A schematic frequency-wavenumber ($\omega - k$) plane at a single azimuth showing the distribution of typical seismic phases and the effect of various filters.

where $\mathbf{k}_0 = \omega_0 \mathbf{s}$, then

$$V(\omega, k_x, k_y) = \delta(\omega - \omega_0, k_x - k_{0_x}, k_y - k_{0_y})$$ 7.

The spatiality of a seismic array, however finite and discrete, allows the estimation of the right side of equation 5 and thus an estimate of the wavenumber and slowness vectors, $\mathbf{k}$ and $\mathbf{s}$, of propagating signals and noise.

Figure 2 shows, in a schematic way, the properties of propagating seismic waves in an $\omega - \mathbf{k}$ plane. Here all types of waves, including the noise, are propagating in the same direction; however, the basic methods of signal enhancement through the frequency-wavenumber decomposition should be clear from the figure. Although not strictly done in practice, the idea is that arrays allow the seismologist to isolate any portion of this plane and then reconstitute the wave form with only those frequencies and wavenumbers of interest. The significance of terms like frequency band pass filter, velocity filter, and "pie-slice" filter should be evident from Figure 2. The reader must remember that this figure represents only one plane through a three-dimensional diagram, thus all of the filters that exploit the velocity differences of the various wave types may use their directional characteristics as well. The $\omega - \mathbf{k}$ diagram also suggests how an array may be used to enhance a seismic phase of a given wavenumber with respect to a second phase propagating simultaneously across the array with a different wavenumber.

In addition to signal enhancement, the seismic array may be extensively used in observational studies of the propagation of seismic waves. The transformation of the seismic field across an array into the $\omega - \mathbf{k}$ field directly yields observations on the velocity, or slowness, and direction of propagation, quantities that may be used in studies of the fine structure of the earth's interior.

In practice we cannot exactly evaluate the right side of equation 5 but only estimate it through various array processing techniques. These are well developed in the literature (Burg 1964, Kelly 1965, 1967, Capon 1969, 1973, Lacoss et al 1969), and the interested reader is directed toward these studies and their references. No extensive development of these rather sophisticated techniques are developed here; only a few simple concepts are set down.

The seismic array may be represented by a suite of N seismometers, each being a sample point in a two-dimensional horizontal plane. The position of these sample points may be determined with respect to some origin by the position vectors $\mathbf{r}_n$. As Kelly (1967) describes, the seismic signal, $v(t)$, of some bandwidth is represented at the origin in terms of its frequency spectrum as

$$v(t) = \int_{-\infty}^{\infty} V(\omega) \exp(i\omega t) d\omega$$ 8.

Assuming nondispersive propagation, the form of the signal at the nth sensor is

$$x_n(t) = v(t - \mathbf{s} \cdot \mathbf{r}_n)$$ 9.

where $\mathbf{s}$ is the two-dimensional slowness vector. Thus

$$x_n(t) = \int_{-\infty}^{\infty} V(\omega) \exp\left[i(\omega t - \mathbf{k} \cdot \mathbf{r}_n) d\omega\right]$$ 10.

where again $\mathbf{k} = \omega\mathbf{s}$. Now if the individual channels from the array are directly summed in the time domain, we obtain the zero-delay beam represented as

$$b(t) = N^{-1} \sum_{n=1}^{N} x_n(t) \qquad 11.$$

Now, using equation 10,

$$b(t) = \int_{-\infty}^{\infty} V(\omega)\exp(i\omega t)N^{-1}\sum_{n=1}^{N}\exp(-i\mathbf{k}\cdot\mathbf{r}_n)\,d\omega \qquad 12.$$

The array spatial sample function may be represented as

$$f(r) = \sum_{n=1}^{N} \delta(\mathbf{r}\cdot\mathbf{r}_n) \qquad 13.$$

where $\delta(\mathbf{r}) = 1$ at $\mathbf{r} = 0$ and $\delta(\mathbf{r}) = 0$ otherwise. The wavenumber spectrum of $f(r)$ may be computed by

$$F(\mathbf{k}) = N^{-1} \sum_{n=1}^{N} \exp(-i\mathbf{k}\cdot\mathbf{r}_n) \qquad 14.$$

this is the wavenumber response of the array. Now equation 12 may be written as

$$b(t) = \int_{-\infty}^{\infty} V(\omega)F(\mathbf{k})\exp(i\omega t)\,d\omega \qquad 15.$$

The interpretation of equation 15 is that each frequency component of the propagating signal is, during the direct summing operation, modulated by the wave-

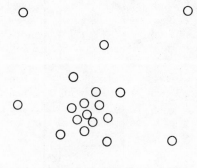

Figure 3   The configuration of the sub-arrays at LASA. Each subarray consists of about 15 single short-period seismometers.

100 km

number response of the array. This modulation depends both on the frequency and slowness of the signal. If appropriate delays had been introduced to the individual channels, say $X_n(t + \mathbf{s} \cdot \mathbf{r}_n)$, before summing, then the sum or beam would represent, in the absence of noise and signal distortion, the original signal. This technique is usually referred to as beam-forming.

It is useful to consider a typical array wavenumber response or rather the array pattern defined as $P(k) = |F(k)|^2$. The instrument configuration of LASA is shown in Figure 3 and the resulting array pattern in Figure 4. This pattern, contoured in decibels down from the peak response, shows the resolution or fidelity of the array in the wavenumber plane for monochromatic waves. The pattern is independent of its position in the wavenumber plane; if the array is steered to a specific wavenumber, $\mathbf{k}_0$, through the introduction of appropriate channel delays, the response of the array will be that shown in Figure 3 centered at $\mathbf{k}_0$. As Kelly (1967) comments, the response of an array to a realistic wide band signal such as equation 8 is not $F(k)$ but some smoothed version of $F(k)$, with smoothing increasing with signal band width.

$V^2(\omega, k_x, k_y)$ may be estimated by forming many beams, filtering the output from

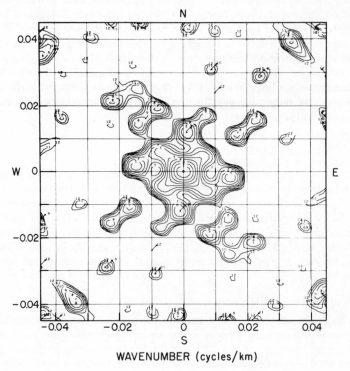

Figure 4 The wavenumber response of the LASA subarray configuration shown in Figure 3.

these beams, and squaring and averaging this output over some time window. Lacoss et al (1969) point out that there are more efficient ways of estimating $V^2(\omega, k_x, k_y)$ through the auto and cross-power spectral densities between elements of the array. However it is computed, the wavenumber observation of the seismic field must be viewed as the convolution of the array response with the true wavenumber composition of that field. Thus array patterns, like that of Figure 4, which approximate at two-dimensional impulse at the origin are desirable. Capon (1969) has developed a "high-resolution" frequency-wavenumber analysis technique which tends to reduce the effects of the convolution of the array pattern with the true wavenumber field. This method effectively deconvolves the observed wavenumber field and the array pattern.

These and other methods cannot be expressed here with the rigor and detail used in the original papers. Additional array techniques may be found in Embree et al (1963), Backus et al (1964), Green et al (1965, 1966), Birtill & Whiteway (1965), Johnson (1966), Capon et al (1967, 1968, 1969), Sengbush & Foster (1968), Schneider & Backus (1968), Capon (1970), and King et al (1973).

There is one result of importance to seismologists which has arisen from these exhaustive studies of various array processing techniques. Due to the incoherent components of both short- and long-period seismic noise over distances of 3.0 and 7.5 km, respectively, in most cases simple delay and sum beam-forming at arrays yields nearly as much signal enhancement as other more elaborate schemes tested (Capon 1973). With seismometer spacings on these intervals the signal to noise ratio varies as roughly $\sqrt{(N)}$, where N is the number of seismometers. This is a happy consequence for seismologists who may use arrays in search of smaller phases but who do not wish to be encumbered with the heavy computational burden that the more sophisticated techniques require.

RAY PARAMETER STUDIES

Figure 5 shows a summary of some published measurements of the ray parameter (p or $dT/d\Delta$) for the P wave in the distance range 0–180°. It is obvious from this figure that the earth beneath the crust may be divided into three major regions on the basis of $dT/d\Delta$. These are the upper mantle, the lower mantle, and the core. Also given in Figure 5 are the approximate maximum depths to which a ray of a given parameter penetrates, giving a general idea of the extent of these regions. In what follows we discuss the various interpretations given in studies of data such as that shown in Figure 5.

The Upper Mantle and Transition Zone

From the ray parameter measurements shown in Figure 5 the natural lower limit of the upper mantle appears to be about 800 km, although most authors speak of a transition zone between 400 and 800 km depth. The term transition refers to various mineralogical phase changes which are proposed between these depths, giving rise to abrupt increases in density and seismic velocity. Byerly (1926, 1935) first noted abrupt changes in the slope of the initial P wave time-distance curves near 17° and

28° in the western United States. Although his measurements came from individual stations, the type of phenomenon he observed may be seen in the array example of Figure 6. These are short-period seismograms of the *P* waves arriving at LASA (see Table 1) from an event at 15° distance. Here an initial phase of low amplitude and long period is followed by a stronger phase propagating across the array at a higher velocity. Clearly at a slightly greater distance this latter phase will become the first arrival and the slope of the time-distance curve of the initial *P* wave will

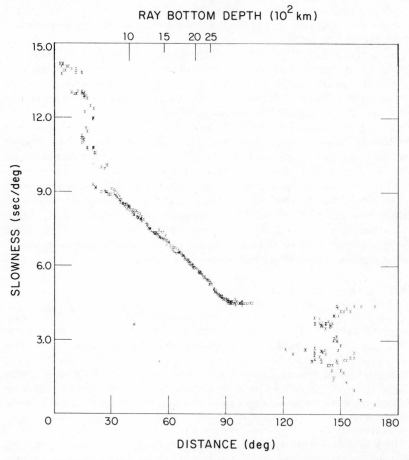

Figure 5 A summary of *P* wave d*T*/dΔ measurements from Johnson (1967, 1969), Corbishley (1970), Doornbos & Husebye (1972), and Davies et al (1971). Waves that bottom in the upper mantle and transition zone arrive between 0 and 30°, those bottoming in the lower mantle between 30 and 90°, and those that penetrate the core between 120 and 180°. The diagram clearly shows the differences in the complexities of these three regions.

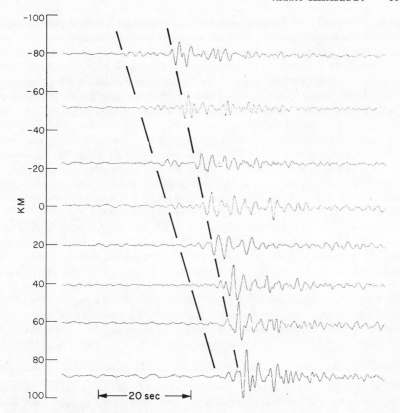

Figure 6 A series of LASA seismograms showing the *P* wave arrivals from an earthquake 15° to the west, off the Oregon coast. The later, stronger arrivals with a higher phase velocity have been refracted back to the surface more steeply, due to a sharp increase in velocity at about 350 km depth.

undergo a sharp change. The reason for this is that rays encountering an abrupt increase in velocity gradient with depth are refracted back toward the surface at an angle more nearly vertical than the shallower rays. The wave fronts of the deeper rays propagate across the array at a greater velocity or decreased apparent showness ($dT/d\Delta$).

Early work by Niazi & Anderson (1965) using TFO (see Table 1) located in central Arizona gave an indication of two relatively sharp decreases in $dT/d\Delta$ which were attributed to *P* wave velocity gradient increases near 320 and 640 km depth. A later study by Johnson (1967) using an extended TFO array confirmed these features in the upper mantle structure of western North America. Kanamori (1967) found $dT/d\Delta$ evidence for an abrupt *P* velocity increase near 375 km depth beneath Japan. Simpson (1973), using the Warramunga array in Australia, put forth a model with similar increases in the 300–400 km and 650–700 km depth ranges. Earlier Simpson et al (1971) had suggested two sharp increases in velocity at 280 and 440 km in the mantle north of Australia but this interpretation does not seem to have held its validity in light of further work (Simpson 1973). Kovach & Robinson (1969), using TFO, found evidence from shear wave $dT/d\Delta$ measurements for high velocity gradients near 400 and 600 km. Thus, although differences exist in the velocities and depths associated with these increased gradients found in various studies, they all are in general agreement and provide perhaps the best direct evidence of the detailed structure of the lower upper mantle or transition zone. These high velocity gradients near the top and base of the transition zone are most likely to be associated with solid-solid phase changes resulting in higher silicate densities than those predicted from simple compression. Some of these predicted changes are experimentally demonstrable in the laboratory. Anderson et al (1972) and Ringwood (1972) review the mineralogical evidence for these phase transitions within the mantle.

Although array measurements of body wave $dT/d\Delta$ cannot directly confirm the existence of a low velocity zone in the upper mantle at between 100–200 km depth, most of the relevant observations in studies cited above are consistent with such an interpretation. The $dT/d\Delta$ measurements are practically constant from 5° to 15° distance and show a second arrival with a lower value beginning at about 10°. The thicknesses of and velocities in the lithosphere and athenosphere seem best fixed by surface wave studies (Kanamori & Press 1970, Dziewonski 1971), although studies of P wave travel times and amplitudes using extensive networks (e.g. Archambeau et al 1969) usually include estimates of the properties of these features.

The Lower Mantle

As can be inferred from Figure 5, seismic velocities in the lower mantle increase more smoothly with depth than those in the upper mantle or transition zone. The nature of this increase is consistent with a homogeneous material compressed under adiabatic conditions. Array $dT/d\Delta$ studies of the lower mantle have attracted many workers including Chinnery & Toksöz (1967, Johnson (1969), Fairborn (1969), Greenfield & Sheppard (1969), Corbishley (1970), and Wright (1970). Despite these efforts the resulting velocity models for the lower mantle show only very slight changes from those proposed as early as 1939 by Jeffreys and Gutenberg. Thus most of the discussion of lower mantle $dT/d\Delta$ measurements has centered on zones exhibiting higher or lower values than some average value for the lower mantle (e.g. Toksöz et al 1967, Chinnery 1969). To point up the difficulty in drawing general conclusions from these "anomalous" $dT/d\Delta$ measurements, Johnson (1969) and Corbishley (1970) show relatively steep $dT/d\Delta$ versus distance gradients in the vicinity of 32°, while Greenfield & Sheppard (1969) and Wright (1970) describe relatively flat gradients at the same distance. The former imply increased velocity-depth gradients at 800 km while the latter have been interpreted as being due to low-velocity zone at the same depth (Wright 1968). It is worth noting that the data of Johnson (1969) from TFO and Greenfield & Sheppard (1969) from LASA sample regions only slightly separated geographically. Further, the data of Corbishly (1970) come from four arrays distributed globally, and it is combined simultaneously in a technique designed to minimize the effects of site corrections (Douglas & Corbishley 1968). It seems clear that although worldwide high- or low-velocity gradients may exist in the lower mantle, the resolving power of the arrays is greater than the lateral or radial homogeneity of many of these "anomalous" zones.

The lower 300 km of the mantle, just above the core boundary, introduces a large scatter in $dT/d\Delta$ measurements at distances of 90–100°. Johnson (1969) found that a correction for the presence of the core-mantle boundary, dependent both on frequency and azimuth, greatly but not totally reduced much of this scatter. Wright (1973) concluded that $dT/d\Delta$ measurements at these distances were a strong function of azimuth and suggested regional variations in the velocity structure of the lower mantle at depths greater than 2700 km as a possible cause. Using travel-time residuals of deep focus earthquakes Julian & Sengupta (1973) have also made a rather compelling case for lateral inhomogeneity in the deep mantle below 2000 km.

The Core

Most of the $dT/d\Delta$ studies of the earth's core have centered on the precursors to the phase which penetrates the inner core, $PKIKP$. Bolt (1962, 1964) suggested that these precursors were due to a layered boundary between the inner and outer core in which the P-wave velocity of the lower fluid outer core (about 10.0 km/sec) rose in two steps, over a radial distance of about 300 km, to a higher velocity (11.3 km/sec) at the top of the solid inner core. The basic framework was more or less confirmed by other workers using travel-time data (Engdahl 1968, Gogna 1968, Husebye & Madariaga 1970). Early array studies of core phases (Hannon & Kovach 1966) seemed to support a yet more complicated outer-inner core boundary suggested by Adams & Randall (1964). More recent studies by Doornbos & Husebye (1972), Doornbos & Vlaar (1973), and Doornbos (1974), using $dT/d\Delta$ measurements at NORSAR, as well as a reinterpretation of published travel times by Cleary & Haddon (1972), have suggested an alternative hypothesis for the precursors to $PKIKP$. These arguments are closely tied to the PKP caustic at an angular distance of about 117° on the core-mantle boundary. The precursor $dT/d\Delta$ measurements and the scatter in their arrival times suggest that the precursors are due to seismic waves diffracted and scattered by this caustic near the core mantle boundary. As Cleary and Haddon state, "a unique feature of scattering from a caustic is that the scattered wave appears as a first arrival in the shadow zone in front of the caustic." Estimates of the radial dimension of the proposed scattering zone just above the core mantle boundary are of the order of a few hundred kilometers.

Although the structure of this transition zone between the outer and inner core may not yet be completely resolved, array detections have shown definitively that the top of the inner core, at a radial distance of about 1216 km from the earth's center, is marked by a sharp discontinuity. From explosions at a distance of about 12°, reflections from the near side of the inner core, $PKiKP$ (Engdahl et al 1970), and the far side, $PKIIKP$ (Bolt 1974), have been observed. Bolt (1974) and Bolt & Qamar (1970) have used these observations to set limits on density, velocity, and attenuation in the inner core. If the inner core is solid it should transmit shear waves generated by compressional waves striking its surface, and these would be recognizable at the earth's surface as *PKJKP*. Julian et al (1972) made a search for this phase using LASA and reported five separate detections with reasonable

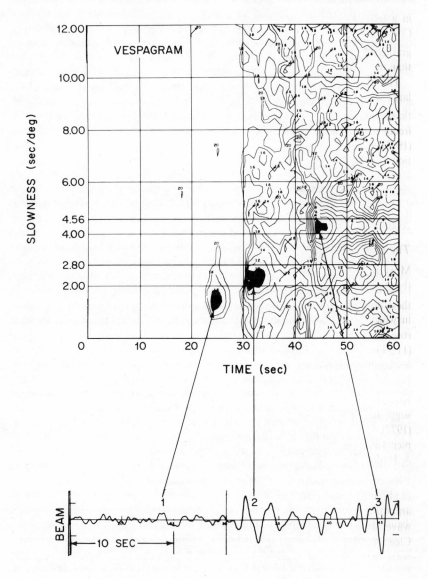

Figure 7   The slowness-time spectrum (Vespagram) for *P* wave arrivals at LASA from an event at 153° distance.

arrival time and $dT/d\Delta$ measurements. If the shear wave velocity in the inner core were constant, a value of 2.95 ± 0.1 km/sec is indicated.

In this latter study and in many of those discussed above, a specific type of array processing, called velocity spectral analysis (VESPA) (Davies et al 1971), was used. In this process the array is used to yield a graphical display seismic amplitude contoured as a function of slowness and time. This allows the separation of phases which arrive at the array at the same time but with different velocities, such as those refracted along various paths through the core. Figure 7 shows an example of a vespagram upon which three distinct core phases and their different velocities are clearly discernible. From an event at about $153°$, the first and fastest phase is $PKIKP$ refracted through the inner core, the second phase is PKP_1 refracted through the middle outer core, and the last phase is PKP_2 refracted more deeply into the outer core. These latter two phases are from the two sides of the PKP caustic discussed above.

FINE AND ANOMALOUS EARTH STRUCTURE

In addition to the $dT/d\Delta$ measurements summarized above, which are in a sense the grossest measurements to be made with an array, much effort has been put into the interpretation of the finer details present in the array observations of the seismic field. These detailed observations include the measurement of variation in amplitude, frequency content, wave form, velocity, and direction or azimuth of approach; not only as anomalies at the array as a whole but also as mutations within the array. These anomalies and variations are perhaps the most interesting and vexing aspects of seismology on which array data has been brought to light. It should be made clear at the onset what these anomalies are and how they are measured. Array travel-time anomalies, for a given path, may consist of two components: an array-wide anomaly which is evident at all the sensors, say because the array is located on thick sedimentary deposits; and intra-array anomalies, arrival-time differences within the array with respect to some standard travel-time table and a reference station within the array. Similarly there are two types of $dT/d\Delta$ and azimuth anomalies. An array $dT/d\Delta$ anomaly exists when the $dT/d\Delta$ measurement, usually by fitting a plane wave to the arrival times, does not agree with that predicted by some reference curve assuming the location of the event is given by some reliable independent means. An array azimuth anomaly exists when the fitted plane wave front is not normal to the great circle between the array and the given epicenter. Intra-array $dT/d\Delta$ and azimuth anomalies are found when the array is subdivided into smaller groups of seismometers or subarrays. All of these anomalies, except the azimuthal ones, must be referenced to some standard travel-time base, usually the Jeffreys-Bullen (J-B) tables, which must be considered non-anomalous and representative of a true worldwide average. Thus the term "array anomaly" by itself is rather meaningless; the variations under study must be well defined. It is incumbent upon the researcher to exactly define the type of anomaly being discussed and to exhaustively justify the interpretation of the source of that anomaly. The latter, in some cases, is rather difficult.

The Crust

As shown by Niazi (1966) and Zengeni (1970), the effect on teleseismic P waves of a plane-dipping crust-mantle boundary beneath an array will be seen as an anomaly in $dT/d\Delta$ and azimuth of approach which varies sinusoidally as a function of azimuth. A correction for such an effect would need to be made to $dT/d\Delta$ measurements before these observations are used in a study of the properties of the lower mantle. Unfortunately, none of the arrays studied to date seem to be given entirely to such a simple interpretation. Early workers such as Otsuka (1966a,b) using the Berkeley array, Johnson (1967, 1969) using TFO, and Greenfield & Sheppard (1969) using LASA found that more complicated variations of the plane-dipping interface model were needed and that even these did not explain all of the observations. As an example, array $dT/d\Delta$ and azimuth anomalies and intra-array travel-time anomalies at LASA have been studied as much as at any other array (Greenfield & Sheppard 1969, Glover & Alexander 1969, Iyer 1971, Engdahl & Felix 1971, Iyer & Healy 1972), yet the interpretations given to these studies hardly present a clear picture of the underlying crustal structure. One feature common to most of the models is a synclinal trend at the base of the crust striking northeast. The axis of this trend is nearly under the array center, and a crustal thickness on the order of 60 km at this axis is implied. An extension of the anomalous zone beneath the crust under LASA into the upper mantle has been suggested by Iyer & Healy (1972) and Engdahl & Felix (1971). While the crust under YKA is thought to be quite simple (Weichert & Whitham 1969), Wright (1973) now asserts that $dT/d\Delta$ anomalies of core phases imply lateral variations in the upper mantle beneath this Canadian array. Lateral upper mantle variations and complex topography at the base of the crust are taken as the major cause of array azimuth and velocity anomalies measured at the Warramunga array (Wright et al 1974).

The one point that arises from this brief discussion is that no clear, simple picture of the crust and upper mantle has been gained from array studies. The primary conclusion is that lateral variations in the crust and upper mantle, at least on the scale of an array, exist beneath most arrays studied. Recently work has been done which attempts to describe the structure under arrays by statistical models rather than strictly deterministic ones. Mack (1969) attributed the short-period P-wave form variations across LASA to the multiple paths of the primary mantle wave due to deep crustal relief. Using the array beam as a master signal he was able to deconvolve individual traces into a series of two or three randomly but closely spaced impulses and then successfully synthesize the trace of the same seismometer due to a second event from the same region. Larner (1970) described the seismic field scattered from a corrugated crust-mantle boundary and was able to predict the trend of intra-array amplitude and travel-time anomalies at LASA based on a single deep (60 km) corrugation under the center of the array, similar to the model of Greenfield & Sheppard (1969). Considerable scatter still existed within this trend, and recently Aki (1973) and Capon (1974) tried to model this scatter in terms of the statistical properties of the variation in P velocity in the crust and upper mantle beneath the array. Chernov's (1960) theory was applied in both studies, although different types of measurements of the LASA data were taken. Aki concluded, for waves of 2 sec period, that fluctuations in amplitude and phase could be explained by 4% random velocity variations, with a correlation distance of about 10 km, existing to a depth of some 60 km. Measuring amplitude and slowness anomalies within the array, Capon finds, at 1.25 sec period, 1.9% velocity fluctuations correlated over a distance of 12 km and existing to a depth of 136 km. This seems to be rather good agreement considering the difficulties involved in this type of analysis and appears to give a promising interpretation of anomalies which exist between individual stations of large and medium aperture arrays.

The Mantle

Other workers have found difficulty in placing the source of array-wide $dT/d\Delta$ and azimuth anomalies within the crust or upper mantle beneath the array. Davies & Sheppard (1972) argue in terms of the "array diagram" of LASA shown in Figure 8. Here in a $dT/d\Delta$ and azimuth space are plotted the mislocation vectors of LASA plane wave measurements (tail of the arrows) with respect to US Geological Survey locations (head of the arrows). The slownesses of the latter are based on the J-B tables. Davies and Sheppard argue that although an array bias exists in Figure 8, about 0.25 sec/deg to the north, the variations shown cannot be explained by any realistic structure beneath the array. The chief argument is that the vectors change too rapidly in certain regions to be caused by any array structure, whose effect would necessarily be smoothed in such a diagram. They further argue that the J-B tables are an adequate reference since any systematic reference bias would show up as pure radial vectors at all azimuths. They conclude that these vectors offer strong evidence for lateral velocity variations in the mantle. In particular they note that the lower mantle, just above the core boundary, is particularly well sampled in Figure 8 and that the rapid changes in the direction and magnitude of these vectors for events beyond 90° is indicative of strong lateral heterogeneity in this region. Wright (1973), previously discussed, and Kanasewich et al (1973) have also presented array evidence for lateral heterogeneity in the lower mantle using Canadian arrays. The latter have suggested that an anomalous zone in the lower mantle beneath Hawaii may have a bearing on hypotheses concerning the source of those islands (see also Needham & Davies 1973). Again the point is made by Figure 8 that the resolving power of arrays is greater than the lateral homogeneity of the earth.

The d Phases

Figure 9 shows the short-period traces of certain subarray sums at LASA recording the P wave and coda from a deep, South American earthquake. The P wave and its surface reflection, pP, are clearly seen on the section of records. Between these two phases are clearly seen at least two coherent arrivals across the array; these arrivals are probably reflections from the upper mantle above the earthquake source.

Bolt et al (1968) initially suggested the nomenclature PdP for precursors to the phase PP which reflects at the earth's surface midway between the source and

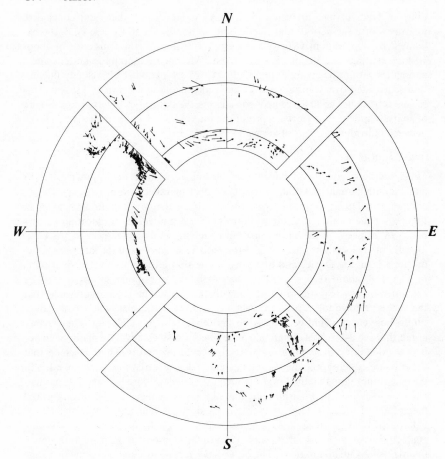

Figure 8 The array diagram of mislocation vectors at LASA in slowness and azimuth space (Davies & Sheppard 1972). The tails of the arrows mark the array location, the heads mark the more accurate network locations. The circles from the center mark 4.0, 5.0, 7.5, and 10 sec/deg slowness values.

receiver. It was suggested that PdP reflected beneath this midpoint at some depth d. Following this notation the intermediate phases seen in Figure 8 would be called $p270P$ and $p150P$. Using single stations, as did Bolt et al (1968), Adams (1968) found precursors to the core phase $PKPPKP$ or $P'P'$ which traces symmetric paths to a station by traversing the mantle and core and reflecting off the far surface of the earth. These $P'P'$ precursors are designated $P'dP'$. These and other d phases have been studied at single stations and arrays by Whitcomb & Anderson (1970), Whitcomb (1973), Wright (1972), and Gutowski & Kanasewich (1974). Within these

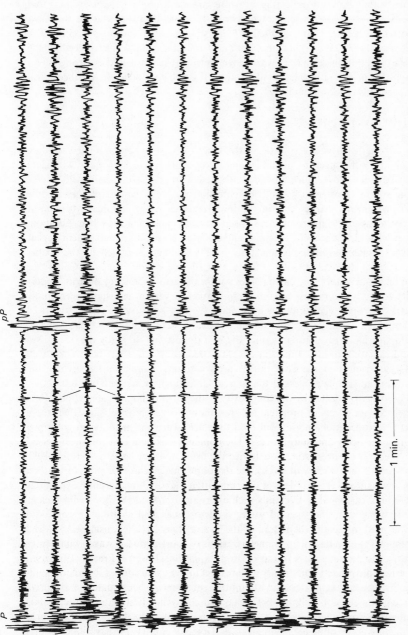

Figure 9 A series of LASA seismograms showing the coherency across the array of weak arrivals between *P* and *pP*. The event location is 650 km beneath western Brazil and these weak arrivals are interpreted as reflections from the upper mantle above the source.

studies, although many reflecting horizons are evident, the region at the rapid velocity increase and phase change near 650 km depth appears to be the most consistent reflector. From detailed $dT/d\Delta$ studies some authors (Wright 1972, Whitcomb 1973) have argued that certain d reflections, those just preceding the surface reflection phase, may represent asymmetric ray paths caused by strong lateral complexity in the crust and shallow upper mantle just beneath the theoretical homogeneous reflection point.

SOURCE STUDIES

Arrays have been used very little in the study of seismic sources. These studies, if they are to be given to unambiguous interpretation, are best made with data recorded over a wide frequency band at sites well distributed azimuthally with respect to the source. Most arrays, built for detection purposes, consist of very narrow band instruments, particularly in the low frequency range, a range where most conventional source interpretations based on wide band data are made. Additionally, at teleseismic distances, the dimensions of most earthquakes make them appear as point sources to an array and thus little advantage is gained by the spatial properties of the array. Some arrays are in seismic regions, such as California, Montana, and Alaska, and earthquakes within the array boundary offer an opportunity to study the radiation from these sources at very near distances at closely spaced sites. One such study has been made by Berg & Lutschak (1973) who used the long-period Alaskan array to measure the tilt field of an intra-array earthquake.

Short-period arrays probably offer the best opportunity to study the 0.5–2.0 sec period radiation from seismic sources. This frequency range has been fairly well neglected in most source work but it surely contains information concerning the details of the seismic process at the source. Unfortunately, as we have seen above, it also is heavily affected by the complexity of the earth through propagation. However, if certain effects of propagation are random between array sites, then these effects of propagation "noise" will tend to cancel when the array beam is formed through the delay and sum process. The contention is here that the short-period array beam is a better estimation, above the usual ambient noise reduction, of the short-period time history of the source than any single sensor. If this is true then it has relevance to the problem of discrimination between underground explosions and natural earthquakes. Certainly such discrimination could be done with greater facility if the same short-period waves used to detect and locate an event could also be used to identify the nature of the source. Lacoss (1969), using LASA, and Anglin (1971) both describe parameterizations of short-period P waves which have been used to discriminate earthquakes from explosions. Filson & Frasier (1972) used five arrays distributed azimuthally with respect to five explosions of different magnitude but at a common site in an attempt to simultaneously estimate the explosive source parameters of each event. The method also yielded estimates of the seismic attenuation along each of the five ray paths which seemed, as did the source parameters, to be quite reasonable values. Douglas et al (1974) have used models of short-period earthquake seismograms and have compared them with recordings at various arrays. By varying the model parameters and the source depth they have successfully matched the model to the observations; notably in the case of an event of questionable origin in East Kazakhstan.

Of course arrays have been used extensively in the measurement of body and surface wave magnitude ratios for the purposes of source identification (Lacoss 1969, Filson & Bungum 1972) but these are more exercises in detection than source studies. The chief value of arrays in source studies, which with present instrumentation lies at shorter periods, has yet to be fully realized or appreciated.

OTHER PROBLEMS

Long-Period Multi-Path Propagation

Evernden (1953, 1954), studying data recorded at a long-period network in central California, recognized that surface waves did not always propagate in true great circle paths and suggested that these deviations were due to lateral variations within the crust and upper mantle. Capon (1970, 1971), applying his high-resolution wave-number techniques to long-period data recorded at LASA, confirmed Evernden's observations concerning the complex nature of surface wave propagation. This phenomenon (called multi-path propagation) is seen at a single station as an amplitude modulation or beating of the long-period seismogram, caused by surface waves of a given period propagating along different paths and thus arriving at the station at different times. Capon measured the azimuth of approach and time of arrival of various wave groups and, in some cases, was able to associate deviations with paths that included reflections and refractions at topographical features. Julian (1973) has attempted to model the complex surface wave propagation pattern observed at LASA from Nevada sources by taking into account the variation in crustal thickness and velocities in crust and upper mantle in the western United States. The effect of the high crustal velocities in the basalts of the Snake River plain on the 10 sec Rayleigh waves was accurately predicted by Julian's model. The azimuth of arrival of the 40 sec waves, east of the great circle path at LASA, was not reproduced by the model and was attributed to unknown complexities in the boundary between the Colorado plateau and the Basin and Range province.

Array Location and Detection

As one may infer from Figure 8, arrays, as they stand, are rather poor at yielding accurate locations of epicenters. Of course the various seismic regions of the world may be calibrated at the array using accurate locations based on other data. A set of station corrections results, functions of azimuth and slowness, which then are applied in subsequent locations to the arrivals recorded at individual sites within an array. In spite of these refinements array locations are seldom accurate to less than one degree, an accuracy routinely obtained by networks of widely dispersed stations. Yet due to the signal enhancement procedures which may be applied, arrays are much better detectors of seismic events than individual stations or networks of stations. In some cases this is done automatically by sequentially phasing the array at various predetermined points in the slowness and azimuth plane and comparing the average power in a short interval of interest with that of a longer, preceding interval.

Obviously a combination of array and network data should yield a larger number of well-located events than either set of data independently. In a recent study Lacoss et al (1974) combined array detections and measurements of $dT/d\Delta$ with network arrival time observations to gain locations for some 996 events occurring within a one-month period. This represented more than 40% more events than reported by any other single array or network routinely locating seismic events within the same time period. In this study the array measurements were included within the normal equations usually solved in the least-squares estimates of epicenter location based on arrival times.

As indicated earlier, arrays and combinations of arrays may be very helpful in identifying phases in the seismic coda of earlier phases. The problem of identifying separate events in the case of interfering surface waves has been studied by Capon & Evernden (1971) and Mack & Smart (1973). The ability of an array to sort out interfering events depends on the relative size and azimuth of approach of the signals at the array and the array response function. For signals of the same size arriving simultaneously from the same azimuth, the array is of little use; however, signals approaching from widely different azimuths may be separated measurably. The separation of mixed short-period phases and automatic short-period phase identification using arrays has been studied by Cohen (1974) and Schlien & Toksöz (1974).

COMMENTS

Within the last decade seismic arrays have added a vast new store of data upon which seismologists may draw. Here we have tried to review some of the inferences already made from this data and how these interpretations have refined our notions of the interior of the earth. Many controversies still exist. There is no doubt that arrays can detect lateral inhomogeneities deep within the earth better than any other tool available to seismologists. The unambiguous interpretation of many anomalous features brought to light by a single array is difficult, which suggests the use of multiple arrays focused on the same anomalous zone as the next obvious step; however, this may not always be possible. In many cases a statistical parameterization of seismic field as a scattered phenomenon, more easily done with array than single-station data, may be more realistic than a classical deterministic interpretation. In any case the variations of the seismogram in space and time have now been clearly defined by the arrays; they have yet to be completely deciphered.

ACKNOWLEDGMENTS

Much of the work discussed here is that of my colleagues in the Seismic Discrimination Group of Lincoln Laboratory and I thank them for an association which has been extremely helpful to me. This work was sponsored by the Advanced Research Projects Agency of the United States Department of Defence.

Literature Cited

Adams, R. 1968. Early reflections of $P'P'$ as an indication of upper mantle structure. *Bull. Seismol. Soc. Am.* 58:1933–48

Adams, R., Randall, M. 1964. Fine structure of the earth's core. *Bull. Seismol. Soc. Am.* 54:1299–1313

Aki, K. 1973. Scattering of P waves under the Montana LASA. *J. Geophys. Res.* 78:1334–46

Anderson, D., Sammis, C., Jordan, T. 1972. *The Nature of the Solid Earth,* ed. E. C. Robertson, 41–66. New York: McGraw. 677 pp.

Anglin, F. 1971. Discrimination of earthquakes and explosions using short period seismic array data. *Nature* 233:51–52

Archambeau, C., Flinn, E., Lambert, D. 1969. Fine structure of the upper mantle, *J. Geophys. Res.* 74:5825–65

Backus, M., Burg, J., Baldwin, D., Bryan, E. 1964. Wide-band extraction of mantle P waves from ambient noise. *Geophysics* 29:672–92

Båth, M. 1968. *Mathematical Aspects of Seismology.* Amsterdam: Elsevier. 415 pp.

Berg, E., Lutschak, W. 1973. Crustal tilt fields and propagation velocities associated with earthquakes. *Geophys. J. Roy. Astron. Soc.* 35:5–29

Birtill, J., Whiteway, F. 1965. The application of phased arrays to the analysis of seismic body waves. *Phil. Trans. Roy. Soc. London Ser. A* 258:421–93

Bolt, B. 1962. Gutenberg's early PKP observations. *Nature* 196:122

Bolt, B. 1964. The velocity of seismic waves near the earth's center. *Bull. Seismol. Soc. Am.* 54:191–208

Bolt, B. 1974. *The detection of PKIIKP and damping in the inner core.* Presented at 69th Ann. Meet. Seismol. Soc. Am., Las Vegas

Bolt, B., O'Neill, M., Qamar, A. 1968. Seismic waves near 110°: is structure in the core or upper mantle responsible? *Geophys. J. Roy. Astron. Soc.* 16:475–87

Bolt, B., Qamar, A. 1970. Upper bound to the density jump at the boundary of the earth's inner core. *Nature* 288:148–50

Bullen, K. 1963. *An Introduction to the Theory of Seismology.* Cambridge, England: Cambridge Univ. Press. 381 pp.

Burg, J. 1964. Three-dimensional filtering with an array of seismometers. *Geophysics* 39:693–713

Byerly, P. 1926. The Montana earthquake of June 28, 1975. *Bull. Seismol. Soc. Am.* 16:209–65

Byerly, P. 1935. The first preliminary waves

of the Nevada earthquake of December 20, 1932. *Bull. Seismol. Soc. Am.* 25:62–80

Capon, J. 1969. High resolution frequency-wavenumber spectrum analysis. *Proc. IEEE* 57:1408–18

Capon, J. 1970. Application of detection and estimation theory to large array seismology, *Proc. IEEE* 58:760–70

Capon, J. 1970. Analysis of Rayleigh wave multi-path propagation at LASA. *Bull. Seismol. Soc. Am.* 60:1701–31

Capon, J. 1971. Comparison of Love and Rayleigh wave multi-pathing at LASA. *Bull. Seismol. Soc. Am.* 61:1327–44

Capon, J. 1973. *Methods in Computational Physics,* 13, ed. B. Bolt, 13:1–59. New York: Academic. 473 pp.

Capon, J. 1974. Characterization of crust and upper mantle structure under LASA as a random medium. *Bull. Seismol. Soc. Am.* 64:235–66

Capon, J., Evernden, J. 1971. Detection of interfering Rayleigh waves at LASA. *Bull. Seismol. Soc. Am.* 61:807–49

Capon, J., Greenfield, R., Kolker, R. 1967. Multidimensional maximum-likelihood processing of a large aperture seismic array. *Proc. IEEE* 55:192–211

Capon, J., Greenfield, R., Kolker, R., Lacoss, R. 1968. Short-period signal processing results for the Large Aperture Seismic Array. *Geophysics* 33:452–72

Capon, J., Greenfield, R., Lacoss, R. 1969. Long-period signal processing results for the Large Aperture Seismic Array. *Geophysics* 34:305–29

Chernov, L. 1960. *Wave Propagation in a Random Medium.* New York: Dover. 168 pp.

Chinnery, M. 1969. Velocity anomalies in the lower mantle. *Phys. Earth Planet. Interiors* 2:1–10

Chinnery, M., Toksöz, M. N. 1967. P wave velocities in the mantle below 700 km. *Bull. Seismol. Soc. Am.* 57:199–226

Cleary, J., Haddon, R. 1972. Seismic wave scattering near the core-mantle boundary: a new interpretation of precursors to PKP. *Nature* 240:549–51

Cohen, T. 1974. Coda suppression capabilities of the beam and mixed signal processor. *Bull. Seismol. Soc. Am.* 64:415–26

Corbishley, D. 1970. Multiple array measurements of the P wave travel-time derivative. *Geophys. J. Roy. Astron. Soc.* 19:1–14

Davies, D. 1973. Seismology with large arrays. *Rep. Progr. Phys.* 36:1233–83

Davies, D., Kelly, E., Filson, J. 1971. Vespa

process for analysis of seismic signals. *Nature Phys. Sci.* 232:8–13

Davies, D., Sheppard, R. 1972. Lateral heterogeneity in the earth's mantle. *Nature* 239:318–23

Department of State Publication 7008. 1960. *Documents on Disarmament 1945–1959.* 2: 1090–1111

Doornbos, D. 1974. Seismic wave scattering near caustics: Observations of PKKP precursors. *Nature* 247:352–53

Doornbos, D., Husebye, E. 1972. Array analysis of PKP phases and their precursors. *Phys. Earth Planet. Interiors* 5: 387–99

Doornbos, D., Vlaar, N. 1973. Regions of seismic wave scattering in the earth's mantle and precursors to PKP. *Nature Phys. Sci.* 243:58–61

Douglas, A., Corbishley, D. 1968. Measurement of $dT/d\Delta$. *Nature* 217:1243–44

Douglas, A., Marshall, P., Yound, J., Hudson, J. 1974. Seismic source in east Kazakhstan. *Nature* 248:743–45

Dziewonski, A. 1971. Upper mantle models from pure path despersion data. *J. Geophys. Res.* 76:2587–2601

Embree, P., Burg, J., Backus, M. 1963. Wide band filtering—the "pie-slice" process. *Geophysics* 28:948–74

Engdahl, E. 1968. *Core phases and the earth's core.* PhD thesis. Saint Louis Univ., St. Louis, Missouri. 197 pp.

Engdahl, E., Felix, C. 1971. Nature of travel time anomalies at LASA. *J. Geophys. Res.* 76:2706–15

Engdahl, E., Flinn, E., Romney, C. 1970. Seismic waves reflected from the earth's inner core. *Nature* 228:852–53

Evernden, J. 1953. Direction of approach of Rayleigh waves and related problems, 1. *Bull. Seismol. Soc. Am.* 43:335–74

Evernden, J. 1954. Direction of approach of Rayleigh waves and related problems, 2. *Bull. Seismol. Soc. Am.* 44:159–84

Fairborn, J. 1969. Shear wave velocities in the lower mantle. *Bull. Seismol. Soc. Am.* 59:1983–2000

Filson, J., Bungum, H. 1972. Initial discrimination results from the Norwegian seismic array. *Geophys. J. Roy. Astron. Soc.* 31:315–28

Filson, J., Frasier, C. 1972. Multi-site estimation of explosive source parameters. *J. Geophys. Res.* 77:2045–61

Glover, P., Alexander, S., 1969. Lateral variations under the LASA and their effect on $dT/d\Delta$ measurements. *J. Geophys. Res.* 74:505–31

Gogna, M. 1968. Travel times of PKP from Pacific earthquakes. *Geophys. J. Roy. Astron. Soc.* 16:489–514

Green, P., Frosch, R., Romney, C. 1965. Principles of an experimental Large Aperture Seismic Array (LASA). *Proc. IEEE* 53:1821–33

Green, P., Kelly, E., Levin, M. 1966. A comparison of array processing techniques. *Geophys. J. Roy. Astron. Soc.* 11: 67–84

Greenfield, R., Sheppard, R. 1969. The Moho depth variations under LASA and their effect on $dT/d\Delta$ measurements. *Bull. Seismol. Soc. Am.* 59:409–20

Gutowski, P., Kanasewich, E. 1974. Velocity spectral evidence of upper mantle discontinuities. *Geophys. J. Roy. Astron. Soc.* 36:21–32

Hannon, W., Kovach, R. 1966. Velocity filtering on seismic core phases. *Bull. Seismol. Soc. Am.* 56:441–54

Husebye, E., Madariaga, R. 1970. The origin of precursors to core waves. *Bull. Seismol. Soc. Am.* 60:939–52

Iyer, H. 1971. Variation of apparent teleseismic P waves across the Large Aperture Seismic Array, Montana. *J. Geophys. Res.* 76:8554–67

Iyer, H., Healy, J. 1972. Teleseismic residuals at the LASA-USGS extended array and their interpretation in terms of crust and upper mantle structure. *J. Geophys. Res.* 77:1503–27

Johnson, L. 1966. *Measurements of mantle velocities of P waves with a large array.* PhD thesis. Calif. Inst. Tech., Pasadena, 107 pp.

Johnson, L. 1967. Array measurements of P velocities in the upper mantle. *J. Geophys. Res.* 72:6309–25

Johnson, L. 1969. Array measurements of P velocities in the lower mantle. *Bull. Seismol. Soc. Am.* 59:973–1008

Julian, B. 1973. Predicting Rayleigh wave propagation behavior from geological and geophysical data. *Mass. Inst. Tech. Lincoln Lab. Semiann. Tech. Sum.* June 1973: 27

Julian, B., Davies, D., Sheppard, R. 1972. PKJKP. *Nature* 235:317–18

Julian, B., Sengupta, M. 1973. Seismic travel time evidence for lateral inhomogeneity in the deep mantle. *Nature* 242: 443–47

Kanamori, H. 1967. Upper mantle structure from apparent velocities of P waves recorded at Wakayama micro-earthquake observatory. *Bull. Earthquake Res. Inst.* 45:657–78

Kanamori, H., Press, F. 1970. How thick is the lithosphere? *Nature* 226:330–31

Kanasewich, E., Ellis, R., Chapman, C., Gutowski, P. 1973. Seismic array evidence for a core boundary source for the

Hawaiian linear volcanic chain. *J. Geophys. Res.* 78:1361–71

Kelly, E. 1965. *Mass. Inst. Tech. Lincoln Lab. Tech. Note* 1965–21

Kelly, E. 1967. *Mass. Inst. Tech. Lincoln Lab. Tech. Note* 1967–30

King, D., Mereu, R., Muirhead, K. 1973. The measurement of apparent velocity and azimuth using adaptive processing techniques of data from the Warramunga seismic array. *Geophys. J. Roy. Astron. Soc.* 35:137–68

Kovach, R., Robinson, R. 1969. Upper mantle structure in the basin and range province, western North America, from the apparent velocities of S waves. *Bull. Seismol. Soc. Am.* 59:1653–65

Lacoss, R. 1969. *Mass. Inst. Tech. Lincoln Lab. Tech. Note* 1969–24

Lacoss, R., Kelly, E., Toksöz, M. 1969. Estimation of seismic noise structure using arrays. *Geophysics* 34:21–38

Lacoss, R., Needham, R., Julian, B. 1974. *Mass. Inst. Tech. Lincoln Lab. Tech. Note* 1974–14

Larner, K. 1970. *Near receiver scattering of teleseismic body waves in layered crust-mantle models having irregular interfaces.* PhD thesis. Mass. Inst. Technol., Cambridge. 274 pp.

Mack, H. 1969. Nature of short-period P-wave signal variations at LASA. *J. Geophys. Res.* 74:3161–70

Mack, H., Smart, E. 1973. Automatic processing of multi-array long-period seismic data. *Geophys. J. Roy. Astron. Soc.* 35:215–24

Needham, R., Davies, D. 1973. Lateral heterogeneity in the deep mantle from seismic body wave amplitudes. *Nature* 244:152–53

Niazi, M. 1966. Correction to apparent azimuth and travel-time gradients for a dipping Mohorovicic discontinuity. *Bull. Seismol. Soc. Am.* 56:491–509

Niazi, M., Anderson, D. 1965. Upper mantle structure of western North America from apparent velocities of P waves. *J. Geophys. Res.* 70:4633–40

Otsuka, M. 1966a. Azimuth and slowness anomalies of seismic waves measured on the central California seismograph array 1. Observations. *Bull. Seismol. Soc. Am.* 56:223–40

Otsuka, M. 1966b. Azimuth and slowness anomalies of seismic waves measured on the central California seismographic array 2. Interpretation. *Bull. Seismol. Soc. Am.* 56:655–76

Ringwood, A. E. 1972. *The Nature of the Solid Earth,* ed. E. C. Robertson, 67–92. New York: McGraw. 677 pp.

Schlien, S., Toksöz, M. N. 1974. Automatic phase identification with one and two large aperture seismic arrays. *Bull. Seismol. Soc. Am.* 64:221–33

Schneider, W., Backus, M. 1968. Dynamic correlation analysis. *Geophysics* 33:105–26

Sengbush, R., Foster, M. 1968. Optimum multichannel velocity filters. *Geophysics* 33:11–35

Simpson, D. 1973. *P velocity structure of the upper mantle in the Australian region.* PhD thesis. Australian National Univ., Canberra. 212 pp.

Simpson, D., Wright, C., Cleary, J. 1971. A double discontinuity in the upper mantle. *Nature Phys. Sci.* 231:201–3

Toksöz, N., Chinnery, M., Anderson, D. 1967. Inhomogeneities in the earth's mantle. *Geophys. J. Roy. Astron. Soc.* 13:31–59

Weichert, D., Whitham, K. 1969. Calibration of the Yellowknife Seismic Array with first zone explosions. *Geophys. J. Roy. Astron. Soc.* 18:461–76

Whitcomb, J. 1973. Asymmetric P'P': An alternative to P'dP' reflections in the uppermost mantle (0 to 110 km). *Bull. Seismol. Soc. Am.* 63:133–43

Whitcomb, J., Anderson, D. 1970. Reflections of P'P' seismic waves from discontinuities in the mantle. *J. Geophys. Res.* 75:5713–28

Wright, C. 1968. Evidence for a low velocity layer for P waves at a depth close to 800 km. *Earth Planet. Sci. Lett.* 5:35–40

Wright, C. 1970. P wave travel time gradient measurements and lower mantle structure. *Earth Planet. Sci. Lett.* 8:41–44

Wright, C. 1972. Array studies of seismic waves arriving between P and PP in the distance range 90° to 115°. *Bull. Seismol. Soc. Am.* 62:385–400

Wright, C. 1973. Array studies of P phases and the structure of the D" region of the lower mantle. *J. Geophys. Res.* 78:4965–82

Wright, C., Cleary, J., Muirhead, K. 1974. The effects of local structure on the short period P wave arrivals recorded at the Warramunga seismic array. *Geophys. J. Roy. Astron. Soc.* 36:295–319

Zengeni, T. 1970. A note on an azimuthal correction for $dT/d\Delta$ for a single dipping plane interface. *Bull. Seismol. Soc. Am.* 60:299–306

ADVANCES IN THE GEOCHEMISTRY OF AMINO ACIDS

�֍10037

Keith A. Kvenvolden
Planetary Biology Division, Ames Research Center, NASA, Moffett Field, California 94035

INTRODUCTION

Amino acids are universal components of all known living systems. These molecules are the building blocks of proteins which serve as enzymes to catalyze biochemical reactions and as structural components of organisms. Their importance in living systems makes amino acids significant in geochemistry wherein the geological fate of biological compounds is one direction of research.

Amino acids are simple molecules (see Appendix) having at a minimum one carboxylic acid and one amino functional group. The presence of these groups permits these compounds to undergo numerous reactions. Although the number of different kinds of amino acids theoretically possible is almost unlimited, only 20 commonly occur in proteins of organisms. Many other amino acids occur naturally in organisms, and because they usually are not found in proteins they are classified as nonprotein amino acids. All of the protein amino acids are α-amino derivatives of carboxylic acids; except for glycine, all contain at least one asymmetric carbon atom (Figure 1).

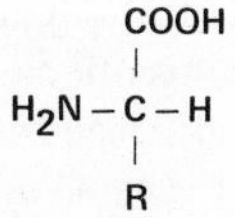

$$
\begin{array}{c}
\text{COOH} \\
| \\
\text{H}_2\text{N} - \text{C} - \text{H} \\
| \\
\text{R}
\end{array}
$$

Figure 1   General structural formula (Fischer projection) for a typical protein amino acid.

Amino acids with one asymmetric carbon atom can exist in two configurations which are mirror images called enantiomers, optical isomers, or optical antipodes. The relative configuration of enantiomers is designated by the symbols D and L, the absolute configuration by (R) and (S). L-Amino acids usually have the (S) configuration, but there are exceptions such as L-cysteine which has the (R) configuration. For this paper the D,L convention will be used. There are four common amino acids that contain two asymmetric carbon atoms, and thus can occur not only as enantiomers but also as diastereomers. They are isoleucine, threonine, hydroxypro-

183

184   KVENVOLDEN

line, and hydroxylysine. Whereas enantiomers of a given amino acid have the same physical properties (except for the ability to rotate plane-polarized light in opposite directions) and cannot be resolved chromatographically without the aid of optically active reagents, diastereomers have different physical properties and can therefore be resolved by simple chromatographic techniques.

All amino acids in protein, except glycine, are optically active and are all in the L-stereoisomeric configuration. D-Amino acids are found in quantitatively small amounts in living organisms, but not in proteins. As optically active amino acids pass from the biosphere to the lithosphere, they lose their optical activity by undergoing interconversions leading eventually to equilibrium mixtures of D- and L-amino acids. This process is known as racemization, when enantiomers are involved, and as epimerization, when diastereomers undergo interconversions at one of their asymmetric carbon atoms.

Modern study of amino acids from an organic geochemical point of view began with the investigations of Abelson (1954), who first pointed out the significance of amino acids in fossil shells and bones. Following this work, Erdman et al (1956) showed that amino acids are present in sediments and sedimentary rocks. Although the amino acid content of soils had been studied since 1908, Stevenson (1954) applied new and quantitative techniques yielding a deeper understanding of the problem. These and other early investigations were first reviewed by Abelson (1963) and later by Hare (1969), who anticipated a number of research areas which, after 1969, have received increasing attention. It is the purpose of this present review to touch only aspects of the past work and mainly to focus on those developments of the last five years that seem particularly significant.

TECHNIQUES

Methods utilized in the analysis of amino acids have greatly improved since consideration was first given to the geochemical occurrences of these compounds (Abelson 1954). Free amino acids have been removed from geochemical samples by means of extraction with water, 80% ethanol, and ammonium acetate while combined or bound amino acids conventionally are removed from samples by hydrolysis with 6 N hydrochloric acid (Degens & Reuter 1964). One of the major problems in amino acid analyses is the removal of inorganic salts which dissolve in the extracting solvents, particularly in 6 N hydrochloric acid, during hydrolysis. Salt removal, or desalting, is commonly accomplished by use of strongly acidic cation-ion exchange resins (Degens & Reuter 1964). In certain specialized situations, salts have been removed by precipitation. For example, Hare (1969) has used hydrogen fluoride to precipitate calcium which dissolves during the hydrolysis of calcareous fossils. Pollock & Miyamoto (1971) have also shown that precipitation techniques can remove interfering ions such as iron, aluminum, calcium, and magnesium.

Identification of individual amino acids recovered from geochemical samples initially was done by paper chromatography (Degens & Reuter 1964). This method of analysis was replaced by ion exchange techniques incorporating sulfonated polystyrene resins (Moore & Stein 1951) when automated systems generally became available (Moore et al 1958, Spackman et al 1958). Hamilton (1963) in particular showed how the automated systems could be improved to increase resolution and sensitivity, and Hare (1969) developed an automated ion exchange system incorporating a gas-drive system with small-bore Teflon columns especially for geochemical purposes. This system is at least three orders of magnitude more sensitive and one order of magnitude faster than the original automated instruments which are available commercially (Hare 1973a). In these ion exchange procedures ninhydrin is reacted with the amino acids producing derivatives which are detected colorimetrically. Recently a fluorometric detection system which uses fluorescamine and provides sensitivity at the picomole level was designed (Udenfriend et al 1972, Hare 1973a).

Ion exchange chromatography has been utilized extensively to separate the naturally occurring diastereomers such as isoleucine and alloisoleucine (see section on stereochemistry). In addition, diastereomeric derivatives of enantiomeric amino acids can be prepared, and these also can be resolved by ion exchange chromatography. In geochemistry this technique has been applied to aspartic acid from bone (Bada & Protsch 1973); the method can be used with other amino acids as well. In this technique the determination of the ratio of the D and L isomers of aspartic acid was based on the separation of the diastereomeric dipeptides obtained by derivatization with N-carboxyanhydride (Manning & Moore 1968). Before derivatization individual amino acids must be separated from one another, and this separation has been accomplished by means of gradient elution with hydrochloric acid from a cation exchange column (Wall 1953). Without separation the dipeptide synthesis of the total amino acid mixture from bone resulted in a large number of peaks, many of which overlapped (Bada & Protsch 1973).

Of historical interest for geochemistry is the fact that the stereochemistry of amino acids from shells was first determined by enzymatic methods (Hare & Abelson 1968). These methods involve oxidases which destroy either the D or the L isomer of certain amino acids and leave one or the other isomer intact. The methods have not been used extensively mainly because contaminating amino acids can be introduced by the oxidases.

By gas chromatographic techniques it is now possible to make rapid and accurate analyses of amino acids. In this technique the volatility of the amino acids must be increased by reacting the compounds with various derivatizing reagents. Gehrke & Stalling (1967) have reviewed the literature on amino acid derivatives for gas chromatography and reported the development of the now commonly used N-trifluoroacetyl-n-butyl ester derivatives. Methods of analysis incorporating this derivative have yielded sensitivities in the nanogram and picogram regions (Zumwalt et al 1971). Casagrande (1970) has applied this derivative in the study of amino acids in peat accumulations in the Everglades. He was able to separate 35 amino acids on two gas chromatographic columns.

Gas chromatography has been quite useful in providing a means of separating and identifying individual amino acid enantiomers. Two approaches have been employed in geochemistry: 1. volatile enantiomeric derivatives of amino acids are chromatographed on optically active stationary phases (Nakaparksin et al 1970a, Koenig et al 1970), and 2. enantiomers are converted to diastereomeric derivatives and chromatographed on various optically inactive stationary phases (Kvenvolden et al 1972). Diastereomeric derivatives of amino acid enantiomers can be formed by the reaction of either their amino or carboxylic acid groups with an optically active reagent. For geochemical investigations esterification of carboxylic acid groups with optically active alcohols, particularly (+)-2-butanol, has been successfully applied (see section on stereochemistry). The method of preparation and resolution of 2-butyl esters of amino acids was developed independently by Gil-Av et al (1965) and Pollock et al (1965). Figure 2 illustrates the separation and resolution of amino acid diastereomeric derivatives that can be achieved by this method.

Ion exchange chromatography and gas chromatography have proven to be important methods for the separation, identification, and quantitation of amino acids, especially the common protein amino acids. With unknown mixtures of naturally occurring and synthetic amino acids these methods alone do not in general provide sufficient information for unequivocal identifications. Mass spectrometry, however, offers quite reliable criteria for structural determination and identification of amino acids; the combination of gas chromatography and mass spectrometry is especially suited for the analysis of unknown amino acids (Gelpi et al 1969, Lawless & Chadha 1971). The identification of nonprotein amino acids in meteorites (Kvenvolden et al 1971) illustrates a particularly successful application of the method of combined gas chromatography/mass spectrometry. The development of computer-controlled gas chromatography/mass spectrometry has led to the technique

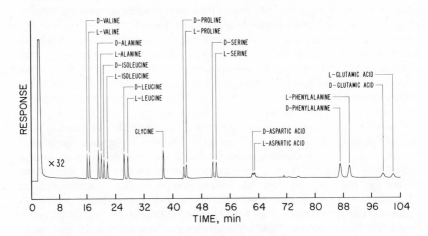

Figure 2 Separation and resolution of a standard mixture of amino acid derivatives. Gas chromatogram shows N-trifluoracetyl-O-acetyl-($\pm$)-2-butyl esters of 10 common amino acids. Chromatography: 46 m × 0.05 cm stainless steel column coated with UCON 75-H-90,000; temperature programmed from 100 to 150°C at 1°C/min (Kvenvolden et al 1972).

of mass fragmentography. The technique has recently been applied to the quantitation of 10 amino acids in soil extracts (Pereira et al 1973), and will undoubtedly find additional applications in the future.

AMINO ACIDS IN FOSSILS

After Abelson (1954) showed that amino acids are common organic constituents of fossil shells and bones, numerous studies followed, and much of this work has been reviewed (Hare 1969). Although most of the amino acids in Recent fossils are likely bound to each other in the form of peptides and proteins, older fossils usually contain increasing proportions of free amino acids, but the total concentration of amino acids decreases with time. Studies of modern and fossil clams of the genus *Mercenaria* illustrate the point. Abelson (1963) showed that most of the amino acids in Recent *Mercenaria* are protein-bound while in Miocene *Mercenaria* the content of protein-bound amino acid is negligible; the amino acids remaining in the Miocene sample are free. These observations show that hydrolysis of proteins is an important geochemical process taking place over geologic time. The extensive studies of *Mercenaria* strongly suggest that peptide bonds are hydrolyzed in geologically short times of 10,000 to 1,000,000 years, and the total amino acid concentration in older fossils is essentially equivalent to the content of free amino acids (Hare 1969). However, a large body of literature, much of it reviewed by Florkin (1969), indicates that peptide bonds may remain intact in fossils for several hundred million years. The work of Akiyama & Wyckoff (1970) supports the idea that the peptide bond can exist for long periods of time. They studied a series of fossil *Pecten* shells ranging in age from Pleistocene through Jurassic. All samples retained amino acids apparently bound as protein and peptide as well as free amino acids. From all these studies it becomes obvious that the long-term fate of the peptide bond is not yet well understood. Additional, careful, and systematic work on a number of fossil specimens is needed to clarify the issue or at least to explain the present discrepancy among results.

One example (Hare & Mitterer 1967) of the evidence supporting the idea that the content of amino acids in fossils diminishes with time is shown in Figure 3. In a modern *Mercenaria* the total amino acid content measured 9505 nmol/g; the total amount of amino acids in a Middle Miocene *Mercenaria* fossil was 408 nmol/g. The most abundant species in the modern shell are aspartic acid, serine, and glycine, while in the Miocene shell the most abundant amino acids are alanine, proline, and glutamic acid. Although the nonprotein amino acid, γ-aminobutyric acids. From all these studies it becomes obvious that the long-term fate of the independent study, Peterson et al (1972) observed that pipecolic acid is not found in modern *Mercenaria* but is an abundant amino acid in Miocene *Mercenaria* as shown by the gas chromatographic record in Figure 4.

Studies of the occurrence of amino acid enantiomers and diastereomers in fossil shells and bones have been increasingly emphasized since the diastereomeric pair, L-isoleucine–D-alloisoleucine, was first recognized (Hare & Mitterer 1967, Hare & Abelson 1968). These kinds of studies are discussed later.

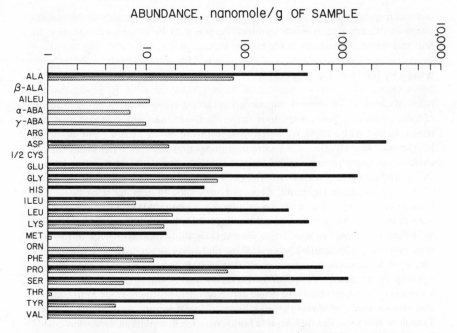

Figure 3 Histogram showing the distribution of amino acids in Modern (solid bars) and Middle Miocene (shaded bars) *Mercenaria* (data obtained from Table 26, Hare & Mitterer 1967).

Another important application of the geochemistry of amino acids has been given by King & Hare (1972a,b). They have determined the amino acid composition of 16 species of planktonic foraminifera isolated from deep-sea cores. The patterns of distribution of the amino acids appear to reflect the genotype, and phylogenetic affinities can be assigned among extinct species living at least 18 m.y. ago. This work represents an excellent example of geochemistry applied to paleotaxonomy and provides a potentially valuable paleobiochemical approach to evolution.

AMINO ACIDS IN SEDIMENTS

Initial interest in the amino acid content of sediments and sedimentary rocks began with the work of Erdman et al (1956). They showed that these compounds could be found in a Recent marine sedimentary deposit, as well as in a marine shale, Anahuac Formation, of Oligocene age. This work, along with Abelson's (1954) work on fossils, established the general ubiquity of amino acids in geochemical samples.

Extensive studies of sediment cores from basins offshore of California provided guides to the changes in concentration of amino acids that take place with depth (4 m) over short spans of geologic time. Degens et al (1961) identified in reducing sediments from Santa Barbara Basin 19 protein amino acids whose total concentration ranged from 815 to 177 ppm of dried sediment. The amino acids were irregularly distributed in the core, and there was no obvious concentration trend. In contrast, amino acids from cores from oxidizing sediments from the San Diego Trough decreased in concentration with depths (Degens et al 1963), a similar trend being observed for amino acids from the Experimental Mohole (Rittenburg et al 1963). Other examples of observations of the general trend of decreasing concentration of amino acids with depth are: (a) sediments off southeast Devon, England, ranging to about 9,000 years in age (Clarke 1967); (b) reducing sediments spanning about 10,000 years from Saanich Inlet, British Columbia (Brown et al 1972); (c) Pleistocene sediments from the Vema Fracture Zone of the Mid-Atlantic Ridge (Aizenshtat et al 1973); (d) Holocene to Oligocene sediments from Japan (Itihara 1973); and (e) Pleistocene sediments from the Cariaco Trench (Hare 1972, 1973b).

In much of the more recent work, increased attention is being given to nonprotein amino acids which may represent products from biological and/or chemical diagenesis. The two compounds, β-alanine and γ-aminobutyric acid, are often identified. Aizenshtat et al (1973) point out that the percentage of these two nonprotein amino acids relative to the total amino acids increases dramatically (13 to 70%) down the sediment column. They suggest that these compounds were derived by decarboxylation of other amino acids such as aspartic and glutamic acids. Itihara (1973) proposed that β-alanine and γ-aminobutyric acid came from biogenic matter produced by microorganisms. In studies of sediments from the Cariaco Trench, Hare (1972, 1973b) found, in addition to the common protein amino acids, several ninhydrin-positive compounds including the nonprotein amino acids γ-aminobutyric acid, diaminopimelic acid, β-alanine, and ornithine. Kemp & Mudrochova (1973) also call attention to the nonprotein amino acids in sediments from Lake Ontario. Another class of nonprotein amino acids that results from diagenetic processes in sediments includes amino acids of the D stereoisomic con-

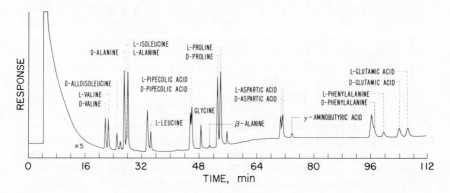

Figure 4 Gas chromatogram showing N-trifluoracetyl-O-acetyl-(+)-2-butyl esters of amino acids from a Miocene *Mercenaria*. Chromatography: 46 m × 0.05 cm stainless steel column coated with Carbowax 20M; temperature programmed from 100 to 160°C at 1°C/min.

figuration. The occurrences of D-amino acids in sediments have been studied in some detail and are described later. As an example of the presence of D-amino acids in marine sediments, Figure 5 shows the chromatographic record of amino acids recovered from foraminifera in sediments from the Caribbean Sea.

Amino acids also occur in the water column above the sediment-water interface. Degens et al (1964) demonstrated that many common amino acids are present in the water, and that the distribution of amino acids differs in the water and in the sediment. Other reports of the occurrences of amino acids in marine and fresh waters are Tatsumoto et al (1961), Pocklington (1971), Clarke et al (1972), and Peake et al (1972).

The possible role of amino acids during sedimentation has been emphasized by Mitterer (1968, 1971a,b, 1972a,b). He has shown that amino acids are present in calcareous oolites, aragonite needles, and calcified tissues of a variety of organisms. The amino acids in these substances apparently play a major role in the process of calcification.

A different kind of approach to the geochemistry of amino acids in sediments has been made by Stevenson & Cheng (1969, 1972). By measuring the amino acid nitrogen in core samples from the Argentine Basin they have noted fluctuations with depth. They believe that these changes with depth are related to alterations in warm and cool climates during Quaternary times. The amino acid nitrogen content of the sample is a measure of the total amino acids present. The fluctuations of amino acid nitrogen with depth in the Argentine Basin are in contrast with the general trend of decreasing total amino acid content with depth observed for many other marine sediments.

Following the first discovery of amino acids in sedimentary rocks by Erdman et al (1956), other rock samples have been shown to contain amino acids. Jones & Vallentyne (1960) found five amino acids in the Eocene Green River Formation. Later Kvenvolden & Peterson (1969) identified 12 amino acids in a sample from the

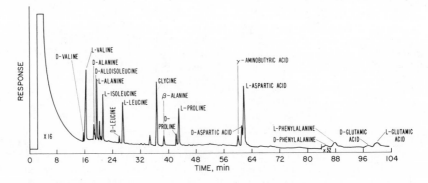

Figure 5   Gas chromatogram of N-trifluoroacetyl-O-acetyl-(+)-2-butyl esters of amino acids from foraminifera in sediments about 120,000 years old in the Caribbean Sea. Chromatography: same as for Figure 2 (Kvenvolden et al 1973).

same formation; they found that both D and L enantiomers of the amino acids were present, with the L enantiomers in greatest abundance. This observation suggested that some of the amino acids in this rock may be attributed to modern contamination because, as explained later, amino acids of Eocene age would be expected to occur in racemic mixtures.

Amino acids have been reported to be associated with highly carbonaceous substances. For example, Casagrande & Given (1974) describe a detailed investigation of amino acids in the peat deposits of the Florida Everglades. Amino acids were also reported to be present in bituminous sediments (Degens & Bajor 1960), anthracite (Heijkenskjöld & Möllerberg 1958), Mountsorrel bitumen (Aucott & Clarke 1966), and Precambrian thucholite (Prashnowsky & Schidlowski 1967). Reports of occurrences of indigenous amino acids in metamorphosed materials such as anthracite or in very old samples such as the thucholite must be viewed with skepticism. For example, serine and threonine are reported to be in the thucholite and associated Precambrian rocks, yet these two amino acids are known to be extremely unstable. Clearly many of the early claims for amino acids in metamorphosed and very ancient samples must be attributed to contamination. Certainly contamination should be considered as a possible source for amino acids found in Precambrian asbestos (Harington & Cilliers 1963). With the availability of techniques to study the stereochemistry of amino acids it should now be possible to reinvestigate some of these unusual occurrences of amino acids and more accurately ascertain their significance.

The ultimate fate of amino acids in sediments and fossils is not clearly understood. In early stages of diagenesis, microorganisms undoubtedly play a significant part in the degradation processes. Nonbiological processes take place during later stages of diagenesis and, as Erdman (1961) proposed, decarboxylation and deamination reactions may account for the formation of low molecular weight hydrocarbons. Support for this idea can be found in the work of Thompson & Creath (1966). They found that part of the light hydrocarbon content of modern and fossil shells can be accounted for as products from amino acids. Another mechanism by which amino acids may disappear from the sediments involves the irreversible uptake of amino acids by kerogen, as originally suggested by Abelson & Hare (1970), or by reaction of amino acids with other organic materials to yield humic acids and eventually kerogen (Abelson & Hare 1971, Hoering 1973).

LABORATORY DIAGENESIS

Some guidelines to the occurrence and fate of amino acids in geological substances have been obtained through laboratory studies wherein the effects of time, temperature, and environment have been evaluated. From the very beginning of geochemical considerations of amino acids, the thermal stability of these compounds was investigated. By heating alanine in solution at various times and temperatures, Abelson (1954) determined that this amino acid is sufficiently stable to remain intact for more than a billion years. This observation suggested that alanine and possibly many other amino acids would be useful chemical fossils for geochemical studies. The order of thermal stabilities for amino acids in solution was later established to be (Abelson 1959):

Relatively stable: alanine, glycine, glutamic, acid, leucine, isoleucine, proline, valine.

Moderately stable: aspartic acid, lysine, phenylalanine.

Relatively unstable: serine, threonine, arginine, tyrosine.

A somewhat similar order of thermal stabilities of amino acids in solution was determined by Vallentyne (1964).

Hare & Mitterer (1969) discovered that, besides the effects of temperature and time, pH also is a significant factor in the stability of some amino acids. Because of the influence of pH and possibly of the matrix on the stability of amino acids, they simulated the diagenesis of amino acids by subjecting amino acid-containing shells to various time-temperature-environment conditions. In the presence of water the shell amino acids degraded, and the order of stability was similar to that determined for amino acids in solution (Abelson 1959). Confirmation and extension of the work by Hare & Mitterer (1969) was provided by Vallentyne (1969), who lists the order of increasing thermal stability of amino acids in shells as serine, threonine, methionine, lysine, tyrosine, phenylalanine, isoleucine, leucine, proline, valine, glycine, alanine, and glutamic acid.

Thermal stabilities of amino acid components of humic acids were determined under oxidative conditions (Khan & Sowden 1971). The order of thermal stabilities followed closely that given previously for amino acids in shells and in solution: least stable were serine and threonine; intermediate in stability were proline, arginine, and lysine; and most stable were glutamic acid, aspartic acid, glycine, alanine, valine, isoleucine, leucine, tyrosine, phenylalanine, and histidine. From all of this work it seems that the relative thermal stability of amino acids has been well established.

Products resulting from thermal stability studies have provided clues to possible

Table 1 Products possibly resulting from the chemical diagenesis of amino acids

Amino acid	Products
Aspartic acid	Malic acid, ammonia
Glutamic acid	γ-aminobutyric acid
Glycine	Methylamine
Alanine	Ethylamine
Serine	Glycine, alanine, ethanolamine
Threonine	Glycine
Phenylalanine	Phenylethylamine
Methionine	Glycine, alanine, ammonia
Arginine	Proline, ornithine, ammonia
Proline	Ammonia
Leucine	Ammonia
Lysine	Pipecolic acid

precursor-product relationships occurring during the chemical diagenesis of amino acids. Table 1 lists some possible diagenetic reactions from thermal reaction kinetics studies (Vallentyne 1964, Povoledo & Vallentyne 1964, Vallentyne 1968). Although Vallentyne (1968) did not identify products, except ammonia, resulting from thermal treatment of lysine, Peterson et al (1972) found by gas chromatography/mass spectrometry that one of the products of thermal treatment is pipecolic acid. This nonprotein amino acid was probably not observed previously because of the low color constant of its ninhydrin derivative. Decomposition and transformations of amino acids also take place biologically (Abelson 1963), and two of the precursor-product relationships which may account for some of the amino acids in geochemical samples are : aspartic acid → β-alanine and α-alanine, and glutamic acid → γ-amino-butyric acid. Both β-alanine and γ-aminobutyric acid are nonprotein amino acids that are commonly observed in geochemical studies. It appears that γ-aminobutyric acid can result from both chemical and biological diagenesis of glutamic acid; β-alanine apparently can be derived from aspartic acid only by biological decarboxylation as emphasized by Bada (1971).

Laboratory simulations of amino acid diagenesis have also been used to study the changes that take place with time in the stereochemistry of amino acids. Research into kinetics of racemization and epimerization appears next.

STEREOCHEMISTRY

Geochemical studies of D and L enantiomers of amino acids have added new dimensions to the understanding of the occurrence and fate of amino acids in terrestrial and extraterrestrial samples. The importance to geochemistry of the stereochemical properties of amino acids was first recognized by Hare & Mitterer (1967) and Hare & Abelson (1968). In studies of the amino acids they observed that, although isoleucine was commonly found in shells ranging in age from modern to Miocene, alliosoleucine was absent in living shell protein but was found to increase in abundance relative to isoleucine in progressively older fossils. Equilibrium mixtures of isoleucine and alloisoleucine were established by Miocene time. Using enzymatic techniques (see section on techniques), Hare & Abelson (1968) were able to follow the change in relative abundances with geologic time of D- and L-amino acids isomers in fossil *Mercenaria* shells. They showed, for example, that L-isoleucine interconverts to D-alloisoleucine and that no appreciable amounts of D-isoleucine or L-alloisoleucine are formed. (Figure 6 illustrates the possible stereochemical structures of isoleucine–alloisoleucine.) From these observations they inferred that the extent of racemization (epimerization in the case of diastereomers such as isoleucine) of amino acids might be applied to ascertaining the age of shells. This idea of using amino acids as a potential chemical geochronometer was expanded upon the next year by Hare & Mitterer (1969), who also suggested that these kinds of studies could be used for paleothermometry and stratigraphic correlations.

The dominant process giving rise to the apparent interconversion of D and L isomers of amino acids in geochemical samples is racemization, and a first-order reversible kinetic model has been used to interpret the data (Hare & Mitterer 1969,

Bada et al 1970, Nakaparksin et al 1970b, Wehmiller & Hare 1971, Bada 1971, Bada & Schroeder 1972). The model was derived as follows:

$$\text{L-amino acid} \underset{k_2}{\overset{k_1}{\rightleftarrows}} \text{D-amino acid} \qquad\qquad\qquad 1.$$

where k_1 and k_2 are the first-order rate constants for interconversion of the L-amino and D-amino acids. For enantiomeric amino acids, $k_1 = k_2$. For diastereomers such as L-isoleucine–D-alloisoleucine, $k_1 \neq k_2$. The rate law for the interconversion (racemization and epimerization) of amino acids is

$$\frac{-d(\text{L})}{dt} = k_1(\text{L}) - k_2(\text{D}) \qquad\qquad\qquad 2.$$

where (L) is the concentration of the L-amino acid and (D) is the concentration of the D-amino acid. The integrated rate expression is

$$\ln\left[1 + \frac{(\text{D})}{(\text{L})}\middle/ 1 - \frac{k_2(\text{D})}{k_1(\text{L})}\right] = \left(1 + \frac{k_2}{k_1}\right)k_1\, t + \text{const} \qquad\qquad 3.$$

A derivation of equation 3 from equation 1 has been given by Bada & Schroeder (1972).

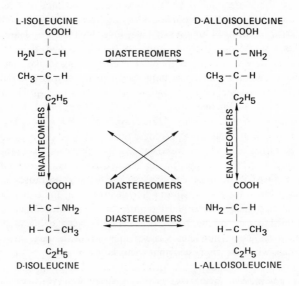

Figure 6   Stereoisomers of isoleucine; absolute configuration is indicated by Fischer projections.

First-order equations have been expressed by Wehmiller & Hare (1971) in a different but equivalent form. They determined the extent of racemization as a fraction of the completion of the reaction. This approach results in the expression:

$$\frac{X_e - X}{X_e} = \exp\left[-\left(1 + \frac{k_1}{k_2}\right)k_2 t \right] \qquad\qquad 4.$$

where X_e is the value of (D)/(D)+(L) at equilibrium and X is the value of (D)/(D)+(L) at any time t. For the enantiomeric amino acids, X_e has a value of 0.50; for L-isoleucine–D-alloisoleucine, $X_e = 0.56$ when k_1/k_2 is chosen to be 1.25. This equilibrium ratio has been found to be between 1.25 and 1.40 from laboratory experiments and from fossil shells (Hare & Mitterer 1967, 1969; Nakaparksin et al 1970b).

Marine sediments have proven useful in testing the idea that the changes in relative ratios of amino acid enantiomers and diastereomers result mainly from racemization and epimerization, and that the reaction follows a first-order reversible kinetic model. Bada et al (1970) showed that isoleucine in a sediment core from near the Mid-Atlantic Ridge apparently had undergone a slow epimerization giving rise to mixtures of isoleucine and alloisoleucine, and that the extent of epimerization increases with increasing depth and age. They assumed that the rate of interconversion of isoleucine in sediments is the same as that in aqueous solution (Bada 1971). With this assumption they calculated sedimentation rates and ages based on the first-order reversible kinetic model. Wehmiller & Hare (1971) and Hare (1971) quickly pointed out that rates of epimerization obtained from kinetic studies of free isoleucine in aqueous solution cannot be used directly in considerations of isoleucine in marine sediments. They showed also that first-order reversible kinetics do not totally describe the reactions taking place in sediments. Linear kinetics seem to apply only to alloisoleucine–isoleucine ratios up to about 0.25 after which the kinetics are non-linear. From these observations they calculated a sedimentation rate about an order of magnitude greater than suggested by Bada et al (1970). Bada & Schroeder (1972) later revised their estimate of sedimentation rate so that it conformed more closely to the value calculated by Wehmiller & Hare (1971). This revision was based on applications of kinetic studies of marine sediments at high temperatures which confirmed the findings of Wehmiller & Hare (1971). For alloisoleucine–isoleucine ratios greater than 0.3–0.4, the epimerization no longer obeys reversible first-order kinetics.

Much attention has been given to the epimerization of isoleucine in marine sediments and fossils. A reason for this attention is that isoleucine and alloisoleucine can be conveniently measured by conventional automated ion exchange chromatography. With the development of specialized gas chromatographic techniques (see section on techniques), it has become possible to measure the relative ratios of many chiral amino acids. For example, Kvenvolden et al (1970a) examined the amino acids in sediments of Saanich Inlet, British Columbia, and showed that the D/L enantiomeric ratios of alanine, valine, leucine, proline, phenylalanine, and glutamic acid increase with sample age (to about 9000 years) apparently because of partial racemization.

In a more detailed study, Kvenvolden et al (1973) extended the original work of Wehmiller & Hare (1971) by showing that the enantiomeric ratios of a number of amino acids from foraminifera in sediments from the Atlantic Ocean and Caribbean Sea increase with depth and age over a span from 40,000 to 2,000,000 years. The changing ratios did not follow first-order reversible rate laws; this observation supports the findings of Wehmiller & Hare (1971), who showed that the epimerization of isoleucine did not obey reversible, first-order kinetics. Figure 7 is a graph of $(X_e - X)/X_e$ vs time from equation 4 for isoleucine and alanine. Clearly, first-order reversible kinetics do not totally describe the actual kinetics of racemization (epimerization) as observed in these natural sediment samples. Each amino acid in the total amino acid fraction of foraminifera from marine sediments has its own characteristic rate of racemization (Kvenvolden et al 1973). Valine, glutamic acid, and leucine racemize at slower rates than do phenylalanine, alanine, aspartic acid, and proline. Isoleucine epimerizes about as slowly as valine racemizes. Free amino acids in the foraminifera are more extensively racemized than are the total amino acids, probably due to metal ion-catalyzed racemization as suggested by Bada (1971) and Bada & Schroeder (1972).

Apparently one reason racemization of amino acids in sediments does not follow reversible first-order kinetics is that amino acids are present in various forms, such as

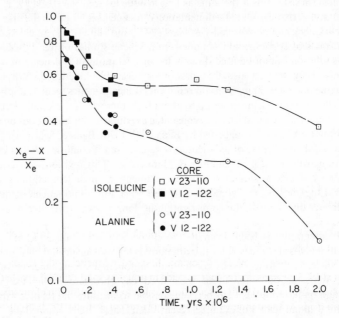

Figure 7   Graph of $(X_e - X)/X_e$ vs time for isoleucine–alloisoleucine $(k_1/k_2 = 1.25)$ and alanine $(k_1/k_2 = 1.0)$ for time range 0 to 2×10^6 years. Data obtained from amino acids in foraminifera from sediments of Core V23-110 (Atlantic Ocean) and V-12-122 (Caribbean Sea) (Kvenvolden et al 1973).

protein, peptide, and free. The amino acids in these forms racemize at very different rates, and the final collection of amino acids examined is a composite from the different sources. In a detailed study of isoleucine–alloisoleucine from five fractions ("free", "peptide", "residue", total sediment, and foraminifera) isolated from cores of the Deep Sea Drilling Project, Bada & Mann (1973) showed a striking difference in rates of epimerization among the fractions. Although the isoleucine–alloisoleucine in the free, peptide, and foraminifera fractions did not follow reversible first-order rate laws, the total sediment and residue fractions were nearly linear. The explanation offered for these results is that the bulk of the amino acids in these cores is in the form of protein in which the isoleucine epimerizes at a very slow rate. Another potential complication indicated by King & Hare (1972c) is that isoleucine in individual species of foraminifera appears to epimerize at different rates.

It becomes obvious from this discussion that many interrelated reactions must affect the stereochemistry of amino acids in sediments. Bada & Schroeder (1972) developed a model of a multistep reaction sequence showing the various reactions isoleucine is likely to undergo in sediments. The scheme is redrawn in Figure 8 to apply to any chiral amino acid. The scheme offers a reasonable assessment of the reaction sequence that may take place in sediments. Although the interconversion of D and L isomers of amino acids appears to be the dominant process influencing D/L ratios of amino acids in sediments, it should be recognized that amino acids of secondary origins may also affect D/L ratios of certain amino acids. For example, alanine may be produced during decomposition of other amino acids such as serine, and the resulting alanine would influence in an unknown manner the final D/L ratio measured for the total alanine in the sample.

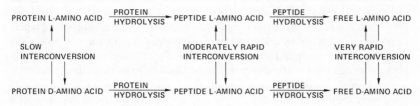

Figure 8   Possible reaction sequence of amino acid enantiomers during diagenesis (adapted from Bada & Schroeder 1972).

In contrast to the nonlinear kinetics exhibited by amino acids from calcareous sediments (Wehmiller & Hare 1971, Bada & Schroeder 1972, Kvenvolden et al 1973) amino acids in bone follow the reversible first-order kinetic model (Bada 1972b, Bada et al 1973a) just as do free amino acids in aqueous solution (Nakaparksin et al 1970b, Bada 1971, 1972a). This observation has led to the application of racemization of amino acids to the dating of fossil bones (Turekian & Bada 1972). By using the racemization of aspartic acid from bones dated by radiocarbon, Bada & Protsch (1973), in a unique approach, calculated the in situ first-order rate constant for the interconversion of aspartic acid enantiomers (see also Bada et al 1974a, Bada et al 1974b). They applied this constant to date other bones, either too old or too small for radiocarbon dating, from the same or similar localities. In addition to geochronologic implications, the racemization of amino acids has also been used to determine paleotemperatures in bones (Bada et al 1973b, Schroeder & Bada 1973). Mitterer (1972a) calculated paleotemperatures based on alloisoleucine–isoleucine ratios from fossil *Mercenaria* shells from Florida and North Carolina. The temperatures obtained placed an upper limit to late Pleistocene glacial temperatures in the area. Besides paleothermometry, Mitterer (1974) also described other applications for amino acid racemization studies: correlation of stratigraphic units, verification of radiocarbon ages, and determination of fluctuations in Pleistocene sea levels.

Geochemical studies of the stereochemistry of amino acids have shown a number of potentially useful areas of application, particularly in geochronology and paleothermometry (Bada & Schroeder 1974). But the reactions taking place in geochemical samples are complicated, and the possibility of contamination is always high. Therefore, although the literature reports successful applications of amino acid racemization in age and temperature determinations, these reports should be considered preliminary and a beginning to more intensified and detailed work.

AMINO ACIDS IN METEORITES

The fall of the Murchison carbonaceous meteorite in Australia in 1969 has provided extraterrestrial material containing indigenous amino acids which undoubtedly originated through abiotic processes (Kvenvolden et al 1970b, Cronin & Moore 1971, Oró et al 1971a). The abiotic nature of these processes is supported by the fact that most of the amino acids discovered are not found in terrestrial proteins and by the observation that asymmetric amino acids are present as racemic mixtures. Evidence for the racemic nature of the amino acids was provided by the gas chromatographic response of diastereomeric derivatives as illustrated in Figure 9 showing the resolution of various amino acids and especially of (DL)-α-amino-n-butyric acid and (DL)-α-alanine from the Murchison meteorite. Continuing work on the amino acids in this meteorite has demonstrated that at least 17 additional amino acids are present; these were identified by structural class and not by specific molecular structures (Lawless 1973).

Prior to their discovery in the Murchison meteorite, amino acids had been reported to be present in a number of carbonaceous and noncarbonaceous meteorites, but interpretations with regard to the significance of these findings were quite uncertain. Degens & Bajor (1962) first found a suite of common protein amino acids in Murray and Bruderheim meteorites. Three principal hypotheses were considered to explain these amino acids: (*a*) they are products of extraterrestrial life, (*b*) they are abiotic compounds from early stages of the solar nebula, or (*c*) they are the result of terrestrial contamination. Although the amino acids found are common in living organisms, these authors rejected hypothesis *a* and suggested that *b* and/or *c* accounted for their observations.

This first work was quickly followed by more detailed investigations. Kaplan et al (1963) found a number of protein amino acids in eight carbonaceous and six non-

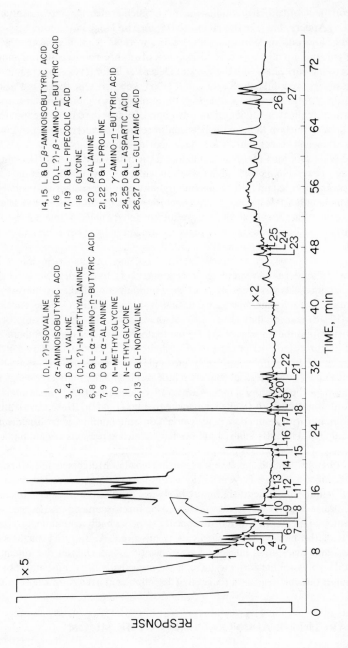

Figure 9 Gas chromatogram of N-trifluoroacetyl-O-acetyl-(+)-2-butyl esters of amino acids from the Murchison meteorite. Separation and resolution of D- and L-α-amino-*n*-butyric acid and D- and L-α-alanine are emphasized. Chromatography: same as for Figure 2.

carbonaceous meteorites. Their results on the Orgueil meteorite were essentially duplicated by Anders et al (1964). Vallentyne (1965), who looked for amino acids in Orgueil and Holbrook meteorites, confirmed the work of Kaplan et al (1963) in showing that amino acids do occur in samples of both carbonaceous and non-carbonaceous meteorites. As with the work of Degens & Bajor (1962), interpretation of these results proved difficult. Kaplan et al (1963) believed that the amino acids resulted from abiotic chemical syntheses with a possible overprint of terrestrial contaminations. Vallentyne (1965), on the other hand, felt that the data at the time did not allow final distinction to be made between terrestrial-extraterrestrial, biogenic-abiogenic, or ancient-recent sources for the meteoritic amino acids.

Detailed considerations of amino acids found on hands and fingers (Hamilton 1965, Oró & Skewes 1965) provided information which clearly showed that most of the amino acids reported up to that time in Orgueil and other meteorites probably resulted from handling (Hayes 1967). This conclusion plus the observation by Oró & Tornabene (1965) that a sample of Murray meteorite contained viable, terrestrial bacteria generally discredited the early reports claiming indigenous amino acids in meteorites.

But it is now known, as mentioned in the first part of this discussion, that carbonaceous chondrites contain indigenous amino acids. In the early work, where indigenous amino acids were present, most were undoubtedly masked by terrestrial contamination. For example, a reexamination of a sample of Murray meteorites by Oró et al (1971b) showed that about 10% of the amino acids are probably of a chemical origin and the remainder can be attributed to biological contamination. Using a freshly prepared interior piece of Murray meteorite, Cronin & Moore (1971) and Lawless et al (1971) found a suite of amino acids which appears to be essentially uncontaminated by any terrestrial sources. These results are qualitatively similar to those reported for the Murchison meteorite (Kvenvolden et al 1970b, Kvenvolden et al 1971). Additional work on Orgueil meteorite by Oró et al (1971b) suggested that most of the amino acids resulted from terrestrial contamination by microorganisms. Lawless et al (1972) found a suite of amino acids which represents contamination superimposed on an indigenous population of molecules.

It is now obvious that the two carbonaceous meteorites, Murchison and Murray, which are classified as C2 (see Hayes 1967 for a summary of the classification system), contain indigenous amino acids that most likely are products of abiotic syntheses. Only one C1 carbonaceous meteorite, the Orgueil, has been examined with techniques available since 1970, and this meteorite appears to be so badly contaminated that the pattern of the indigenous amino acids is obscured. A definitive picture of indigenous amino acids in C1 meteorites must await the finding of uncontaminated specimens. The only C3 meteorite to be reported on recently is Allende, and it contains no indigenous amino acids at the level of detection employed (Cronin & Moore 1971).

THE SEARCH FOR AMINO ACIDS ON THE MOON

In 1969 geochemists had their first opportunity to examine samples returned from the Moon, and amino acids were among the several classes of organic compounds sought by four different groups of investigators (Ponnamperuma et al 1970, Oró et al 1970a, Fox et al 1970, Nagy et al 1970a). Unfortunately, starting with the first reports of these investigations until the present time, a consensus has never been obtained with regard to the occurrence and significance of amino acids from the Moon. Following the initial publications, more detailed papers concerning Apollo 11 samples appeared. For example, Murphy et al (1970) and Nagy et al (1970b) reported the common amino acids glycine and alanine (total concentration 68 ppb), while Hare et al (1970) reported the presence of these two compounds plus aspartic acid, glutamic acid, serine, and threonine (total concentration about 50 ppb). Gehrke et al (1970) and Oró et al (1970b), on the other hand, failed to find amino acids.

Disagreement continued as results were obtained from Apollo 12 lunar samples. Harada et al (1971) reported amino acids at a total concentration level of 69 ppb, but Nagy et al (1971), Oró et al (1971c), and Gehrke et al (1971) did not detect amino acids above the general background levels of their techniques. Analyses of samples from the Apollo 14 mission yielded more satisfactory results. Modzeleski et al (1973), Fox et al (1972), and Gehrke et al (1972) all found amino acids in a specially collected Apollo 14 sample but at total concentrations below 10 ppb. Other Apollo 14 sample extracts examined by Fox et al (1972) contained up to 37 ppb total amino acid. Amino acids were also detected in extracts from Apollo 15 and 16 missions but the total concentrations were generally lower than in extracts from samples of earlier missions (Fox et al 1973, Gehrke et al 1973).

From the extensive work done since 1969 it seems definitely established that amino acids are present in low concentrations in acid-hydrolyzed extracts of lunar fines. Glycine and alanine are most commonly observed, and the total concentration of amino acids reported has not exceeded 70 ppb. The identity of the amino acids has been established through both ion exchange and gas chromatographic techniques. There has never been sufficient material available, however, to confirm identifications by mass spectrometry. The exact significance of these amino acids remains a mystery. There are several possibilities: 1. The amino acids could represent laboratory contamination. This explanation probably can be eliminated because of the strict controls used by the investigative teams. 2. The amino acids could be contaminants reaching the sample between the time it was collected and the time that it was analyzed in the chemical laboratories. Contamination of this kind is very difficult to monitor. Modzeleski et al (1973) attempted to check one aspect of this potential problem by examining amino acids on the astronaut's glove; they concluded that the amino acids in lunar samples could not have arisen from the glove. 3. The amino acids could have been synthesized on the Moon by rocket exhaust. This idea has been investigated by Fox et al (1970), who concluded that this source for amino acids was unlikely. 4. The amino acids are indigenous to the Moon. This theory is not considered likely because of the inclement conditions that exist at the lunar surface (Sagan 1972); any indigenous amino acids would likely be destroyed. 5. The amino acids could be derived during laboratory extraction and hydrolysis procedures from precursor molecules which are indigenous to the Moon or which are produced at the lunar surface through, for example, solar-wind interactions. This last explanation, first suggested by Harada et al (1971), has some appeal, but until the specific identity of the postulated precursors is obtained the idea must remain a working hypothesis.

AMINO ACIDS IN SPACE RESEARCH

In the design of experiments to search for extraterrestrial life by means of techniques involving organic chemistry, the ability to distinguish between biogenic and abiogenic molecules is very important. In recent investigations of the geochemistry and cosmochemistry of amino acids, data have been obtained from four sources: 1. organisms of the biosphere, 2. modern and ancient sediments on Earth, 3. carbonaceous meteorites, and 4. laboratory experiments simulating prebiotic chemical processes. From the data, criteria (Table 2) have been developed to characterize amino acid molecules with regard to their biological or nonbiological origins (Kvenvolden 1973). These criteria are less restrictive than those originally established by Hare (1969) based on the isomeric compositions of only isoleucine. With these criteria the significance of amino acids found on Mars, for example, should be more correctly ascertained.

CONTAMINATION

One of the great difficulties in the study of the geochemistry of amino acids is the persistent problem of contamination. Amino acids are ubiquitous, and it is often a challenge to prove that amino acids are indigenous to the samples under investigation. It is an even greater challenge to provide convincing evidence that the amino acids in geochemical samples are both indigenous and syngenous, meaning

Table 2 Criteria for mode of origin of amino cids

	Biogenic	Abiogenic
Structure	α-amino carboxylic acids dominant with occasional β- and γ-amino carboxylic acids in ancient deposits of biologic origin	α-amino carboxylic acids common along with β- and γ-amino carboxylic acids. N-alkyl (methyl and ethyl) amino acids present
Enantiomers	L-amino acids most abundant. Abundance of D-amino acids increases with age. Amino acids older than 15×10^6 yr should occur as racemic mixtures.	Racemic mixtures of amino acid enantiomers
Composition	Protein amino acids dominate. Ancient amino acids may include nonprotein amino acids such as β-alanine, γ-amino-butyric acid, and alloisoleucine	Mixtures of both protein and nonprotein amino acids. Most amino acid isomers of low carbon number present
Distribution	Variable. Glutamic acid and aspartic acid common in young amino acid populations; glycine and alanine common in ancient amino acids	Variable. Glycine and α-aminoisobutyric acid possibly dominant

that the amino acids were incorporated into the sample at the same time that the original sample was formed.

Contamination can occur any time during the history of a sample from lithification and fossilization to the final analysis in the laboratory. That amino acids in finger-prints of persons handling the sample were a possible serious source of contamination was recognized by Hamilton (1965), Oró & Skewes (1965), and Hare (1965). They pointed out that the characteristic patterns of fingerprint contamination always included high concentrations of both serine and glycine. The level of contamination of individual amino acids from this source could be as much as 10^{-7} mole. Rash et al (1972) reviewed numerous sources of laboratory contamination and showed that nanogram to microgram amounts of amino acids are found in reagents, latex gloves, masslin fibres, fingerprints, skin, dandruff, hair, saliva, dust, and cigarette smoke. Amino acids are often found in the hydrochloric acid reagents commonly used in amino acid analyses (Hare 1965, Rash et al 1972), and it appears that distillation of the hydrochloric acid will not remove the contaminating amino acids (Wolman & Miller 1971). Hydrochloric acid with less than 0.2 nmole of amino acid per liter can be prepared, however, by condensing gaseous hydrochloric acid with liquid nitrogen and transferring the gas through a vacuum line to water which has been carefully distilled in glass (Wolman & Miller 1971). Sources of contamination within the analytical laboratory can usually be identified, and the contamination minimized. What happens to the samples before they reach the laboratory cannot usually be well controlled.

A dramatic example of contamination in geochemical studies is the work before 1970 in which numerous protein amino acids were reported to be present in meteorites, as discussed previously. It is now generally agreed that the reported amino acids came from fingerprints (Oró & Skewes 1965) and/or possibly terrestrial bacteria (Oró & Tornabene 1965).

The danger of contamination always exists when the concentration of the indigenous amino acids in small. An example of possible natural contamination is the finding by Schopf et al (1968) of amino acids in three different samples of chert—Bitter Springs, Gunflint, and Fig Tree—representing points in time 1, 2, and 3 b.y. ago, respectively. Great precautions were taken in preparing these samples to assure that any surficial contamination was removed and that the laboratory procedures had sufficiently low analytical blanks to permit significant results to be obtained. The amino acids found were interpreted to be indigenous and syngenous to the cherts. After this work was completed, new techniques in organic geochemistry were applied to determine the stereochemistry of amino acid enantiomers (see section on stereochemistry). Abelson & Hare (1969) studied in detail the amino acids in the Gunflint chert and concluded, on the basis of their thermal history and stereo-chemistry, that the compounds were of recent origin. Kvenvolden et al (1969) determined by gas chromatographic techniques that the amino acids in Fig Tree chert have the L configuration only, a result supporting what was observed by Abelson & Hare (1969), who had used enzymatic techniques to show that the amino acids in Gunflint chert were also of the L configuration.

It is generally believed that billion-year-old amino acids, if they survive this great span of time at all, should be present as racemic mixtures (both D and L enantiomers of each amino acid equally abundant). Studies of amino acids in Tertiary sediments and shells certainly suggest that amino acids of Mesozoic age and older should occur as racemic mixtures (Hare & Abelson 1968, Wehmiller & Hare 1971). The finding of only L-configured amino acids in the Precambrian samples indicates that natural, as yet unspecified, contamination has probably taken place.

If this interpretation with regard to the significance of amino acids found in very ancient rocks is correct, then the stereochemistry of amino acids becomes useful in evaluating the extent of contamination of amino acids in geochemical samples. Whereas geologically young samples should show ratios of D- to L-amino acids of less than one, in older samples (greater than 15 m.y. old) the amino acid enantiomers and diastereomers should be at equilibrium. This limit of 15 m.y. may be pushed back in time as additional studies are made, particularly on noncalcareous samples, but it is doubtful that unequivocal evidence will ever be found for nonracemic amino acids in sediments of Mesozoic age and older. Thus claims for the presence of indigenous amino acids in fossils of the Mesozoic, Paleozoic, and Precambrian can be tested by determining their stereochemistry.

Stereochemistry can be used in another way to evaluate whether a sample is contaminated. Studies by Kvenvolden et al (1973) and Bada et al (1973a) have shown that during racemization each amino acid has its own characteristic rate. While amino acids such as valine and leucine racemize slowly, aspartic acid and alanine racemize quickly. The order for the racemization rates in bone, for example, is aspartic acid > alanine = glutamic acid > isoleucine $\cong$ leucine (Bada et al 1973a). If the extent of racemization of amino acids in an unknown sample matches this order, the sample is considered to be uncontaminated and of use for geochronology and paleothermometry.

The hazards of contimination in geochemical studies of amino acids are great. With new knowledge and techniques it may be wise to reevaluate some of the previous work, especially where very low concentrations of amino acids have been reported.

SUMMARY

During the last five years important advances have been made in understanding the geochemistry of amino acids. Some of these advances were made possible through the availability of new techniques, particularly gas chromatography/mass spectrometry, which permitted the identification of uncommon amino acids such as those found in carbonaceous meteorites. New techniques also increased sensitivity in analyses like those applied to the search for amino acids in samples from the Moon.

Much attention was focused on studies concerned with the stereochemistry of amino acids. The ability to measure the extent of racemization or epimerization of amino acids in various samples led to the development of potentially valuable tools for geochronology and paleothermometry. The finding of racemic mixtures of amino acids in carbonaceous meteorites provided evidence that these compounds were not biological in origin, at least not modern terrestrial contaminations.

The effects of natural diagenetic processes can now be generalized for geochemical studies where changes in amino acid content and stereochemistry are observed as a function of increasing time: 1. the concentration of total amino acids diminishes, 2. the more stable amino acids become dominant, 3. nonprotein amino acids become more obvious as products of degradation, and 4. the distribution of amino acid enantiomers and diastereomers reaches equilibrium. Laboratory studies of diagenesis support these generalizations. The ultimate fate of amino acids may be conversion to simple hydrocarbons or incorporation into humus and kerogen.

Contamination has been recognized as a potential problem in geochemical studies of amino acids. Fortunately, stereochemical studies can aid in determining the extent of possible contamination. Also, increased knowledge of the overall geochemistry of amino acids now permits better assessment to be made of the results obtained.

The long-term fate of the peptide bond is not clearly defined because of apparently conflicting experimental results and interpretations. Systematic studies with the new techniques should resolve this controversial problem in the geochemistry of amino acids.

Appendix Amino acid structures

GLYCINE† (GLY)	$\begin{array}{c} \text{COOH} \\ \| \\ H_2N - C - H \\ \| \\ H \end{array}$	**GLUTAMIC ACID†** (GLU)	$\begin{array}{c} \text{COOH} \\ \| \\ H_2N - C^* - H \\ \| \\ (CH_2)_2 \\ \| \\ \text{COOH} \end{array}$
ALANINE† (ALA)	$\begin{array}{c} \text{COOH} \\ \| \\ H_2N - C^* - H \\ \| \\ CH_3 \end{array}$	**CYSTINE†** (CYS)	$\begin{array}{cc} \text{COOH} & \text{COOH} \\ \| & \| \\ H_2N - C^*-H & H_2N-C^* - H \\ \| & \| \\ H_2C - S & - S - CH_2 \end{array}$
VALINE† (VAL)	$\begin{array}{c} \text{COOH} \\ \| \\ H_2N - C^* - H \\ \| \\ H_3C - CH \\ \| \\ CH_3 \end{array}$	**CYSTEINE†** (CYSH or ½CYS)	$\begin{array}{c} \text{COOH} \\ \| \\ H_2N - C^* - H \\ \| \\ CH_2 \\ \| \\ SH \end{array}$
LEUCINE† (LEU)	$\begin{array}{c} \text{COOH} \\ \| \\ H_2N - C^* - H \\ \| \\ CH_2 \\ \| \\ H_3C - CH \\ \| \\ CH_3 \end{array}$	**METHIONINE†** (MET)	$\begin{array}{c} \text{COOH} \\ \| \\ H_2N - C^* - H \\ \| \\ (CH_2)_2 \\ \| \\ S \\ \| \\ CH_3 \end{array}$
ISOLEUCINE† (ILEU)	$\begin{array}{c} \text{COOH} \\ \| \\ H_2N - C^* - H \\ \| \\ H_3C - C^* - H \\ \| \\ C_2H_5 \end{array}$	**PHENYLALANINE†** (PHE)	$\begin{array}{c} \text{COOH} \\ \| \\ H_2N - C^* - H \\ \| \\ CH_2 \\ \bigcirc \end{array}$
SERINE† (SER)	$\begin{array}{c} \text{COOH} \\ \| \\ H_2N - C^* - H \\ \| \\ CH_2 \\ \| \\ OH \end{array}$		
THREONINE† (THR)	$\begin{array}{c} \text{COOH} \\ \| \\ H_2N - C^* - H \\ \| \\ H - C^* - OH \\ \| \\ CH_3 \end{array}$	**TYROSINE†** (TYR)	$\begin{array}{c} \text{COOH} \\ \| \\ H_2N - C^* - H \\ \| \\ CH_2 \\ \bigcirc \\ OH \end{array}$
PROLINE† (PRO)	$\begin{array}{c} \text{COOH} \\ \| \\ HN - C^* - H \end{array}$	**LYSINE†** (LYS)	$\begin{array}{c} \text{COOH} \\ \| \\ H_2N - C^* - H \\ \| \\ (CH_2)_4 \\ \| \\ NH_2 \end{array}$
HYDROXYPROLINE† (HYPRO)	$\begin{array}{c} \text{COOH} \\ \| \\ HN - C^* - H \\ \\ OH \end{array}$		
ASPARTIC ACID† (ASP)	$\begin{array}{c} \text{COOH} \\ \| \\ H_2N - C^* - H \\ \| \\ CH_2 \\ \| \\ \text{COOH} \end{array}$	**HYDROXYLYSINE†** (HYLYS)	$\begin{array}{c} \text{COOH} \\ \| \\ H_2N - C^* - H \\ \| \\ (CH_2)_2 \\ \| \\ HO - C^* - H \\ \| \\ CH_2 \\ \| \\ NH_2 \end{array}$

Appendix (*continued*)

HISTIDINE† (HIS)

$$COOH$$
$$|$$
$$H_2N - C^* - H$$
$$|$$
$$CH_2$$

(imidazole ring, NH / N)

ARGININE† (ARG)

$$COOH$$
$$|$$
$$H_2N - C^* - H$$
$$|$$
$$(CH_2)_3$$
$$|$$
$$NH$$
$$|$$
$$C=NH$$
$$|$$
$$NH_2$$

TRYPTOPHANE† (TRY)

$$COOH$$
$$|$$
$$H_2N - C^* - H$$
$$|$$
$$CH_2$$

(indole ring, N–H)

ALLOISOLEUCINE

$$COOH$$
$$|$$
$$H - C^* - NH_2$$
$$|$$
$$H_3C - C^* - H$$
$$|$$
$$C_2H_5$$

CITRULLINE (CIT)

$$COOH$$
$$|$$
$$H_2N - C^* - H$$
$$|$$
$$(CH_2)_3$$
$$|$$
$$NH$$
$$|$$
$$C=O$$
$$|$$
$$NH_2$$

ORNITHINE (ORN)

$$COOH$$
$$|$$
$$H_2N - C^* - H$$
$$|$$
$$(CH_2)_3$$
$$|$$
$$NH_2$$

DIAMINOPIMELIC ACID (DAPA)

$$COOH$$
$$|$$
$$H_2N - C^* - H$$
$$|$$
$$(CH_2)_3$$
$$|$$
$$H_2N - C^* - H$$
$$|$$
$$COOH$$

α-AMINOISOBUTYRIC ACID (α-AIB)

$$COOH$$
$$|$$
$$H_2N - C - CH_3$$
$$|$$
$$CH_3$$

α-AMINO-n-BUTYRIC ACID (α-ABA)

$$COOH$$
$$|$$
$$H_2N - C^* - H$$
$$|$$
$$C_2H_5$$

β-AMINOISOBUTYRIC ACID (β-AIB)

$$COOH$$
$$|$$
$$H_3C - C^* - H$$
$$|$$
$$CH_2$$
$$|$$
$$NH_2$$

β-AMINO-n-BUTYRIC ACID (β-ABA)

$$COOH$$
$$|$$
$$CH_2$$
$$|$$
$$H_2N - C^* - H$$
$$|$$
$$CH_3$$

γ-AMINOBUTYRIC ACID (γ-ABA)

$$COOH$$
$$|$$
$$(CH_2)_3$$
$$|$$
$$NH_2$$

β-ALANINE (β-ALA)

$$COOH$$
$$|$$
$$(CH_2)_2$$
$$|$$
$$NH_2$$

ISOVALINE

$$COOH$$
$$|$$
$$H_2N - C^* - CH_3$$
$$|$$
$$C_2H_5$$

NORVALINE

$$COOH$$
$$|$$
$$H_2N - C^* - H$$
$$|$$
$$(CH_2)_2$$
$$|$$
$$CH_3$$

N-METHYLALANINE

$$COOH$$
$$|$$
$$H_3C - N - C^* - H$$
$$\quad\; |\qquad |$$
$$\quad\; H \quad CH_3$$

N-METHYLGLYCINE

$$COOH$$
$$|$$
$$H_3C - N - CH_2$$
$$\quad\; |$$
$$\quad\; H$$

N-ETHYLGLYCINE

$$COOH$$
$$|$$
$$H_5C_2 - N - CH_2$$
$$\qquad |$$
$$\qquad H$$

PIPECOLIC ACID

$$COOH$$
$$|$$
$$HN - C^* - H$$

(piperidine ring)

†PROTEIN AMINO ACID
*ASYMMETRIC CARBON ATOM

208   KVENVOLDEN

Literature Cited

Abelson, P. H. 1954. Organic constituents of fossils. *Carnegie Inst. Wash. Yearb.* 53: 97–101

Abelson, P. H. 1959. Geochemistry of organic substances. In *Researches in Geochemistry*, ed. P. H. Abelson, 1:79–103. New York: Wiley. 511 pp.

Abelson, P. H. 1963. Geochemistry of amino acids. In *Organic Geochemistry*, ed. I. A. Breger, 431–55. New York: Macmillan. 658 pp.

Abelson, P. H., Hare, P. E. 1969. Recent amino acids in the Gunflint chert. *Carnegie Inst. Wash. Yearb.* 67:208–10

Abelson, P. H., Hare, P. E. 1970. Uptake of amino acids by kerogen. *Carnegie Inst. Wash. Yearb.* 68:297–303

Abelson, P. H., Hare, P. E. 1971. Reactions of amino acids with natural and artificial humus and kerogens. *Carnegie Inst. Wash. Yearb.* 69:327–34

Aizenshtat, Z., Baedecker, M. J., Kaplan, I. R. 1973. Distribution and diagenesis of organic compounds in JOIDES sediment from Gulf of Mexico and western Atlantic. *Geochim. Cosmochim. Acta* 37:1881–98

Akiyama, M., Wyckoff, R. W. G. 1970. The total amino acid content of fossil pecten shells. *Proc. Nat. Acad. Sci. USA* 67:1097–1100

Anders, E. et al 1964. Contaminated meteorite. *Science* 146:1157–61

Aucott, J. W., Clarke, R. H. 1966. Amino acids in the Mountsorrel bitumen, Leicestershire. *Nature* 212:61–63

Bada, J. L. 1971. Kinetics of the nonbiological decomposition and racemization of amino acids in natural waters. Nonequilibrium Systems in Natural Water Chemistry. *Am. Chem. Soc. Advan. Chem. Ser.* 106:309–31

Bada, J. L. 1972a. Kinetics of racemization of amino acids as a function of pH. *J. Am. Chem. Soc.* 95:1371–73

Bada, J. L. 1972b. The dating of fossil bones using the racemization of isoleucine. *Earth Planet. Sci. Lett.* 15:223–31

Bada, J. L., Kvenvolden, K. A., Peterson, E. 1973a. Racemization of amino acids in bones. *Nature* 245:308–10

Bada, J. L., Luyendyk, B. P., Maynard, J. B. 1970. Marine sediments: dating by the racemization of amino acids. *Science* 170:730–32

Bada, J. L., Mann, E. H. 1973. Racemization of isoleucine in cores from Leg 15, Site 148. *Initial Reports of the Deep Sea Drilling Project* 20:947–51

Bada, J. L., Protsch, R. 1973. Racemization reaction of aspartic acid and its use in dating fossil bones. *Proc. Nat. Acad. Sci. USA* 70:1331–34

Bada, J. L., Protsch, R., Schroeder, R. A. 1973b. The racemization reaction of isoleucine used as a paleotemperature indicator. *Nature* 241:394–95

Bada, J. L., Schroeder, R. A. 1972. Racemization of isoleucine in calcareous marine sediments: kinetics and mechanism. *Earth Planet. Sci. Lett.* 15:1–11

Bada, J. L., Schroeder, R. A. 1974. Amino acid racemization reactions and their geochemical implications. *Naturwissenschaften* 61:In press

Bada, J. L., Schroeder, R. A., Carter, G. F. 1974a. New evidence for the antiquity of man in North America deduced from aspartic acid racemization. *Science* 184:791–93

Bada, J. L., Schroeder, R. A., Protsch, R., Berger, R. 1974b. Concordance of collagen-based radiocarbon and aspartic acid racemization ages. *Proc. Nat. Acad. Sci. USA* 71:914–17

Brown, F. S., Baedecker, M. J., Nissenbaum, A., Kaplan, I. R. 1972. Early diagenesis in a reducing fjord, Saanich Inlet, British Columbia—III. Changes in organic constituents of sediment. *Geochim. Cosmochim. Acta* 36:1185–1203

Casagrande, D. J. 1970. Gas-liquid chromatography of thirty-five amino acids and two amino sugars. *J. Chromatogr.* 49:537–40

Casagrande, D. J., Given, P. H. 1974. Geochemistry of amino acids in some Florida peat accumulations—I. Analytical approach and total amino acid concentrations. *Geochim. Cosmochim. Acta* 38:419–34

Clarke, M. E., Jackson, G. A., North, W. J. 1972. Dissolved free amino acids in southern California coastal waters. *Limnol. Oceanogr.* 17:749–58

Clarke, R. H. 1967. Amino acids in Recent sediments off southeast Devon, England. *Nature* 213:1003–5

Cronin, J. R., Moore, C. B. 1971. Amino acid analyses of the Murchison, Murray, and Allende carbonaceous chondrites. *Science* 172:1327–29

Degens, E. T., Bajor, M. 1960. Die Verteilung von Aminosäuren in bituminösen Sedimenten und ihre Bedeutung fur die Kohlen —und Erdölgeologie. *Glueckauf* 96:1525–34

Degens, E. T., Bajor, M. 1962. Amino acids and sugars in the Bruderheim and Murray meteorite. *Naturwissenschaften* 49:605–6

Degens, E. T., Emery, K. O., Reuter, J. H.

1963. Organic materials in recent and ancient sediments—III : Biochemical compounds in San Diego Trough, California. *Neues Jahrb. Geol Palaentol. Monatsh.* 5 : 231–48

Degens, E. T., Prashnowsky, A., Emery, K. O., Pimenta, J. 1961. Organic materials in recent and ancient sediments—II. Amino acids in marine sediments of Santa Barbara Basin, California. *Neues Jahrb. Geol. Palaeontol. Monatsh.* 8 :413–26

Degens, E. T., Reuter, J. H. 1964. Analytical techniques in the field of organic geochemistry. In *Advances in Organic Geochemistry,* ed. U. Colombo, G. D. Hobson, 377–402. New York : Macmillan. 488 pp.

Degens, E. T., Reuter, J. H., Shaw, D. N. F. 1964. Biochemical compounds in offshore California sediments and sea waters. *Geochim. Cosmochim. Acta* 28 : 45–66

Erdman, J. G. 1961. Some chemical aspects of petroleum genesis as related to the problem of source bed recognition. *Geochim. Cosmochim. Acta* 22 : 16–36

Erdman, J. G., Marlett, E. M., Hanson, W. E. 1956. Survival of amino acids in marine sediments. *Science* 124 : 1026

Florkin, M. 1969. Fossil shell "conchiolin" and other preserved biopolymers. In *Organic Geochemistry-Methods and Results,* ed. G. Eglinton, M. T. J. Murphy, Chap. 20 : 498–520. Berlin : Springer. 828 pp.

Fox, S. W., Harada, K., Hare, P. E. 1972. Amino acid precursors in lunar fines from Apollo 14 and earlier missions. *Proc. Third Lunar Sci. Conf., Geochim. Cosmochim. Acta Suppl. 3,* 2 : 2109–18

Fox, S. W., Harada, K., Hare, P. E. 1973. Accumulated analyses of amino acid precursors in returned lunar samples. *Proc. Fourth Lunar Sci. Conf., Geochim. Cosmochim. Acta Suppl. 4,* 2 : 2241–48

Fox, S. W., Harada, K., Hare, P. E., Hinsch, G., Mueller, G. 1970. Bio-organic compounds and glassy microparticles in lunar fines and other materials. *Science* 167 : 767–70

Gehrke, C. W., Stalling, D. L. 1967. Quantitative analysis of the twenty natural protein amino acids by gas-liquid chromatography. *Separ. Sci.* 2 : 101–38

Gehrke, C. W. et al 1970. Carbon compounds in lunar fines from Mare Tranquillitatis—III. Organosiloxanes in hydrochloric acid hydrolysates. *Proc. Apollo 11 Lunar Sci. Conf., Geochim. Cosmochim. Acta Suppl. 1,* 2 : 1845–56

Gehrke, C. W. et al 1971. A search for amino acids in Apollo 11 and 12 lunar fines. *J. Chromatogr.* 59 : 305–19

Gehrke, C. W. et al 1972. Amino acid analyses of Apollo 14 samples. *Proc. Third Lunar Sci. Conf., Geochim. Cosmochim. Acta Suppl. 3,* 2 : 2119–29

Gehrke, C. W. et al 1973. Extractable organic compounds in Apollo 15 and 16 lunar fines. *Proc. Fourth Lunar Sci. Conf., Geochim. Cosmochim. Acta Suppl. 4,* 2 : 2249–59

Gelpi, E., Koenig, W. A., Gibert, J., Oró, J. 1969. Combined gas chromatography-mass spectrometry of amino acid derivatives. *J. Chromatogr. Sci.* 7 : 604–13

Gil-Av, E., Charles, R., Fischer, G. 1965. Resolution of amino acids by gas chromatography. *J. Chromatogr.* 17 : 408–10

Hamilton, P. B. 1963. Ion exchange chromatography of amino acids—a single column, high resolving, fully automatic procedure. *Anal. Chem.* 35 : 2055–64

Hamilton, P. B. 1965. Amino acids on hands. *Nature* 205 : 284–85

Harada, K., Hare, P. E., Windsor, C. R., Fox, S. W. 1971. Evidence for compounds hydrolyzable to amino acids in aqueous extracts of Apollo 11 and Apollo 12 lunar fines. *Science* 173 : 433–35

Hare, P. E. 1965. Amino acid artifacts in organic geochemistry. *Carnegie Inst. Wash. Yearb.* 64 : 232–35

Hare, P. E. 1969. Geochemistry of proteins, peptides, and amino acids. In *Organic Geochemistry—Methods and Results,* ed. G. Eglinton, M. T. J. Murphy, Chap. 18 : 438–63. Berlin : Springer. 828 pp.

Hare, P. E. 1971. Effect of hydrolysis on the racemization rate of amino acids. *Carnegie Inst. Wash. Yearb.* 70 : 256–58

Hare, P. E. 1972. Amino acid geochemistry of a sediment core from the Cariaco Trench. *Carnegie Inst. Wash. Yearb.* 71 : 592–96

Hare, P. E. 1973a. Fluorescent method for the analysis of amino acids and peptides. *Carnegie Inst. Wash. Yearb.* 72 : 701–4

Hare, P. E. 1973b. Amino acids, amino sugars, and ammonia in sediments from the Cariaco Trench. *Initial Reports of the Deep Sea Drilling Project* 20 : 941–42

Hare, P. E., Abelson, P. H. 1968. Racemization of amino acids in fossil shells. *Carnegie Inst. Wash. Yearb.* 66 : 526–28

Hare, P. E., Harada, K., Fox, S. W. 1970. Analyses for amino acids in lunar fines. *Proc. Apollo 11 Lunar Sci. Conf., Geochim. Cosmochim. Acta Suppl. 1,* 2 : 1799–1803

Hare, P. E., Mitterer, R. M. 1967. Nonprotein amino acids in fossil shells. *Carnegie Inst. Wash. Yearb.* 65 : 362–64

Hare, P. E., Mitterer, R. M. 1969. Laboratory simulation of amino-acid diagenesis in

210   KVENVOLDEN

Harington, J. S., Cilliers, J. J. le R. 1963. A possible origin of the primitive oils and amino acids isolated from amphibole asbestos and banded ironstone. *Geochim. Cosmochim. Acta* 27:411–18

Hayes, J. M. 1967. Organic constituents of meteorites—a review. *Geochim. Cosmochim. Acta* 31:1395–1440

Heijkenskjöld, F., Möllerberg, H. 1958. Amino acids in anthracite. *Nature* 181: 334–35

Hoering, T. C. 1973. A comparison of melanoidin and humic acid. *Carnegie Inst. Wash. Yearb.* 72:682–90

Itihara, Y. 1973. Amino acids in the Cenozoic sediments of Japan. *Pac. Geol.* 6:51–63

Jones, J. D., Vallentyne, J. R. 1960. Biogeochemistry of organic matter—I. Polypeptides and amino acids in fossils and sediments in relation to geothermometry. *Geochim. Cosmochim. Acta* 21:1–34

Kaplan, I. R., Degens, E. T., Reuter, J. H. 1963. Organic compounds in stony meteorites. *Geochim. Cosmochim. Acta* 27: 805–34

Kemp, A. L. W., Mudrochova, A. 1973. The distribution and nature of amino acids and other nitrogen-containing compounds in Lake Ontario surface sediments. *Geochim. Cosmochim. Acta* 37:2191–2206

Khan, S. U., Sowden, F. J. 1971. Thermal stabilities of amino acid components of humic acids under oxidative conditions. *Geochim. Cosmochim. Acta* 35:854–58

King, K. Jr., Hare, P. E. 1972a. Amino acid composition of planktonic foraminifera: a paleobiochemical approach to evolution. *Science* 175:1461–63

King, K. Jr., Hare, P. E. 1972b. Amino acid composition of the test as a taxonomic character for living and fossil planktonic foraminifera. *Micropaleontology* 18:285–93

King, K. Jr., Hare, P. E. 1972c. Species effects in the epimerization of L-isoleucine in fossil planktonic foraminifera. *Carnegie Inst. Wash. Yearb.* 71:596–98

Koenig, W. A., Parr, W., Lichtenstein, H. A., Bayer, E., Oró, J. 1970. Gas chromatographic separation of amino acids and their enantiomers: non-polar stationary phases and a new optically active phase. *J Chromatogr. Sci.* 8:183–86

Kvenvolden, K. A. 1973. Criteria for distinguishing biogenic and abiogenic amino acids—preliminary considerations. *Space Life Sci.* 4:60–68

Kvenvolden, K. A., Lawless, J. G., Ponnamperuma, C. 1971. Nonprotein amino acids in the Murchison meteorite.

Proc. Nat. Acad. Sci. USA 68:486–90

Kvenvolden, K. A., Peterson, E. 1969. Amino acid enantiomers in Green River Formation oil shale. *Geol. Soc. Am. Abstr. with Programs.* P. 7:132

Kvenvolden, K. A., Peterson, E., Brown, F. S. 1970a. Racemization of amino acids in sediments from Saanich Inlet, British Columbia. *Science* 169:1079–82

Kvenvolden, K. A. et al 1970b. Evidence for extraterrestrial amino-acids and hydrocarbons in the Murchison meteorite. *Nature* 228:923–26

Kvenvolden, K. A., Peterson, E., Pollock, G. E. 1969. Optical configuration of amino-acids in Precambrian Fig Tree chert. *Nature* 221:141–43

Kvenvolden, K. A., Peterson, E., Pollock, G. E. 1972. Geochemistry of amino acid enantiomers: gas chromatography of their diastereomeric derivatives. In *Advances in Organic Geochemistry 1971*, ed. H. R. von Gaertner, H. Wehner, 387–401. Braunschweig: Pergamon. 736 pp.

Kvenvolden, K. A., Peterson, E., Wehmiller, J., Hare, P. E. 1973. Racemization of amino acids in marine sediments determined by gas chromatography. *Geochim. Cosmochim. Acta* 37:2215–25

Lawless, J. G. 1973. Amino acids in the Murchison meteorite. *Geochim. Cosmochim. Acta* 37:2207–12

Lawless, J. G., Chadha, M. S. 1971. Mass spectral analysis of C_3 and C_4 aliphatic amino acid derivatives. *Anal. Biochem.* 44:473–85

Lawless, J. G., Kvenvolden, K. A., Peterson, E., Ponnamperuma, C., Jarosewich, E. 1972. Evidence for amino-acids of extraterrestrial origin in the Orgueil meteorite. *Nature* 236:66–67

Lawless, J. G., Kvenvolden, K. A., Peterson, E., Ponnamperuma, C., Moore, C. 1971. Amino acids indigenous to the Murray meteorite. *Science* 173:626–27

Manning, J. M., Moore, S. 1968. Determination of D- and L-amino acids by ion exchange chromatography as L-D and L-L dipeptides. *J. Biol. Chem.* 243:5591–97

Mitterer, R. M. 1968. Amino acid composition of organic matrix in calcareous oolites. *Science* 162:1498–99

Mitterer, R. M. 1971a. Influence of natural organic matter on $CaCO_3$ precipitation. In *Carbonate Cements,* ed. O. Bricker, 252–58. Baltimore: Johns Hopkins Univ. Press

Mitterer, R. M. 1971b. Comparative amino acid composition of calcified and non-calcified polychaete worm tubes. *Comp. Biochem. Physiol.* 38B:405–9

Mitterer, R. M. 1972a. Calcified proteins in

the sedimentary environment. In *Advances in Organic Geochemistry 1971*, ed. H. R. von Gaertner, H. Wehner, 441–52. Braunschweig: Pergamon. 736 pp.

Mitterer, R. M. 1972b. Biogeochemistry of aragonite mud and oolites. *Geochim. Cosmochim. Acta* 36:1407–22

Mitterer, R. M. 1974. Pleistocene stratigraphy in southern Florida based on amino acid diagenesis in fossil *Mercenaria*. *Geology* 2:425–28

Modzeleski, V. E. et al 1973. Carbon compounds in pyrolysates and amino acids in extracts of Apollo 14 lunar samples. *Nature Phys. Sci.* 242:50–52

Moore, S., Spackman, D. H., Stein, W. H. 1958. Chromatography of amino acids on sulfonated polystyrene resins—an improved system. *Anal. Chem.* 30:1185–90

Moore, S., Stein, W. H. 1951. Chromatography of amino acids on sulfonated polystyrene resins. *J. Biol. Chem.* 192:663–81

Murphy, M. E. et al 1970. Analysis of Apollo 11 lunar samples by chromatography and mass spectrometry: pyrolysis products, hydrocarbons, sulfur, amino acids. *Proc. Apollo 11 Lunar Sci. Conf., Geochim. Cosmochim. Acta Suppl. 1,* 2:1879–90

Nagy, B. et al 1970a. Organic compounds in lunar samples: pyrolysis products, hydrocarbons, amino acids. *Science* 167:770–73

Nagy, B. et al 1970b. Carbon compounds in Apollo 11 lunar samples. *Nature* 225:1028–32

Nagy, B. et al 1971. Carbon compounds in Apollo 12 lunar samples. *Nature* 232:94–98

Nakaparksin, S., Birrell, P., Gil-Av, E., Oró, J. 1970a. Gas chromatography with optically active stationary phases: resolution of amino acids. *J. Chromatogr. Sci.* 8:177–82

Nakaparksin, S., Gil-Av, E., Oró, J. 1970b. Study of the racemization of some neutral α-amino acids in acid solution using gas chromatographic techniques. *Anal. Biochem.* 33:374–82

Oró, J., Gibert, J., Lichtenstein, H., Wikström, S., Flory, D. A. 1971a. Amino-acids, aliphatic and aromatic hydrocarbons in the Murchison meteorite. *Nature* 230:105–7

Oró, J., Nakaparksin, S., Lichtenstein, H., Gil-Av, E. 1971b. Configuration of amino-acids in carbonaceous chondrites and a Pre-cambrian chert. *Nature* 230:107–8

Oró, J. et al 1971c. Abundances and distribution of organogenic elements and compounds in Apollo 12 lunar samples. *Proc. Second Lunar Sci. Conf., Geochim. Cosmochim. Acta Suppl. 2,* 2:1913–25

Oró, J., Skewes, H. B. 1965. Free amino-acids on human fingers: the question of contamination in microanalysis. *Nature* 207:1042–45

Oró, J., Tornabene, T. 1965. Bacterial contamination of some carbonaceous meteorites. *Science* 150:1046–48

Oró, J. et al 1970a. Organogenic elements and compounds in surface samples from the Sea of Tranquility. *Science* 167:765–67

Oró, J. et al 1970b. Organogenic elements and compounds in type C and D lunar samples from Apollo 11. *Proc. Apollo 11 Lunar Sci. Conf., Geochim. Cosmochim. Acta Suppl. 1,* 2:1901–20

Peake, E., Baker, B. L., Hodgson, G. W. 1972. Hydrogeochemistry of the surface waters of the MacKenzie River drainage basin, Canada—II. The contribution of amino acids, hydrocarbons and chlorins to the Beaufort Sea by the MacKenzie River system. *Geochim. Cosmochim. Acta* 36:867–83

Pereira, W. E., Hoyano, Y., Reynolds, W. E., Summons, R. E., Duffield, A. M. 1973. The simultaneous quantitation of ten amino acids in soil extracts by mass fragmentography. *Anal. Biochem.* 55:236–44

Peterson, E., Shum, S., Kvenvolden, K. A. 1972. Pipecolic acid in *Mercenaria* of Pleistocene and Micocene ages. *Geol. Soc. Am. Abstr. with Programs* 4:627

Pocklington, R. 1971. Free amino acids dissolved in North Atlantic Ocean waters. *Nature* 230:374–75

Pollock, G. E., Miyamoto, A. K. 1971. A desalting technique for amino acid analysis of use in soil and geochemistry. *Agr. Food Chem.* 19:104–7

Pollock, G. E., Oyama, V. I., Johnson, R. D. 1965. Resolution of racemic amino acids by gas chromatography. *J. Gas Chromatogr.* 3:174–76

Ponnamperuma, C. et al 1970. Search for organic compounds in the lunar dust from the Sea of Tranquillity. *Science* 167:760–62

Povoledo, D., Vallentyne, J. R. 1964. Thermal reaction kinetics of the glutamic acid-pyroglutamic acid system in water. *Geochim. Cosmochim. Acta* 28:731–34

Prashnowsky, A. A., Schidlowski, M. 1967. Investigation of Precambrian thucholite. *Nature* 216:560–63

Rash, J. J. et al 1972. GLC of amino acids: a survey of contamination. *J. Chromatogr. Sci.* 10:444–50

Rittenburg, S. C. et al 1963. Biogeochemistry of sediments in experimental mohole. *J. Sediment. Petrology* 33:140–72

Sagan, C. 1972. The search for indigenous

lunar organic matter. *Space Life Sci.* 3: 484–89

Schopf, J. W., Kvenvolden, K. A., Barghoorn, E. S. 1968. Amino acids in Precambrian sediments: an assay. *Proc. Nat. Acad. Sci. USA* 59:639–46

Schroeder, R. A., Bada, J. L. 1973. Glacial-postglacial temperature difference deduced from aspartic acid racemization in fossil bones. *Science* 182:479–82

Spackman, D. H., Stein, W. H., Moore, S. 1958. Automatic recording apparatus for use in the chromatography of amino acids. *Anal. Chem.* 30:1190–1206

Stevenson, F. J. 1954. Ion exchange chromatography of the amino acids in soil hydrolysates. *Soil Sci. Soc. Am. Proc.* 18:373–77

Stevenson, F. J., Cheng, C.-N. 1969. Amino acid levels in the Argentine Basin sediments: correlation with Quaternary climatic changes. *J. Sediment. Petrology*, 39:345–49

Stevenson, F. J., Cheng, C.-N. 1972. Organic geochemistry of the Argentine Basin sediments: carbon-nitrogen relationships and Quaternary correlations. *Geochim. Cosmochim. Acta* 36:653–71

Tatsumoto, M., Williams, W. T., Prescott, J. M., Hood, D. W. 1961. Amino acids in samples of surface sea water. *J. Mar. Res.* 19:89–95

Thompson, R. R., Creath, W. B. 1966. Low molecular weight hydrocarbons in Recent and fossil shells. *Geochim. Cosmochim. Acta* 30:1137–52

Turekian, K. K., Bada, J. L. 1972. The dating of fossil bones. In *Calibration of Hominoid Evolution*, ed. W. W. Bishop,

J. A. Miller, 171–85. New York: Scottish Academic Press

Udenfriend, S. et al 1972. Fluorescamine: a reagent for assay of amino acids, peptides, proteins, and primary amines in the picomole range. *Science* 178:871–72

Vallentyne, J. R. 1964. Biogeochemistry of organic matter—II. Thermal reaction kinetics and transformation products of amino compounds. *Geochim. Cosmochim. Acta* 28:157–88

Vallentyne, J. R., 1965. Two aspects of the geochemistry of amino acids. In *The Origins of Prebiological Systems and of Their Molecular Matrices*, ed. S. W. Fox, 105–25. New York: Academic. 482 pp.

Vallentyne, J. R. 1968. Pyrolysis of proline, leucine, arginine and lysine in aqueous solution. *Geochim. Cosmochim. Acta* 32: 1353–56

Vallentyne, J. R. 1969. Pyrolysis of amino acids in Pleistocene *Mercenaria* shells. *Geochim. Cosmochim. Acta* 33:1453–58

Wall, J. S. 1953. Simultaneous separation of purines, pyrimidines, amino acids, and other nitrogenous compounds by ion exchange chromatography. *Anal. Chem.* 25: 950–53

Wehmiller, J., Hare, P. E. 1971. Racemization of amino acids in marine sediments. *Science* 173:907–11

Wolman, Y., Miller, S. L. 1971. Amino acid contamination of aqueous hydrochloric acid. *Nature* 234:548–49

Zumwalt, R. W., Kuo, K., Gehrke, C. W. 1971. A nanogram and picogram method for amino acid analysis by gas-liquid chromatography. *J. Chromatogr.* 57: 193–208

PRECAMBRIAN PALEOBIOLOGY: PROBLEMS AND PERSPECTIVES

×10038

J. William Schopf
Department of Geology, University of California, Los Angeles,
Los Angeles, California 90024

PROLOGUE: OF PARADIGMS AND MULTIPLE WORKING HYPOTHESES

In 1890, T. C. Chamberlin, the distinguished American geologist and educator, wrote a classic paper entitled "The method of multiple working hypotheses." The paper has been repeatedly reprinted in various forms (most recently in the May 7, 1965, issue of *Science*). On numerous occasions I have returned to this paper, struck by what I regarded initially as its rather curious central theme. In brief, Chamberlin argues that a scientist should take serious measures to avoid undue parental affection for a newly constructed theory (or paradigm, to use Kuhnian phraseology); that such affection too often leads the unwary to embrace concepts that are at best partial, and at worst erroneous, views of reality; and that to achieve the desired goal of intellectual objectivity, a scientist should erect a set of alternative working hypotheses that will serve to temper his enthusiasm for his favorite "intellectual offspring." On first reading, this theme seemed flawed—how could a scientist generalize effectively, and how could he have the courage to put forth new but possibly radical and unpopular concepts unless he felt, and presumably felt deeply, that he really had discovered the Rosetta stone of his science? Would not such affection add to, rather than detract from, the workings of science?

It seems to me now, however, that Chamberlin's message is well taken. To make progress in science, the acceptance of a given set of interpretations is no doubt required, but only up to a point; the threshold of reasonability can be passed rapidly if one begins to press the data into a preconceived set of paradigmatic pigeonholes. It seems likely that this threshold has been surpassed in recent interpretations of the Precambrian fossil record. If this has in fact occurred, it is probably a function of the newness of the field. With few data available, it becomes relatively easy to construct uncomplicated, plausible, and seemingly correct hypotheses. Certainly, it could be a result of the style of evidence and argument necessitated by historical science, where the data are always incomplete. Much of the evidence can be interpreted only in a qualitative, rather subjective, fashion, and "proof" by analogy is commonplace. A careful reading of the literature suggests that

213

virtually all major interpretations have been consistent with the evidence at hand. That, however, is the crux of the issue; there is a substantive difference between those data that are merely *consistent* with a given construct and those that *compel* its acceptance, a distinction not always explicit in recent big picture syntheses. (At the same time, however, it should be noted that consumers of the Precambrian "product," and especially textbook writers, have exhibited a decided tendency to overlook the various cautionary phrases that are generously sprinkled throughout much of the writing in this field.)

The purpose of this paper, then, is to separate the merely consistent from the compelling, to segregate that which could have been the case from that which is "known." Certainly, this is a highly personal assessment; it would be naive to suggest that these views represent a consensus of current thought, even (or perhaps especially) of active workers in the field. I am convinced, however, that an exercise of this type is much needed. It is important that a long, critical look at the available evidence be taken before tentative working hypotheses become entrenched as ruling dogma.

INTRODUCTION

As is evident from inspection of the dates of publications listed in Tables 1–4, Precambrian paleobiology is a young, rapidly growing field. More than two-thirds of the 75 papers cited have been published since 1969 and more than 95% since 1960. While this listing is not at all exhaustive, it is reasonably representative; with few exceptions, important studies in the field have been carried out during the past decade. A related index of the relative youth of the subject is summarized in Table 1; all but one of the 20 well-preserved microbiotas now known from the Precambrian were first reported during the past decade, 60% have been discovered since 1973, and fully 40% were first reported during the six months prior to July 1974. From perusal of other geological literature it is clear that this recent flush of interest in the Precambrian is by no means limited to paleobiology. Increasingly, earth scientists generally are turning their attention to those processes and events that molded the evolving planet during the earliest seven-eighths of earth history. This apparently significant development in the geosciences is reflected partly by the establishment during the past year of yet another new scientific journal, *Precambrian Research,* and by the recasting and retitling of a second, now known as *Origins of Life,* to handle the expected influx of data.

Limitations of the Early Fossil Record

I have discussed elsewhere (Schopf 1970, 1974) what seem to me to be the major limitations intrinsic to studies of Precambrian life. Three rather general aspects, however, deserve reemphasis.

1. Primary among such limitations is the youth of the field. Although there are certainly more workers now than ever before (worldwide, perhaps 150 active micropaleontologists, "stromatoliphiles," and organic geochemists), and although data are accumulating at an increasingly rapid pace, much of the published informa-

tion is preliminary, not yet followed up by detailed analyses nor confirmed by multiple independent investigations. For example, a monographic treatment of only two of the 20 Precambrian microbiotas listed in Table 1 has been published as yet, and the studies have not been exhaustive in either case. Extended descriptions of three of the other microfloras listed are currently in press or in preparation. For the remaining 15 microbiotas—three-quarters of those now known—published data are in general limited to reports several paragraphs in length, illustrated with a few photomicrographs. These short reports accomplish the crucial job of establishing the existence of the assemblages but they provide only limited insight into their composition, variability, ecologic setting, and evolutionary status. Although rewarding, the detailed studies now needed are tedious and time-consuming. Useful data will no doubt become increasingly available over the next few years. In the meantime, however, understanding of the course and timing of early evolutionary advance must be based on the rather meager information currently available about these biotas. For this reason, if for no other, it is difficult to place great confidence in the possible prescience of the various speculative overviews of early evolution set forth in recent years (e.g. Cloud 1968, 1974, Margulis 1970, Schopf 1970 and in press, Barghoorn 1971).

2. As summarized in Table 1, all but three of the diverse Precambrian microbiotas now known have been preserved by silica permineralization in a distinctive and rather uncommon facies—fine-grained, primary or very early diagenetic cherts occurring in association with, and often partially replacing, algal–laminated, commonly stromatolitic, carbonates. In a sense, emphasis on this single facies has been a fortunate historical accident. In the early 1950s, Stanley Tyler made the first important discovery of Precambrian microfossils in just such stromatolitic cherts occurring along the northern shore of Lake Superior in the Proterozoic Gunflint Iron-Formation; subsequent publications (Tyler & Barghoorn 1954, Barghoorn & Tyler 1965) established that the Gunflint microflora was diverse and remarkably well preserved. The paleobiological potential of the stromatolitic cherty facies was thus recognized early, a potential that has been fruitfully exploited by several subsequent workers (e.g. Schopf, Cloud, Licari, Hofmann, and Muir).

In retrospect, it now appears to be particularly fortunate that cherts have received a disproportionately large share of attention during the formative stages of the field. Two principal objectives must be accomplished in micropaleontologic studies of early sediments: first, it must be established that the putative microfossils are actually biogenic, rather than being nonbiologic artifacts, and second, it must be demonstrated that they are actually of Precambrian age, rather than being subsequently introduced contaminants. Cherts provide answers to both questions: since cellular preservation in the siliceous matrix is often excellent, the biogenicity of detected microstructures can usually be established beyond doubt. Equally important, since cherts are readily studied in petrographic thin sections, the indigenous nature, and thus the Precambrian age of the permineralized microorganisms, can be clearly demonstrated. Despite these advantages, advantages which are in general not shared by other rock types (calcareous concretions such as "coal balls" being the single exception that comes to mind), this predominantly unifacies approach is

Table 1   Precambrian microbiotas (microfossil assemblages that are diverse, well preserved,

Status of study	Approximate age (m.y.)	Geologic unit	Locality
b	~ 725 (600 > 850)	Hector Fm. (Windermere Grp.)	southwestern Alberta, Canada
a	~ 725 (600 > 800)	Tindir Grp. (Upper Dolomitic Sandstone and Shale Mem.?)	east-central Alaska
a	~ 800 (600 > 1000)	Upper Chuar Grp. (Kwagunt Fm., Walcott Mem.)	eastern Grand Canyon, Arizona
a	~ 850 (600 > 1000)	Lower Chuar Grp. (Galeros Fm., Carbon Canyon Mem.)	eastern Grand Canyon, Arizona
c	~ 900 (790 > 1370)	Bitter Springs Fm. (Loves Creek Mem.)	southern Northern Territory, Australia
a	~ 900 (740 > 1340)	Auburn Dolomite	southern South Australia
a	~ 950 (740 > 1340)	cf Myrtle Springs Fm. (Burra Grp.)	east-central South Australia
b	~ 1000 (740 > 1340)	Skillogalee Dolomite and cf Skillogalee Dolomite (Burra Grp.)	east-central and southern South Australia
a	~ 1150 (740 > 1340)	River Wakefield Grp. (Blyth Dolomite)	southern South Australia
a	~ 1200 (1175 > 1225)	Dismal Lakes Grp.	western Northwest Territories, Canada (Great Bear Lake Region)
b	~ 1300 (1200 > 1400)	Beck Spring Dolomite (Pahrump Grp.)	southeastern California
a	~ 1400 (1340 > 1430)	Vempalle Fm. (Cuddapah Grp.)	south-central Andhra Pradesh, India
a	~ 1500 (1100 > 1800)	Bungle Bungle Dolomite	northern Western Australia
a	~ 1600 (1510 > 1700)	Amelia Dolomite (McArthur River Grp.)	northeastern Northern Territory, Australia
a	~ 1650 (1550 > 1750)	Paradise Creek Fm.	northwestern Queensland, Australia
a	~ 1800 (1750 > 2200)	Belcher Grp. (lower carbonate unit)	southern Hudson Bay, Canada
c	~ 1900 (1750 > 2200)	Gunflint Iron Fm. (Lower Algal Chert Mem.)	southern Ontario, Canada
a	~ 2000 (1750 > 2200)	Kasegalik Fm. (Belcher Grp.)	southern Hudson Bay, Canada
a	~ 2000 (1750 > 2500)	Pokegama Quartzite	northeastern Minnesota
a	~ 2250 (2220 > 2300)	Transvaal Dolomite (Olifants River Grp.)	eastern Transvaal, South Africa

a = preliminary report only; in need of detailed study.
b = monographic treatment in press or in preparation.
c = monographic treatment published.

and of established biogenicity and Precambrian age)

Fossiliferous lithology	First published report	Subsequent important reports
black shale	Cloud & Licari 1968	Moorman 1974
siliceous shale and calcareous mudstone	Allison & Moorman 1973	
cherty pisolites in carbonate	Schopf, Ford & Breed 1973	
chert lenses in algal laminated carbonate	Schopf et al 1974	
cherts interbedded with algal laminated carbonate and in partially silicified carbonate stromatolites	Barghoorn & Schopf 1965	Schopf 1968, Schopf & Blacic 1971, Schopf 1972, Schopf 1974
cherts interbedded with carbonate	Schopf et al 1974	
cherts interbedded with algal laminated carbonate and in partially silicified carbonate stromatolites	Schopf et al 1974	
cherts interbedded with algal laminated carbonate and in partially or wholly silicified carbonate stromatolites	Schopf & Barghoorn 1969	Schopf & Fairchild 1973, Fairchild, in preparation
cherts interbedded with carbonate	Schopf et al 1974	
chert lenses in algal laminated carbonate	Schopf et al 1974	
cherts interbedded with algal laminated carbonate and in partially or wholly silicified carbonate stromatolites	Cloud et al 1969	Gutstadt & Schopf 1969 Licari 1974
cherts in partially silicified carbonate stromatolites	Schopf et al 1974	
cherts in partially silicified carbonate stromatolites	Diver 1974	
cherts in partially silicified carbonate stromatolites	Croxford et al 1973	Muir 1974
cherts in partially silicified carbonate stromatolites	Licari, Cloud & Smith 1969	Licari & Cloud 1972
chert lenses in algal laminated carbonate	Hofmann & Jackson 1969	
bedded, algal laminated chert and chert stromatolites, in places interbedded with carbonate	Tyler & Barghoorn 1954	Barghoorn & Tyler 1965, Cloud 1965, Licari & Cloud 1968, Edhorn 1973
chert lenses in algal laminated carbonate	Hofmann 1974	
chalcedonic chert-filled reentrants into pre-Pokegama erosion surface	Cloud & Licari 1972	
carbonate stromatolite	Nagy, Zumberge & Nagy 1973	Nagy 1974, MacGregor et al 1974

not without drawbacks. Specifically, it remains to be established whether the degree of diversity and level of evolutionary development represented in these permineralized biotas reflect accurately the nature of their comtemporaneous biospheres. Extra-polations based primarily on a single, perhaps ecologically restricted facies (required for in situ precipitation of silica), are obviously tenuous. Geologic evidence indicates that stromatolitic cherty facies were substantially more widespread during the Precambrian, prior to the evolution of the siliceous protists (diatoms, radiolarians), parazoans, and plants (e.g. *Equisetum*, grasses) that comprise the biologic com-ponents of the Phanerozoic silica cycle, than they were in subsequent geologic time. Still, there can be no doubt that a biased and quite incomplete picture of even the earliest Phanerozoic biosphere would be obtained if only the cherty carbonate facies were studied; certainly, this could also be true of the Precambrian. To date, terrigenous, moderate to deep water, and especially clastic facies have yet to be adequately sampled in the Precambrian (representative reports are summarized in Table 3). As suggested previously (Schopf et al 1973, Schopf et al 1974), the inter-bedded, fossiliferous, pisolitic cherts, shales, and cherty stromatolitic carbonates of the Proterozoic Chuar Group of the Grand Canyon provide an excellent opportunity to investigate these questions, questions which must be resolved before the under-standing of early evolution (and of the possible use of microfossils for bio-stratigraphic correlation in the Precambrian) can be regarded as more than speculative inference.

3. In the mid-1960s, there was widespread optimism that the newly developed analytical techniques of organic geochemistry would provide definitive evidence of the biochemical and physiological capabilities of Precambrian life. However, as is perhaps common following the first applications of a new approach to the solution of a difficult problem, subsequent studies have shown that the approach is less applicable, and the problem more complex, than was initially appreciated. The enthusiasm of a decade ago has yet to be rewarded.

Recently, McKirdy (1974) skillfully reviewed the status of organic geochemistry in Precambrian research. The current situation appears to be as follows: (*a*) both non-extractable (kerogen) and extractable (amino acids, fatty acids, hydrocarbons, porphyrins, etc) organic matter occur in Precambrian sedimentary rocks; (*b*) the total amount of such organic matter is small (generally comprising 0.05–0.40% of the rock, by weight), with the particulate, insoluble kerogen fraction predominant (very commonly comprising well over 95% of the total); (*c*) although there is good evidence that the kerogen occurring in such rocks was deposited syngenetically with the enclosing sediments and is thus of Precambrian age, kerogen is a geochemically altered, complex, polymeric material that as yet has yielded only limited information of paleobiologic significance; and (*d*) it has proven very difficult to establish that the extractable components (which are of readily interpretable biologic significance) are syngenetic with original sedimentation, rather than having been emplaced during later geologic time. The central difficulty is that no specific chemical tests are available to indicate with certainty the geologic age of extracted organic matter. In the absence of such tests—coupled with the realization that all rocks are somewhat permeable, that soluble organic matter is carried by ground waters, and that an enormous time period is available for the secondary emplacement of such compounds in Pre-cambrian strata—interpretations based on the extractable components of ancient sediments are virtually certain to be ambiguous (Schopf 1970, Smith et al 1970). To date, probably the most promising results from studies of Precambrian organic mat-ter have come from analyses of the carbon isotopic composition of kerogen (Schopf et al 1971, Oehler et al 1972). These studies suggest that autotrophs, capable of fixing CO_2, may have been extant more than 3000 m.y. ago (see discussion of the Archean record, which follows).

THE PRECAMBRIAN FOSSIL RECORD

In broad overview, the Precambrian rock record can be divided temporally into two major phases: the *Archean* record, dating from the time of deposition of the oldest known sediments (the approximately 3760 m.y.-old Isua Iron-Formation of West Greenland described by Moorbath et al 1973) to the first widespread appearance of cratonal, platform-type deposits (2500 m.y. ago); and the *Proterozoic* record, extend-ing from the close of the Archean to the beginning of the Phanerozoic (600 m.y. ago). It is becoming increasingly apparent that these two phases of the rock record were characterized by decidedly differing tectonic styles. The Archean was apparently a time of rather widespread volcanic and plutonic activity. Deposits of this age include the major greenstone, metabasaltic belts of the world, which comprise arcuate, synclinal troughs of geochemically primitive, predominantly volcanogenic sediments (Glickson 1970, 1971). In contrast, the Proterozoic is characterized by typically cratonal sediments of the type deposited on submerged continental margins, and by less primitive igneous rocks. In major respects, the tectono-sedimentary framework of the Proterozoic appears to have been fundamentally comparable to that of later geologic time (Sidorenko 1969).

Significantly, and perhaps not surprisingly, this major geologic division is paralleled in the known Precambrian fossil record. Evidence of Archean life is notably meager, and, as discussed below, much of the evidence now commonly accepted appears to have been misinterpreted. However, with the onset of widespread cratonal sedimentation and the approximately synchronous appearance of widespread stromatolitic carbonates near the beginning of the Proterozoic, the quality and quantity of paleobiologic evidence markedly increase. It remains to be established whether this parallelism reflects a cause-effect relationship (i.e. the meagerness of the Archean fossil record being a function of the sparcity of preserved fossiliferous facies) or whether the apparent break at the Archean-Proterozoic "boundary" reflects a major biological event (such as the origin and rapid diversification of oxygen-producing photoautotrophs).

In keeping with the principal aim of this paper—to segragate that which is "known" from that which at present can only be speculated—I will first consider the relatively well-established Proterozoic fossil record; for convenience, and also to reflect possible stages of evolutionary advance, the discussion is divided into three segments encompassing time periods of roughly equal duration. Subsequently, I will consider available evidence of Archean life.

Proterozoic I: 2500 to 1800 m.y. Ago

Geologically, this earliest third of the Proterozoic encompasses the first widespread appearance of shelf-type sediments (2500 m.y.); the earliest known evidence of large-scale, presumably continental glaciation (2200 m.y., Gowganda and Bruce tillites); the last apparent occurrence of "abundant, easily oxidized, detrital pyrite and uraninite in stream deposits" (2300–2100 m.y.; Cloud 1974, p. 58); virtually the last, and volumetrically the greatest, phase of deposition of banded iron-formations (2200–1900 m.y.; Goldich 1973); and the apparently earliest appearance of oxidized, terrestrial, red beds (2000 m.y.; Cloud 1974). Biologically, Proterozoic I includes the oldest known occurrence of widespread carbonate stromatolites (2300 m.y.; Table 2); the oldest known structurally preserved microbiota (2250 m.y.; Table 1); and the highly diverse and rather intensively studied microfossil assemblage of the Gunflint cherts (1900 m.y.; Table 1).

ATMOSPHERIC EVOLUTION   What can be said with confidence regarding the environment and life of this time? Many workers, but most prominently Preston Cloud, have argued that the geologic evidence indicates a major transition in atmospheric oxygen content (from a nearly anoxic condition to an oxygen content of perhaps 1% of the present level) during this period. Two aspects of this matter need be considered: first, does evidence require that such a transition actually occurred during the Precambrian, and second, if so, is there compelling evidence for its occurrence during Proterozoic I?

In answer to the first question, the relevant data seem to point to the early occurrence of just such a transition in atmospheric composition. First of all, the biosynthetic pathways, physiological capabilities, and to a limited degree the morphological characteristics (e.g., heterocysts in blue-green algae; J. W. Schopf, in press) of modern microorganisms appear to reflect evolutionary derivation from an initial stock of obligate anaerobes that gave rise during the Precambrian to microorganisms capable of both coping with and utilizing atmospheric oxygen. The most logical and perhaps the only reasonable interpretation of this evolutionary sequence is that it paralleled the development of the earth's environment (see J. W. Schopf, in press, and Margulis 1970, for reviews of the very extensive literature on this subject). Second, a large body of experimental data (Ponnamperuma & Gabel 1968) indicates that, as currently envisioned, the early Precambrian abiotic synthesis of living systems could have occurred only under essentially anoxic conditions. Third, computations of the mass of oxygen and of carbon currently sequestered in the biosphere, lithosphere, atmosphere, and oceans suggest that the oxygen of the present and past atmospheres of the earth could have been entirely produced photosynthetically over geologic time by blue-green algae, protists, and plants (Holland 1962). If these calculations are valid, they indicate that the environment quite likely was virtually anoxic during the early Precambrian, prior to the origin of this type of biological photosynthesis, and that atmospheric oxygen content must therefore have increased prior to the late Precambrian appearance of aerobic metazoans and metaphytes. Based on these and other lines of reasoning, it seems safe to assume that the postulated transition did in fact occur. The time of its occurrence, however, is somewhat less certain.

In recent years (most recently in 1974), Cloud has repeatedly made a case for interpreting the apparent near synchronization (about 2000 m.y. ago) of the cessation of deposition of banded iron-formations and the onset of terrestrial red bed sedimentation as an indication of an important rise in atmospheric oxygen content. Very briefly, he postulates (a) that the iron of banded iron-formations was dispersed in solution in the ferrous state in an oxygen-poor environment; (b) that it was precipitated (often as relatively thin layers continuous throughout large basins) as ferric or ferro-ferric oxides because it reacted with oxygen produced photosynthetically by primitive precursors of blue-green algae; (c) that iron was required as an oxygen-acceptor by these primitive algae (hypothesized to have lacked a suitably advanced internal oxygen-mediating enzyme system) to depress local oxygen concentrations to a tolerable level; and (d) that the subsequent evolution of effective oxygen-mediating enzymes freed these photoautotrophs from dependence on external oxygen-depressors and led to a marked increase in biomass and subsequent burial, thus also increasing net oxygen production and "sweeping the oceans free of iron forever" (evidenced in the rock record by the last major phase of deposition of banded iron-formations). This resulted in the generation of a relatively oxygenic atmosphere (evidenced by the approximately synchronous onset, about 2000 m.y. ago, of terrestrial red bed sedimentation). He supports his argument by further noting that easily oxidized detrital pyrite and uraninite have been reported in stream deposits older, but not younger, than the postulated time of transition. In apparent agreement with this model, it can be additionally noted that the onset of relatively oxygenic conditions could have resulted in an abrupt decrease in ambient surface temperature (Sagan & Mullen 1972), a development that might be reflected by the glacial deposits of Proterozoic I.

In several ways, this is an attractive model. It unifies a large number of seemingly disparate observations into an apparently plausible and uncomplicated whole. At the same time, however, aspects of the model are obviously rather speculative. For example, since the oxygen liberated by photoautotrophs is produced intracellularly, and since it is the intracellular biochemical "machinery" that requires protection from oxidation, it is difficult to envision the development of oxygen-producing photoautotrophy without the prior or concurrent development of at least a primitive type of internal oxygen-mediating system. Certainly, as Cloud suggests, the external oxidation of ferrous iron could have aided in maintenance of microaerophilic conditions in the vicinity of such microorganisms. However, since banded iron-formations are reported to have been deposited as early as 3760 m.y. ago (Moorbath et al 1973), the model would appear to require a time lag of some 1800 m.y. (roughly three times as long as the Phanerozoic) for the evolution of "advanced" oxygen-depressing enzymes from their primitive precursors. Although I know of no data that completely rule out this possibility, it demands an exceedingly and probably unreasonably slow rate of evolution. This is perhaps especially notable, since the model seems also to suggest that the many steps of abiotic and biotic evolution bridging the gap between nonliving organic matter and the earliest oxygen-producing microorganisms must have occurred within a time period less than half as long (e.g. between 4500 and 3760 m.y. ago). The postulated dependence of stromatolite-forming microorganisms on ambient ferrous iron seems also somewhat difficult to reconcile with the occurrence of well-developed stromatolites, some containing cellularly preserved microorganisms, in nonferruginous (principally carbonate) horizons and formations deposited earlier than 2000 m.y. ago (Tables 1 and 2).

Nevertheless, the model makes the important point that some type of special interrelationship may have existed between the production of oxygen and its consumption in the deposition of banded iron-formations. However, this relationship might have been passive rather than active. If so, the oxygen would not necessarily need to have been entirely the product of biological photosynthesis; other oxygen sources, such as the photodissociation of water vapor induced by uv light, could have played an important role, especially prior to the origin of oxygen-producing photoautotrophs, and especially if such deposits are a result of oceanic upwelling as suggested by Holland (1973). In such an alternative working hypothesis, up to about 2000 m.y. ago, oxidation of aqueous ferrous iron (and of other environmental constituents such as H_2S, CO, possibly NH_3, etc) could have served as an effective oxygen sink. The concentration of atmospheric oxygen might have remained depressed until oxygen-producing photoautotrophs had become established and widespread (evidenced by the earliest widespread stromatolites?), and the transition from banded iron-formation to red bed deposition could be interpreted, as Cloud has suggested, as reflecting the advent of relatively oxygenic conditions. The major difference between this scenario and the foregoing model is that the occurrence of banded iron-formation would be evidence consistent with but not necessarily indicative of the presence of oxygen-producing photoautotrophs (or, indeed, of the necessary existence of any type of biological activity).

In sum, it seems apparent that a transition in atmospheric oxygen content, resulting at least partially from the cumulative effects of oxygen-producing biological photosynthesis, occurred during Precambrian time; geologic data suggest rather strongly (but are probably insufficient to establish firmly) that the transition occurred about 2000 m.y. ago. Among the unsolved portions of the puzzle: the time spanned by this transition—whether it was part of a long-term continuum or occurred as an abrupt step-function; its quantitative aspects—the partial pressure of atmospheric oxygen prior to and following the event; the relationship of the transition to Proterozoic glaciation—whether causative or fortuitous; and the role of photoautotrophs in the deposition of banded iron-formation—whether active and required or passive and essentially superfluous.

PALEOBIOLOGY   I have previously, and rather recently, reviewed in detail the fossil record of Proterozoic I (discussion of the Middle Precambrian record; Schopf 1970); here I will discuss only those aspects of particular import that have come to light since that time.

The oldest known microfossils of Proterozoic I (which also comprise the oldest diverse microbiota now known in the geologic record) are those recently reported from carbonate stromatolites of the approximately 2250 m.y. old Transvaal Dolomite of South Africa (Table 1); both septate filaments (Figure 1b) and colonial unicells (Figure 1c,d), interpreted as blue-green algae (L. Nagy 1974, MacGregor et al 1974), occur in the deposit. Although these appear to me to be the oldest bona fide

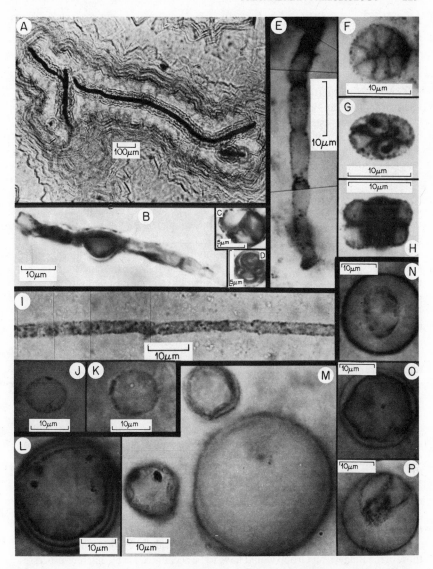

Figure 1 Optical photomicrographs showing mineralogic pseudofossils (*a*) and organically preserved filamentous (*b, e, i*) and spheroidal (*c, d, f–h, j–p*) microfossils in petrographic thin sections of Precambrian carbonates (*b–d*) and cherts (*a, e–p*); *e* and *i* show composite photomicrographs. *a*: Dodguni chert, southern India (? 2450 m.y.). *b–d*: Transvaal Dolomite, eastern Transvaal, Africa (~2250 m.y.). *e–h*: Dismal Lakes Grp., Northwest Territories, Canada (~1200 m.y.). *i–k*: Vempalle Fm., southern India (~1400 m.y.). *l–p*: Galeros Fm., Arizona (~850 m.y.).

microfossils now known from the Proterozoic, microfossil-like objects of possibly greater age (perhaps 2450 m.y. old) have been reported by Gowda (1970) from the Dodguni Cherts of South India and have been interpreted by Khan (1973) as cyanophytes. After studying these Dodguni "microfossils," however, I have become convinced that they are mineralogic artifacts (pseudofossils), composed predominantly of iron minerals that became concentrated into tubular (Figure 1a) and irregular microstructures during cyrstallization of the fibrous chalcedonic matrix. [This interpretation suggests that some and perhaps all of the filaments recently reported by Fairchild et al (1973) from the Devonian Caballos Novaculite of Texas, also occurring in fibrous chalcedonic chert, may be similarly nonbiologic. These two examples illustrate well the difficulties involved in interpretation of microstructures occurring in secondary, fibrous chalcedony. Fortunately, such difficulties are only rarely encountered in studies of microstructures occurring in primary, or early diagenetic, microcrystalline cherts, which is the type of preservation (permineralization of organic-walled microfossils in fine-grained quartz) exhibited by the majority of known Precambrian microbiotas (Table 1).]

During the past few years, a substantial number of stromatolitic sediments older than 2000 m.y. have been detected (Table 2). Two generalizations stand out: (1) only two stromatolitic units have been reported from rock sequences thought to have been deposited prior to the beginning of Proterozoic I (linestones of the Bulawayan Group, the age of which seems relatively well defined, and carbonates of the Steeprock Group, the age of which is not well known and which could conceivably be slightly less than the minimum of 2500 m.y. suggested in Table 2); (2) in both South Africa and Western Australia, the two areas of the world in which the geochronology of early stromatolitic units is well known, stromatolites first become abundant and widespread in sediments immediately younger than about 2300 m.y. (this may also be true for North America, although the geochronology is not as well-defined). It seems possible that the apparent paucity of older stromatolites may simply be a function of the known rock record; carbonate facies, in which stromatolites are normally (but not exclusively) preserved, are apparently relatively rare, due to nondeposition or nonpreservation, in terrains older than about 2300 m.y. However, it is also conceivable that the rather sudden appearance of abundant, widespread stromatolites might reflect the occurrence of a major evolutionary advance at about this time, for example, the origin, or at least the rapid diversification, of stromatolite-forming microorganisms, or the development of biochemical processes that enabled such organisms to flourish at the sediment-water interface and to precipitate and bind calcium carbonate. Were this the case, stromatolites older than about 2300 m.y. and especially those of the Bulawayan Group would have to be accounted for by different processes or by the activities of an earlier, presumably more primitive biologic group.

What is known about stromatolite-forming microorganisms? Probably the best known and most studied modern stromatolites are those occurring at Hamelin Pool near Shark Bay, on the west coast of Australia, first described in detail by Logan (1961). In major respects—their calcareous composition, internal laminar organization, size, external form, growth habit, and response to the physical environment—these modern stromatolites bear considerable resemblance to their Precambrian

Table 2 Stromatolites > 2000 million years in age[a]

Approximate age (× 10⁶ yr)	Geologic units	Locality	Form "Genera"	Reference
~2000 (1850 > 2185)	Wyloo Grp. (Duck Creek Dolomite; Mount Bruce Supergroup)	west-central Western Australia (Hamersley Basin)	*Pilbaria; Patomia*	Walter 1972
~2050 (1900 > 2100)	Hamersley Grp. (Wittenom Dolomite and Carawine Dolomite; Mount Bruce Supergroup)	west-central Western Australia (Hamersley Basin)	*"Collenia"*	Hunty 1964, 1967
~2100 (1950 > 2220)	Pretoria Grp. (Houtenbek Fm., Vermont Fm., Machadodorp Volcanics, Boven Shale; Transvaal "Sequence")	eastern Transvaal, South Africa	several unnamed forms	Button 1971
~2200 (2090 > 2290)	Fortescue Grp. (Tumbiana Pisolite and Pillingina Tuff = Mount Jope Volcanics; Mount Bruce Supergroup)	west-central Western Australia (Hamersley Basin)	*Alcheringa; Gruneria*	Walter 1972
~2250 (2220 > 2300)	Olifants River Grp. (Transvaal Dolomite; Transvaal "Sequence")	northern Cape Province, South Africa	*Katerina, Pilbaria*(?), and several unnamed forms	Truswell & Eriksson 1973, Walter 1972, Cloud & Semikhatov 1969
~2275 (2220 > 2300)	Wolkberg Grp. (Abel Erasmus Volcanics; Transvaal "Sequence")	eastern Transvaal, South Africa	several unnamed forms	Button 1973
~2300 (2220 > 2400)	Ventersdorp Grp. (Rietgat Fm.)	Orange Free State, South Africa	*"Collenia," "Cryptozoon,"* and *Gruneria*	Winter 1963, 1965, Walter 1972
~2600 (2500 > 2800)	Steeprock Grp. (Steeprock Limestone)	Steep Rock Lake, Ontario, Canada	*"Cryptozoon"*	Rothpletz 1916, Jolliffe 1955, Hofmann 1971
~3100 (3000 > 3300)	Bulawayan Grp. ("Zwankendaba Limestone")	southwestern Rhodesia, Africa	several unnamed forms	MacGregor 1941. Schopf et al 1971

[a] Other stromatolitic units possibly of comparable age: Belcher Grp. (1750 > 2200), Hofmann & Jackson 1969; Gunflint Iron Fm. (1750 > 2200), Hofmann 1969; Nash Fm. (1650 > 2340), Knight 1968; Biwabik Iron Fm. (1750 > 2200), Cloud & Semikhatov 1969; Denault Fm. (1800 > 2440), Donaldson 1963; Koolpin Fm. (1830 > 2440), Walter 1972.

analogues; moreover, it is now well established that both modern and fossil stromatolites are products of microbial activity. The calcareous laminae of modern stromatolites are produced by two processes, sometimes combined: (1) entrapment and binding of particulate carbonate minerals in mucilagenous sheaths and gelatinous mats formed at the stromatolite growth surface by a microbial community which is usually dominated by filamentous blue-green algae; and by (2) precipitation of calcium carbonate as a result of the photosynthetic removal of CO_2 from waters immediately adjacent to actively growing surfaces. Sequential layers are deposited largely as a result of the phototactic mobility of members of the microbial community. As the overlying layer of detrital and precipitated mineral matter progressively thickens, filamentous members of the community glide upward toward the light (leaving behind a mucilagenous residual layer) and establish a new growth surface. Evidence indicates that stromatolites of Proterozoic I were produced in a similar fashion; where preserved (e.g. stromatolites of the Gunflint Iron-Formation), former growth surfaces can be seen to have been composed of interwoven, anastomosing masses of predominantly filamentous microorganisms (see Schopf 1970 for photomicrographs of these and other early microfossils). In addition, in many fossil stromatolites, carbonate laminae of detrital, precipitated, or combined origins can be readily distinguished. Thus, it can be deduced that the predominant stromatolite-formers of Proterozoic I were filamentous, mat-building microorganisms that were probably mucilagenous, phototactic, and mobile; they were able to fix CO_2 (thus being at least partial autotrophs) and precipitate calcium carbonate, and they flourished at the sediment-water interface. Based on such reasoning and on analogy with modern stromatolite-formers, it has been almost universally assumed that these organisms were blue-green algae. Certainly, all of the evidence cited above is consistent with such an interpretation. Is it, however, compelling? Alternative hypotheses come to mind. For example, it seems equally conceivable that these organisms were members of a now extinct stock, similar to and possibly precursors of the essentially "modern" cyanophytes that are well known from the later Precambrian. Alternatively, the evidence might be interpreted as indicating that these organisms were related to gliding flexibacteria (which, although photoautotrophic, do not produce oxygen as a photosynthetic byproduct) of the types that occur in some modern algal mat communities. Two additional factors must be considered: is there morphologic evidence supporting the presumed cyanophytic affinities of these microorganisms? And, is there strong evidence for oxygen-producing, rather than bacterial, photosynthesis? On both counts, it seems to me that the evidence is as yet equivocal.

The postulated cyanophytic affinity of the dominant filaments of Proterozoic I has been based principally on the occurrence of relatively large intercalary cells interpreted (Licari & Cloud 1968, Cloud & Licari 1972, L. Nagy 1974, Schopf, in press) as heterocysts and akinetes, two specialized cell types that are characteristic of many modern blue-green algae. If critically viewed, however, the published photomicrographs of these structures could alternatively be interpreted as illustrating preservational or diagenetic artifacts. Probably the most convincing of such structures now known is the enlarged ("akinete"-like) cell shown in Figure 1b recently reported (L. Nagy 1974) from the Transvaal Dolomite; however, only this single example has been illustrated from the formation and additional data are needed to establish that structures of this type exhibit a size range, pattern of distribution, and an abundance consistent with the proposed interpretation. Similarly, heterocysts in modern blue-green algae usually exhibit a regular pattern of distribution, have distinct organic "plugs" at polar pores, and are commonly rather abundant in a single trichome; none of these characteristics is clearly demonstrated in known Precambrian specimens (including that reported from the upper Proterozoic Skillogalee Dolomite by Schopf & Barghoorn 1969). In the absence of such evidence, the proposed cyanophytic affinity of early Proterozoic filaments seems open to question; no other uniquely blue-green algal features, i.e. morphological characters not known also to occur in extant bacterial groups, have been reported from these assemblages. Indeed, a substantial number of the described taxa of Proterozoic I are morphologically unlike any known members of the modern biota (see figures in Barghoorn & Tyler 1965 and in Hofmann 1971b). In view of these considerations, it is possible that a fresh, potentially less confining approach may be needed in the classification of these microfossils, one that recognizes that early Proterozoic organisms might be only distantly related phylogenetically to modern microbes.

Although it has been readily assumed, based on circumstantial evidence, that the CO_2-fixing photoautotrophs of Proterozoic I were oxygen-producers, there appear to be no direct data that so indicate; evidence commonly cited (their occurrence in stromatolites; their morphologic similarity to modern cyanophytes; their presumed role in the deposition of banded iron-formations), while consistent with this interpretation, is not compelling. Resolution of this problem could hinge on understanding of the oxygen concentration at the sediment-water interface. If such concentrations were low, stromatolites might have been built by anaerobic communities dominated by photoautotrophic bacteria; in a living condition, the microbes in such populations would have been stratified as a function of their metabolic characteristics and the availability of organic and inorganic substrates. However, if oxygen concentrations at the sediment-water interface were relatively high, as in present algal mats, the growth surface would have been oxygenic and the population would have been stratified as a function of both oxygen concentration and metabolic requirements; in a living condition, aerobes (including oxygen-producers) would occur only in the upper reaches of such communities. Evidence of such stratification and of possible differences among stromatolitic communities of various ages, evidence that could well be detectable in permineralized stromatolites, has yet to be reported from the Precambrian. A second approach to the problem is suggested by results of carbon isotopic analyses of modern blue-green algal mat communities which indicate that aerobic and anaerobic decomposition of organic matter can result in differing isotope effects. Specifically, Behrens and Frishman (1971) report that when buried in the anaerobic zone of algal mats, organic carbon surviving bacterial degradation is enriched in ^{13}C by about two parts per thousand relative to carbon subjected to diagenesis under oxidizing conditions. Such differences are easily measurable and are potentially preservable in the kerogen fraction of ancient stromatolites. Studies of these types could yield new insight into the biological composition and physiological capabilities of stromatolitic communities, possibly providing a better basis for deciphering the timing and nature of the transitional phase in the development of the earth's oxygenic atmosphere.

Proterozoic II: 1800 to 1200 m.y. Ago

In recent years, the known fossil record of this middle third of the Proterozoic has substantially increased in quality and, especially, in quantity (Table 1); more than half of the microbiotas now known were undetected in 1969 when I prepared an earlier summary of this interval (Schopf 1970). Interestingly, the tentative interpretations then drawn seem generally quite consistent with the newly acquired data. This apparent consonance, however, may not be especially meaningful; all of the assemblages now known from Proterozoic II are in need of additional detailed study (except perhaps for the Beck Spring microbiota of which Licari's monographic treatment is currently in press) and for two of the microbiotas, those of the Vempalle Formation (Figure 1i–k) and the Dismal Lakes Group (Figure 1e–h), the photomicrographs here included are the first published. [This is true of virtually all of the other figures as well; except for the photomicrographs of the Transvaal microfossils (Figure 1b–d), the figures also include the first published pictures of the Dodguni pseudofossils (Figure 1a) and of microfossils from the newly discovered microbiotas of the Galeros Formation (Figure 1l–p), the Blyth Dolomite (Figure 2), the Myrtle Springs Formation (Figure 3a–c), and the Auburn Dolomite (Figure 3d–g).] Thus, as I suggested in the earlier review, "inferences here drawn from the fragmentary data available should be regarded as being of a most tentative sort" (Schopf 1970, p. 320).

The prime inference, and the only inference that I will here discuss at length, is that Proterozoic II may encompass the time of origin of the nucleated, eukaryotic level of cellular organization. With but few exceptions (e.g. the dinoflagellates, regarded in some quarters as "mesokaryotic"), all known living systems can be readily categorized as either *prokaryotic*—forms such as bacteria and blue-green algae that divide by fission rather than mitosis and that lack membrane-bound organelles such as nuclei and plastids—or *eukaryotic*— the protists, fungi, plants, and animals of the extant biota, all of which exhibit the basic similarity in cellular design of having various functional units packaged into discrete organelles, all of which exhibit mitotic cell division, and most of which are sexual, exhibiting both meiosis and syngamy during their life cycles.

In current taxonomies, this dichotomy in cellular organization is regarded as the principal division point in the evolutionary hierarchy; it is clear that prokaryotes and eukaryotes have long comprised major, independently evolving stocks. The probable evolutionary relationship between these two stocks is perhaps somewhat less certain. Two relationships seem conceivable: the groups might be of separate, independent origins, unrelated phylogenetically and sharing biochemical characteristics only because they originated from the same or similar starting materials within a similar environment and have subsequently evolved effective means for coexistence; or, apparently far more probably, the eukaryotic cell could be an evolutionary derivative of a prokaryotic ancestor (or ancestors, if a symbiotic hypothesis of the style championed by Margulis in 1970 proves ultimately acceptable). In either case (and I know of no serious supporters of the former possibility), there is no doubt that both groups became established during the Precambrian. Unlike modern members of

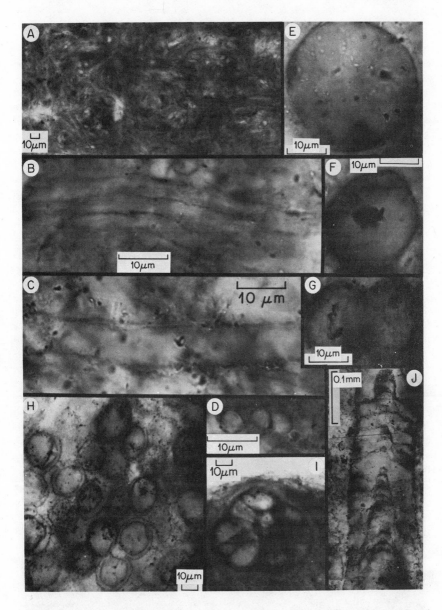

Figure 2 Optical photomicrographs showing organically preserved filamentous (*a–d*) and spheroidal (*e–i*) microfossils and a cylindrical "mini-stromatolite-like" structure (*j*) in petrographic thin sections of Precambrian black cherts from the lower River Wakefield Group (? Blyth Dolomite; ∼1150 m.y.) of South Australia; *j* is a composite photomicrograph. *a* shows a mat of interwoven, sheath-like filaments of the type shown at higher magnification in *b* and *c* (longitudinal sections) and *d* (transverse sections). Organically laminated "ministromatolite-like" structures of the type shown in *j* occur in colony-like layers up to 5 cm thick and composed of several hundred close-packed cylinders; the biologic group responsible for formation of these curious structures is unknown.

the "Prokaryota," which includes anaerobic and microaerophilic forms in addition to aerobes, extant eukaryotic cells are fundamentally oxygen-requiring (see Margulis 1970 for a detailed discussion and remarks about apparent exceptions). It is therefore

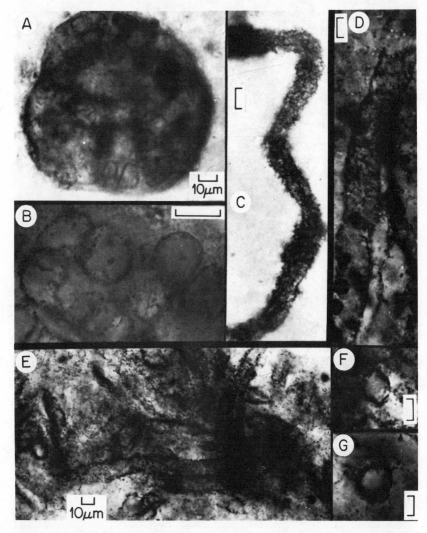

Figure 3 Optical photomicrographs showing organically preserved colonial unicells (*a, b*) and sheath-like filaments (*c–e*, longitudinal sections; *f, g*, transverse sections) in petrographic thin sections of Precambrian black chert from South Australia. *a–c*: Myrtle Springs Fm. (~950 m.y.). *d–g*: Auburn Dolomite (~900 m.y.). Lines for scale represent 10 μm; *c* shows a composite photomicrograph.

difficult to envision the advent of the "Eukaryota" prior to the development of oxygenic conditions; the banded iron-formation to red bed transition would thus seem to suggest an earliest reasonable date. Available fossil evidence is consistent with such a suggestion; assured eukaryotes are unknown from sediments older than 2000 m.y. At the same time, however, the fossil record is equivocal as to the actual time of origin of nucleated organisms. Although microfossils interpreted by some workers as possibly of eukaryotic affinities have been reported from Proterozoic II sediments (and, indeed, from units deposited during the latter third of Proterozoic I as well), compelling evidence of this advanced cell type extends back only to about 900 m.y. ago.

Microfossils suggested as possible early eukaryotes fall into three general categories: (1) objects resembling the statocysts of chrysophycean algae, reported from the approximately 1300 m.y.-old Beck Spring microflora (Cloud et al 1969, Licari 1974); (2) organic-walled, tubular, branching structures of relatively large diameter (15–60 μm), occurring in both the Beck Spring Dolomite (Licari 1974) and the somewhat younger Skillogalee Dolomite (Fairchild & Schopf 1974) that have been interpreted as siphonaceous green algae (Cloud 1974); and (3) solitary, clumped, and colonial unicells, some of relatively large size (> 50 μm), containing dense organic bodies interpreted by many workers as possible remnants of organelles. Cells containing such organelle-like bodies have been reported to occur in the Gunflint (Barghoorn 1971, Edhorn 1973), Belcher (Hofmann & Jackson 1969), Amelia (Muir 1974), Bungle Bungle (Diver 1974), Beck Spring (Cloud et al 1969), Dismal Lakes (Figure 1g; Schopf et al 1974), Blyth (Figure 2f; Schopf et al 1974), Skillogalee (Schopf & Fairchild 1973), Bitter Springs (Schopf 1968, 1974), and Galeros (Figure 1m; Schopf et al 1974) microbiotas (see Table 1 for ages of these assemblages).

Of these three categories of possible eukaryotes, the putative chrysophytes seem least convincing. Chrysophytes and chrysophyte cysts are elsewhere unreported from sediments older than Cretaceous (Loeblich 1974); comparison of Precambrian structures (perhaps 1200 m.y. older) with such cysts thus seems dubious. Since interpretation of these microstructures as eukaryotes is based solely on this comparison, it similarly is open to question. Although more plausibly of eukaryotic origin, the branching tubular structures exhibit an irregular branch pattern, rather variable diameters, and pinching and swelling over short distances in an apparently unsystematic fashion. These characteristics are not normally exhibited by siphonaceous chlorophytes. The evidence at hand does not seem to rule out the possibility that such tubes might reflect the activities of boring cyanophytes or might be remnants of sheaths that originally enclosed masses of blue-green algal trichomes. In contrast, at least some members of the third category—cells containing organelle-like bodies—seem convincingly eukaryotic. As discussed elsewhere (Awramik et al 1972, Schopf 1974), however, to be accepted as evidence of eukaryotic organization, such bodies must be demonstrated to be actually organellar in origin rather than merely being remnants of collapsed, "coalified" cytoplasm. In addition, to date, the detailed investigations needed to adequately establish this point—comparative statistical studies of modern and fossil algal populations and studies of artificially silicified modern algae supported by data from optical microscopy, transmission electron microscopy, and chemical analyses—have been carried out on but one taxon of one Precambrian microbiota (*Glenobotrydion* of the 900 m.y.-old Bitter Springs assemblage; Schopf 1972, 1974, Oehler 1973). Due to lack of preservation of cellular detail, it is doubtful that investigations of this type could be fruitfully applied to the possible eukaryotes in several of the fossil assemblages now known (e.g. those preserved in the Gunflint, Beck Spring, Dismal Lakes, and Skillogalee cherts). Nevertheless, assuming that neither structurally complex microscopic eukaryotes nor trace or body fossils of megascopic eukaryotes are discovered, such studies will be required before the history of eukaryotic organization can be traced with confidence into Proterozoic II (judging from published photomicrographs, a prime candidate for such studies is the microflora of the Bungle Bungle Dolomite described by Diver in 1974). Thus, although available evidence seems *consistent* with the proposition that eukaryotes were probably extant as early as 1300 m.y. (Beck Spring microbiota) or 1500 m.y. ago (Bungle Bungle microbiota), and may have existed even earlier, *compelling* evidence dates only from the time of deposition of the approximately 900 m.y.-old Bitter Springs cherts.

Proterozoic III : 1200 to 600 m.y. Ago

Of the three subdivisions of the Proterozoic here recognized, the paleobiology of this latter third of the Era is best known. As is summarized in Table 1, half of the diverse, well-preserved microbiotas thus far detected in the Precambrian occur within this 600 m.y.-long interval. Perhaps most importantly for the present discussion, this interval includes the most studied and probably best understood of all assemblages yet discovered, that of the Bitter Springs Formation of central Australia.

I have recently summarized in detail the apparent evolutionary implications of the abundant, diverse, and exceptionally well-preserved microbiota of the Bitter Springs cherts (Schopf 1972). Based on this assemblage, and on other microfossils now known from Proterozoic III (e.g. Figure 1*l*–*p*, Figure 2*a*–*i*, Figure 3), the following generalizations seem warranted:

(1) The abundant, widespread, stromatolitic communities of this time were composed predominantly of blue-green algae. Unlike the microbiotas of Proterozoic I—which included a relatively large proportion of organisms of unknown affinities and which were dominated by presumably prokaryotic microorganisms that bear resemblance to, but might have been only distantly allied with, modern blue-green algae—those of Proterozoic II and especially Proterozoic III contained diverse assemblages of filamentous (apparently including oscillatoriacean, nostocacean, and rivulariacean) and coccoid (entophysalidacean, pleurocapsacean, but predominantly chroococcacean) cyanophytes.

(2) Unicellular eukaryotes, apparently of chlorophycean and/or rhodophycean affinities (Schopf & Blacic 1971), had become established at least as early as 900 m.y. ago (Schopf 1972, 1974). A tetrahedral tetrad of spore-like cells and isolated unicells bearing surficial trilete-like marks, presumed to be of meiotic origin (Schopf & Blacic 1971), have been detected in deposits of this age. These data strongly suggest (but do not prove, since the tetrad could be the rearranged products of mitosis or fission, and the apparent triletes could be preservational artifacts) that at least some

Table 3 Precambrian microfossils < 2500 m.y. in age (exclusive of fossils occurring in stromatolitic microbiotas)

Types of microfossils	Age range
"sporomorphs"; "palynomorphs"; acritarchs; isolated algal and fungal(?) filaments, etc.	primarily late Precambrian; most < 900 million years, many < 750 million years.

Geographic distribution	Lithologies
North America; South America; India; U.S.S.R.; China; Europe; Scandinavia; Australia.	principally shales and graywackes; some carbonates and metasediments; few cherts.

References
Reviews: Glaessner 1962, 1966, Timofeev 1966, 1969, 1973, Downie 1967; J. M. Schopf 1969, J. W. Schopf 1970. *Recent reports*: Timofeev 1970, Fournier-Vinas & Debat 1970, Vidal & Röshoff 1971, Cloud & Germs 1971, Pacltová 1972, Binda 1972, Konzalová 1972, 1973, Shimron & Horowitz 1972, Ford & Breed 1973, LaBerge 1973, White 1974.

of these unicellular eukaryotes were sexual, having life cycles composed of alternating haploid and diploid phases.

(3) In addition to benthic, stromatolitic communities of microorganisms, planktonic unicellular algae (represented in the fossil record by acritarchs, "palynomorphs," etc) were diverse, widespread, and abundant during Proterozoic III (Table 3).

(4) It is now known with certainty that megascopic, multicellular organisms had evolved prior to the close of the Precambrian. Fossils of megascopic algae, apparently between about 800 and 600 m.y. in age, occur in Norway (Spjeldnaes 1963), South-West Africa (Glaessner 1963), and the Soviet Union (Vologdin 1966, Gnilovskaja 1971). Metazoan trace and body fossils are known from latest Precambrian (700 to 600 m.y. in age) sediments of South Australia and elsewhere (Glaessner & Wade 1966, Glaessner 1969, 1971). In addition, megascopic structures interpreted as remains of eukaryotic algae (Walter et al 1974) and possibly metazoans (Glaessner 1969; for a conflicting interputation, see Cloud 1973) have been reported from strata perhaps as old as 1300 m.y. Although possibly biogenic, these older structures are of uncertain origin and are separated from established metaphytes and metazoans by a time period nearly as long as that encompassed by the entire Phanerozoic. If these structures are in fact remnants of metaphytes or metazoans, it can be anticipated that additional evidence will be obtained from sediments of intermediate geologic age.

(5) Little is known with certainty regarding the timing and nature of processes and events resulting in the late Precambrian emergence of megascopic life. However, as new data have rapidly accumulated in recent years, workers have become increasingly confident in interpreting the known late Precambrian fossil record "as is"; earlier

approaches to the classic question of the origin of the Metazoa, approaches that relied heavily on special causes and ad hoc assumptions to explain the apparent paucity of Precambrian animal life, seem increasingly unattractive. If this confidence is justified, it seems likely that the ultimate solution to the problem will be multi-faceted, involving a mix of evidence from both neobiology (i.e. genetics, bio-chemistry, developmental biology, ecology) and paleobiology, and that it will take into account the interactions between, and the possible coevolution of, late Pre-cambrian metazoans and metaphytes (Schopf et al 1973b, Stanley 1973).

The Archean Record

As is summarized in Figure 4, the Proterozoic-Archean "boundary" (at about 2500 m.y.) corresponds roughly to the first appearance of widespread stromatolites (2300 m.y.) and diverse microbiotas (2250 m.y.); only two stromatolitic horizons have been reported from the Archean (Table 2) and sediments of the Era are not known to contain diverse, well-preserved microbiotas similar to those relatively widespread in the Proterozoic (Table 1). This evident dichotomy probably does not reflect a marked difference in emphasis and activity on the part of researchers; the successes in Proterozoic paleobiology have certainly stimulated ongoing study of deposits of

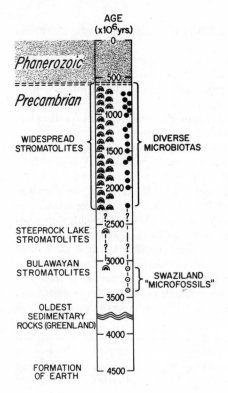

Figure 4 Geologic column summarizing the temporal distribution of stromatolites, microbiotas, and fossil-like micro-structures now known from sediments of Archean (3760 < 2500 m.y.) and Pro-terozoic (2500 < 600 m.y.) age.

that age, but a number of geologists, micropaleontologists, and geochemists have also made a concerted effort to uncover evidence of Archean life. It would probably be premature to interpret this dichotomy as being real, i.e. as reflecting a basic difference in the abundance, distribution, or nature of living systems on either side of the "boundary," rather than, for example, being due to geologic effects such as differing degrees of preservation or styles of tectonism that might tend to obscure the earlier record. At the same time, however, if the current trend continues unchecked, if evidence of Archean life continues to prove elusive in coming years, possibilities of this type would merit serious consideration.

The most commonly cited evidence of Archean life is the occurrence of micro-fossil-like objects, primarily detected in sediments of the Swaziland Supergroup (Fig Tree and Onverwacht Groups) of the eastern Transvaal, South Africa, that has been reported in the literature since the mid-1960s. Three varieties of such objects have been described: (1) rod-shaped (and possibly spheroidal) bacterium-like bodies; (2) filamentous, microscopic thread-like structures; and (3) coccoidal, unicell-like microstructures.

BACTERIUM-LIKE MICROSTRUCTURES   At this writing, I regard the bacterium-like bodies (*Eobacterium isolatum*) which Barghoorn and Schopf (1966) reported as being demonstrably indigenous to the Swaziland sediments (following revision of Swaziland stratigraphy, they should now be regarded as occurring in cherts of the upper Onverwacht Swartkoppie Formation, rather than the Fig Tree cherts as originally reported). These structures are almost certainly composed of organic matter; I interpret them as probably biogenic. It should be recognized, however, that these minute (< 1 μm) structures are of very simple morphology, and that their assessment as probable microfossils is a somewhat subjective matter, based on analogy with the morphology of modern microbes and on the supposition that the shape and form of these structures (especially their smooth, circular, cross-sectional outline) would be difficult to explain by solely nonbiologic processes. Moreover, although bits of organic matter of similar size and form can be discerned optically in petrographic thin sections of the Swaziland cherts, these structures were studied principally in surface replicas by transmission electron microscopy. This and similar techniques however, have led to the report of a rather large number of microstructures that although interpreted by various workers as "Precambrian microfossils" now appear to me (cf Cloud 1973) to be either artifacts and assuredly nonbiologic, or to be recent, and primarily bacterial, contaminants (e.g. Schopf et al 1965, Cloud et al 1965, Jackson 1967, Oberlies & Prashnowsky 1968, Schidlowski 1968, 1970, Prashnowsky & Oberlies 1972, Dungworth & Schwartz 1972). Thus, at present, *Eobacterium* is probably best regarded as suggestive, rather than compelling, evidence of Archean life.

FILAMENTOUS "MICROFOSSILS"   A number of workers have reported the occurrence of "filaments," "filamentous bodies," "fibrillar structures," etc, from sediments of Archean age. Judging from the known Proterozoic fossil record, the search for older filamentous microfossils would appear promising; moreover, the existence of well-laminated Archean stromatolites (Table 2) provides at least presumptive evidence for the early appearance of the filamentous habit (Schopf et al 1971). Nevertheless, none of the reports of Archean filamentous microfossils that have appeared to date seem compelling.

The most recent and probably the most noteworthy report of Archean filaments is that of Brooks, Muir & Shaw (1973). These workers report the rare occurrence in Onverwacht cherts of "filamentous bodies," 4–8 μm in diameter, which they regard as "undoubtedly the remains of living organisms." However, of the four structures they illustrate, two (their Figures 1e and 1f) are short, rather nondescript, and do not seem obviously biogenic; a third (their Figure 1h), although apparently biogenic, is not demonstrably indigenous to the chert; and the fourth (their Figure 1g) is puzzling, since it appears to occur in a petrographic thin section (rather than in etched chert as is indicated in the text), is more than twice as long (about 55 μm) as the maximum length (20 μm) indicated in the text for Onverwacht filaments, and bears considerable resemblance to filamentous pseudofossils—organic matter occurring in quartz-filled microfractures—reported by Pflug (1967) from the overlying Fig Tree cherts.

Two other reports of Archean filaments merit attention. In 1968, Engel and his colleagues reported the occurrence of carbonaceous filaments "of diverse morphology, many remarkably lifelike" in the Onverwacht sediments. Although these forms are described as being "abundant," only one such form is figured (their Figure 2b) and, as Engel et al indicate, the figured object is difficult to interpret. In the absence of additional data from these workers concerning Onverwacht filaments, and despite their initial interpretation that "many" of these forms "appear to be true fossils," it now appears that all of these structures (rather than only a portion, as the authors suggested) are best regarded as "carbonaceous material of accidental form, distributed unevenly and also in regular geometric patterns along the bedding planes by sedimentary and diagenetic processes." Second, Pflug (1967) has recorded the occurrence of unbranched, filamentous microstructures in acid-resistant residues of shales of the overlying Fig Tree Group; he regards these forms as similar to the modern blue-green alga *Trichodesmium*. None of the forms figured, however, is evidently cellular ; all but two are quite short and not obviously filliform; and the two relatively long filaments (his Plate III, Figures 13 and 14, and 9 and 15) are regarded by Pflug as questionable, and of "problematic" nature. Objects found in thin sections and regarded by Pflug as being similar to these filamentous microstructures appear to me to be "accidental" aggregations of organic matter.

In sum, none of these reported microstructures provides compelling evidence of the existence of filamentous Archean microorganisms. Although it is possible and perhaps likely that such organisms were extant prior to the Proterozoic, the only available evidence that seems reasonably suggestive of their earlier presence is the occurrence of laminated Archean stromatolites.

SPHEROIDAL "MICROFOSSILS"   The third variety of Archean "microfossils"—coccoidal, unicell-like microstructures—is similarly difficult to interpret. During recent years, a large number of such microscopic spheroids, variously interpreted as "spheroidal microfossils," "alga-like microfossils," or "fossil-like microstructures," have been reported from Swaziland sediments (Pflug 1966, 1967, Schopf & Barghoorn

1967, Engel et al 1968, B. Nagy & L. Nagy 1969, Pflug et al 1969, Brooks 1971, Brooks & Muir 1971, L. Nagy 1971, Dungworth & Schwartz 1972, Brooks, Muir & Shaw 1973). For a variety of reasons, however, it is difficult to meaningfully synthesize results of these many studies. As is summarized in Table 4, fossil-like objects of this type (cf *Archaeosphaeroides barbertonensis*) have been reported from at least four horizons in the Swaziland Supergroup (the Sheba, Swartkoppie, Kromberg, and Theespruit Formations, together spanning a stratigraphic interval of about 13,000 m). However, results of studies of these horizons cannot easily be compared since the geographic and stratigraphic provenience of samples investigated are often not clearly indicated (Table 4). In addition, results of studies of various rock types (cherts, shales, and "siderites"), prepared and examined by a variety of techniques (e.g. petrographic thin sections, acid-resistant residues, and powdered samples studied by optical and electron microscopy), have rather commonly been lumped together in published descriptions such that it is difficult and sometimes impossible to determine which objects were found in which lithologies by use of which technique.

Thus, the data available are generally insufficient to resolve conflicting interpretations. For example, Nagy and Nagy (1969) have published histograms showing the size distribution of approximately 400 microstructures, about half of which range in diameter from 7 to 95 μm (Table 4, number 6) and occur in "a sedimentary zone in the upper Onverwacht," and the remainder of which occur in "another nearby sedimentary zone in the upper Onverwacht" and range in diameter from 11 to 193 μm (Table 4, number 7). [Unfortunately, the lithology and exact geographic and stratigraphic source of the rocks studied are not indicated, nor is it evident how the samples were prepared and examined, by what technique the measurements were obtained, or, most importantly, on what basis it was (presumably) determined that all of the objects measured were of carbonaceous rather than of mineralic composition.] In apparent conflict with these results, Brooks, Muir and Shaw (1973) specifically cite the report of Nagy and Nagy and state that "contrary to some earlier results, the size distribution [of spheroids] in a single sample [of Onverwacht sediment] tends to fall within a narrow range," indicated as being between 7 and 20 μm; these workers, however, present no statistical data in support of this contention and do not indicate on what evidence, other than observation of an unspecified number of microstructures, it is based. In sum, it is at present difficult to sort out the stratigraphic, lithologic, and geographic relationships among the numerous microstructures that have been reported from Swaziland strata, and to resolve the apparently conflicting testimony of workers who have reported these occurrences. Clearly, there is pressing need for a carefully executed, monographic treatment of the Swaziland Supergroup and its fossil-like microstructures.

Based on the data currently available, however, what can be said regarding the possible biogenicity of these Archean spheroids? Are these oldest "fossils," fossils? To approach this problem, I have investigated the possibility of devising a series of simple statistical tests to distinguish between populations of unicellular microorganisms, both modern (e.g. *Chlorella, Gloeocapsa, Tetraspora*) and fossil (e.g. Bitter Springs unicells containing organelle-like bodies and *Huroniospora* of the Gunflint

Table 4   Modern unicellular algae, fossil unicells, and "fossil-like" spheroids

Approximate age (m.y.)	Material	Number of Specimens measured	Diameter Range (μm)	Diameter Mode (μm)
	MODERN UNICELLULAR ALGAE			
Modern	*Chlorella* sp.	426	1.5–5	2–3
Modern	*Gloeocapsa* sp.	338	2.5–5.5	3–5
Modern	*Tetraspora* sp.	292	6–12	8–10
	LATE PRECAMBRIAN UNICELLS			
~900	Bitter Springs Fm.—cells with organelle-like "spots"	519	5.5–15.5	8–10
	MIDDLE PRECAMBRIAN UNICELLS			
~1900	Gunflint Iron Fm.—*Huroniospora* spp.	805	1.1–15.8	2–5
	ABIOTIC "ORGANIZED ELEMENTS"			
~4500	Orgueil carbonaceous meteorite—organic spheroids	268	2–36	7–11
	EARLY PRECAMBRIAN SPHEROIDS			
	Fig Tree Group			
~3100	1. Sheba Fm.?	19	10–36	16–20
~3100	2. Sheba Fm.?	7	26–39	36
~3100	3. Fm.?	not indicated	<30	...
~3100	4. Fm.?	not indicated	...	~20
	Onverwacht Group			
~3200	5. Swartkoppie Fm.	28	15–24	18–21
~3200	6. "Upper Onverwacht," Fm.?	199	7–95	9–15
~3200	7. "Upper Onverwacht," Fm.?	203	11–193	polymodal?
~3200	8. Swartkoppie Fm.	not indicated	15–20	...
~3200	9. Kromberg Fm.	not indicated	15–20	...
~3200?	10. Fm.?	not indicated	20–50	...
~3400	11. Theespruit Fm.? (Zone SF-77)	192	6–74	polymodal?
~3400	12. Theespruit Fm.	not indicated	7–10	...
	Swaziland Supergroup			
~3150	"Swaziland-1"	54	10–39	16–20
~3150	"Swaziland-2"	456	7–193	polymodal?
~3400	"Swaziland-3"	192	6–74	polymodal?

cherts), and populations of unicell-like objects known to be of abiotic origin (e.g. the carbonaceous "organized elements" of the Orgueil meteorite). These tests could then be used to elucidate the nature of the Swaziland spheroids. I will discuss in detail elsewhere the results of this study (J. W. Schopf, in preparation), results that are partially summarized in Table 4; here, it seems sufficient to briefly outline conclusions that now seem indicated. Three size parameters have been studied: (1) the standard deviation, a measure of the dispersion from the mean spheroid diameter of the population; (2) a measure of the approach of the population to an arithmetic normal, Gaussian distribution, or a combination of several such distributions; and (3) a of Spheroids

Mean (μm)	Standard deviation (μm)	(%)	Sample preparation	References
2.7	0.7	24	water mount	Schopf, this report
4.2	0.5	12	water mount	Schopf, this report
9.1	1.4	15	water mount	Schopf, this report
9.3	1.9	21	thin sections of chert	Schopf 1974 (Figure 16)
5.3	2.5	46	thin sections of chert	Barghoorn & Tyler 1965 (Figure 9) (*right*)
10.9	5.1	46	acid-resistant residues	Rossignol-Strick & Barghoorn 1971 (Figure 1)
18.7	7.3	39	acid-resistant residues of shale	Pflug 1967 (Plate I, Figures 8, 9, 12–15, 22, 23; Plate II, Figures 5, 12–16, 18–21, 23–28, 32)
32.7	4.6	66	thin sections of chert	Pflug 1967 (Plate I, Figures 11, 26–28)
...	...	...	not indicated	Engel et al 1968, p. 1008
...	...	...	acid-resistant residues of chert	Dungworth & Schwartz 1972, p. 701
18.7	1.8	10	thin sections of chert	Schopf & Barghoorn 1967 (Figure 5)
20.2	11.9	59	powdered preparation?	Nagy & Nagy 1969 (Figure 2)
55.6	37.4	67	powdered preparation?	Nagy & Nagy 1969 (Figure 3)
...	...	...	not indicated	Brooks, Muir & Shaw 1973, p. 213
...	...	...	not indicated	Brooks, Muir & Shaw 1973, p. 213
...	...	...	acid-resistant residues of chert	Dungworth & Schwartz 1972, p. 701
27.2	14.0	51	not indicated; from chert	Engel et al 1968 (Figure 4)
...	...	...	not indicated	Brooks, Muir & Shaw 1973, p. 213
20.3	6.8	33	as indicated above for 1+2+5	as indicated above for 1+2+5
36.0	31.6	88	as indicated above for 1+2+5+6+7	as indicated above for 1+2+5+6+7
27.2	14.0	51	as indicated above for 11	as indicated above for 11

measure designed to determine whether the end points of the size range of the population are biologically "reasonable" (i.e. whether they are compatible with the pattern of distribution resulting from vegetative division in populations of extant unicellular algae). As can be seen from inspection of Table 4, homogeneous, vegetatively dividing populations (the three modern algae and, apparently, the Bitter Springs unicells) exhibit a relatively low standard deviation (i.e. 12–24%). Application of the three tests indicates that the Gunflint population is heterogeneous, apparently being composed of several discrete biologic components each of which approaches a Gaussian distribution; thus, although it is similar to the abiotic population in having a high standard deviation (46%), the Gunflint unicells differ in this important respect from the Orgueil "organized elements." In contrast, the Swaziland spheroids (e.g. "Swaziland-2," a combination of all available measurements for spheroids from the Fig Tree and upper Onverwacht, and "Swaziland-3," the available data for lower Onverwacht spheroids) are similar in their pattern of size distribution to the abiotic "organized elements," a similarity reflected by each of the three parameters examined. It would be premature and possibly erroneous to interpret this degree of similarity as establishing that the Swaziland spheroids are entirely nonbiologic. It does seem to suggest, however, that the Archean populations are not composed of a single biologic group or assemblage of closely similar biologic groups; that if all of the objects in these populations are of a single origin, that origin is almost certainly not biological; that these populations are composed of a morphologically heterogeneous mixture of entities which may be of heterogeneous origin; and that if these populations are composed solely of organic spheroids (rather than including an admixture of spheroidal mineral grains), they could be, at least in part, of abiotic (or possibly prebiotic) origin. Based on the data now available, the Swaziland spheroids should not be regarded as constituting firm evidence of Archean life.

EVIDENCE OF ARCHEAN LIFE   Although direct evidence is scanty, the existence of biologic systems during the Archean seems indicated by three independent lines of reasoning.

First, as is indicated above, calcareous stromatolites are known to occur in at least two Archean units (Table 2). The biogenic interpretation of these structures appears to be firmly grounded; they are essentially indistinguishable from younger stromatolites that are of demonstrable biogenicity and there seems no reason to suspect that they could have resulted solely from inorganic, accretionary processes.

Second, the insoluble organic matter (kerogen) occurring in unmetamorphosed Archean sediments exceeding 3000 m.y. in age (i.e. carbonaceous material that seems certain to have been deposited at the time of original sedimentation) has a carbon isotopic composition comparable with that of demonstrably biologic organic matter of younger geologic age and is distinctly different from that of coexisting and younger inorganic, carbonate carbon (Schopf et al 1971, Oehler et al 1972). Although relatively little is known regarding the isotopic composition and mass distribution of carbon in the various Archean reservoirs, it seems highly difficult to account for the isotopic composition of Archean kerogen without the intervention of biologic activity.

And third, the occurrence both of relatively abundant, diverse, and morphologically complex microorganisms, and of widespread stromatolites in deposits of early Proterozoic age, evidences an episode of prior Archean evolution. Although there are no a priori means of estimating accurately the time encompassed by this earlier evolutionary development, the Bulawayan stromatolites (Schopf et al 1971) and the carbon isotopic data (Oehler et al 1972) suggest that living systems probably originated earlier than 3300 m.y. ago. The paths traversed during the earliest phase of biologic activity (for example, whether the Archean could have been characterized by a multiplicity of sequential origins, evolutions, and extinctions of primitive biospheres, and whether abiotic and biotic organic systems could have coexisted for geologically significant periods of time) and, indeed the time and mode of origin of terrestrial living systems are presently problems that seem far from solution.

PATHS OF FUTURE RESEARCH

As is perhaps all too evident in the foregoing discussion, discoveries of the past decade (and in some cases their possible misinterpretation) have posed nearly as many questions as they have helped to answer; much more remains to be learned. No doubt future studies will continue the ongoing attempts to resolve obvious, outstanding problems, some of which—such as the classic question of the origin of the Metazoa—have defined acceptable solution for more than a century. Judging from developments of the past ten years, however, one can also expect the discovery of important new questions; the time of origin of eukaryotic organization and of the secondary development of such features as meiosis and sexuality, for example, were not recognized as paleobiologic problems of major import until a very few years ago. And finally, based on the experience of the Apollo program it seems safe to predict that the techniques and approach of Precambrian paleobiology could play an important role in entirely new ventures, such as the search for evidence of life in returned Martian samples, a prospect that seems certain to come to fruition within this century and possibly as early as 1988.

Perhaps the most salient observation that comes to mind at the present time, however, is that the study of Precambrian paleobiology appears to have reached a threshold; questions regarding the existence of Precambrian life, questions that only a decade ago were popular and pressing, are no longer raised. The question now is not whether evidence of early life exists, but what does it say? What does it tell us of the timing and nature of the early evolutionary sequence? The field is rapidly progressing from a formative phase of widespread, hopeful speculation into a more mature phase of critical evaluation leading to well-founded hypotheses.

SUMMARY AND CONCLUSIONS

1. In recent years, there has been a striking increase of interest in the paleobiology of the Precambrian. Of the 20 well-preserved microbiological communities now known from sediments of pre-Phanerozoic age, all but one have been discovered during the past decade (Table 1); well over half of these assemblages have been discovered since 1973 and fully 40% were first reported during the six months prior to July 1974. Precambrian microbiotas, in toto spanning a time range (2250–725 m.y.) nearly three times as long as that encompassed by the entire Phanerozoic, are now known from Australia, Canada, the United States, India, and Africa. These cellularly preserved microfossil assemblages comprise the foundation on which current understanding of the course and timing of early evolutionary advance must be based.

2. Of the 20 microbiotas now known, however, only two—the well-known assemblages of the Gunflint Iron-Formation (Barghoorn & Tyler 1965) and the

Bitter Springs Formation (Schopf 1968, Schopf & Blacic 1971)—have received monographic treatment. Similar extended descriptions of three other biotas (those of the Hector Formation, the Skillogalee Dolomite, and the Beck Spring Dolomite) are currently in press or are in preparation (Table 1). Thus, a minimum of 15 microbiotas—three-quarters of all assemblages now known—are in need of additional, detailed investigation. Moreover, even the two best known biotas are as yet incompletely documented; a variety of unusual and potentially significant taxa remain to be formally described from the Gunflint assemblage (Schopf 1963, Barghoorn 1971) and several profusely fossiliferous localities of the Bitter Springs cherts have yet to be studied in detail (J. W. Schopf, unpublished).

3. Virtually all of these important microbiological communities have been preserved by permineralization in a distinctive and rather uncommon facies—fine-grained, primary or very early diagenetic cherts, occurring in association with (or in part replacing) algal-laminated, commonly stromatolitic, carbonates (Table 1). It remains to be established whether the degree of diversity and level of evolutionary development represented in these assemblages reflect accurately the nature of their contemporary biospheres. Until this matter is resolved, and until an adequate number of permineralized microbiotas have been studied in detail, speculations regarding early evolution will remain just that—speculative inferences founded on incomplete and possibly misleading evidence.

4. Organic geochemical investigations of ancient sediments, studies which only a few years ago appeared to hold promise for rapid elucidation of important biochemical and physiological aspects of early evolution, have been beset by major difficulties. Careful studies have demonstrated that many ancient sediments contain both extractable and insoluble organic components. At present, however, it is "extremely difficult (if not impossible)" (McKirdy 1974, p. 108) to establish that the extractable components (amino acids, fatty acids, porphyrins, hydrocarbons, etc.) date from the time of sedimentation, rather than having been introduced (e.g. via ground waters) in the relatively recent geologic past. In contrast, the insoluble carbonaceous component (kerogen) appears generally to be syngenetic with Precambrian sedimentation; however, the insoluble kerogen polymer is chemically quite complex and, with the possible exception of carbon isotopic analyses, its study has yet to yield definite insight into the biochemical and physiological capabilities of early life.

5. In broad brush outline, the known Precambrian fossil record can be divided temporally into two major phases: the *Archean* record, dating from the time of deposition of the oldest known sediments (3760 m.y.) to the approximately synchronous first appearance of widespread cratonal, platform-type sediments (2500 m.y.) and associated stromatolitic carbonates (2300 m.y.; Table 2); and the *Proterozoic* record, extending from the close of the Archean to the first appearance of widespread invertebrate metazoans that marked the beginning of the Phanerozoic (600 m.y.). Although these major phases are of roughly equal duration, and although both have been rather intensively investigated by paleobiologists, geochemists, and geologists over the past decade, the phases differ markedly in both the quality and quantity of evidence of biological activity they are now known to contain (Figure 1).

6. Despite the studies of recent years, evidence of Archean life has remained meager, difficult to interpret and in many cases unconvincing. In retrospect, it now appears likely that some "early" (mid-1960) interpretations may have been partly in error: although it remains possible that organic compounds extracted from Archean sediments date, at least in part, from the time of sedimentation, available evidence is less than compelling and interpretations based on such components (e.g. interpretation of the occurrence of pristane, phytane, and porphyrins as evidence of chlorophyll and, therefore, of photosynthetic life) are subject to question. Similarly, well-preserved microbiotas are still unknown from the Archean and the simple, rather nondescript "microfossils" that have been detected bear considerable resemblance to nonbiologic objects (Table 4) and are thus of questionable biogenicity. In sum, little is known regarding the antiquity, morphology, affinities, physiological variability, ecology, or evolution of Archean life.

7. Although direct evidence is scanty, the existence of biologic systems during the Archean seems indicated: (a) carbonate stromatolites, essentially indistinguishable from demonstrably biogenic Proterozoic bioherms, occur in at least two Archean units (Table 2; Figure 1); (b) insoluble organic matter, evidently syngenetic with the enclosing sediments and having a carbon isotopic composition consistent with a biological origin, occurs in Archean deposits at least as old as 3300 m.y.; (c) banded iron-formations, possibly (but probably not necessarily) reflecting the presence of biologically–generated free oxygen, occur in the oldest sedimentary sequence now known (3760 m.y.) and are locally well developed in the Archean; and (d) the occurrence of relatively abundant, diverse, and morphologically complex fossil microorganisms in deposits of early Proterozoic age evidences an episode of prior Archean evolution (although there is, however, no a priori means of estimating accurately the time encompassed by this earlier evolutionary development). It seems evident, therefore, that life originated during the Archean. Although the processes involved are as yet incompletely defined, studies of the past two decades strongly suggest that the abiotic synthesis of living systems may have been a rather rapid (10^3–10^6 yr) and not uncommon event on the primitive earth. It may therefore be surmised that living systems probably originated on several or more likely many occasions. (It also seems conceivable that the Archean could have been characterized by a multiplicity of sequential origins, evolutions, and extinctions of primitive biospheres, and that biotic and abiotic organic systems could have coexisted for geologically significant periods of time.)

8. In notable contrast with studies of the Archean, paleobiologic investigations of the Proterozoic have met with considerable success (Tables 1 and 3). Bearing in mind the limitations outlined above, the available data appear to indicate that morphologically and biochemically complex (genetically "modern") microorganisms had become established as early as 2300 m.y. ago; that stromatolitic, microbial biocoenoses of this time and the later Precambrian were based on filamentous photoautrotrophs; that the evolution of the earth's atmosphere, especially the advent of relatively oxygenic conditions perhaps 2000 m.y. ago, had a marked effect on the course and rate of early biotic diversification; that the Proterozoic biosphere, and indeed that of the entire Precambrian, was dominated by microscopic and for the

most part apparently prokaryotic forms of life; that the development of the megascopic, multicellular, eukaryotic level of organization was a relatively recent innovation, possibly occurring about 800–700 m.y. ago; and finally, that the present biota represents the most recent expression of an unbroken evolutionary continuum that extends back through the geologic past to a time equal to and probably greater than half the currently accepted age of the earth.

9. Extrapolating from the progress of the past decade, it seems probable that ongoing investigations of the early fossil record will focus on a rather unusual mix of classic problems, as yet unsolved, and novel questions only recently perceived. Problems of particular moment include: (a) Precambrian biostratigraphy—can paleobiologic data provide a solid framework for biostratigraphic division of the Precambrian? (b) Archean paleobiology—when did life begin and by what paths did it evolve during the earliest half of earth history? (c) Taxonomy and phylogeny—was the Precambrian actually "The Age of Blue-Green Algae" (J. W. Schopf, in press), or is a fresh and potentially less confining approach needed in the classification of primitive fossil microbes? (d) Secular variations—when and at what rate might such factors as day-length (Pannella 1972, Rosenberg & Jones 1974), surface temperature (Sagan & Mullen 1972), tectonic style, atmospheric composition, and ocean chemistry have changed during the Precambrian? How might such variations have affected the evolving biota? (e) Organic geochemistry—can kerogen yield evidence of evolving biochemistries? What fraction of the organic matter extractable from ancient sediments dates from the time of original sedimentation? (f) Major evolutionary events—what was the timing and nature of major evolutionary innovations that occurred during the Precambrian? Were these biologic innovations directly coupled to important environmental changes? Finally, it is evident that the past decade has witnessed the development, application, refinement, and improvement of techniques and approaches applicable to the search for the earliest traces of life on this planet. It is not difficult to envision that within the foreseeable future, and possibly as early as the next decade, these same methods and their derivatives could find application in the search for evidence of past or present life on Mars and elsewhere within the accessible reaches of our solar system.

ACKNOWLEDGMENTS

The views here expressed have evolved gradually over the past several years (and will, I hope, where applicable be regarded as superseding those expressed in J. W. Schopf, in press, a paper now nearly three years out of date). I acknowledge with thanks the influence of my students, both graduate and undergraduate, who have convinced me by their searching questions and, on occasion, by their perhaps too ready acceptance of speculative interpretations that a reappraisal of the known early fossil record is in order.

Field work in South Australia leading to discovery of the new microbiotas here illustrated was carried out in collaboration with the South Australian Department of Mines. Fossiliferous samples of the lower River Wakefield Group (Figure 2) were collected in 1968 with the assistance of B. G. Forbes, W. V. Preiss, and D. McKirdy (now a graduate student at the Australian National University, Canberra). In 1969,

Forbes forwarded additional fossiliferous chert samples, including those of the Auburn Dolomite (Figure 3d–g), and in 1973, field work carried out with the assistance of Preiss and T. R. Fairchild (of UCLA) resulted in discovery of the Myrtle Springs microflora (Figure 3a–c). As a result of other field work done in 1973, fossiliferous cherts were discovered in the Vempalle Formation (Figure 1l–k), collected in collaboration with K. N. Prasad of the Geological Survey of India; in the Galeros Formation (Figure 1l–p), collected with the assistance of R. J. Horodyski (of UCLA) and W. J. Breed of the Museum of Northern Arizona; and in the Dismal Lakes Group (Figure 1e–h), collected by Horodyski and J. A. Donaldson of Carleton University, Ottawa, Canada. Samples of Dodguni Chert (Figure 1a) were generously provided by S. S. Gowda, and the photomicrographs of the Transvaal micro-fossils (Figure 1b–d), by L. A. Nagy. I much appreciate the interest and assistance of all of these individuals.

Field work and laboratory research leading to the preparation of this paper have been supported by NSF Grant GB 37257 (Systematic Biology Program), by NASA Grant NGR 05–007–407, and by a Fellowship awarded by the John Simon Guggenheim Memorial Foundation.

Literature Cited

Allison, C. W., Moorman, M. A. 1973. Microbiota from the late Proterozoic Tindir Group, Alaska. *Geology* 1:65–68

Awramik, S. M., Golubić, S., Barghoorn, E. S. 1972. Blue-green algal cell degradation and its implications for the fossil record. *Geol. Soc. Am. Abstr. with Programs* 4:438

Barghoorn, E. S. 1971. The oldest fossils. *Sci. Am.* 224:30–42

Barghoorn, E. S., Schopf, J. W. 1965. Microorganisms from the late Precambrian of central Australia. *Science* 150:337–39

Barghoorn, E. S., Schopf, J. W. 1966. Microorganisms three billion years old from the Precambrian of South Africa. *Science* 152:758–63

Barghoorn, E. S., Tyler, S. A. 1965. Microorganisms from the Gunflint chert. *Science* 147:563–77

Behrens, E. W., Frishman, S. A. 1971. Stable carbon isotopes in blue-green algal mats. *J. Geol.* 79:94–100

Binda, P. L. 1972. Preliminary observations on the palynology of the Precambrian Katanga Sequence, Zambia. *Geol. Mijnbouw* 51:315–19

Brooks, J. 1971. Some chemical and geochemical studies on sporopollenin. *Sporopollenin*, ed. J. Brooks et al, 351–407. New York: Academic

Brooks, J., Muir, M. D. 1971. Morphology and chemistry of the organic insoluble matter from the Onverwacht Series Pre-

cambrian chert and the Orgueil and Murray carbonaceous meteorites. *Grana* 11:9–14

Brooks, J., Muir, M. D., Shaw, G. 1973. Chemistry and morphology of Precambrian microorganisms. *Nature* 224:215–17

Button, A. 1971. Early Proterozoic algal stromatolites of the Pretoria Group, Transvaal Sequence. *Trans. Geol. Soc. S. Afr.* 74:201–10

Button, A. 1973. Algal stromatolites of the early Proterozoic Wolkberg Group, Transvaal Sequence. *J. Sediment. Petrology* 43:160–67

Cloud, P. E. Jr. 1965. Significance of the Gunflint (Precambrian) microflora. *Science* 148:27–35

Cloud, P. E. Jr. 1968. Pre-Metazoan evolution and the origins of the Metazoa. *Evolution and Environment*, ed. E. T. Drake, 1–72. New Haven, Conn.: Yale Univ. Press

Cloud, P. 1973. Pseudofossils: a plea for caution. *Geology* 1:123–27

Cloud, P. 1974. Evolution of ecosystems. *Am. Scientist* 62:54–66

Cloud, P. E. Jr. Germs, A. 1971. New pre-Paleozoic nannofossils from the Stoer Formation (Torridonian), northwest Scotland. *Geol. Soc. Am. Bull.* 82:3469–74

Cloud, P. E. Jr., Gruner, J. W., Hagen, H. 1965. Carbonaceous rocks of the Soudan

Iron Formation (Early Precambrian). *Science* 148:1713–16

Cloud, P. E. Jr., Licari, G. R. 1968. Morphological criteria for biogeochemical processes. *Abstr. Geol. Soc. Am. Ann. Meet., Mexico City*: 57

Cloud, P., Licari, G. R. 1972. Ultrastructure and geologic relations of some two-aeon old nostocacean algae from northeastern Minnesota. *Am. J. Sci.* 272:138–49

Cloud, P. E. Jr., Licari, G. R., Wright, L. A., Troxel, B. W. 1969. Proterozoic eucaryotes from eastern California. *Proc. Nat. Acad. Sci. USA* 62:623–31

Cloud, P. E. Jr., Semikhatov, M. A. 1969. Proterozoic stromatolite zonation. *Am. J. Sci.* 267:1017–61

Croxford, N. J. W., Janecek, J., Muir, M. D., Plumb, K. A. 1973. Microorganisms of Carpentarian (Precambrian) age from the Amelia Dolomite McArthur Group, Northern Territory, Australia. *Nature* 245: 28–30

De la Hunty, L. E. 1964. Balfour Downs, W. A. *Explan. Notes Geol. Surv. W. Aust. 1:250,000 Geol. Ser.,* Sheet SF/51–9. 23 pp.

De la Hunty, L. E. 1967. Explanatory notes on the Robertson 1:250,000 geological sheet, Western Australia. *Rep. Geol. Surv. W. Aust., 1967/4.* 36 pp.

Diver, W. L. 1974. Precambrian microfossils of Carpentarian age from Bungle Bungle Dolomite of Western Australia. *Nature* 247:361–62

Donaldson, J. A. 1963. Stromatolites in the Denault Formation, Marion Lake, Coast of Labrador, Newfoundland. *Bull. Geol. Surv. Can.* 102. 33 pp.

Downie, C. 1967. The geologic history of the microplankton. *Rev. Palaeobot. Palynology* 1:269–81

Dungworth, G., Schwartz, A. W. 1972. Kerogen isolates from the Precambrian of South Africa and Australia: analysis for carbonised micro-organisms and pyrolysis gas liquid chromatography. *Advances in Organic Geochemistry 1971,* ed. H. R. v. Gaertner, H. Wehner, 699–706. New York: Pergamon

Edhorn, A. 1973. Further investigations of fossils from the Animikie, Thunder Bay, Ontario. *Proc. Geol. Ass. Can.* 25:37–66

Engel, A. E. J., Nagy, B., Nagy, L. A., Engel, C. G., Kremp, G. O. W., Drew, C. M. 1968. Alga-like forms in Onverwacht Series, South Africa: oldest recognized life-like forms on earth. *Science* 161:1005–8

Fairchild, T. R., Schopf, J. W. 1974. A late Precambrian stromatolitic microflora from Boorthanna, South Australia. *Am. J. Bot.* 61(5, Suppl.): 15 (Abstr.)

Fairchild, T. R., Schopf, J. W., Folk, R. L. 1973. Filamentous algal microfossils from the Caballos Novaculite, Devonian of Texas. *J. Paleontol.* 47:946–52

Ford, T. D., Breed, W. J. 1973. The problematical Precambrian fossil *Chuaria. Palaeontol.* 16:535–50

Fournier-Vinas, C., Debat, P. 1970. Présence de microorganismes dans les terrains métamorphiques précambriens (schistes X) de l'ouest de la Montagne Noire. *Bull. Soc. Géol. Fr.* (7), XII, no. 2:351–55

Glaessner, M. F. 1962. Pre-Cambrian fossils. *Biol. Rev. Cambridge Phil. Soc.* 37:467–94

Glaessner, M. F. 1963. Zur Kenntnis der Nama-Fossilien Südwest-Afrikas. *Ann. Naturhist. Mus. Wien.* 66:113–20

Glaessner, M. F. 1966. Precambrian paleontology. *Earth-Sci. Rev.* 1:29–50

Glaessner, M. F. 1969. Trace fossils from the Precambrian and basal Cambrian. *Lethaia* 2:369–93

Glaessner, M. F. 1971. Geographic distribution and time range of the Ediacara Precambrian fauna. *Geol. Soc. Am. Bull.* 82:509–14

Glaessner, M. F., Wade, M. 1966. The late Precambrian fossils from Ediacara, South Australia. *Palaeontol.* 9:599–628

Glikson, A. Y. 1970. Geosynclinal evolution and geochemical affinities of Early Precambrian systems. *Tectonophysics* 9:397–433

Glikson, A. Y. 1971. Primitive Archaean element distribution patterns: chemical evidence and geotectonic significance. *Earth Planet. Sci. Lett.* 12:309–20

Gnilovskaja, M. B. 1971. Drevneyshiye vodnyye rasteniya venda Russkoy platformy (pozdiydokembriy) [The oldest aquatic plants of the Vendian of the Russian Platform (late Precambrian)]. *Paleontol. Zh.* 1971(3):101–7 [*Paleontol. J.* 5:372–78]

Goldich, S. S. 1973. Ages of Precambrian banded iron-formations. *Econ. Geol.* 68:1126–34

Gowda, S. S. 1970. Fossil blue-green algae and fungi from the Archean Complex of Mysore, South India. *Proc. (Abstr.), Int. Symp. Taxon. Biol. Blue-Green Algae, 1st, Madras, India,* Jan. 8–13, 1970

Gutstadt, A. M., Schopf, J. W. 1969. Possible algal microfossils from the late pre-Cambrian of California. *Nature* 223:165–67

Hofmann, H. J. 1969. Stromatolites from the Proterozoic Animikie and Sibley Groups, Ontario. *Geol. Surv. Can. Pap.* 68–69. 77 pp.

Hofmann, H. J. 1971. Precambrian fossils,

pseudofossils and problematica in Canada. *Geol. Surv. Can. Bull. 189.* 146 pp.

Hofmann, H. J. 1971b. Polygonomorph acritarch from the Gunflint Formation (Precambrian), Ontario. *J. Paleontol.* 45: 522–24

Hofmann, H. J. 1974. Mid-Precambrian prokaryotes(?) from the Belcher Islands, Canada. *Nature* 249: 87–88

Hofmann, H. J., Jackson, G. D. 1969. Precambrian (Aphebian) microfossils from Belcher Islands, Hudson Bay. *Can. J. Earth Sci.* 6: 1137–44

Holland, H. D. 1962. Model for the evolution of the earth's atmosphere. *Petrologic Studies. A Volume to Honor A. F. Buddington,* ed. A. E. J. Engel et al, 447–77. New York: Geol. Soc. Am.

Holland, H. D. 1973. The oceans: a possible source of iron in iron-formations. *Econ. Geol.* 68: 1169–72

Jackson, T. A. 1967. Fossil actinomycetes in middle Precambrian glacial varves. *Science* 155: 1003–5

Jolliffe, A. W. 1955. Geology and iron ores of Steep Rock Lake. *Econ. Geol.* 50: 373–98

Khan, M. 1973. Algae through the ages. *Acta Bot. Indica* 1: 55–67

Konzalová, M. 1972. Some new microorganisms from the Bohemian Precambrian (upper Proterozoic). *Čas. Mineral. Geol.* 17: 267–72

Konzalová, M. 1973. Algal colony and rests of other microorganisms in the Bohemian Upper Proterozoic. *Věstn. Ústřed. Ústavu Geol.* 48: 31–33

Knight, S. H. 1968. Precambrian stromatolites, bioherms and reefs in the lower half of the Nash Formation, Medicine Bow Mountains, Wyoming. *Contrib. Geol.* 7(2): 73–116

LaBerge, G. L. 1973. Possible biological origin of Precambrian iron-formations. *Econ. Geol.* 68: 1098–1109

Licari, G. R. 1974. Paleontology and paleoecology of the Proterozoic Beck Spring Dolomite in eastern California. *J. Paleontol.* In press

Licari, G. R., Cloud, P. 1972. Prokaryotic algae associated with Australian Proterozoic stromatolites. *Proc. Nat. Acad. Sci. USA* 69: 2500–4

Licari, G. R., Cloud, P. E. Jr. 1968. Reproductive structures and taxonomic affinities of some nannofossils from the Gunflint Iron Formation. *Proc. Nat. Acad. Sci. USA* 59: 1053–60

Licari, G. R., Cloud, P. E. Jr., Smith, W. D. 1969. A new chroococcacean alga from the Proterozoic of Queensland. *Proc. Nat. Acad. Sci. USA* 62: 56–62

Loeblich, A. R. Jr. 1974. Protistan phylogeny as indicated by the fossil record. *Taxon* 23: 277–90

Logan, B. W. 1961. Cryptozoon and associated stromatolites from the Recent, Shark Bay, western Australia. *J. Geol.* 69: 517–33

MacGregor, A. M. 1941. A pre-Cambrian algal limestone in Southern Rhodesia. *Trans. Geol. Soc. S. Afr.* 43: 9–16

MacGregor, I. M., Truswell, J. F., Eriksson, K. A. 1974. Filamentous algae from the 2300 m.y. old Transvaal Dolomite. *Nature* 247: 538–40

Margulis, L. 1970. *Origin of Eukaryotic Cells.* New Haven, Conn.: Yale Univ. Press. 349 pp.

McKirdy, D. M. 1974. Organic geochemistry in Precambrian research. *Precambrian Res.* 1: 75–137

Moorbath, S., O'Nions, R. K., Pankhurst, R. J. 1973. Early Archaean age for the Isua Iron Formation, West Greenland. *Nature* 245: 138–39

Moorman, M. 1974. Microbiota of the late Proterozoic Hector Formation, southwestern Alberta, Canada. *J. Paleontol.* 48: 524–39

Muir, M. D. 1974. Microfossils from the middle Precambrian McArthur Group, Northern Territory, Australia. *Origins Life* 5: 105–18

Nagy, B., Nagy, L. A. 1969. Early pre-Cambrian Onverwacht microstructures: possibly the oldest fossils on earth? *Nature* 223: 1226–29

Nagy, L. A. 1971. Ellipsoidal microstructures of narrow size range in the oldest known sediments on earth. *Grana* 11: 91–94

Nagy, L. A. 1974. Transvaal stromatolite: first evidence for the diversification of cells about 2.2 × 10^9 years ago. *Science* 183: 514–16

Nagy, L. A., Zumberge, J. E., Nagy, B. 1973. Early Precambrian life: problems and significance. *Program, Int. Conf. Origin Life, 4th, Barcelona, Spain.* Session II, Pap. 65 (Abstr.)

Oberlies, F., Prashnowsky, A. A. 1968. Biogeochemische und elektronenmikroskopische Untersuchung präkambrischer Gesteine. *Naturwissenschaften* 55: 25–28

Oehler, D. Z. 1973. *Carbon isotopic and electron microscopic studies of organic remains in Precambrian rocks.* PhD thesis. Univ. Calif., Los Angeles. 123 pp.

Oehler, D. Z., Schopf, J. W., Kvenvolden, K. A. 1972. Carbon isotopic studies of organic matter in Precambrian rocks. *Science* 175: 1246–48

Pacltová, B. 1972. *Palaeocryptidium* Deflandre from the Proterozoic of

248 SCHOPF

Bohemia. *Čas. Mineral. Geol.* 17:357–63

Pannella, G. 1972. Paleontological evidence on the earth's rotational history since Early Precambrian. *Astrophys. Space Sci.* 16:212–237

Pflug, H. D. 1966. Structured organic remains from the Fig Tree Series of the Barberton Mountain Land. *Econ. Res. Unit Inform. Circ. 28,* Univ. Witwatersrand, Johannesburg, South Africa. 14 pp.

Pflug, H. D. 1967. Structured organic remains from the Fig Tree Series (Precambrian) of the Barberton Mountain Land (South Africa). *Rev. Palaeobot. Palynology* 5:9–29

Pflug, H. D., Meinel, W., Neumann, K. H., Meinel, M. 1969. Entwicklungstendenzen des frühen Lebens auf der Erde. *Naturwissenschaften* 56:10–14

Ponnamperuma, C., Gabel, N. W. 1968. Current status of chemical studies on the origin of life. *Space Life Sci.* 1:64–96

Prashnowsky, A. A., Oberlies, F. 1972. Über Lebenszeugnisse im Präkambrium Afrikas und Südamerikas. *Advances in Organic Geochemistry 1971,* ed. H. R. v. Gaertner, H. Wehner, 683–98. New York: Pergamon

Rosenberg, G. D., Jones, C. B. 1974. Approaches to chemical periodicities in molluscs and stromatolites. *Growth Rhythms and History of the Earth's Rotation,* ed. G. D. Rosenberg, S. K. Runcorn. New York: Wiley. In press

Rossignol-Strick, M., Barghoorn, E. S. 1971. Extraterrestrial abiogenic organization of organic matter: the hollow spheres of the Orgueil meteorite. *Space Life Sci.* 3:89–107

Rothpletz, A. 1916. Über die systematische Deutung und die stratigraphische Stellung der ältesten Versteinerungen Europas und Nordamerikas mit bedonderer Berücksichtigung der Cryptozoen und Ooolithe. II. Über *Cryptozoon, Eozoom,* und *Atikokania. Abh. Bayer. Akad. Wiss., Math-Phys. Kl.* 28(4). 92 pp.

Sagan, C., Mullen, G. 1972. Earth and Mars: evolution of atmospheres and surface temperatures. *Science* 177:52–56

Schidlowski, M. 1968. Untersuchungen an Kohliger Substanz aus dem Präkambrium Südafrikas. *Umsch. Wiss. Techn.* 18:566–67

Schidlowski, M. 1970. Elektronenoptische Identifizierung zellartiger Mikrostrukturen aus sem Präkambrium des Witwatersrand-Systems (>2.15 Mrd. Jahre). *Paläont. Z.* 44:128–133

Schopf, J. M. 1969. Early Paleozoic palynomorphs. *Aspects of Palynology,* ed. R. H. Tschudy, R. A. Scott, 163–92. New York: Wiley-Interscience

Schopf, J. W. 1968. *A paleontological study of the Gunflint microfossil assemblage.* A. B. thesis. Oberlin College, Oberlin, Ohio. 31 pp.

Schopf, J. W. 1968. Microflora of the Bitter Springs Formation, late Precambrian, central Australia. *J. Paleontol.* 42:651–88

Schopf, J. W. 1970. Precambrian microorganisms and evolutionary events prior to the origin of vascular plants. *Biol. Rev. Cambridge Phil. Soc.* 45:319–52

Schopf, J. W. 1972. Evolutionary significance of the Bitter Springs (late Precambrian) microflora. *Proc. Int. Geol. Congr., 24th, Sect. 1, Precambrian Geol.,* Montreal: 68–77

Schopf, J. W. 1974. The development and diversification of Precambrian life. *Origins Life* 5:119–35

Schopf, J. W. Paleobiology of the Precambrian: the age of blue-green algae. *Evolutionary Biology,* ed T. Dobzhansky, M. K. Hecht, W. C. Steere, 7:1–43. New York: Plenum. In press

Schopf, J. W., Barghoorn, E. S. 1967. Algalike fossils from the early Precambrian of South Africa. *Science* 156:508–12

Schopf, J. W., Barghoorn, E. S. 1969. Microorganisms from the late Precambrian of South Australia. *J. Paleontol.* 43:111–18

Schopf, J. W., Barghoorn, E. S., Maser, M. D., Gordon, R. O. 1965. Electron microscopy of fossil bacteria two billion years old. *Science* 149:1365–67

Schopf, J. W., Blacic, J. M. 1971. New microorganisms from the Bitter Springs Formation (late Precambrian) of the north-central Amadeus Basin, Australia. *J. Paleontol.* 45:925–60

Schopf, J. W., Fairchild, T. R. 1973. Late Precambrian microfossils: a new stromatolitic biota from Boorthanna, South Australia. *Nature* 242:537–38

Schopf, J. W., Ford, T. D., Breed, W. J. 1973. Microorganisms from the late Precambrian of the Grand Canyon, Arizona. *Science* 179:1319–21

Schopf, J. W., Haugh, B. N., Molnar, R. E., Satterthwait, D. F. 1973b. On the development of the metaphytes and metazoans. *J. Paleontol.* 47:1–9

Schopf, J. W., Horodyski, R. J., Fairchild, T. R., Donaldson, J. A. 1974. Late Precambrian microfossils: discovery of four new stromatolitic biotas. *Am. J. Bot.* 61(5, Suppl.):19 (Abstr.)

Schopf, J. W., Oehler, D. Z., Horodyski, R. J., Kvenvolden, K. A. 1971. Biogenicity and significance of the oldest known stromatolites. *J. Paleontol.* 45:477–85

Shimron, A. E., Horowitz, A. 1972. Precambrian organic microfossils from Sinai. *Pollen Spores* 14:333–42

Sidorenko, A. V. 1969. A unified geohistoric approach to the study of the Precambrian and post-Precambrian. *Dokl. Akad. Nauk SSSR* 186:36–38

Smith, J. W., Schopf, J. W., Kaplan, I. R. 1970. Extractable organic matter in Precambrian cherts. *Geochim. Cosmochim. Acta* 34:659–75

Spjeldnaes, N. 1963. A new fossil (*Papillomembrana* sp.) from the upper Precambrian of Norway. *Nature* 200:63–64

Stanley, S. M. 1973. An ecological theory for the sudden origin of multicellular life in the late Precambrian. *Proc. Nat. Acad. Sci. USA* 70:1486–89

Timofeev, B. V. 1966. Micropaleophytological study of ancient rock sequences. *Izd. Akad. Nauk SSSR*, Leningrad (in Russian). 240 pp.

Timofeev, B. V. 1969. Sphaeromorphida of the Proterozoic. *Izd. Akad. Nauk SSSR*, Leningrad (in Russian). 146 pp.

Timofeev, B. V. 1970. Sphaeromorphidia géants dans le Précambrien avancé. *Rev. Palaeobot. Palynology* 10:157–60

Timofeev, B. V. 1973. Microphytofossils of the Precambrian of the Ukraine. *Izd. Akad. Nauk SSSR*, Leningrad (in Russian). 100 pp.

Truswell, J. F., Eriksson, K. A. 1973. Stromatolitic associations and their palaeoenvironmental significance: a reappraisal of the lower Proterozoic locality from the northern Cape Province, South Africa. *Sediment. Geol.* 10:1–23

Tyler, S. A., Barghoorn, E. S. 1954. Occurrence of structurally preserved plants in pre-Cambrian rocks of the Canadian Shield. *Science* 119:606–8

Vidal, G. Röshoff, K. 1971. Organic remains in metasedimentary and metatuffitic rocks of the Vetlanda Series, South Sweden. A preliminary report. *Geol. Fören. Stockholm Förh.* 93:775–78

Vologdin, A. G. 1966. Ob otkrytii ostatkov giganticheskikh sifoney v drevnikh sloyakh Timanskogo kryazha [Discovery of the remains of gigantic siphonaceous algae in the ancient strata of the Timan Range] *Dokl. Akad. Nauk SSSR* 169:672–75 [*Dokl. Acad. Sci. USSR, Earth Sci. Sect.* 169:209–13]

Walter, M. R. 1972. Stromatolites and the biostratigraphy of the Australian Precambrian and Cambrian. *Palaeontol. Ass. London, Spec. Pap. Palaeontol., No. 11.* 190 pp.

Walter, M. R., Oehler, J. H., Oehler, D. Z. 1974. Megascopic algae 1,300 million years old from the Belt Supergroup, Montana: a reinterpretation of Walcott's *Helminthoidichnites*. *J. Paleontol.* In press

White, M. 1974. Microfossils from the late Precambrian Altyn Formation of Montana. *Nature* 247:452–53

Winter, H. de la R. 1963. Algal structures in the sediments of the Ventersdorp System. *Trans. Geol. Soc. S. Afr.* 66:115–21

Winter, H. de la R. 1965. *The stratigraphy of the Ventersdorp System in the Borthaville District and adjoining areas.* PhD thesis. Univ. Witwatersrand, Johannesburg, South Africa

VOLCANIC ROCK SERIES AND TECTONIC SETTING

✳10039

Akiho Miyashiro
Department of Geological Sciences, State University of New York at Albany,
Albany, New York 12222

CLASSIFICATION OF BASALTS AND THREE MAIN VOLCANIC ROCK SERIES

Basalts are the most widespread and abundant type of volcanic rock on the earth. Daly (1914) and Bowen (1928) regarded magmas of basaltic composition as the parents of many or most igneous rocks. Many petrographic provinces have a volcanic rock series starting from basalt, as shown for example in Figure 1. Other provinces contain a number of different rock series, each starting from basalts. Bowen discussed possible mechanisms of the derivation of diverse igneous rocks from the same basaltic magma. Andesite, dacite, and rhyolite were regarded as representing residual liquids in successive stages of fractional crystallization. Under some unusual conditions of crystallization differentiation, residual magmas could become highly alkalic. Thus, Daly and Bowen presumed the uniformity of parental basaltic magmas and the derivation of diverse volcanic rocks, depending on certain later conditions.

Kennedy (1933) pointed out that there exist two different kinds of parental basaltic magmas which show contrasting differentiation trends regardless of the conditions of crystallization: one is toward saturation and oversaturation with SiO_2 (that is, a nonalkalic trend), whereas the other is toward an increasing degree of undersaturation with SiO_2 (that is, an alkalic trend).

In the nonalkalic trend, the residual magma tends to become higher in SiO_2, Na_2O, and K_2O content and in FeO^*/MgO ratio, and lower in MgO with advancing fractional crystallization. (Here FeO^* means total iron as FeO.) However, a dispute existed between Bowen (1928) and Fenner (1929) about the behavior of FeO (or FeO^*). Bowen considered that residual magma becomes progressively lower in FeO (and FeO^*), as is represented by the basalt-andesite-dacite-rhyolite series in orogenic belts (Figure 1), whereas Fenner maintained that residual magma becomes higher in FeO (and FeO^*) in the normal course of crystallization of basaltic magma. The existence of Fenner's trend was conclusively demonstrated by Wager & Deer (1939) in their study of the Skaergaard intrusion in Greenland, and was confirmed by many later authors in studies of other doleritic and gabbroic intrusions in stable

251

continents, e.g. the Palisade sill in New Jersey and New York (Walker 1940, Walker 1969) and the Stillwater complex in Montana (Hess 1960, Wager & Brown 1967). Thus, the nonalkalic (subalkalic) trends are classified into two series: the calc-alkalic series (Bowen trend) and the tholeiitic series (Fenner trend).

The main constituent minerals of ordinary basaltic rocks are olivine, pyroxenes, plagioclase, and nepheline. The albite component of plagioclase is intermediate between nepheline composition $NaAlSiO_4$ and SiO_2, so the major feature of the mineral composition of basaltic rocks can be represented by a tetrahedron with normative minerals quartz, nepheline, olivine, and diopside at the corners (Figure 2). Normative hypersthene falls on the quartz-olivine edge. Albite falls on the quartz-nepheline edge, and normative anorthite is added to albite to make plagioclase. The above-mentioned major trends in volcanic rocks are discussed below with reference to this diagram.

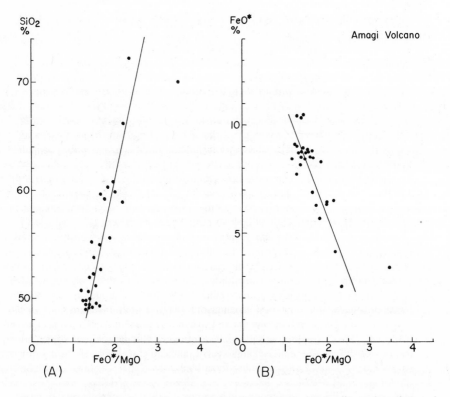

Figure 1 Compositional variation in the basalt-andesite-dacite-rhyolite series of Amagi Volcano, Japan (Kurasawa 1959). FeO* means total iron as FeO. The FeO*/MgO ratio is used here as a measure of the degree of fractional crystallization. The rocks with low FeO*/MgO ratios are basalts.

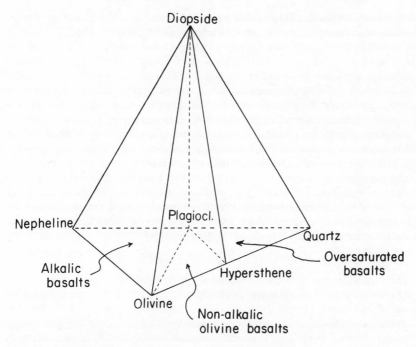

Figure 2 Classification of basalt rocks in terms of normative minerals. Modified from Yoder & Tilley (1962).

Tholeiitic Series (Fenner Trend)

Tholeiitic (TH) magmas contain normative hypersthene together with olivine or quartz, and therefore plot on the right side of the olivine-plagioclase-diopside plane in Figure 2. The crystallization of olivine, plagioclase, and clinopyroxene moves the composition of the residual magma toward the right, and the residual magma may pass the hypersthene-plagioclase-diopside plane. Then, olivine stops crystallizing and reacts with the residual magma to form Ca-poor pyroxene (orthopyroxene or pigeonite). As a consequence, olivine is commonly rimmed by Ca-poor pyroxene, and the series forms: tholeiitic basalt → andesite → dacite → rhyolite. Usually such TH series volcanic rocks contain no hornblende nor biotite.

Calc-alkalic Series (Bowen Trend)

Calc-alkalic (CA) series basaltic magmas, like TH series magmas, contain normative hypersthene together with olivine or quartz, thus plotting on the right side of the olivine-plagioclase-diopside plane in Figure 2. The early stage of crystallization of CA series magmas resembles that of TH series magmas, resulting in an analogous series: basalt → andesite → dacite → rhyolite. Olivine shows a peritectic-reaction relation to Ca-poor pyroxene (usually orthopyroxene in this case). The major difference between the TH and CA series lies in the more rapid increase of SiO_2 with advancing crystallization, and in the lower FeO* content of the CA series.

Kennedy (1955) and Osborn (1962) ascribed the difference between the TH and CA series to a difference in oxygen fugacity of magmas. Under a higher oxygen fugacity, much magnetite would crystallize from a relatively early stage, leading to depletion of Fe in the residual magma and to complementary enrichment in SiO_2; this should form the CA trend. On the other hand, under a lower oxygen fugacity, crystallization of magnetite would be delayed, and hence Fe is not so strongly removed from magma as it is in the CA series; this should form the TH series.

Ti and V are elements characteristically concentrated in magnetite along with Fe, and show behaviors similar to that of FeO* in fractional crystallization of TH and CA series magmas (Miyashiro and Shido, in press). In later stages of crystallization of the CA series, hornblende and biotite commonly form. Hence, Bowen's (1928) well-known reaction series—olivine → pyroxenes → hornblende → biotite—holds for the CA series.

Alkalic Series

The most widespread and abundant kind of alkalic basalts are alkali olivine basalts, which have only a small percentage of normative nepheline. They plot on the left side of, and close to, the olivine-plagioclase-diopside plane in Figure 2. Poldervaart (1964) stated that some basaltic rocks slightly on the right side of this plane show a crystallization trend similar to that of alkali olivine basalt magma, probably because natural clinopyroxene contains a small amount of hypersthene molecule. Crystallization of olivine, plagioclase, and clinopyroxene from such magmas changes the residual magma toward the left in Figure 2, resulting in the formation of the rock series: basalt → hawaiite → mugearite → trachyte (and phonolite).

Recent experimental studies suggest that alkali basaltic magmas with various degrees of SiO_2 undersaturation may form as primary magmas through partial melting of the upper mantle at great depths (Green 1970). In that case, a number of different alkalic rock series starting from such magmas may exist.

GRAPHICAL DISTINCTION BETWEEN ALKALIC AND NONALKALIC SERIES

If we compare rocks with the same SiO_2 content, rocks of alkalic series usually have higher $Na_2O + K_2O$ and lower CaO contents than those of nonalkalic series (TH and CA). Thus, alkalic and nonalkalic series may be distinguished with Harker-type variation diagrams with SiO_2 on the abscissa (Harker 1909). This method was adopted particularly by Kuno (1959, 1966) and Macdonald & Katsura (1964). Macdonald & Katsura demonstrated that the fields of alkalic and tholeiitic rocks in Hawaii can be approximately divided by a straight line in an SiO_2 vs $Na_2O + K_2O$ diagram (Figure 3), and Kuno (1959, 1966) found that the boundary between the alkalic and nonalkalic (TH and CA) volcanic rocks in and around Japan lies close to the above straight line in such a diagram (Figure 3).

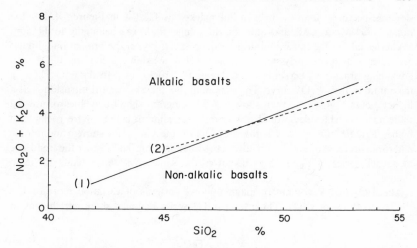

Figure 3   Alkali contents of alkalic and nonalkalic basaltic rocks. Curve 1 is the boundary between the alkalic and tholeiitic basalts in Hawaii (Macdonald & Katsura 1964), and curve 2, the boundary between alkalic and nonalkalic volcanic rocks in Japan (Kuno 1959, 1966).

GRAPHICAL DISTINCTION BETWEEN THOLEIITIC AND CALC-ALKALIC SERIES

Wager & Deer (1939) demonstrated the usefulness of triangular diagrams with MgO, FeO, and $Na_2O + K_2O$ at the corners (FMA diagrams) for the distinction between the TH and CA series. Because of the surface or near surface changeability of the oxidation state of iron, some later authors used $FeO + Fe_2O_3$ or FeO^* in place of FeO, as shown in Figure 4. In such a diagram, the early and middle stages of crystallization of typical TH series are represented by curves approximately parallel to the $MgO-FeO^*$ side, and in the late stage, a sharp turn occurs toward the $Na_2O + K_2O$ corner to indicate more silicic rocks (e.g. granophyre). On the other hand, rocks of typical CA series show trends nearly perpendicular to the $MgO-FeO^*$ side. A complete gradation exists between typical TH and typical CA series.

Miyashiro (1974) proposed another method of graphical distinction between the TH and CA series as shown in Figure 5. In typical TH series, such as in the Skaergaard intrusion, the SiO_2 content remains nearly constant or decreases slightly during the early stage of fractional crystallization. In other TH series, such as in Miyake-jima, Kilauea, and Tofua (Figure 5), the SiO_2 content increases slowly with advancing fractional crystallization, i.e. with increasing FeO^*/MgO. On the other hand, the magmas of the CA series show more rapid increases of the SiO_2 content with fractional crystallization, as indicated by curves for Asama and Amagi in the same figure. The TH and CA series are defined as having a gentler and a steeper slope, respectively, than the broken line marked as TH/CA in Figure 5A. For the range FeO*/MgO > 2.0, and only for this range, the rocks belonging to TH and CA series fall on the lower and upper side, respectively, of the broken line. Figure 5B indicates that the early stage of typical TH series shows increasing FeO* with advancing fractional crystallization. Here, the rocks of Miyake-jima show a maximum in their FeO* curve. The maximum for the Skaergaard intrusion is at a higher FeO*/MgO value than shown in this diagram. The trend lines of abyssal tholeiites and Macauley would also show a maximum, if there were rocks with higher FeO*/MgO ratios. On the other hand, the CA series show a monotonic and rapid decrease of FeO* with advancing fractionation. The steeper the trend line of a rock series in Figure 5A, the lower the same rock series tends to lie in Figure 5B.

The TiO_2 and V content of magma shows patterns of variation more or less similar to those of FeO*. Thus, typical TH series magmas show enrichment in

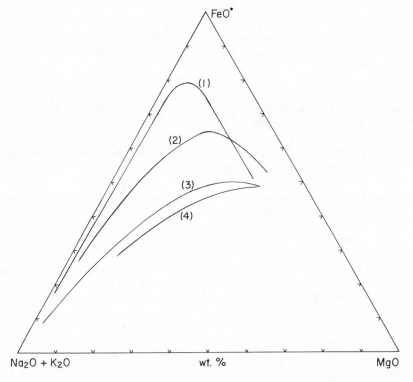

Figure 4   MgO: FeO*: ($Na_2O + K_2O$) diagram. Curve 1 represents the compositional change of the magmatic liquid in the Skaergaard intrusion (typical TH series), curves 2 and 3, the average TH and CA series, respectively, in the Izu-Hakone region in Japan, curve 4, the CA series in Amagi Volcano (Figure 1).

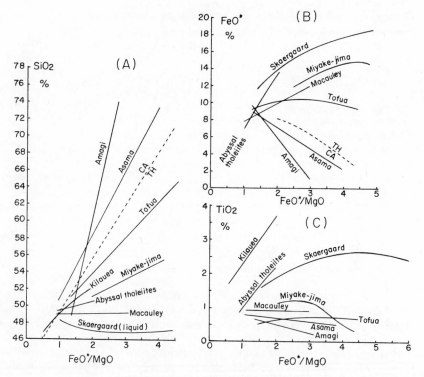

Figure 5   Variations of SiO_2, FeO*, and TiO_2 content of igneous rocks with increasing FeO*/MgO. The rocks of Amagi and Asama Volcanoes (Japan) belong to the CA series, and those of the other areas to the TH series. Tofua Island is in the Tonga Arc, Miyake-jima in the Izu-Bonin Arc, and Macauley Island in the Kermadec Arc (see Miyashiro 1974, Table 1). Kilauea Volcano is in Hawaii and the Skaergaard intrusion is in Greenland. (The curves for Amagi were taken from Figure 1.)

TiO_2 with a maximum during fractional crystallization, whereas CA series magmas show monotonically decreasing TiO_2 content as shown in Figure 5C.

RELATIONSHIP BETWEEN VOLCANIC ROCK SERIES AND TECTONIC SETTING

There is a close relationship between volcanic rock series and tectonic settings, as summarized in Table 1. The CA series is characteristic of orogenic belts (island arcs and active continental margins). On the other hand, TH and alkalic series occur in all major tectonic settings. Alkalic series rocks are the major constituents of oceanic islands and some plateau basalts. TH series rocks are predominant in mid-oceanic ridges and other plateau basalts. Tholeiites are the most abundant volcanic rocks on earth.

Table 1 Volcanic rock series in various tectonic settings

	CA	TH	Alkalic
1. Orogenic belts:			
active continental margins	+ +[a]	(+)	(+)
mature island arcs	+ +	+ +	(+)
immature island arcs	(+)	+ +	
2. Stable continents:			
plateau basalts	(+)	+ +	+ +
3. Seamounts and intra-oceanic islands:			
ordinary seamounts			+ +
St. Helena and many other islands			+ +
Hawaii		+ +	+
Iceland and Galapagos		+ +	+ +
4. Mid-oceanic ridges:			
ordinary submarine parts		+ +	
fracture zones		+	+

[a] + + abundant, + scarce, (+) occasionally present.

VOLCANIC ROCK SERIES IN ISLAND ARCS AND ACTIVE CONTINENTAL MARGINS

Immature island arcs are made up of small volcanic islands built on thin oceanic-type crusts. They are entirely or predominantly composed of TH series rocks, possibly accompanied by a small amount of CA series rocks, e.g. Kermadec and Tonga Arcs (Figures 6 and 7).

Well-developed (mature) island arcs, such as the Northeast Japan arc, are made up of larger islands having thicker continental-type crusts with granitic rocks. The volcanic rocks there usually include basalts, andesites, and dacites of both the TH and CA series. As shown in Figures 6–8, the average SiO_2 content and the proportion of CA series rocks among all the volcanic rocks tend to increase with greater thickness of crust, i.e. with advancing development of continental-type crust (Miyashiro 1974).

Well-developed island arcs commonly have wide volcanic belts, across which the $Na_2O + K_2O$ and K_2O content and the $(Na_2O + K_2O)/Al_2O_3$ ratio tend to increase toward the adjacent continent, i.e. away from the adjacent trench (or the ocean). Thus, the volcanic rocks in the trench-side zone of such an island arc volcanic belt are TH and CA series rocks with low $Na_2O + K_2O$ content, whereas those in the continental-side zone of the same belt are TH and CA series rocks with higher $Na_2O + K_2O$ content, occasionally accompanied by a small amount of alkalic series rocks at the continental-side margin of the zone (Figure 18 of Miyashiro 1974). The basalts of the TH and CA series in the continental-side zone, being high in Al_2O_3, were called *high-alumina basalt* by Kuno (1960). This regular

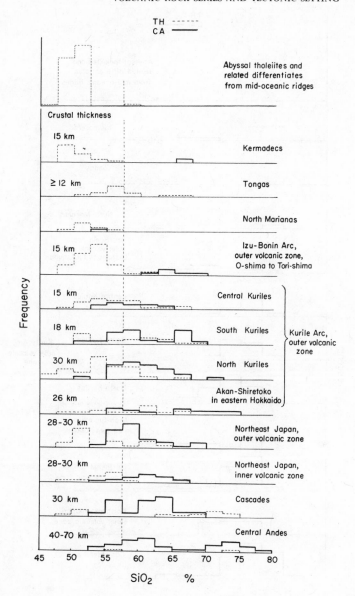

Figure 6 Frequency distributions of SiO_2 percentages in abyssal tholeiites and volcanic rocks in island arcs and active continental margins. Arcs and continental margins are shown in an approximate order of increasing crustal thickness (Miyashiro 1974 and in press).

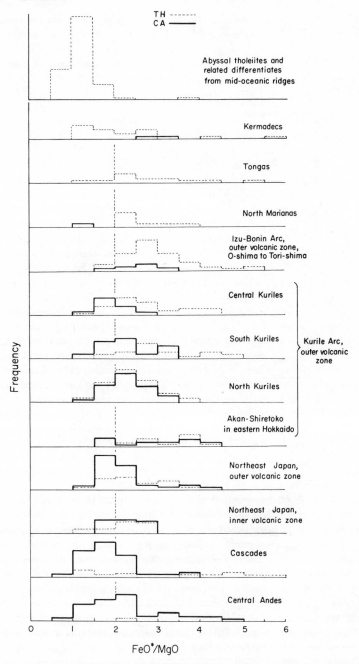

Figure 7 Frequency distributions of FeO*/MgO ratios in abyssal tholeiites and volcanic rocks in island arcs and active continental margins (Miyashiro 1974 and in press).

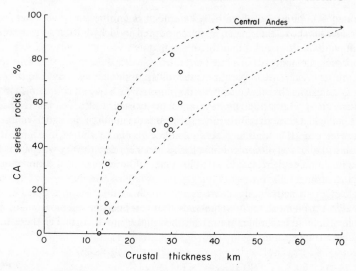

Figure 8   Relationship between the crustal thickness and the percentage of CA series rocks among all the volcanics in island arcs and active continental margins (Miyashiro 1974).

compositional variation across the arc has been regarded by many authors as being genetically related to the increasing depth of seismic foci toward the continent (Kuno 1959, 1966, Sugimura 1968, Dickinson & Hatherton 1967). If magmas are generated along the plane of seismic foci, the deeper origin of magmas on the continental side could increase the alkalic composition. However, magmas may be

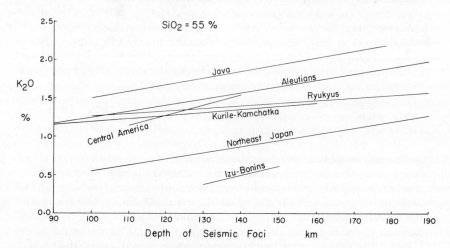

Figure 9   Relation between the depth of seismic focii and the K_2O contents of volcanic rocks with 55% SiO_2 for several island arcs (Nielson & Stoiber 1973). As K_2O content increases with SiO_2, the values at 55% SiO_2 are used.

generated not along the plane of seismic focii but rather in some places in the upper mantle above that plane. Water produced by dehydration of a descending slab of plate would rise through the upper mantle and might cause partial melting of upper mantle materials in some places. Because the descending slab dips toward the continent, such sites of partial melting in the upper mantle may tend to become deeper, increasing the alkalic composition toward the adjacent continent.

Dickinson & Hatherton (1967) claimed the existence of a definite relationship between the K_2O content of volcanic rocks and the depth of the underlying seismic plane. However, the relation differs in different arcs, as shown in Figure 9. The K_2O and $Na_2O + K_2O$ content of volcanic rocks appears to vary with such factors as the rate of plate convergence and the degree of development of continental-type crust (Miyashiro 1974, pp. 341–47).

Volcanoes on active continental margins such as the Cascades and the Andes are made up of mostly CA series rocks. The proportion of more silicic rocks, such as dacite and rhyolite, among all the volcanic rocks is higher there than in island arcs (Figures 6–8).

PLATEAU BASALTS ON CONTINENTS

Many plateau basalts, including the Columbia River Basalts in North America and the Deccan Basalts in India, are composed mainly of tholeiites with subordinate alkalic rocks. Other plateau basalts, including the Basin-Range Basalts in North America and the Patagonian Basalts in South America, are mildly alkalic. The Scottish Tertiary province includes plateau basalts of both TH and alkalic series in large quantities, accompanied by some amount of CA series silicic rocks, which were probably formed by contamination by, or melting of, the silicic continental-type basement (Moorbath & Bell 1965, Miyashiro & Shido, in press).

The majority of rocks in plateau basalts are basaltic so far as their SiO_2 content and color indices are concerned. For that reason the plateau basalts were widely misunderstood as being compositionally homogeneous, nearly undifferentiated rocks. Actually the FeO^*/MgO ratio shows a wide range of variation (usually 1–3.5), suggesting a considerable degree of differentiation (Kuno 1969). Some rocks with more advanced fractional crystallization also occur.

VOLCANIC ROCK SERIES IN SEAMOUNTS AND OCEANIC ISLANDS

Seamounts and most oceanic (intra-oceanic) islands are apparently composed of basalts of alkalic series. The most widespread rock series is alkali olivine basalt → trachyte, occasionally accompanied by phonolite. More strongly undersaturated series, including such rocks as melilite-olivine-nephelinite, also occur. Some exceptional oceanic islands are made up almost entirely of feldspathoidal rocks, e.g. Tahiti in the Southern Pacific and Trinidade east of the Brazilian continent.

Daly (1927) and Chayes (1972) considered that in oceanic islands basalt and trachyte are widespread but rocks intermediate between them are relatively rare.

Probably, however, this bimodal frequency distribution is not real, but rather results from sampling bias and inappropriate graphical representation. A careful study of the alkalic series in Saint Helena by Baker (1968) has shown that the amount of rocks decreases monotonically with an increasing degree of fractional crystallization.

Tholeiites occur in addition to alkalic rocks in some oceanic islands, including Hawaii, Galapagos, and Iceland. In each volcano in Hawaii, the main phase of activity is represented by eruption of an enormous amount of tholeiite. In the declining stage, mildly alkalic rocks are erupted. At the end, a small amount of strongly undersaturated alkalic rocks may be erupted (Macdonald & Katsura 1964).

Iceland straddles the Mid-Atlantic Ridge. Tholeiites occur in a zone which may be regarded as a continuation of the axis of the Mid-Atlantic Ridge, whereas alkalic volcanics occur on both sides of this zone (Jakobsson 1972). A small amount of andesite (icelandite) and rhyolite of the TH series occurs in association with basalts. The basalts, andesite, and rhyolite have practically the same Sr^{87}/Sr^{86} ratios (O'Nions & Grönvold 1973), which are distinctly higher than those of abyssal tholeiites (Hart et al 1973). This is consistent with the idea that Iceland represents a hot spot whose magmas differ from abyssal tholeiite magma in origin, and that the diversity of volcanic rocks in Iceland is mainly due to crystallization differentiation of the hot-spot magmas and not to contamination by, or melting of, an old silicic basement.

Volcanism in oceanic islands and stable continents may be a manifestation of hot spots. Though the volcanic rocks in oceanic islands generally resemble those of stable continents, there are some regular chemical differences between them. The

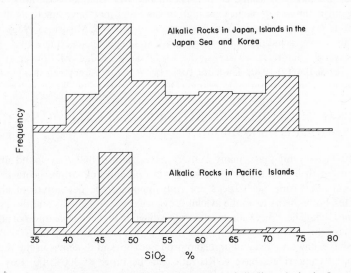

Figure 10 Frequency distribution of SiO_2 percentages of alkalic rocks in the Japan-Korea region as compared with that of alkalic rocks in intra-Pacific islands (Yagi 1959).

K_2O contents of tholeiites in oceanic islands are usually in the range 0.0–0.6%, whereas those on stable continents are usually higher (0.5–1.5%). The "alkalic" rocks in oceanic islands are mostly undersaturated with SiO_2, whereas those in island arcs and active and stable continents contain considerable proportions of both undersaturated and oversaturated rocks, as is understood from the frequency distributions of SiO_2 percentages shown in Figure 10.

Tholeiites in stable continents, oceanic islands, and mid-oceanic ridges tend to have higher TiO_2 contents than those of island arcs (Miyashiro, in press).

VOLCANIC ROCK SERIES IN MID-OCEANIC RIDGES

The hard surface of mid-oceanic ridges is mostly made up of tholeiites which have normative olivine and are characteristically low in K_2O ($<0.40\%$), as was pointed out by Engel et al (1965). Such rocks have been called *oceanic tholeiite* or *abyssal tholeiite*. The latter term is preferred here because the former was used by some authors for all tholeiitic rocks in oceanic regions, including oceanic island tholeiites.

Abyssal tholeiites show a relatively small extent of crystallization differentiation (Miyashiro et al 1970). The range of their FeO^*/MgO ratios is narrower than those of basaltic rocks in other tectonic settings. In abyssal tholeiites and related differentiates, FeO^*/MgO ratios >2.0 are very rare (Figure 7). In island arcs and continental regions, a majority or a large proportion of basaltic rocks have FeO^*/MgO >2.0. Abyssal tholeiites are made up mainly of plagioclase, olivine, clinopyroxene, and magnetite. Plagioclase is the first mineral to crystallize in abyssal tholeiite magmas high in Al_2O_3, whereas olivine is the first mineral to crystallize in those low in Al_2O_3 (Shido et al 1971). In the Mid-Atlantic Ridge north of the Azores, many abyssal tholeiites are high in CaO and may have augite as the first mineral to crystallize (Hekinian & Aumento 1973, Shido & Miyashiro 1973).

Alkalic rocks, including alkali olivine basalt and nepheline-bearing gabbro, occur in transverse fracture zones across the Mid-Atlantic Ridge (Melson et al 1967, Honnorez & Bonatti 1970). Tholeiitic rocks occur in other fracture zones. Volcanic seamounts, superposed on mid-oceanic ridges, are made up of alkalic rocks.

ABYSSAL THOLEIITES VS OTHER NONALKALIC BASALTS; THE OPHIOLITE PROBLEM

For the last several years, many authors have claimed that they found abyssal tholeiites or their metamorphic derivatives in ophiolites of various geologic ages. It appears that some such rocks are truly metamorphic derivatives of abyssal tholeiites, but in many cases the available evidence was ambiguous and discussions were mistaken. Therefore, I summarize major diagnostic chemical features of abyssal tholeiites.

Fresh (i.e. unweathered and unmetamorphosed) abyssal tholeiites in the present-day mid-oceanic ridges have K/Rb ratios in the range of 300–2100 (Kay et al 1970, Shido et al, in press). Although some tholeiitic rocks in island arcs and stable continents show K/Rb ratios as high as 1000–1100 (Jakeš & White 1970),

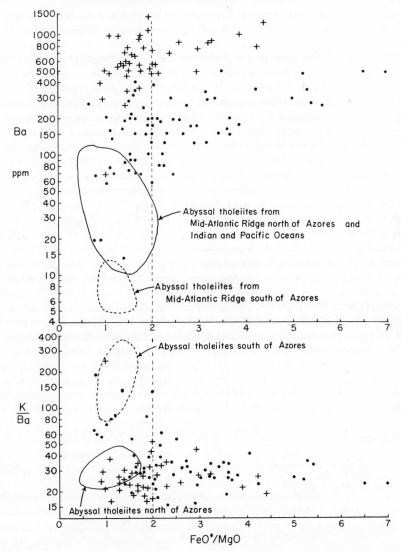

Figure 11 Ba content and K/Ba ratios of volcanic rocks from island arcs and active continental margins plotted against FeO*/MgO ratios. The composition fields of abyssal tholeiites are outlined for comparison (Fumiko Shido and A. Miyashiro, unpublished). Abyssal tholeiites from the Mid-Atlantic Ridge segments north and south of the Azores have different composition fields (Shido & Miyashiro 1973, Shido and Miyashiro, unpublished).

the majority of tholeiitic and other volcanic rocks in such tectonic settings appear to have K/Rb lower than 800.

Ba content and K/Ba ratio are also useful to some extent in the distinction between abyssal tholeiites and other nonalkalic volcanic rocks, as shown in Figure 11.

Fresh abyssal tholeiites have $K_2O < 0.40\%$ (usually $<0.30\%$), and usually $Na_2O > 2.0\%$ ($>2.5\%$ except in the northern part of the Mid-Atlantic Ridge, as shown by Shido & Miyashiro 1973). In other words, a combination of low K_2O and high Na_2O content is characteristic of abyssal tholeiites. Thus it is a mistake to regard, as some authors have done, abyssal tholeiites as being characteristically low in all alkalies. Some fresh island arc tholeiites show $K_2O < 0.30\%$ (e.g. Kuno 1960, Table 1; 1966, Table 1). Such island arc tholeiites, however, are usually characterized by relatively low Na_2O content (usually $<2.0\%$).

Alkalies and alkaline earths are usually mobile during secondary processes such as weathering, alteration, and metamorphism. It is likely that most volcanic rocks in ophiolites were subjected to some degree of such secondary processes. Thus identification of abyssal tholeiites in ophiolites based on K_2O, Na_2O, Ba, K/Rb, and/or K/Ba is not reliable. Examination of the $(Na_2O + K_2O)$ vs Na_2O/K_2O relations in basaltic rocks both in ophiolites found in high pressure metamorphic terranes and in spilites gives evidence that K_2O content usually decreases in low temperature changes (Miyashiro, in press). Low content of K_2O therefore cannot be used as evidence for an abyssal tholeiite origin.

Iron (FeO*) and MgO are not as mobile as alkalies, and so give more reliable criteria. Almost all of the abyssal tholeiites fall in the FeO*/MgO range of 0.8–2.0, whereas large proportions of island arc basalts and plateau basalts show FeO*/MgO ratios above as well as below 2.0 (Figure 7). In many island arc volcanoes

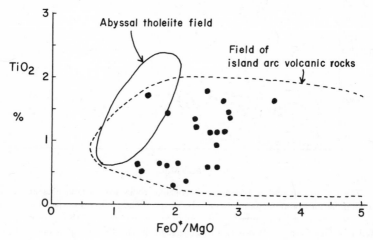

Figure 12   Comparison of TiO_2 content in abyssal tholeiites and island arc volcanic rocks (TH and CA). Dots represent low-K tholeiites of island arcs ($K_2O < 0.40\%$).

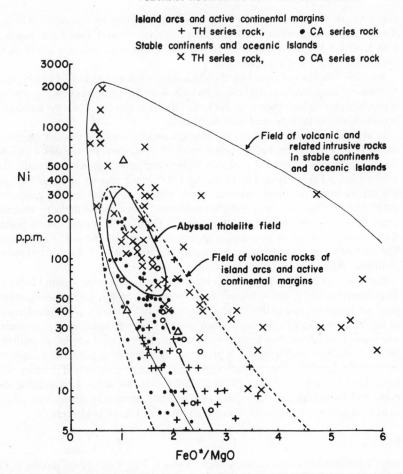

Figure 13   Ni content in abyssal tholeiites and volcanic rocks in island arcs, active continental margins, stable continents, and oceanic islands. Triangles indicate abyssal tholeiites plotting outside the area shown here as the abyssal tholeiite field.

composed almost entirely of tholeiites, predominant rocks show FeO*/MgO > 2.0, e.g. volcanic islands Oshima and Mt. Fuji in the Izu-Bonin Arc. Thus, the presence of a high proportion of rocks with FeO*/MgO > 2.0 suggests that the rocks are not abyssal tholeiites (Miyashiro 1973).

TiO_2 is also relatively immobile during secondary changes. Figure 5 indicates that the abyssal tholeiites show a higher rate of increase of FeO* and TiO_2 with increasing FeO*/MgO than do island arc volcanic rocks. This may be used as a diagnostic feature, if analyses of a series of rocks are available.

The composition fields of abyssal tholeiites and island arc volcanic rocks (TH and CA) are shown in the FeO^*/MgO vs TiO_2 diagram of Figure 12. Though the two fields widely overlap, each still has a characteristic area. Island arc tholeiites with $K_2O < 0.40\%$ are plotted in order to show that most (but not all) of them fall outside the abyssal tholeiites field.

Abyssal tholeiites and island arc tholeiites show similar ranges of SiO_2, although island arc volcanic rocks as a whole commonly show SiO_2 ranges extending toward a much higher value (Figure 6). SiO_2 is relatively mobile in many secondary processes, and so must be used with due caution.

In abyssal tholeiites and island arc volcanic rocks, Cr and Ni content decreases regularly with increasing FeO^*/MgO (Figure 13). Therefore, the use of Cr and Ni for the distinction of abyssal tholeiites from island arc volcanics is virtually the same as using FeO^*/MgO. The ranges of FeO^*/MgO, and so of Cr and Ni, in abyssal tholeiites are much narrower than, and are included in, the ranges of FeO^*/MgO, Cr, and Ni, respectively, in island arc volcanics. Some tholeiitic volcanic rocks in oceanic islands and stable continents show a Ni content much higher than is found in island arc volcanics or abyssal tholeiites with the same FeO^*/MgO. This may be useful in identifying volcanic rocks formed in oceanic islands and stable continents (Miyashiro & Shido, in press).

Rocks of the CA series are formed in island arcs and continental regions (Table 1). The occurrence of CA series volcanic rocks in some ophiolitic complexes therefore gives an important clue to their origin (Miyashiro 1973). On the other hand, rocks of the TH and alkalic series are formed in almost any tectonic setting, so determination of the origin of such rocks requires detailed chemical studies, as outlined above. Another paper (Miyashiro, in press) will show that some ophiolitic complexes contain CA and TH series volcanics whereas others contain only TH series volcanics, and still others contain TH and alkalic series volcanics. Furthermore, the paper will show that some ophiolites were probably created in island arcs, others in mid-oceanic ridges, and still others in hot spots and stable continents.

Literature Cited

Baker, I. 1968. Intermediate oceanic volcanic rocks and the "Daly gap." *Earth Planet. Sci. Lett.* 4:103–6

Bowen, N. L. 1928. *The Evolution of the Igneous Rocks.* Princeton, NJ: Princeton Univ. Press. 332 pp. (Reprinted 1956. New York: Dover)

Chayes, F. 1972. Silica saturation in Cenozoic basalt. *Phil. Trans. Roy. Soc. London* A 271:285–96

Daly, R. A. 1914. *Igneous Rocks and their Origin.* New York & London: McGraw-Hill. 563 pp.

Daly, R. A. 1927. The geology of Saint Helena Island. *Proc. Am. Acad. Arts Sci.* 62:No. 2

Dickinson, W. R., Hatherton, T. 1967. Andesitic volcanism and seismicity around the Pacific. *Science* 157:801–3

Engel, A. E. J., Engel, C. G., Havens, R. G. 1965. Chemical characteristics of oceanic basalts and the upper mantle. *Geol. Soc. Am. Bull.* 76:719–34

Fenner, C. N. 1929. The crystallizations of basalts. *Am. J. Sci. 5th Ser.* 18:225–53

Green, D. H. 1970. The origin of basaltic and nephelinitic magmas. *Trans. Leicester Lit. Phil. Soc.* 64:26–54

Harker, A. 1909. *The Natural History of Igneous Rocks.* New York: Macmillan. 384 pp. (Reprinted 1965. New York: Hafner)

Hart, S. R., Schilling J. -G., Powell, J. L. 1973. Basalts from Iceland and along the Reykjanes Ridge: Sr isotope geochemistry. *Nature Phys. Sci.* 246:104–7

Hekinian, R., Aumento, F. 1973. Rocks from the Gibbs fracture zone and the Minia seamount near 53° N in the Atlantic Ocean. *Mar. Geol.* 14:47–72

Hess, H. H. 1960. Stillwater igneous complex, Montana. *Geol. Soc. Am. Mem.* 80. 230 pp.

Honnorez, J., Bonatti, E. 1970. Nepheline gabbro from the Mid-Atlantic Ridge. *Nature* 228:850–52

Jakeš, P., White, A. J. R. 1970. K/Rb ratios of rocks from island arcs. *Geochim. Cosmochim. Acta* 34:849–56

Jakobsson, S. P. 1972. Chemistry and distribution pattern of Recent basaltic rocks in Iceland. *Lithos* 5:365–86

Kay, R., Hubbard, N. J., Gast, P. W. 1970. Chemical characteristics and origin of oceanic ridge volcanic rocks. *J. Geophys. Res.* 75:1585–1613

Kennedy, G. C. 1955. Some aspects of the role of water in rock melt. *Geol. Soc. Am. Spec. Pap.* 62:489–504

Kennedy, W. Q. 1933. Trends of differentiation in basaltic magmas. *Am. J. Sci. 5th Ser.* 25:239–56

Kuno, H. 1959. Origin of Cenozoic petrographic provinces of Japan and surrounding areas. *Bull. Volcanol. Ser. 2,* 20:37–76

Kuno, H. 1960. High-alumina basalt. *J. Petrol.* 1:121–45

Kuno, H. 1966. Lateral variation of basalt magma type across continental margins and island arcs. *Bull. Volcanol.* 29:195–222

Kuno, H. 1969. Plateau basalts. *Am. Geophys. Union Geophys. Monogr.* 13:495–501

Kurasawa, H. 1959. Petrology and chemistry of the Amagi volcanic rocks, Izu Peninsula, Japan (in Japanese). *Chikyu-kagaku* No. 44:1–18

Macdonald, G. A., Katsura, T. 1964. Chemical composition of Hawaiian lavas. *J. Petrol.* 5:82–133

Melson, W. G., Jarosewich, E., Cifelli, R., Thompson, G. 1967. Alkali olivine basalt dredged near St. Paul's Rocks, Mid-Atlantic Ridge. *Nature* 215:381–82

Miyashiro, A. 1973. The Troodos ophiolitic complex was probably formed in an island arc. *Earth Planet. Sci. Lett.* 19:218–24

Miyashiro, A. 1974. Volcanic rock series in island arcs and active continental margins. *Am. J. Sci.* 274:321–55

Miyashiro, A., Shido, F., Ewing, M. 1970. Crystallization and differentiation in abyssal tholeiites and gabbros from mid-

oceanic ridges. *Earth Planet. Sci. Lett.* 7:361–65

Moorbath, S., Bell, J. D. 1965. Strontium isotope abundance studies and rubidium-strontium age determinations on Tertiary igneous rocks from the Isle of Skye, Northwest Scotland. *J. Petrol.* 6:37–66

Nielson, D. K., Stoiber, R. E. 1973. Relationship of potassium content in andesitic lavas and depth to the seismic zone. *J. Geophys. Res.* 78:6887–92

O'Nions, R. K., Grönvold, K. 1973. Petrogenetic relationships of acid and basic rocks in Iceland. Sr-isotopes and rare-earth elements in late and post-glacial volcanics. *Earth Planet. Sci. Lett.* 19:397–409

Osborn, E. F. 1962. Reaction series for subalkaline igneous rocks based on different oxygen pressure conditions. *Am. Min.* 47:211–26

Poldervaart, A. 1964. Chemical definition of alkali basalts and tholeiites. *Geol. Soc. Am. Bull.* 75:229–32

Shido, F., Miyashiro, A. 1973. Compositional difference between abyssal tholeiites from north and south of the Azores on the mid-Atlantic Ridge. *Nature Phys. Sci.* 245:59–60

Shido, F., Miyashiro, A., Ewing, M. 1971. Crystallization of abyssal tholeiites. *Contr. Min. Petrol.* 31:251–66

Sugimura, A. 1968. *Spatial relations of basaltic magmas in island arcs.* In *Basalts,* ed. H. H. Hess, A. Poldervaart, 2:537–71. New York: Interscience

Wager, L. R., Brown, G. M. 1967. *Layered Igneous Rocks.* San Francisco: Freeman. 588 pp.

Wager, L. R., Deer, W. A. 1939. The petrology of the Skaergaard intrusion Kangerdlugssuaq, East Greenland. *Medd. on Grønland* 105:No. 4

Walker, F. 1940. Differentiation of the Palisade diabase, New Jersey. *Geol. Soc. Am. Bull.* 51:1059–1106

Walker, K. R. 1969. The Palisades sill, New Jersey: a reinvestigation. *Geol. Soc. Am. Spec. Pap.* 111. 178 pp.

Yagi, K. 1959. Petrochemistry of the Cenozoic alkali rocks of Japan and surrounding areas (in Japanese). *Kazan Ser. 2* 3:63–75

Yoder, H. S. Jr., Tilley, C. E. 1962. Origin of basalt magmas: an experimental study of natural and synthetic rock systems. *J. Petrol.* 3:342–532.

PLATE TECTONICS: COMMOTION IN THE OCEAN AND CONTINENTAL CONSEQUENCES

×10040

Clement G. Chase
Department of Geology and Geophysics, University of Minnesota,
Minneapolis, Minnesota 55455

Ellen M. Herron
Lamont-Doherty Geological Observatory, Palisades, New York 10964

William R. Normark
U. S. Geological Survey, Menlo Park, California 94025[1]

Writing a review paper about plate tectonics is in a sense like attempting to review the science of geology. This new set of ideas about the large-scale behavior of the outer parts of the earth has had an enormous impact on the thinking of earth scientists in the past few years. The theory of plate tectonics provides unifying concepts that tie together great masses of geological and geophysical information whose relationships were previously obscure. The simplicity and exactness of the theory has made possible successes in quantitative prediction of geological phenomena that were previously only vaguely, if at all, understood.

Faced with a large and burgeoning literature concerned with plate tectonics and its applications, we perforce must restrict ourselves to examining a limited selection of the available work. To this end, we have chosen to sketch briefly the underlying theory of plate tectonics and some of the objections that have been raised, then to examine the present and past plate motions in oceanic areas, and finally to discuss some of the implications of plate theory for the geology of the continents. In our selection from the available literature, there are no papers published prior to 1960. This reflects merely the tremendous outpouring of recent literature, and we intend no slight to the long and distinguished history of geologic thought on which the conceptual edifice of plate tectonics has been erected.

---

[1] Research done at Department of Geology and Geophysics, University of Minnesota, Minneapolis, Minnesota 55455.

THE THEORY OF PLATE TECTONICS

The fundamental postulates of plate tectonics, independently formulated by McKenzie & Parker (1967) and Morgan (1968), are remarkably simple. The first is that the surface of the earth is covered by a number of large rigid spherical caps, or plates, that undergo significant deformation only around their edges. The second is that these plates are in motion relative to one another. A philosophical case can be made that plate tectonics, sensu strictu, is just the geometric and kinematic relationships of plate motions. The elements of how the plate motions are expressed in any particular environment on the earth, such as divergent or convergent plate boundaries, then lie in the more general realm of geology and geophysics.

After the concept of sea-floor spreading was advanced by Hess (1962) and Dietz (1961), an essential step in the development of plate tectonics was the suggestion by Vine & Matthews (1963) that the linear patterns of magnetic anomalies observed in the world's oceans could have been generated by the continuous emplacement of hot material along an axis of crustal divergence and the subsequent magnetization of this material in a direction parallel to the earth's magnetic field. Periodic reversals of the geomagnetic field would produce a series of positively and negatively magnetized crustal blocks which would result in a linear pattern of magnetic anomalies symmetric about the axis of spreading, a mid-oceanic ridge. The hypothesis of Vine & Matthews has been abundantly confirmed, and remains one of the most important tools for measuring plate motions past and present.

Wilson (1965a) came forth with the concept of the transform fault, which was implicitly a plate boundary parallel to the direction of relative motion. Wilson's hypothesis of motion on transform faults was shown to be consistent with seismological data (Sykes 1968). Ironically, Bullard, Everett & Smith (1965) used the mathematics of rotation of rigid plates on a sphere in reconstructing position of the continents around the Atlantic prior to drift, without pursuing the global implications of such an operation.

Support for the plate tectonic hypothesis advanced by McKenzie & Parker (1967) and Morgan (1968) quickly followed. The Cenozoic magnetic anomaly patterns in the Pacific, Atlantic, and Indian Oceans were identified and correlated (Pitman, Herron & Heirtzler 1968, Dickson, Pitman & Heirtzler 1968, Le Pichon & Heirtzler 1968). A reversal time scale for the entire Cenozoic Era based on these magnetic data was presented by Heirtzler et al (1968) and independently by Vine (1968). Le Pichon (1968) then summarized all the marine magnetic data and determined the instantaneous relative directions and rates of motion for six large crustal plates. These vectors agreed well with similar vectors determined seismologically by Isacks, Oliver & Sykes (1968). Much of the impetus of plate tectonics came from this early predictive success.

The main outline of plate tectonic theory as applied to the oceans has not changed much since the summary by Isacks, Oliver & Sykes (1968). As we shall see later, the emphasis in new ideas has shifted to the consequences for continental geology. The plates, collectively termed lithosphere, are approximately 80 km thick in oceanic areas (Press 1972, Walcott 1970, Sclater & Francheteau 1970) and up to twice as thick under the continents (Press 1972, Walcott 1970); they move about on the underlying, more easily deformed, asthenosphere. Elongate belts of shallow seismicity (Barazangi & Dorman 1969) mark the boundaries of the plates, where major geologic activity is concentrated.

Where the plates move apart, asthenospheric material wells up in the gap between the receding edges and creates new oceanic lithosphere. The top of the crustal portion of this lithosphere preserves in linear magnetic anomalies the reversals of the geomagnetic field, providing a powerful tool for deducing the rates and history of sea-floor spreading. Where plates just slide past each other, strike-slip transform faults record faithfully the local direction of relative motion. Where plates approach, if at least one of them is oceanic, it descends into the underlying asthenosphere in the process of subduction, and leads to the creation of volcano-plutonic arcs on the overlying plate. Where both are continental, very complicated thrust mountain belts ensue. Since a given plate may contain both oceanic and continental lithosphere (Figure 1), the older idea of continental drift is contained within plate tectonics.

By claiming that the theory of plate tectonics is successful, we do not mean to imply that all earth scientists are happy with it. Objections to the theory, based on the interpretation of the data and the implications of the theory have been raised, notably by Beloussov (1970), Wesson (1972), and Meyerhoff (for example, Meyerhoff & Meyerhoff 1972a,b). Jeffreys (1970) has also criticized the theory on the grounds that the upper mantle should be too stiff to allow large-scale motion. Although some of the objections presented may be valid and most are worthy of serious consideration, seismological, geological, and marine magnetic and bathymetric data provide almost overwhelming evidence in support of the existence of lithospheric plates and continental drift (Vine & Hess 1970, McKenzie 1972).

THE KINEMATICS OF THE MAJOR PLATES

It is important to emphasize again that the theory of plate tectonics is kinematic in nature, dealing with the motions of the plates and not necessarily with the reasons for the motions. The question of a driving mechanism, although of great interest, is thus divorced from the success of the theory. Although there is as yet no general agreement on what the causative forces are, there fortunately is substantial agreement as to what the present-day relative motions of the plates are.

Present-Day Relative Motions

Since the instantaneous relative motion of any two rigid plates on the surface of a sphere is an infinitesimal rotation that can be described by an angular velocity vector, the quantitative expression of the motions is very simple. Taking any two plates and calling them A and B, the motion of A with respect to B is completely described by the relative angular velocity vector $_A\omega_B$. If we know $_A\omega_B$ in advance, the local relative velocity $_A\mathbf{v}_B$ at any point $\mathbf{r}$ on the boundary between plate A

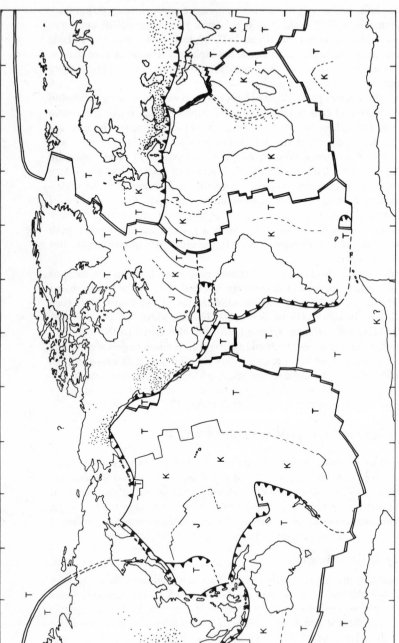

Figure 1 Modern plate boundaries and the age of oceanic crust. The plate boundaries are marked by heavy lines: double lines indicate spreading centers; single lines mark the location of transform faults; toothed lines indicate convergent boundaries, with the teeth pointing down the subducted plate. The stippled areas are continental regions of diffuse seismicity, in which the continental lithosphere probably is not behaving as a rigid plate. The light solid and dashed lines in the oceans are the boundaries, determined from marine magnetic lineations, between Jurassic (J), Cretaceous (K), and Tertiary (T) oceanic crust. The data were compiled from a variety of the sources listed in the references. The map projection is Miller's modified Mercator, which can represent high latitudes but introduces considerable distortion there.

and B is given by the simple relationship $_A\mathbf{v}_B = {_A\omega_B} \times \mathbf{r}$. These relative angular velocity vectors, or poles of relative rotation, can be determined by using rates of local plate separation deduced from the linear marine magnetic anomalies and the directions of local relative plate motion provided by transform fault trends and slip vectors of mechanism solutions for shallow earthquakes.

If we now expand our set of hypothetical plates to three, A, B, and C, the vector relationship $_A\omega_C = {_A\omega_B} + {_B\omega_C}$ (McKenzie & Parker 1967) ensues. Thus, knowing the motions on two plate boundaries of three interacting plates, we can deduce the motion on the third plate boundary. Le Pichon (1968) defined six major plates, found poles of rotation for the five plate pairs for which sea-floor spreading data were then available, and deduced the motions for the other pairs of plates by vector circuits like those above. However, in this approach there is no guarantee that a rotation vector measured directly for two plates will be consistent with one deduced by vector circuits, that is, that the motions will be internally self-consistent.

More recent, more sophisticated motion analyses have avoided this difficulty. Morgan (1972) has published a set of world plate motions adjusted to be self-consistent. Chase (1972, and 1974, in preparation) and Minster et al (1974) have ensured self-consistency by incorporating constraints imposed by the vector circuits into least-square-error methods of finding all the poles of a set of interacting plates simultaneously, using all the available data at once. These results also incorporate more plates than Le Pichon's analysis (Figure 1). The encouraging thing about the results is that the most recent sets of plate motions (Chase, 1974 in preparation, Minster et al 1974) agree to within the limits of confidence in almost all cases, even though different data sets and different least-squares fit criteria were used. An indication of the essential correctness of the plate tectonics assumption and geometry is given by the fact that the motion found for the Pacific plate relative to the North American plate was essentially the same in three cases: in case 1, only information from the Pacific-North American plate boundary was considered; in case 2, all worldwide data were used; and in case 3, all worldwide data *except* Pacific-North American were considered (Chase, 1974, in preparation).

The simultaneous inversion approach of Minster et al (1974) and Chase (1974, in preparation) has also shown that there is statistically significant evidence of current relative motion between the North and South American plates, which may help to explain the complicated tectonics of the Caribbean area (Figure 1).

Chase (1972) has calculated that 3 km^2 of lithosphere is created annually by sea-floor spreading on the boundaries of the larger plates. Assuming an earth of constant radius, an equal amount must be subducted annually. If the average thickness of the volcanic oceanic second layer is 1.4 km (Shor & Raitt 1969), then the total discharge of basaltic magma is about 4 km^3 yr^{-1}. Thus formation of the oceanic crust represents the dominant form of vulcanism on the earth. Furthermore, this rate of lithospheric creation confirms the estimate of Sclater & Francheteau (1970) that heat released by formation of hot oceanic lithosphere is almost half the total conductive heat loss through the surface of the earth.

On a more regional scale, it can be shown (Chase 1972) that the Pacific basin is shrinking by about 0.5 km^2 yr^{-1}, even though it contains the fastest spreading ridges in the world. Opening of the Atlantic and Indian Oceans is thus partly at the expense of the Pacific, whose bordering subduction zones are slowly approaching one another.

The Search for the Absolute—Hotspots

By now, the reader must have noticed the emphasis on the word "relative" when discussing plate motions. This is because the data used in determining present-day kinematics all refer to the motion of one plate relative to its neighbor. Yet, in relating plate motions to paleomagnetic and paleoclimatic evidence, we need a more absolute frame of reference, namely, motion of the plates relative to the spin axis of the earth. Wilson (1965b) proposed that the linear, aseismic volcanic island chains were created by motion of the crust over a "hotspot" in the underlying mantle. The Hawaiian and Emperor volcanic lineaments in the North Pacific (Jackson, Silver & Dalrymple 1972) are the best examples of these hotspot traces, although most active hotspots occur very close to the axis of the mid-ocean ridge system. Morgan (1971, 1972) has suggested that the hotspots are the surface expression of thermal plumes, fixed with respect to and rising from the deep mantle. Subsequently, the mantle plume hypothesis has led to the generation of a considerable body of literature.

If the hotspots do indeed give us a reference framework fixed to the deep mantle and presumably to the earth's spin axis, then our problem of absolute motions is solved. The most recent analyses of present-day plate kinematics seem to be generally consistent with hotspots that are relatively fixed within a short time span, since a single rotation added to the relative plate motions can model successfully the trends of most of the hotspot traces proposed on various plates (Morgan 1972, Minster et al 1974).

However, although the idea of hotspots seems to remain valid in many cases, recent work has cast doubt on their fixity. Molnar & Atwater (1973) have shown that relative motion between hotspots of the order of 2 cm/yr is required in a reconstruction of plate positions of 38 mybp. Burke, Kidd & Wilson (1973) have demonstrated that two groups of hotspots in the Atlantic have moved relative to one another at about the same rate. In modeling present-day instantaneous plate motions, Chase (1974, in preparation) has found that the mean minimum relative movement of 20 suggested hotspots is about 0.5 cm/yr, and that certain hotspot traces imply relative hotspot motion of from 1–2 cm/yr, in the most favorable case. In view of this evidence for nonfixity of hotspots, perhaps the suggestion of McDougall (1971) that hotspots are located within the asthenospheric zone of counterflow to plate motions should be examined in more detail.

More severe problems have arisen with some proposed hotspot traces. Sclater & Klitgord (1973) have demonstrated that the Carnegie Rise and the Cocos Ridge in the eastern Pacific cannot both be traces of the same hotspot as suggested by Morgan (1971), because of differences in age of the underlying oceanic crust. Even more embarrassing, Deep Sea Drilling Project results have shown (Scientific Staff 1974) that in the Late Cretaceous, volcanism was taking place simultaneously along a 1270 km long section of the Line Islands chain. Thus the Line Islands cannot represent the trace of plate motion over a localized melting anomaly. This makes it difficult to maintain the interpretation that the Line Islands represent the same episode of northward motion of the Pacific plate that is proposed to explain the Emperor seamount chain (Morgan 1972). Clearly there is need for much more work on the linear island and seamount chains and other proposed hotspot traces in order to properly evaluate their origin.

HISTORY OF THE OCEAN BASINS

The linear marine magnetic anomalies formed by sea-floor spreading (Vine & Matthews 1963) during intervals of frequent geomagnetic reversals give us the simplest and most powerful technique for deciphering the history of oceanic crust. By 1968, the Cenozoic magnetic anomaly patterns observed in the Pacific, Atlantic, and Indian Oceans had been identified and correlated (Pitman, Herron & Heirtzler 1968, Dickson, Pitman & Heirtzler 1968, Le Pichon & Heirtzler 1968).

During the past few years, papers have been published which essentially complete the basic mapping of magnetic anomalies in the major ocean basins. Although many areas are not yet well surveyed, and although no lineated anomalies were generated in the oceanic crust during the Cretaceous and Jurassic periods of constant magnetic polarity (Helsley & Steiner 1968, McElhinny & Burek 1971), the basic framework of the late Mesozoic and Cenozoic evolution of the ocean basins is now known. It is a sign of the impermanence of oceanic lithosphere that no in situ oceanic crust older than Jurassic has been documented, although the age of perhaps about 80% of the oceanic area is now well established.

The Pacific Ocean

Although the Cenozoic magnetic lineations in the Pacific north of 20°N and south of 30°S were among the first to be mapped (Vacquier, Raff & Warren 1961, Menard & Atwater 1968, Pitman, Herron & Heirtzler 1968), the pattern of magnetic anomalies in the eastern Pacific between 20°N and 30°S has only recently been described. A complex lineation pattern has been identified (Herron 1972) in which the present East Pacific Rise is shown to be a young feature, much of it less than 20 m.y. old. Until about 9 m.y. ago, a northwest trending spreading center was active between 10 and 36°S, within what is now the Nazca plate. Magnetic anomalies, the presence of a topographic rise (Menard, Chase & Smith 1964), abnormally high heat flow values (Langseth 1969), and the relative youth of sediments recovered at the base of cores taken near the axis of the fossil ridge (Burckle et al 1967) all support the existence of this fossil ridge within the Nazca plate.

Immediately north of the equator, holes drilled during Leg 9 of the Deep Sea Drilling Program (Scientific Staff 1970a) reached basalt at four sites between 107 and 122°W. Spreading half-rates of 12 to 13 cm yr^{-1} were computed for the interval between 10 and 38 m.y. ago. These spreading rates at least twice those measured elsewhere in the Pacific from magnetic data for the same time interval. Herron (1972) suggested that a fossil spreading center which ceased spreading about 10 mybp had been crossed near 2°N, 115°W, and a fortuitous pattern of sampling had given the appearance of fast continuous spreading. Heat flow and sediment distribution also support the idea of an old spreading center at 115°W.

In the western North Pacific, Larson, Smith & Chase (1972) mapped an east-west trending magnetic lineation set (the Phoenix lineations) along the equator near 180°E. Larson & Chase (1972) studied the shapes of these anomalies and managed to correlate them to two previously mapped lineation sets: the Japanese and Hawaiian lineations (Uyeda & Vacquier 1968, Hayes & Pitman 1970). A Deep Sea Drilling hole on the Phoenix lineations (Scientific Staff 1971a) showed that they were Early Cretaceous in age, and thus reflect the Mesozoic pattern of spreading in the Pacific.

Reconstruction of the geometry associated with the three lineation sets implies that two triple points and five spreading centers divided the Late Mesozoic Pacific Ocean into four plates. Larson & Chase suggested that the northern triple point moved north, then jumped to the southeast, followed by evolution into the Cenozoic pattern found off the west coast of North America (Menard & Atwater 1968). The Emperor trough may have formed initially by conversion of a spreading center into a transform fault during the Late Cretaceous. The southern triple point migrated rapidly to the south-southeast along a line closely parallel to the Eltanin fracture zone.

Because they contain paleomagnetic information, the marine magnetic anomalies have value beyond their very important role in deciphering the relative motions of sea-floor spreading. Larson & Chase (1972) employed the shapes of the three Mesozoic lineation sets in the Pacific to determine the paleomagnetic pole position for the Pacific plate at around 110 mybp. The pole position indicated that all the anomalies originated some 35 to 40° south of their present locations. This agrees with the northward drift of the Pacific plate suggested by independent paleomagnetic data from Pacific seamounts of Cretaceous age (Uyeda & Richards 1966, Francheteau et al 1970).

The fact that sedimentation rates in the open ocean are very high immediately under the equatorial current systems gives a test of northward drift for the Pacific that is independent of paleomagnetic methods. Deep Sea Drilling Project results (Scientific Staff 1971a) combined with seismic reflection profile data (Menard 1972, Winterer 1973) show that the Eocene equatorial bulge in Pacific sediments is now found well north of the equator. The rates of northward motion calculated in this fashion (Winterer 1973) show that the most rapid northward movement of the Pacific plate occurred before Eocene time.

Marine magnetic anomalies are also a potent tool for working out the history of geomagnetic field reversals. They were used by Heirtzler et al (1968) to establish the Cenozoic sequence of reversals beyond 4.5 m.y. ago. Larson & Pitman (1972) extended the reversal time scale well into the Mesozoic, showing that the geomagnetic reversal sequence derived for the Hawaiian lineation set of the Pacific matched the Keathley lineation pattern in the North Atlantic. The stratigraphic age of the lineations could then be estimated using two Deep Sea Drilling Project holes: one on the younger end of the Phoenix lineations, and one on the older side of the Keathley sequence (Scientific Staff 1970b, 1971b). There are difficulties with attaching absolute ages to the stratigraphic ages (Baldwin, Coney & Dickinson 1974), but the entire Mesozoic reversal sequence has been assigned to the interval between 110 and 160 m.y. ago (Larson 1974). The frequent reversals of the period are bracketed by the Cretaceous and Jurassic magnetic quiet zones. One inference that has been drawn from the Mesozoic reversal time scale is that spreading rates in both the Atlantic and Pacific were significantly higher during the Middle and Late Cretaceous quiet zone than either before or after (Larson & Pitman 1972, Larson 1974).

The Atlantic Ocean

In many ways, the history of the Atlantic Ocean is easier to decipher than that of the Pacific. The original configuration of the continents surrounding the Atlantic before its opening is well known from matching of the continental outlines (Bullard, Everett & Smith 1965). In contrast to the Pacific, the four major plates surrounding the Atlantic (North American, South American, African, and Eurasian) include continental crust, and therefore more easily accessible geological evidence on the Mesozoic movements. However, the spreading rates in the Atlantic have been in general slower than in the Pacific, and the precise data on relative motions given by magnetic anomalies has proved much more difficult to extract.

Two basically different anomaly identifications in the North Atlantic were proposed by Pitman, Talwani & Heirtzler (1971) and Williams & McKenzie (1971) for the area north of the Azores-Gibraltar Ridge. Although their identifications agree for the better-developed anomalies which occur beyond number 20 of the Pitman, Herron & Heirtzler (1968) nomenclature system, the anomaly identified as number 13 by Pitman, Talwani & Heirtzler (1971) is much closer to the ridge axis than that identified by Williams & McKenzie (1971). Pitman and his co-workers had to infer a near cessation in the opening of the North Atlantic between 10 and 38 m.y. ago. In contrast, Williams & McKenzie proposed a pattern of relatively constant spreading rates during the same time interval. This latter model was adopted by Pitman & Talwani (1972) in their detailed analysis of the opening of the North Atlantic. In their study, they reconstructed the history of the North Atlantic during the last 80 m.y. by computing poles of relative rotation and angular displacements for various chosen intervals and then fitting together lineations of the same age from opposite sides of the ridge. Their study is of importance not only because it demonstrates the precision with which the separation of Europe and Africa from North America can be described using magnetic anomalies, but also because it provides a key to understanding the complex development of the Mediterranean Sea and the circum-Mediterranean orogenic belts (Dewey et al 1973).

By combining what is known about the sequence of magnetic reversals in the Mesozoic and continental geologic evidence, the time of opening of the Atlantic ocean is now fairly well known. In the North Atlantic, formation of oceanic crust began during the Jurassic magnetically quiet period, or about 180 mybp (Pitman & Talwani 1972, Larson & Pitman 1972). In the South Atlantic, Mesozoic magnetic lineations have been identified off the southern tip of Africa (Larson & Ladd 1973) that place the age of opening at about 135 mybp, or 45 m.y. later than the North Atlantic.

The Indian Ocean

In contrast to the situation in the North Atlantic, existence of a halt in spreading proposed for the central Indian Ocean ridge (Le Pichon & Heirtzler 1968) has stood up to close scrutiny of the magnetic anomaly patterns. In an extensive and detailed examination of the topographic, magnetic, and seismic data from the Indian Ocean, McKenzie & Sclater (1972) and Fisher, Sclater & McKenzie (1971) have reconstructed the history of about two thirds of its area, back to Late Cretaceous. The present pattern of spreading on the central Indian ridge began, in its northeast-southwest direction, about 36 m.y. ago. Although the anomalies can be identified only back to about number 5 (10 m.y. old on the time scale of Heirtzler et al 1968), the present pole of rotation found by Fisher, Sclater & McKenzie (1971) for the Indian-Somalian plate motion brings the Chagos-Laccadive ridge snugly up against the Mascarene Plateau if allowed enough time to displace the plate by 20°. The bathymetry of the fracture zones supports this interpretation. The Chagos-Laccadive ridge plus the southern part of the Mascarene Plateau then forms a long straight feature, which was apparently almost inactive in the period between 36 and 51 m.y. ago. Prior to that time it had acted as a long transform fault, offsetting in a right lateral sense the earlier north-south spreading which generated anomalies 23 through 30 with an east-west trend.

The Wharton Basin, that part of the Indian Ocean east of the Ninetyeast ridge, is also Cretaceous in age (Scientific Staff 1972). Sclater & Fisher (1974) have identified Early Tertiary and Late Cretaceous magnetic anomalies in the Wharton Basin, establishing that the Ninetyeast ridge served as a transform fault by which the east-west spreading center in the central Indian Ocean was offset far to the north, passing north of Australia. Thus, in the vicinity of India, oceanic crust created on the north side of the spreading center is preserved, while the Wharton basin is floored by crust formed on the south side of the ridge, and the spreading center east of the Ninetyeast ridge has been subducted beneath the Java-Sumatra arc. Apparently, this large offset vanished with the separation of Australia and Antarctica (Sclater & Fisher 1974).

Behind Arc Spreading

Another recent development in plate tectonics, due largely to the efforts of Karig (1970, 1971a,b, 1972) is the discovery that the basins found immediately behind the andesite volcanoes of Pacific island arcs are extensional in origin. As opposed to the more continuous nature of sea-floor spreading and subduction, this is an episodic process, in which the constructional volcanic ridge of the arc periodically splits, leaving behind a remnant arc of characteristic structure separated from the still active part by a spreading basin. The Lau Basin (Karig 1970) is an example of such a basin, and seismic studies by Barazangi & Isacks (1971) and Aggarwal, Barazangi & Isacks (1972) show that a zone of high shear wave attenuation and low seismic velocity in the upper mantle underlie that basin. This is consistent with spreading in the area, and is favorable to Karig's (1971b) model of a thermal diapir rising from the subduction zone and causing the extension. The Mariana arc also has an active extensional basin behind it (Karig 1971a), and the Philippine

Sea can reasonably be interpreted as consisting of a series of such basins and remnant arcs, now past their active stage (Karig 1972). The Sea of Japan (Packham & Falvey 1971) has similar characteristics and may be another example of now inactive behind arc spreading.

Barker (1970) has mapped a pattern of magnetic anomalies behind the Scotia Arc that indicates that spreading at a half rate of 2.7 cm yr^{-1} has been going on for the last 7 to 8 m.y. From the pattern of earthquake epicenters in the region (Barazangi & Dorman 1969) and the configuration of the spreading center, it can be shown that a small plate, Barker's "Sandwich" plate, occupies an area on the north side of the Antarctic-South American plate boundary (Figure 1), and is both generating and consuming Atlantic Ocean floor. The strange and apparently independent motion of this small plate may provide some insight into the forces that move plates in more normal situations.

IMPLICATIONS FOR CONTINENTAL GEOLOGY

A liberal application of plate tectonic concepts to solve geologic problems within continental areas is fraught with more pitfalls than such attempts within oceanic areas. This is primarily due to the much longer and more complicated histories of many continental areas. Whereas oceanic lithosphere disappears within 200 m.y., continental regions persist at the earth's surface and may experience repeated episodes of tectonism during subsequent plate motions and collisions. Unfortunately, misuse of plate tectonic concepts to explain rather local continental tectonic events in the past is leading to some confusion in the literature; indeed, many of the objections to plate theory are based on these erroneous applications. To avoid such pitfalls, we will review only examples of active and/or well-documented plate-boundary events associated with continental areas and then discuss the broader problem of continental accretion as a result of plate motions. In such cases, the rather simpler histories of the sea floor may often be used to help unravel the sequence of events within the adjacent continental region: for example, the true extent of crustal shortening within Tethyan belt may be learned more readily through the history of plate motions in and around the Atlantic and Indian Oceans than from detailed land mapping in the Himalayan or Alpine mountains (Smith 1971, Dewey et al 1973). For delineating plate tectonic histories of continental areas where the adjacent, coeval sea floor has been subducted, it is necessary to specifically define lithologic affinities thought to be diagnostic of the margin type (transform, subduction, or spreading) and to systematically apply these criteria; the attempt by Dewey et al (1973) in reconstructing the Alpine system illustrated the potential of such techniques.

Rifted and Transform Margins

Spreading across a newly formed plate boundary, which dissects a continental block, eventually may form a normal ocean basin such as the Atlantic Ocean. The basic characteristics of the resulting continental margins are determined early in the rifting process before any substantial amount of new sea floor is formed.

The geometry of typical oceanic spreading centers can be recognized even in some continental rifts which did not evolve into ocean basins. The mid-continent gravity high of central North America is a series of sharply bounded, en-echelon, positive gravity anomalies associated with basalts and gabbros of Keweenawan age. Chase & Gilmer (1973) show that the gravity pattern agrees rather precisely with that of a plate tectonic rift. This implies that the continental lithosphere behaved as rigid plates during the Late Precambrian, about 1.1 b.y. ago. They further suggest that this Keweenawan rift probably exhibits an intermediate stage of structural development between the East African Rift Valleys and actual sea-floor spreading as seen in the southern Red Sea.

The Red Sea and the Gulf of California are two well-studied examples of continental rifting which have resulted in the formation of oceanic crust but which differ in character because the direction of separation relative to the trend of the rift basin is markedly different for each. Plate separation across the Red Sea is roughly normal to its long axis and occurred in two stages. Uplift and extension in early or pre-Miocene time resulted in crustal thinning. As seen in reflection profiles, normal faulting resulting in graben topography developed as the surface expression of crustal extension (Lowell & Genik 1972, Ross & Schlee 1973). Regional volcanism accompanied this stage of rifting, but oceanic crustal areas may not have been formed. Deposition of Miocene evaporites and Pliocene marine strata in the newly formed Red Sea basin preceded the second stage of plate separation, which began in Pliocene-Pleistocene time (Ross & Schlee 1973). Since then, sea-floor spreading of about 1 cm/yr has produced a narrow axial trough of oceanic character.

At the mouth of the Gulf of California, profiles across the Baja California continental margin and perpendicular to the axis of spreading also show well-developed horst-graben structures (Normark & Curray 1968). Late Neogene sediments deposited after formation of some of these faults are disrupted by them, suggesting continued movement or readjustment as the adjacent Gulf floor formed. Turbidite deposition along the tip of the peninsula has already formed a small continental rise; it is very easy to envisage this area as an immature version of the western North Atlantic continental margin.

Within the Gulf of California, however, the continental margin structure results from a combination of spreading and transform motion (Sykes 1968). Individual transform faults, which have a more westerly trend than the Gulf itself, connect relatively short spreading segments marked by prominent basins. However, symmetrical magnetic anomalies are found only at the very southern end of the Gulf where the East Pacific Rise still exists topographically (Moore & Buffington 1968, Larson et al 1968, Larson et al 1972, Klitgord et al 1974). The crust underlying the Gulf of California changes gradually from normal oceanic at the south to relatively thin continental at the north (Phillips 1964). As continental separation occurs, rapid clastic sedimentation in the northern and central basins of the Gulf precludes the formation of normal oceanic crust (Moore 1973) and the sea-floor spreading, magnetic-anomaly patterns cannot form (Klitgord et al 1974, Larson et al 1972).

The initiation of rifting between North America and Africa in the Triassic was marked by a series of events which Dewey et al (1973) consider indicative of rifted margins. These include the intrusion of alkalic magmas, development of graben structures with local, thick accumulations of clastic sediments, then extensive outpouring of flood basalts. Extensive evaporite sequences appear to coincide with initial rifting (Pautot, Auzende & Le Pichon 1970). About 180 m.y. ago, active sea-floor spreading began to form the North Atlantic Ocean basin (Pitman & Talwani 1972) and continental margin subsidence is marked by the development of a carbonate miogeoclinal sequence (Dewey et al 1973, Dietz et al 1970). Continued deposition from Jurassic to Recent has produced the thick, prograding sedimentary prisms of geosynclinal proportions of a mature rifted margin (Hallam 1971, Rona 1970, 1973, Dietz 1972).

Convergent Margins

The subduction or consumption of oceanic lithosphere along trenches has been documented in several ways. Less obvious, however, is the ultimate effect on the edge of the overriding continental plate during extended periods of convergent motion. Paradoxically, several modes of continental accretion are important along these compressive (subduction) margins. Growth of the andesitic volcanic arc with its associated intrusive roots and voluminous sediments derived from erosion of both can account for substantial crustal thickening as seen along the Andean margin of South America (James 1971). Dickinson (1973) has shown that the width of arc-trench gaps is proportional to the duration of subduction. This growth reflects both the outward migration of the subduction zone away from crustal rocks in the central region of the arc-trench gap, as well as an inward (continentward) migration of the axis of the volcanic arc. Island arcs and major volcanic ridges carried into a subduction zone on the back of an oceanic plate would result in a rather catastrophic style of growth of a compressional margin as these features would not be subducted due to their relatively low density.

Direct accretion of the margin also can occur along the lower trench wall whenever sufficient sediment is shed into the trench. Sediment sources include the adjacent volcanic arc, pelagic or abyssal-plain deposits covering the downgoing plate, or longitudinal transport by turbidity currents moving along the axis of the trench. Piper et al (1973) review these sedimentation processes for the eastern Aleutian trench where the rate of deposition approaches 3 km/m.y. during glacial periods (von Huene & Kulm 1973). Results from the Deep Sea Drilling Program on the lower continental slope along the trench (von Huene & Kulm 1973) document deformation and uplift of young slope sediments. The upper part of the section at Site 181 has been slightly deformed, tilted, and uplifted perhaps as fast as 2 cm/yr in Recent time. Sediments below this section (and separated by a fault or unconformity) underwent intense deformation about 1 m.y. ago and exhibit compaction equivalent to overburden pressures corresponding to a depth of burial of 1.5 km, which is much greater than their present burial of 170 to 370 m. These deformed sediments were deposited at depths corresponding to the lower slope but may not be simply trench sediments scraped off during subduction of the Pacific plates (von Huene & Kulm 1973). In the case of the Java Trench, where the slope of the lower trench wall is rather low, reflection profiles suggest an imbricate thrust structure as well as folding in a wedge of sediments up to 10 km thick which underlies the trench slope and overlies the gently dipping surface of the descending slab (Beck & Lehner 1974).

Off the northwestern United States, the Gorda and/or Juan de Fuca plate is being subducted along the continental margin, although no Benioff zone or topographic trench exists. A broad continental rise/abyssal plain turbidite section is ponded between the eastern flank of the Juan de Fuca spreading center and the continental margin of the Cascadia Basin. Along the southern end of the Juan de Fuca plate, Silver (1969, 1971) presented evidence for deformation and accretion of these ponded turbidite sediments onto the margin. Along the Washington margin, Silver (1972) later obtained reflection profiles which convincingly illustrate this process of deformation of Cascadia Basin sediments against the continental slope. Individual reflecting horizons maintain structural continuity across the basin-slope boundaries; undeformed subbottom reflectors within the basin sequence can be traced into gentle folds which are expressed as low ridges along the lower slope. Silver (1972) further suggests that deformation has progressed westward in time although it is not a continuous process. Many of the faults which bound these folded strata show that vertical offset is west (or basin) side up. An indurated Pleistocene claystone recovered from one of these ridges suggests burial of sediment before being uplifted and exposed on the ridge. This mode of accretion and uplift of basin sediments along the lower continental slope is consistent with the results of the Deep Sea Drilling Project on the Oregon continental margin immediately to the south (von Huene & Kulm 1973).

Western United States

Even in continental areas that are relatively young and well mapped, the determination of plate movements in adjacent oceanic areas can put needed restraints on models for the continental geologic history. For example, Atwater (1970) interpreted the history of plate motions in the northeastern Pacific using primarily the marine magnetic anomaly pattern of the sea floor and related these motions to the pattern of late Cenozoic tectonism in the western United States. Off southern California about 30 m.y. ago, a central portion of the Farallon plate was completely subducted, and the North American and Pacific plates came in contact. Atwater (1970) assumed the relative motion of the plates at that time was such that the old subduction boundary (along the continental margin) became a locus of transform motion. The relative motion between the North American and Pacific plates during the Tertiary has since been determined solely through the use of sea-floor spreading magnetic data used to calculate the motion across a circuit of plates including North America-African-Indian-Antarctica-Pacific (Atwater & Molnar 1974). This is an excellent example of the use of plate tectonic principles coupled with oceanic magnetic data to solve continental geologic problems, in this case, to put limits on the direction, times, and rate of motion along a continental transform, the San Andreas fault system. Provenance studies on sediments from Deep Sea Drilling Program cores through the outer Delgada deep-sea fan also demonstrate a relative motion between the Pacific and North American plates which agrees rather nicely with the motion determined through spreading data (Hein

1973). Since 30 m.y. ago, this transform motion has moved eastward into the continent affecting progressively more continental terrain (Johnson & Normark 1974). The extension seen in the Great Basin province may in part be a result of the basic transform motion between American and Pacific lithospheric plates (Atwater 1970).

Even if Atwater's model for the western United States is basically correct, it seems likely that extension in the Great Basin province began before the Pacific and American plates came in contact. This portion of the continent was probably undergoing behind-the-arc spreading similar to the western Pacific marginal basins (Karig 1970, 1971a,b, 1972, Packham & Falvey 1971), although there is little agreement on the mechanism(s) which may be involved (Scholz et al 1971, Lipman et al 1971). Further south, an extensive survey of the transform fault system (and their fracture zone extensions) in the Gulf of California clearly shows that the rifted topography associated with plate motions recorded in the symmetrical magnetic anomalies at the mouth of the Gulf disrupts extensive mid-Miocene to Pliocene marine sediments deposited in a proto-Gulf (Moore 1973). Karig & Jensky (1972) feel the opening of a proto-Gulf trough in Mid- to Late-Miocene is part of a volcano-tectonic rift zone associated with behind-the-arc extension during active subduction of the Farallon plate. In addition to the marine basinal sediments, other evidence for this extensional zone includes a Mid- to Late-Tertiary ignimbrite complex stretching from the US border to near the mouth of the Gulf along the Sierra Madre Occidental east of the Gulf, and the development in pre-Pliocene time of north/northwest trending tilted faulted blocks along the eastern margin of Baja California, which disrupt the ignimbrite complex on the Mexican mainland.

The recognition of past plate boundaries even in young continental areas can be rather difficult. A case in point is the Pacific margin of the Baja California peninsula which was a locus of subduction until 12–15 mybp (Atwater 1970). From then until 4–5 m.y. ago, when the present Gulf of California began to form (Atwater 1970, Larson et al 1968, Moore 1973), transform motion of the San Andreas trend prevailed. The main evidence for the subduction episode is: 1. a nearly continuous trough along the base of the continental slope, 2. sediment folded against the base of slope over a limited area, and 3. Franciscan-like assemblages along the coast. The period of transform motion resulted in the formation of multiple northwest-trending ridges bounded on the landward side by near-vertical faults; these ridges are underlain by probably Franciscan equivalents. Along the lower half of the peninsula, post-Miocene sediments are ponded behind these ridges. These post-Miocene (i.e. post-transfer of Baja California to the Pacific plate) sediments deposited on the margin have been deformed after the plate boundary shifted to the Gulf (Normark 1971). This continuing tectonic activity is occurring on the northwest-trending faults and is gradually masking the earlier tectonic signatures.

EUSTACY AND OROGENY

The youthful age of the present sea floor (Jurassic to Recent) shows that oceanic crust is recycled rapidly by subduction, but the marine magnetic data discussed above demonstrate that the rate of creation of oceanic crust has not been constant through time. Under the accepted assumption that the earth has not been increasing in surface area during this period, several speculative implications for continental geology have been put forward that are not specific to plate margins. Rapid growth of a spreading center will result in an effective decrease of the volume of its associated ocean basin (Menard 1964). The concurrent rapid growth along several spreading centers may result in a significant eustatic rise in sea level and an extensive marine transgression of continental areas such as occurred during the Late Cretaceous (Pitman & Hays 1973). Rona (1973) even suggests that the rate of sediment accumulation at continental margins reflects the rate of sea-floor spreading as both appear to increase together. Since increased rates of spreading also imply increased rates of subduction, some speculate further that increased rates of plutonic activity and continental orogenic episodes in general are necessarily related to increased spreading rates (Johnson 1971, Larson & Pitman 1972).

CONTINENTAL ACCRETION

Two related and vitally interesting problems now being reconciled within a plate tectonic framework are the concepts of geosynclinal sedimentation and the general processes of continental growth. Dewey & Bird (1970) attempt to match the varietal geosynclines (i.e. paralio-, lepto-, exo-, etc) spacially and/or temporally within compressional (subduction) plate boundary settings. Dickinson (1971a,b) wants to deemphasize the somewhat confusing classification developed for ancient geosynclines and orogenic cycles and prefers to construct models based on modern arc-trench environments. Dickinson (1971b) suggests that arc-trench systems include two sequences generally regarded as eugeosynclinal: the magmatic arc, which includes andesitic extrusive rocks intruded by comagmatic granitic plutons, and the trench, where oceanic crust, pelagic sediments, and turbidite sequences are severely deformed resulting in chaotic melanges with characteristic ophiolite affinities. The arc-trench gap and any basin formed behind a volcanic arc are the sites of thick, elongate, sedimentary prisms contemporary with magmatic activity in the volcanic arc (Dickinson 1971b). The bulk of the clastic material is derived from the volcanic arc, interbedded volcanic flows are few, and such sequences remain relatively untectonized. These areas resemble some ancient miogeosynclinal assemblages. As described above, there are a variety of mechanisms for continental accretion at convergent margins.

On the other hand, Dietz (1972) considers that the modern equivalent of a classical mio-eu-geo(syn)clinal pair is found in a mature continental shelf-slope-rise sedimentary prism. A concurrently active volcanic arc is not required in this model. Subsequent crustal shortening and mountain building episodes leading to continental accretion result when the adjacent sea floor is subducted; this often results in a continent-continent collision as an intervening ocean basin is consumed and two such shelf-slope-rise prisms are collapsed together.

All of these geosynclinal models result in growth of the continental area; the major differences concern the immediate sources for crustal building blocks. It has been suggested that plate tectonic processes have been responsible for continental accretion during the last 3×10^9 years (Dewey & Horsfield 1970, Dickinson & Luth 1971).

CONCLUSION

In the few years since it was proposed, the theory of plate tectonics has had a profound and positive effect on thinking about global geologic processes. Many unresolved problems remain: how far into the geologic past the plate tectonics model can reasonably be applied; just what the driving mechanisms actually are; to what extent the model is relevant to vertical motions of the crust; and many others. Our understanding of the geologic processes that accompany relative motions at the surface of the earth may well change in the future, but it seems safe to say that the reality of such motions can no longer be ignored. Whatever the eventual scientific fate of the theory, it is clear that it will never again be possible to neglect the two thirds of the earth that lies beneath the ocean in studying the one third that happens to be dry land.

Literature Cited

Aggarwal, Y. P., Barazangi, M., Isacks, B. 1972. P and S travel time in the Tonga-Fiji region: a zone of low velocity in the uppermost mantle behind the Tonga island arc. *J. Geophys. Res.* 77:6427–34

Atwater, T. 1970. Implications of plate tectonics for the Cenozoic tectonics of western North America. *Geol. Soc. Am. Bull.* 81:3513–36

Atwater, T., Molnar, P. 1974. Relative motion of the Pacific and North American plates deduced from sea-floor spreading in the Atlantic, Indian, and South Pacific Oceans. In: Proceedings of the Conference on Tectonic Problems of the San Andreas Fault System, ed. R. L. Kovach, A. Nur. *Stanford Univ. Publ. Geol. Sci.* 13:136–48

Baldwin, B., Coney, P. J., Dickinson, W. R. 1974. Dilemma of a Cretaceous time scale and rates of sea-floor spreading. *Geology* 2:267–70

Barazangi, M., Dorman, J. 1969. World seismicity maps compiled from ESSA, Coast and Geodetic Survey epicenter data, 1961–1967. *Seismol. Soc. Am. Bull.* 59:369–80

Barazangi, M., Isacks, B. 1971. Lateral variations of seismic wave attenuation in the upper mantle above the inclined earthquake zone of the Tonga island arc: deep anomaly in the upper mantle. *J. Geophys. Res.* 76:8493–8516

Barker, P. 1970. Plate tectonics of the Scotia Sea region. *Nature* 228:1293–96

Beck, R. H., Lehner, P. 1974. Oceans, new frontier in exploration. *Am. Assoc. Petrol. Geol. Bull.* 58:376–95

Beloussov, V. V. 1970. Against the hypothesis of ocean floor spreading. *Tectonophysics* 9:489–511

Bullard, E. C., Everett, J. E., Smith, A. G. 1965. The fit of the continents around the Atlantic. In: Symposium on Continental Drift. *Phil. Trans. Roy. Soc. London Ser. A* 258:41–51

Burckle, L. H., Ewing, J., Saito, T., Leyden, R. 1967. Tertiary sediment from the East Pacific Rise. *Science* 157:537–40

Burke, K., Kidd, W. S. F., Wilson, J. T. 1973. Relative and latitudinal movement of Atlantic hotspots. *Nature* 245:133–38

Chase, C. G. 1972. The *n*-plate problem of plate tectonics. *Geophys. J.* 29:117–22

Chase, C. G., Gilmer, T. H. 1973. Precambrian plate tectonics: the midcontinent gravity high. *Earth Planet. Sci. Lett.* 21:70–78

Dewey, J. F., Bird, J. M. 1970. Plate tectonics and geosynclines. *Tectonophysics* 10:625–38

Dewey, J. F., Horsfield, B. 1970. Plate tectonics, orogeny, and continental growth. *Nature* 225:521–25

Dewey, J. F., Pitman, W. C., Ryan, W., Bonnin, J. 1973. Plate tectonics and the evolution of the Alpine system. *Geol. Soc. Am. Bull.* 84:3137–80

Dickinson, W. R. 1971a. Clastic sedimentary sequences deposited in shelf, slope and trough settings between magmatic arcs and associated trenches. *Pac. Geol.* 3:15–30

Dickinson, W. R. 1971b. Plate tectonic models of geosynclines. *Earth Planet. Sci. Lett.* 10:165–74

Dickinson, W. R. 1973. Widths of modern arc-trench gaps proportional to past duration of igneous activity in associated magmatic arcs. *J. Geophys. Res.* 78:3376–89

Dickinson, W. R., Luth, W. C. 1971. A model for plate tectonic evolution of mantle layers. *Science* 174:400–4

Dickson, G. O., Pitman, W. C. III, Heirtzler, J. R. 1968. Magnetic anomalies in the south Atlantic and sea-floor spreading. *J. Geophys. Res.* 73:2087–2100

Dietz, R. S. 1961. Continent and ocean basin evolution by spreading of the sea floor. *Nature* 190:854–57

Dietz, R. S. 1972. Geosynclines, mountains, and continent building. *Sci. Am.* 226(3):30–38

Dietz, R. S., Holden, J. C., Sproll, W. P. 1970. Geotectonic evolution and subsidence of Bahama platform. *Geol. Soc. Am. Bull.* 81:1915–28

Fisher, R. L., Sclater, J. G., McKenzie, D. P. 1971. Evolution of the central Indian Ridge, western Indian Ocean. *Geol. Soc. Am. Bull.* 82:553–62

Francheteau, J., Harrison, C. G. A., Sclater, J. G., Richards, M. L. 1970. Magnetization of Pacific seamounts: a preliminary polar curve for the northeastern Pacific. *J. Geophys. Res.* 75:2035–61

Hallam, A. 1971. Mesozoic geology and the opening of the North Atlantic. *J. Geol.* 79:129–57

Hayes, D. E., Pitman, W. C. III 1970. *Magnetic Lineations in the North Pacific,* ed. J. D. Hayes. *Geol. Soc. Am. Mem.* 126:291–314

Hein, J. R. 1973. Increasing rate of movement with time between California and the Pacific plate: from Delgada submarine fan source areas. *J. Geophys. Res.* 78:7752–62

Heirtzler, J. R., Dickson, G. O., Herron, E. M., Pitman, W. C. III, Le Pichon, X. 1968. Marine magnetic anomalies, geomagnetic field reversals and motions of the ocean floor and continents. *J. Geophys. Res.* 73:2119–36

Helsley, C. E., Steiner, M. B. 1968. Evidence for long periods of normal magnetic polarity in the Cretaceous period. *Earth Planet. Sci. Lett.* 5:325–32

Herron, E. M. 1972. Sea floor spreading and the Cenozoic history of the east central Pacific. *Geol. Soc. Am. Bull.* 83:1671–92

Hess, H. H. 1962. History of ocean basins. In *Petrologic Studies: A Volume to Honor A. F. Buddington,* ed. A. E. J. Engel, H. L. James, B. F. Leonard, 599–620. New York: Geol. Soc. Am.

Isacks, B., Oliver, J., Sykes, L. R. 1968. Seismology and the new global tectonics. *J. Geophys. Res.* 73:5855–99

Jackson, E. D., Silver, E. A., Dalrymple, G. B. 1972. Hawaii-Emperor chain and its relation to Cenozoic circum-Pacific tectonics. *Geol. Soc. Am. Bull.* 83:601–18

James, D. E. 1971. Plate tectonics model for the evolution of the Central Andes. *Geol. Soc. Am. Bull.* 82:3325–46

Jeffreys, H. 1970. Imperfections of elasticity and continental drift. *Nature* 225:1007–8

Johnson, J. D., Normark, W. R. 1974. Neogene tectonic evolution of the Salinian block, west-central California. *Geology* 2:11–14

Johnson, J. G. 1971. Timing and coordination of orogenic epeirogenic, and eustatic events. *Geol. Soc. Am. Bull.* 82:3263–98

Karig, D. E. 1970. Ridges and basins of the Tonga-Kermadec island arc system. *J. Geophys. Res.* 75:239–54

Karig, D. E. 1971a. Structural history of the Mariana island arc system. *Geol. Soc. Am. Bull.* 82:323–44

Karig, D. E. 1971b. Origin and development of marginal basins in the western Pacific. *J. Geophys. Res.* 76:2542–61

Karig, D. E. 1972. Remnant arcs. *Geol. Soc. Am. Bull.* 83:1057–68

Karig, D. E., Jensky, W. 1972. The proto-Gulf of California. *Earth Planet. Sci. Lett.* 17:169–74

Klitgord, K. D., Mudie, J. D., Bischoff, J. L., Henyey, T. 1974. Magnetic anomalies in the northern and central Gulf of California. *Geol. Soc. Am. Bull.* 85:815–20

Langseth, M. G. Jr. 1969. The flow of heat from the earth and its global distribution at the surface. *Am. Inst. Aerons. & Astronauts, AIAA Paper 69-589. Thermophys. Conf., 4th, San Francisco*

Larson, R. L. 1974. An updated time scale of magnetic reversals for the Late Mesozoic (abstr.) *Eos, Trans. Am. Geophys. Union* 55:236

Larson, R. L., Chase, C. G. 1972. Late Mesozoic evolution of the western Pacific. *Geol. Soc. Am. Bull.* 83:3627–44

Larson, R. L., Ladd, J. W. 1973. Evidence for the opening of the South Atlantic in the Early Cretaceous. *Nature* 246:209–12

Larson, R. L., Menard, H. W., Smith, S. M.

1968. The Gulf of California: A result of ocean-floor spreading and transform faulting. *Science* 161:781–84

Larson, R. L., Mudie, J. D., Larson, P. A. 1972. Magnetic anomalies in the southern and middle Gulf of California. *Geol. Soc. Am. Bull.* 83:3361–68

Larson, R. L., Pitman, W. C. III. 1972. World-wide correlation of Mesozoic magnetic anomalies and its implications. *Geol. Soc. Am. Bull.* 83:3645–62

Larson, R. L., Smith, S. M., Chase, C. G. 1972. Magnetic lineations of Early Cretaceous age in the western equatorial Pacific Ocean. *Earth Planet. Sci. Lett.* 15:315–19

Le Pichon, X. 1968. Sea-floor spreading and continental drift. *J. Geophys. Res.* 73: 3661–98

Le Pichon, X., Heirtzler, J. R. 1968. Magnetic anomalies in the Indian Ocean and sea floor spreading. *J. Geophys. Res.* 73: 2101–17

Lipman, P. W., Prostka, H. J., Christiansen, R. L. 1971. Evolving subduction zones in the western United States, as interpreted from igneous rocks. *Science* 174:821–25

Lowell, J. D., Genik, G. J. 1972. Sea-floor spreading and structural evolution of the southern Red Sea. *Am. Assoc. Petrol. Geol. Bull.* 56:247–59

McDougall, I. 1971. Volcanic island chains and sea floor spreading. *Nature* 231:141–44

McElhinny, M. W., Burek, P. J. 1971. Mesozoic paleomagnetic stratigraphy. *Nature* 232:98–101

McKenzie, D. P. 1972. Plate tectonics. *The Nature of the Solid Earth,* ed. E. C. Robertson, 323–60. New York: McGraw-Hill. 677 pp.

McKenzie, D. P., Parker, R. L. 1967. The North Pacific: an example of tectonics on a sphere. *Nature* 216:1276–80

McKenzie, D. P., Sclater, J. G. 1972. The evolution of the Indian Ocean since the Late Cretaceous. *Geophys. J.* 24:437–528

Menard, H. W. 1964. *Marine Geology of the Pacific.* New York: McGraw-Hill. 271 pp.

Menard, H. W. 1972. History of the ocean basins. *The Nature of the Solid Earth,* ed. E. C. Robertson, 440–62. New York: McGraw-Hill. 677 pp.

Menard, H. W., Atwater, T. M. 1968. Changes in direction of sea floor spreading. *Nature* 219:463–67

Menard, H. W., Chase, T. E., Smith, S. M. 1964. Galapagos Rise in the southeastern Pacific. *Deep Sea Res.* 11:233–42

Meyerhoff, A. A., Meyerhoff, H. A. 1972a. "The new global tectonics": major incon-

sistancies. *Am. Assoc. Petrol. Geol. Bull.* 56:269–336

Meyerhoff, A. A., Meyerhoff, H. A. 1972b. "The new global tectonics": age of linear magnetic anomalies of ocean basins. *Am. Assoc. Petrol. Geol. Bull.* 56:337–59

Minster, J. B., Jordon, T. H., Molnar, P., Haines, E. 1974. Numerical modelling of instantaneous plate tectonics. *Geophys. J.* 36:541–76

Molnar, P., Atwater, T. 1973. Relative motion of hotspots in the mantle. *Nature* 246:288–91

Moore, D. G. 1973. Plate edge deformation and crustal growth, Gulf of California structural province. *Geol. Soc. Am. Bull.* 84:1883–1905

Moore, D. G., Buffington, E. C. 1968. Transform faulting and growth of the Gulf of California since the late Pliocene. *Science* 161:1238–41

Morgan, W. J. 1968. Rises, trenches, great faults, and crustal blocks, *J. Geophys. Res.* 73:1959–82

Morgan, W. J. 1971. Convection plumes in the lower mantle. *Nature* 230:42–43

Morgan, W. J. 1972. Plate motions and deep mantle convection. *Geol. Soc. Am. Mem.* 132:7–22

Normark, W. R. 1971. Transfer of lower Baja California, Mexico, from North American to Pacific lithospheric plates. *Abstr. Ann. Meet. Cordillera Sect. Geol. Soc. Am.,* p. 172

Normark, W. R., Curray, J. R. 1968. Geology and structure of the tip of Baja California, Mexico. *Geol. Soc. Am. Bull.* 79:1589–1600

Packham, G. H., Falvey, D. A. 1971. A hypothesis for the formation of marginal seas in the Western Pacific. *Tectonophysics* 11:79–109

Pautot, G., Auzende, J. M., Le Pichon, X. 1970. Continuous deep salt layer along North Atlantic margins related to early phase of rifting. *Nature* 227:351–54

Phillips, R. P. 1964. Seismic refraction studies in Gulf of California. *Marine Geology of the Gulf of California. Am. Assoc. Petrol. Geol. Mem.* 3:90–121

Piper, D. J. W., von Huene, R., Duncan, J. R. 1973. Late Quaternary sedimentation in the active eastern Aleutian Trench. *Geology* 1:19–22

Pitman, W. C., Hays, J. D. 1973. Upper Cretaceous spreading rates and the great transgression. *Geol. Soc. Am. Abstr. Programs* 5:768 (Abstr.)

Pitman, W. C., Herron, E. M., Heirtzler, J. R. 1968. Magnetic anomalies in the Pacific and sea floor spreading. *J.*

Geophys. Res. 73:2069–85

Pitman, W. C., Talwani, M. 1972. Sea floor spreading in the North Atlantic. *Geol. Soc. Am. Bull.* 83:619–46

Pitman, W. C. III, Talwani, M., Heirtzler, J. R. 1971. Age of the North Atlantic from magnetic anomalies. *Earth Planet. Sci. Lett.* 7:195–200

Press, F. 1972. The earth's interior as inferred from a family of models. *The Nature of the Solid Earth*, ed. E. C. Robertson, 147–71. New York: McGraw-Hill. 677 pp.

Rona, P. A. 1970. Comparison of continental margins of eastern North America at Cape Hatteras and northwestern Africa at Cap Blanc. *Am. Assoc. Petrol. Geol. Bull.* 54:129–57

Rona, P. A. 1973. Relations between rates of sediment accumulation on continental shelves, sea-floor spreading, and eustacy inferred from the central North Atlantic. *Geol. Soc. Am. Bull.* 84:2851–74

Ross, D. A., Schlee, J. 1973. Shallow structure and geologic development of the southern Red Sea. *Geol. Soc. Am. Bull.* 84:3827–48

Scholz, C. H., Barazangi, M., Sbar, M. L. 1971. Late Cenozoic evolution of the Great Basin, Western United States, as an ensialic interarc basin. *Geol. Soc. Am. Bull.* 82:2979–90

Scientific Staff. 1970a. Deep Sea Drilling Project: Leg 9. *Geotimes* 15(4):11–13

Scientific Staff. 1970b. Deep Sea Drilling Project: Leg 11. *Geotimes* 15(7):14–16

Scientific Staff. 1971a. Deep Sea Drilling Project: Leg 16. *Geotimes* 16(6):12–14

Scientific Staff. 1971b. Deep Sea Drilling Project: Leg 17. *Geotimes* 16(9):12–14

Scientific Staff. 1972. Deep Sea Drilling Project: Leg 22. *Geotimes* 17(6):15–17

Scientific Staff. 1974. Leg 33, Deep Sea Drilling Project: testing a hot-spot theory. *Geotimes* 19(3):16–20

Sclater, J. G., Fisher, R. L. 1974. Evolution of the East-Central Indian Ocean, with emphasis on the tectonic setting of the Ninetyeast Ridge. *Geol. Soc. Am. Bull.* 85:683–702

Sclater, J. G., Francheteau, J. 1970. The implications of terrestrial heat flow observations on current tectonic and geochemical models of the crust and upper mantle of the Earth. *Geophys. J.* 20:509–42

Sclater, J. G., Klitgord, K. D. 1973. A detailed heat flow, topographic, and magnetic survey across the Galapagos spreading center at 86°W. *J. Geophys. Res.* 78:6951–76

Shor, G. G. Jr., Raitt, R. W. 1969. Explosion seismic refraction studies of the crust and upper mantle in the Pacific and Indian Oceans. *The Earth's Crust and Upper Mantle*, ed. P. J. Hart, 225–30. *Am. Geophys. Union Geophys. Monogr.* 13. 735 pp.

Silver, E. A. 1969. Late Cenozoic underthrusting of the continental margin off northernmost California. *Science* 166:1265–66

Silver, E. A. 1971. Transitional tectonics and late Cenozoic structure of the continental margin off northernmost California. *Geol. Soc. Am. Bull.* 82:1–22

Silver, E. A. 1972. Pleistocene tectonic accretion of the continental slope off Washington. *Mar. Geol.* 13:239–49

Smith, A. G. 1971. Alpine deformation and the oceanic areas of the Tethys, Mediterranean, and Atlantic. *Geol. Soc. Am. Bull.* 82:2039–70

Sykes, L. R. 1968. Seismological evidence for transform faults, sea-floor spreading, and continental drift. *The History of the Earth's Crust*, ed. R. A. Phinney, 120–50. Princeton, NJ: Princeton Univ. Press. 244 pp.

Uyeda, S., Richards, M. L. 1966. Magnetization of four Pacific seamounts near the Japanese Islands. *Tokyo Univ. Earthquake Res. Inst. Bull.* 44:179–213

Uyeda, S., Vacquier, V. 1968. Geothermal and geomagnetic data in and around the island arc of Japan. The Crust and Upper Mantle of the Pacific Area, ed. L. Knopoff, C. L. Drake, P. J. Hart, 349–66. *Am. Geophys. Union Geophys. Monogr.* 12. 522 pp.

Vacquier, V., Raff, A. D., Warren, R. E. 1961. Horizontal displacements in the floor of the northeastern Pacific Ocean. *Geol. Soc. Am. Bull.* 72:1251–58

Vine, F. 1968. Magnetic anomalies associated with sea-floor spreading. *The History of the Earth's Crust*, ed. R. A. Phinney, 73–89. Princeton, NJ: Princeton Univ. Press. 244 pp.

Vine, F. J., Hess, H. H. 1970. Sea-floor spreading. *The Seas*, ed. A. E. Maxwell et al, 4: Part 2, 587–622. New York: Wiley. 664 pp.

Vine, F. J., Matthews, D. H. 1963. Magnetic anomalies over oceanic ridges. *Nature* 199:947–49

von Huene, R., Kulm, L. D. 1973. Tectonic summary of Leg 18. In *Initial Reports of the Deep Sea Drilling Project*, ed. L. D. Kulm, R. von Huene et al, XVIII. Wash. DC: GPO

Walcott, R. I. 1970. Flexural rigidity, thickness, and viscosity of the lithosphere. *J.*

Geophys. Res. 75:3941–54

Wesson, P. S. 1972. Objections to continental drift and plate tectonics. *J. Geophys. Res.* 80:185–97

Williams, C. A., McKenzie, D. P. 1971. The evolution of the northeast Atlantic. *Nature* 232:168–73

Wilson, J. T. 1965a. A new class of faults and their bearing on continental drift.

Nature 207:343–47

Wilson, J. T. 1965b. Evidence from ocean islands suggesting movement in the earth. *Roy. Soc. London Phil. Trans. Ser. A* 258:145–65

Winterer, E. L. 1973. Sedimentary facies and plate tectonics of equatorial Pacific. *Am. Assoc. Petrol. Geol. Bull.* 57:265–82

HIGH TEMPERATURE CREEP   ×10041
OF ROCK AND MANTLE VISCOSITY

Johannes Weertman

Departments of Materials Science and Geological Sciences and Materials Research Center, Northwestern University, Evanston, Illinois 60201

Julia R. Weertman

Department of Materials Science, Northwestern University, Evanston, Illinois 60201

INTRODUCTION

Some 20 years ago, the field of glaciology began a renaissance when laboratory–determined creep properties of ice were used as the basis of theoretical analyses of the mechanics of glaciers, ice sheets, ice caps, and ice shelves (Lliboutry 1965, Paterson 1969). The high temperature creep behavior of ice is described by the power law creep equation, an equation which this review considers in detail. Solutions to ice flow problems based on the power law creep equation provide a means of evaluating other, approximate solutions based on mathematically more convenient but physically less realistic laws, such as the linear viscous creep law or the stress-strain curve of a perfectly plastic solid. As a result of this work, we have today a rather sophisticated understanding of the flow of ice in glaciers and other large ice masses, and a body of knowledge that is very useful for recognizing important new problems.

Glacier mechanics offers a paradigm for future progress in the understanding of deformation and flow in the earth's mantle. The starting point for following the glaciological example is the determination of the creep properties of rock. In the last few years a number of creep experiments, which are reviewed in this article, have been carried out on various rocks. However, great difficulties exist in following exactly the path of the glaciologist. The solid earth scientist can make only a shrewd guess about what sort of rock is in the mantle, whereas the glaciologist knows his material is ice; moreover, he can obtain samples of it from any part of existing glaciers, ice sheets, etc. The solid earth scientist knows that no laboratory creep experiment can possibly be carried out at geologic strain rates (nor at all the possible conditions of temperature and pressure that exist in the mantle). The glaciologist on the other hand can carry out laboratory experiments at creep rates that are of the same magnitude and at the same stress levels occurring in the more

293

rapidly deforming parts of glaciers and ice sheets. His laboratory data thus overlap the field data. Because most of the flow of a glacier occurs in those regions where the strain rate is greatest, for all practical purposes the theoretical glaciologist need not make any extrapolation of the laboratory data.

Before the solid earth scientist can use the laboratory data on the creep of rock, he must extrapolate his results to geologic strain rates and mantle conditions. The theory of the creep of crystalline solids must be considered in order to make as sensible an extrapolation as possible. In this article we review results from the theory of the creep of crystalline solids that should be considered when laboratory observations are extrapolated to mantle conditions.

Reviews of the creep of crystalline solids are given by Garofalo (1965), Kennedy (1963), and Sherby & Burke (1967); reviews of creep theories based on dislocation motion are given by Weertman & Weertman (1965) and Weertman (1968, 1974); and reviews of the creep of mantle rock are given by Kirby & Raleigh (1973), Wesson (1972), Stocker & Ashby (1973), Gordon (1965, 1967), McKenzie (1968), and Weertman (1970). The reader's attention is directed to two books on the flow and deformation of rocks: an early volume edited by Griggs & Handin (1960) and a very recent one edited by Heard et al (1972) which is dedicated to David Griggs, who founded the present-day field of the high temperature creep of rocks.

It is to be understood in this paper that the stress and strain are considered to be those appropriate for a uniaxial tension or compression test.

TYPES OF CREEP AND CREEP MECHANISMS

This section briefly reviews the different types of creep and the principal mechanisms producing the creep. It is important to realize that creep behavior varies with stress level, with the ratio of temperature to melting temperature, and with the length of time the stress has been applied. Failure to appreciate this fact has led to confusion and erroneous arguments in some of the older literature on deformation of the earth's mantle.

Creep Diagram

The creep diagram (see Weertman & Weertman 1965, Weertman 1968) shown in Figure 1 gives roughly the conditions of temperature and stress under which each of the three principal types of creep is observed in laboratory tests. Figure 1 is a temperature and stress field plotted as homologous temperature T/T_m versus the normalized stress σ/μ, where T is temperature measured in degrees absolute, T_m is the melting temperature (or the solidus temperature), σ is the applied stress, and μ is the shear modulus.

The theoretical strength of a crystal is of the order of $\sigma \sim \mu/10$. Thus the largest normalized stress shown in the creep diagram is the value $\sigma/\mu = 1/10$. Large scale movement and multiplication of crystal dislocations occurs above a critical stress σ_c that is given by

$$\sigma_c \sim \mu b \sqrt{\rho} \qquad\qquad 1.$$

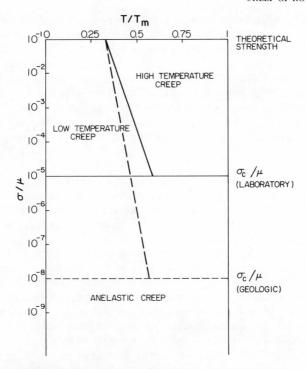

Figure 1   The creep diagram. Homologous temperature (T/T_m) and normalized stress (σ/μ) fields in which the three principal types of creep exist. (Nabarro-Herring creep is not considered in this diagram.) Solid lines stand for laboratory experiments and dashed lines for geologic strain rates. After Weertman & Weertman (1965) and Weertman (1968).

where b is the length of the Burgers vector of the crystal dislocations and ρ is the dislocation density of the material before the stress is applied. By annealing the crystalline material at a high temperature ρ is reduced in value. If the annealing time is long by human standards the factor $b\sqrt{\rho}$ is of the order of 10^{-5}. Hence, as shown in Figure 1, $\sigma_c \sim 10^{-5}\mu$. For material annealed at high temperature for periods of time that are long by geological standards the dislocation density ρ will be reduced, say, by three to six orders of magnitude. Thus a value of $\sigma_c \sim 10^{-7}\mu$ to $10^{-8}\mu$ would not be unreasonable as an estimate of the stress level above which considerable movement and multiplication of crystal dislocations occurs in rock subjected to long term stresses at high temperatures.

Low Temperature Creep

At relatively low temperatures ($T < T_m/3$), dislocations in crystals move only on their glide planes. Diffusional mass transport of atoms is required to cause movement in a direction perpendicular to the glide plane (Such movement, in a direction normal to the glide plane, is called dislocation climb.) At low temperatures diffusion processes take place at such slow rates that the dislocation climb can be ignored. If a crystal is stressed at low temperatures at a level greater than that given by Equation 1, and if mechanisms such as the Peierls stress are too weak to hold up the dislocations, dislocation multiplication and movement on the slip planes will occur on a large scale. As the dislocation density increases, the interactions between the dislocations make it more difficult for further dislocation movement and multiplication to take place. On rather general arguments it can be shown that in this situation (e.g. see Weertman & Weertman 1965) the strain ε is given by

$$\varepsilon = \varepsilon_e + \varepsilon_p + \varepsilon_o \log(1 + vt) \qquad\qquad 2.$$

where ε_e is the elastic and ε_p is the plastic strain produced immediately upon application of stress, t is time measured from the moment of application of the stress, and ε_o and v are, to a relatively weak extent, temperature- and stress-dependent constants. The logarithmic creep law appears to have been first proposed by Phillips (see Sully 1949) near the turn of the century.

Equation 2 shows that at low temperatures the creep strain obeys a logarithmic time dependence. This type of time variation is well verified by laboratory experiments. The value of the constant ε_o is small ($\varepsilon_o \approx 0.01$), and thus even after a geologic time interval the amount of creep strain is relatively small in magnitude. Lomnitz (1956) first demonstrated that at a relatively low temperature, the creep of rock (granite) follows the logarithmic creep equation. In geological and geophysical literature Equation 2 is thus known as the Lomnitz creep equation. As is well known, Jeffreys (1972) has persistently used Lomnitz's creep law to argue against the existence of convection in the earth's mantle. Runcorn (1972) has pointed out that Jeffrey's use of the logarithmic creep equation misapplies a low temperature creep law to a high temperature situation.

Misra & Murrell (1965) have carried out an extensive series of transient creep tests on many rock types (anhydrite, dolomite, sandstone, marble, microgranodiorite, and peridotite). Their tests show that at relatively low temperatures (< 200–$300°C$), the creep curves obey Equation 2. Figure 2 shows a plot of some of their data on microgranodiorite. At higher temperatures the transient creep curves that they measured no longer follow the logarithmic equations. Instead, as was to be expected, the transient creep curve is described by the function βt^m of Equation 3 in the next section.

High Temperature Creep

At higher temperatures ($T > T_m/3$) and at stresses greater than σ_c, the number of degrees of freedom of dislocation movement is increased by one because the dislocations can climb at appreciable rates. As a consequence, dislocations of opposite character are free to move toward each other and annihilate each other. This recovery process offsets the hardening that causes the low temperature creep to progress at ever decreasing rates. (Note: the creep rate, $d\varepsilon/dt$, predicted by Equation 2 is equal to $\varepsilon_o v/(1 + vt) \approx \varepsilon_o/t$.) The amount of creep strain is unlimited at high temperatures. As fast as dislocations are annihilated more dislocations are created

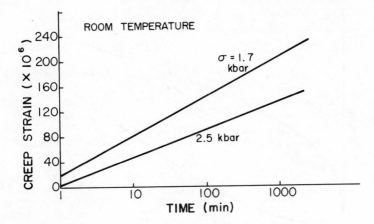

Figure 2 Creep strain vs log-time plot of creep of a microgranodiorite. Data from Misra & Murrell (1965).

at the dislocation sources. Thus, after the stress has been applied for some time, the creep rate $\dot{\varepsilon} = d\varepsilon/dt$ settles down to a value that does not vary with time.

Before a steady-state creep rate is reached in high temperature creep, a period of transient creep is observed in which the creep rate varies with time. In material with a relatively high initial dislocation density and in which no mechanisms exist that appreciably hinder the motion of dislocations on their slip planes, the transient portion of the curve of strain versus time is characterized by a decelerating strain rate. Strain in this region of the curve is given by the Andrade equation

$$\varepsilon = \beta t^m + \dot{\varepsilon}_s t \qquad\qquad 3.$$

where $\dot{\varepsilon}_s$ is the steady state creep rate, β is a constant, and m is another constant whose value ranges between about 1/3 and 2/3. We have dropped out of Equation 3, as we shall do in all the equations to follow, the elastic strain ε_e and the instantaneous plastic strain ε_p. The strain ε hereafter represents the time-dependent creep strain and not the total strain.

In material with a low initial dislocation density the transient portion of the creep curve is one of accelerating flow, if mechanisms exist that slow down the glide motion of dislocations across their slip planes. (Because the initial dislocation density is small, such mechanisms must rely on something other than dislocation interactions.) Figure 3 shows schematically a creep curve with accelerating transient creep. This figure also contains a creep curve with decelerating transient creep. The accelerating transient creep is produced during the period when the dislocation density increases from its initial low value to a much greater, steady-state value (Alexander & Haasen 1968). It is curious that the transient creep curves of ice can be of either type, depending on the orientation of the ice crystal with respect to the stress axis (Weertman 1973).

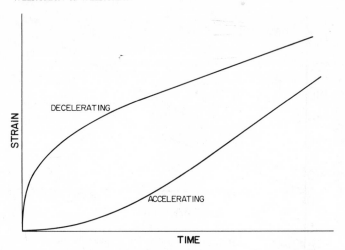

Figure 3   Schematic creep curves of creep strain vs time.

Anelastic Creep

The third type of creep indicated in the creep diagram is anelastic creep. This type of creep is recoverable. If the stress is removed the creep strain decreases in value until it is equal to zero. Under static loading the anelastic creep strain usually is given by the equation

$$\varepsilon = \sum_n \{(\varepsilon_n \sigma/\mu)[1 - \exp(-t/t_n)]\} \qquad 4.$$

where ε_n and t_n are constants. There are numerous mechanisms that do not involve dislocations with relaxation times t_n that can account for Equation 4. Jackson & Anderson (1970) have recently given an extensive review of these mechanisms as applied to rocks. Because $\varepsilon_n < 0.1$, the anelastic creep strain is small compared with the elastic strain. Anelastic strain creep mechanisms that do not involve dislocations operate at stress levels greater than σ_c as well as below σ_c.

Anelastic creep accounts for the damping losses suffered by seismic waves that travel through the mantle. The internal friction is described as $Q^{-1} = (1/2\pi)\Delta W/W$, where W is the total amount of elastic energy stored per unit volume and ΔW is the energy dissipated per unit volume per stress cycle. The anelastic damping is given by

$$Q^{-1} = \sum_n [\varepsilon_n f t_n/(1 + f^2 t_n^2)] \qquad 5.$$

where f is the frequency of the seismic wave.

Dislocation movement also contributes to anelastic creep. At stresses smaller than σ_c, dislocation multiplication and large-scale movement cannot take place within time periods that are short compared to the time required to anneal a dislocation structure into a lower dislocation density configuration. In this situation the only dislocation motion that occurs is a slight bowing-out of dislocation segments that are pinned either at dislocation nodes or by impurity atoms. If the temperature is low, the direction of the dislocation motion is in the slip plane. If the temperature is high, the bowing motion also can take place in the climb direction.

POWER LAW CREEP

The remainder of this paper focuses on steady-state creep. The creep rate produced by dislocation motion, whether in steady-state creep or not, is given by the well-known equation

$$\dot{\varepsilon} = \alpha_0 b\rho v \qquad\qquad 6.$$

where v is the average dislocation velocity and α_0 is a constant ($\alpha_0 \approx 1/4$). In steady-state conditions the dislocation density ρ is given by (Weertman 1974)

$$\rho = (\beta_0 \sigma/\mu b)^2 \qquad\qquad 7.$$

where $\beta_0 \approx 1$. Equation 7 is obtained on the assumption that the dislocation density will approach the value at which the average internal stress field produced by the dislocations is of the same order as the applied stress. Equation 7 is moderately well established by experiment on a number of materials.

Figure 4 shows a plot by Kohlstedt & Goetze (1974) of dislocation density data obtained on high temperature creep specimens of olivine. This figure is a log-log plot of the stress σ versus $\mu b\sqrt{\rho}$. Also shown in the figure is the theoretical curve for $\beta_0 = 1$. The data and theory agree within a factor of 2.

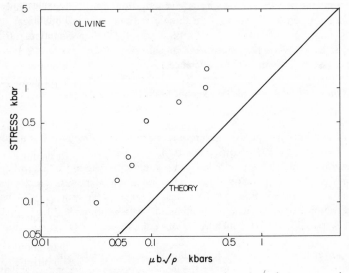

Figure 4   Log-log plot of applied stress vs the expression $\mu b\sqrt{\rho}$ for creep specimens of olivine. After Kohlstedt & Goetze (1974).

Glide-Controlled Creep

It is possible that the dislocation velocity is determined by diffusional processes controlling the glide motion of a dislocation (a number are listed in Weertman & Weertman 1965). The classic example of this kind of diffusional process is the drift of a pinning impurity atmosphere, the Cottrell cloud, with a moving dislocation. The dislocation glide velocity is given by

$$v = \gamma(D/b)(\mu\Omega/kT)(\sigma/\mu)^3 \qquad\qquad 8.$$

where Ω is the atomic volume ($\Omega \approx b^3$), k is Boltzmann's constant, D is a diffusion constant that is appropriate to the particular pinning mechanism, and γ is a dimensionless constant whose value depends on the particular mechanism that controls the glide motion. Combining Equations 6, 7, and 8 gives

$$\dot\varepsilon_s = \alpha(D/b^2)(\mu\Omega/kT)(\sigma/\mu)^3 \qquad\qquad 9.$$

for the steady-state glide-controlled creep rate. The equation for the dimensionless constant is $\alpha = \alpha_0 \beta_0^2 \gamma$. A reasonable lower limit (Weertman 1970) on the value of α is estimated to be $\alpha \approx 0.0015$.

Equation 9 is a power law creep equation. That is, the steady-state creep rate is given by an equation of the form $\dot\varepsilon_s = $ (constant) $(\sigma/\mu)^n$, where n is a constant.

Dislocation Climb-Controlled Creep

Suppose creep is controlled by dislocation climb. (We consider the dragging of jogs by screw dislocations also to be a climb process.) In the absence of ad hoc assumptions all dislocation climb theories of steady-state creep (Weertman 1968, 1974) predict that the dislocation velocity is given by Equation 8 and the steady-state creep rate is given by Equation 9. In these equations D is the coefficient of self-diffusion and the constant $\alpha \approx 0.01$ to 0.1. These dislocation climb creep theories include those of Nabarro, Ivanov & Yanushkevich; Blum, Barrett & Nix; and ourselves (see Weertman 1974 for references).

From the ad hoc assumption that the number of dislocation sources does not depend on stress, it is possible (Weertman 1968) to derive the creep equation

$$\dot\varepsilon_s = \alpha'(D/b^{7/2}M^{1/2})(\mu\Omega/kT)(\sigma/\mu)^{9/2} \qquad\qquad 10.$$

where M is the number of dislocation sources per unit volume and α' is a constant whose value is in the range $0.015 < \alpha' < 0.33$.

Equations 9 and 10 are power law creep equations. The most "natural" power for a power law creep equation is 3. The derivation of powers greater than 3, such as the power 4.5 in Equation 10 as well as power 6 (Weertman 1968), has not been accomplished so far without making ad hoc assumptions. (Many crystalline materials, particularly pure metals, do obey a power law creep equation with $n = 4.5$ to 5.5)

The power law breaks down for metals at stresses greater than about $\mu/1000$. Garofalo (1965) has shown that over the whole stress range the empirical equation

$$\dot\varepsilon_s = AD\{\sin h(\sigma/\sigma_0)\}^n \qquad\qquad 11.$$

describes the creep data well. In this equation A is a constant and $\sigma_0 \sim \mu/1000$. The theory which led to Equation 10 can explain the breakdown of power law creep (Weertman 1974). (The stress level at which power law breakdown occurs is larger than the magnitude of the deviator stresses presumed to exist in the earth's mantle.) If creep is glide-controlled, the power law will break down at stresses that are sufficiently large to tear the dislocations away from impurity pinning atmospheres or from whatever entity prevents the dislocations from gliding freely on their slip planes.

If the steady-state creep is controlled by dislocation climb (or by the subgrain Nabarro-Herring mechanism discussed in the next section), and if the power n of power law creep is equal to 3, power law creep should not break down until the stress level is close to the theoretical strength ($\sigma > \mu/10$) of the material (Weertman 1974).

NABARRO-HERRING CREEP

Dislocation motion is not essential to produce steady-state creep. Steady-state creep can be produced directly by the mass transport of atoms by diffusional processes from one grain boundary to another. Nabarro (1948) proposed this mechanism and Herring (1950) made a complete analysis of it. The steady-state creep equation that can be derived from this mechanism is

$$\dot{\varepsilon}_s = \alpha^*(D/L^2)(\sigma\Omega/kT) \qquad\qquad 12.$$

where L is the average grain diameter and α^* is a dimensionless constant ($\alpha^* \approx 5$). The term D again is the coefficient of self-diffusion. Equation 12 predicts that the creep rate is proportional to the stress.

If subgrains exist within the grains the term L in Equation 12 should be set equal to the subgrain size. Experimentally it is observed for metals that the subgrain size L is given by

$$L = L_0(\mu/\sigma) \qquad\qquad 13.$$

where the constant $L_0 \approx 5 \times 10^{-9}$ m. [In Weertman (1968) the value of L_0 is given erroneously as equal to 5×10^{-6} cm rather than 5×10^{-6} mm. The creep rate predicted from the subgrain mechanism in this reference should be increased by a factor of 100.] If Equations 12 and 13 are combined, Equation 9 again is obtained; in this case $\alpha = \alpha^* b^2/L_0^2 \approx 0.013$. Thus the Nabarro-Herring equation when applied to subgrains and the creep equation derived from dislocation motion are essentially the same, and they both predict approximately the same creep rates.

Raleigh & Kirby (1970) have suggested that Equation 13 be used to estimate the stress that naturally deformed rock was subjected to before it reached the earth's surface. Their measurements on experimentally deformed olivine give a value of L_0 of $\approx 5 \times 10^{-7}$ m, a value 100 times larger than is observed for metals. If $L_0 = 5 \times 10^{-7}$ m for rock, the subgrain Nabarro-Herring creep mechanism gives a creep rate that is about two orders of magnitude smaller than the creep rate of the mechanisms involving dislocation climb.

The Nabarro-Herring creep mechanism is based on the mass transport of atoms through the grains. Coble (1963) has suggested that mass transport also can take place along the grain boundaries through grain boundary diffusion. The effect of grain boundary diffusion can be taken into account (Stocker & Ashby 1973) in Equation 12 by setting $D = D_V[1 + (\pi w/L)(D_B/D_V)]$, where w is the effective thickness of a grain boundary for diffusion, D_V is the volume diffusion coefficient, and D_B is the grain boundary diffusion coefficient.

If a fluid phase is present in the rock, mass transport can occur through this phase (Stocker & Ashby 1973, Elliott 1973), producing an enhanced creep rate.

TEMPERATURE AND PRESSURE DEPENDENCE

Temperature and pressure affect the creep rate primarily through the diffusion coefficient D. In material consisting of only one atomic species the diffusion coefficient is given by

$$D = D_0 \exp(-Q/kT) \exp(-PV/kT) \qquad 14.$$

where Q is the activation energy for diffusion, P is the hydrostatic pressure (a positive quantity in compression), V is the activation volume for diffusion, and D_0 is a constant whose value is of the order of 10^{-5} to 10^{-2} m^2/sec.

It has been established empirically (Sherby & Simnad 1961, Shewmon 1963) for tests carried out at atmospheric pressure that $Q/k = gT_m$, where g is a dimensionless constant whose value varies with crystal structure. For metals g equals 15 to 25. We suggested (Weertman 1970) that this relationship be generalized to

$$(Q + PV)/k = gT_m \qquad 15.$$

where T_m is now the melting temperature under the pressure P. Equation 15 implies that at moderate pressures $V = (g/k)(dT_m/dP)$. This relationship is satisfied approximately for the limited data that are available (Weertman 1970). Equation 15 is useful for estimating the creep properties of the deep mantle. At best, laboratory experiments can measure Q and V only at relatively moderate pressures. At very high pressures the quantities Q and V are not constants but are functions of P.

In material made up of more than one atom species the diffusion coefficient D is replaced by a suitably averaged quantity of the diffusion coefficients of the different species. Herring's (1950) expression for D can be approximated (Stocker & Ashby 1973, Weertman 1968; in Equation 4b of the latter reference the terms D_i and X_j in the denominator should be interchanged) as:

$$D = D_1 D_2 D_3 \dots D_n/(X_1 D_2 D_3 \dots D_n + D_1 X_2 D_3 \dots D_n + \cdots + D_1 D_2 D_3 \dots X_n) \qquad 16.$$

where D_1, D_2, etc, are the diffusion coefficients, and X_1, X_2, etc, are the fractional concentrations of the different species $(X_1 + X_2 + \cdots + X_n = 1)$.

Goetze & Kohlstedt (1973) have performed a very useful experiment for determining the value of D for dislocation climb in olivine. By directly measuring the climb velocity through transmission electron microscope observations, they were able to obtain approximate estimates of D_0 and Q. Their values are listed in Table 1.

Table 1 Diffusion constant data for olivine[a]

D_0	Q	g
3 $m^2 s^{-1}$	135 kcal/mole	31

[a] Data of Goetze & Kohlstedt (1973) obtained by the measurement of the climb of dislocations in olivine. The value of g is calculated taking $T_m = 2160°K$.

These data undoubtedly represent values corresponding to the self-diffusion of oxygen atoms in the olivine lattice. Misener (1972) has measured the metallic diffusion in forsterite-fayalite diffusion couples [forsterite ($Mg_2 Sio_4$) has a melting temperature of 1890°C; fayalite ($Fe_2 Sio_4$) has a melting temperature of 1200°C] at temperatures in the range of 900–1100°C. Misener found an activation energy of 47 ± 4 kcal/mole and a value of $D_0 = 3.4 \times 10^{-7}$ $m^2 s^{-1}$ for the interdiffusion of the metal atoms. Buening & Buseck (1973) have measured the Fe-Mg inter-diffusion coefficients in single crystals of olivine (San Carlos). At temperatures below 1125°C the values of the diffusion constants for the metal ions are approximately $D_0 \approx 3.7 \times 10^{-11}$ $m^2 s^{-1}$ and $Q \approx 29$ kcal/mole. At temperatures above 1125°C the constants have values of approximately $D_0 \approx 1.7 \times 10^{-6}$ $m^2 s^{-1}$ and $Q \approx 58$ kcal/mole. It appears unlikely that the climb of dislocations in olivine is controlled by the diffusion of the metal atoms.

CREEP MEASUREMENTS ON IGNEOUS ROCKS

This section summarizes recent results from laboratory experiments on igneous rocks. Only igneous rocks are considered in this article because of the obvious application of results obtained from them to the problem of the flow of the earth's mantle. [For the reader interested in other kinds of rocks we point out that $\dot{\varepsilon}_s$ for ice is given by a power law creep equation with power $n \approx 3$. The value of α required to fit the data of ice to Equation 9 ($\alpha \approx 0.3$) is consistent with the dislocation climb or the subgrain Nabarro-Herring creep mechanism. Heard & Raleigh (1972) find that Yule marble obeys a power law creep equation with $n \approx 8$. The stresses used in their experiments are in the region where power law creep should break down. Carter & Heard (1970) obtained similar results on halite (NaCl) and found $n \approx 7$ to 8. Later work of Heard (1972) on halite gave $n = 5.5 \pm 0.4$. This latter value of n is in the range ($n \approx 4.5$) observed for pure metals.]

Recent published data on the high temperature creep properties of igneous rocks are compatible with the creep equation

$$\dot{\varepsilon}_s = C\sigma^n \exp(-gT_m/T) = C\sigma^n \exp(-Q/kT) \qquad 17.$$

where C is a constant. (The experiments were carried out under a confining pressure. It is to be understood that the stress σ in Equation 17 represents the quantity $\sigma_1 - \sigma_3$, where σ_1 and σ_3 are principal stress components.) Table 2 lists experimentally determined values of the terms C, n, g, and Q. Also listed in this table

Table 2 Experimental values of power law creep equation parameters

Rock or Mineral	n	g	Q (kcal/mole)[a]	C (s^{-1}kbar^{-n})	$\dot{\varepsilon}_s \exp(Q/kT)$ or $\dot{\varepsilon}_s \exp(gT_m/T)$ (at $\sigma = 10$ bars) s^{-1}	Temperature Range (°K)	Stress Range (kbars)[b]	Reference
Dunite[c]								
wet	3.18 ±0.18	31.8	93.1 ±2.5	3.4×10^8	150	1000 to 1600	1 to 10	Post & Griggs (1973), Post (1973)
dry			130			1000 to 1600	1 to 10	Post & Griggs (1973), Post (1973)
wet	2.4 ±0.2		79.9 ±7.5	6.2×10^6	98	1200 to 1600	1 to 10	Carter & Ave'Lallemant (1970)
dry	4.8 ±0.4		119.8 ±16.6	1.2×10^{10}	30	1200 to 1600	1 to 10	Carter & Ave'Lallemant (1970)
Lherzolite[d]								
wet	2.3 ±0.3		79.8 ±9.4	3.2×10^7	800	1200 to 1500	0.6 to 4	Carter & Ave'Lallemant (1970)
Olivine[e] (dry)	~3	29	125±5	4.2×10^{11}	4.2×10^5	1700 to 2000	0.05 to 1.5	Kohlstedt & Goetze (1974)

Material	n					T	$\dot{\varepsilon}$ range	Reference
Olivine (dry)[f]	3					1373	3 to 10	Kirby & Raleigh (1973)
Lherzolite[g]	5	26.4	106 ±20	10^8	0.01	1300 to 1400	5 to 10	Raleigh & Kirby (1970)
Theory[h]	3			2.6×10^{12}	2.6×10^6			
Quartz[h]	3.64 ±0.18	31.6 ±3.5				520 to 720	0.5 to 14	Balderman (1974)

[a] 1 kcal/mole = 4.19 kJ mole^{-1}.

[b] 1 kbar = 10^9 dyne/cm^2 = 100 MN m^{-2}.

[c] Mt. Burnett dunite of average grain size of 1 mm. Composition is 98% olivine $[(Mg_{0.92}Fe_{0.08})_2 SiO_4]$, 1% diopside, 1% chromite (Carter & Ave'Lallemant 1970, Goetz & Brace 1972).

[d] Lherzolite from Australia of average grain size of 0.5 mm. Composition is 60 to 70% olivine $[(Mg_{0.92}Fe_{0.08}) SiO_4]$, 20-30% enstatite, 5-10% diopside, <1% spinel.

[e] Single crystals $(Mg_{0.92}Fe_{0.08})_2 SiO_4$ from San Carlos (Arizona) peridotite. Value of g calculated by taking $T_m = 2160°K$.

[f] From Mt. Burnett dunite.

[g] Composition not given.

[h] Calculated from Equation 9 using $\alpha = 0.3$ (experimental value found for ice), D_0 and Q from Table 1, $\Omega \approx b^3/2$, $b = 6.98$ nm and $\mu = 791$ kbar [values of b and μ compiled by Stocker & Ashby (1973)]. In Table 2 and the Figures, the strain rate $\dot{\varepsilon}_s$ and the stress σ are considered to be those measured in uniaxial compression or tension tests. For shear tests, the shear stress and strain rate are changed to the equivalent uniaxial stress and strain rate by use of the expressions $\sigma = 3^{1/2}\sigma_{sh}$, where σ_{sh} is the shear stress, and $\dot{\varepsilon}_s = 2\dot{\varepsilon}_t/3^{1/2} = \dot{\varepsilon}_e/3^{1/2}$, where $\dot{\varepsilon}_t$ is the tensor shear strain rate and $\dot{\varepsilon}_e$ is the engineering shear strain rate.

are values of the quantity $\dot{\varepsilon}_s \exp(Q/kT)$ or $\dot{\varepsilon}_s \exp(gT_m/T)$ calculated for the stress $\sigma = 10$ bars (1 MN m^{-2}).

Figure 5 shows a log-log plot of creep rate vs stress, from data of Kohlstedt & Goetze (1974) on olivine. The creep rates were normalized to a temperature of 1673°K (1400°C) through the relationship $\dot{\varepsilon}_{s1} \exp(Q/kT_1) = \dot{\varepsilon}_{s2} \exp(Q/kT_2)$, where $T_1 = 1673°$K, T_2 is the test temperature, and $\dot{\varepsilon}_s$ is the measured creep rate at the test temperature. The value of Q is that listed in Table 2. A straight line with slope of three has been drawn through the data points. In Figure 6 the same plot is made without the actual data points being shown. In addition Figure 6 shows results of

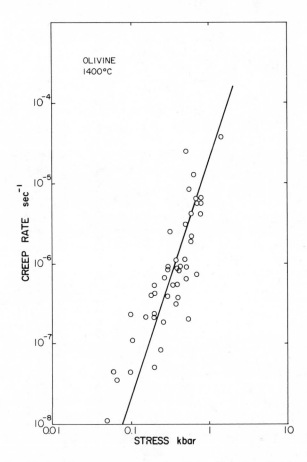

Figure 5 Log-log plot of creep rate vs stress for olivine. Data from Kohlstedt & Goetze (1974). Creep rates taken at different temperatures are normalized to a creep rate at 1400°C through use of a creep activation energy of 125 kcal/mole.

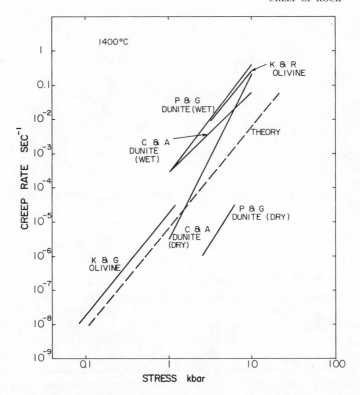

Figure 6   Log-log plot of creep rate vs stress for olivine and dunite. Creep rates normalized to a creep rate at 1400°C through use of creep activation energies listed in Table 2. K&G: Kohlstedt & Goetze (1974), C&A: Carter & Ave'Lallemant (1970), K&R: Kirby & Raleigh (1973) (we have assumed that $Q = 125$ kcal for these data), Theoretical: calculated with data listed in Table 1 and footnote h of Table 2, P&G: Post & Griggs (1973) and Post (1973).

the experiments of Carter & Ave'Lallemant (1970), Post & Griggs (1973), and Post (1973) on dunite and Kirby & Raleigh (1973) on olivine. The values of Q determined by the different investigators were used to normalize the value of $\dot{\varepsilon}_s$ of their data in Figure 6 to a creep rate appropriate for a temperature of 1400°C.

The creep rate predicted by Equation 9 for $\alpha = 0.3$ and the value of D given in Table 1 also is plotted in Figure 6. The theoretical and observed creep rates agree to within about an order of magnitude. The power n of the theoretical curve in Figure 6 is equal to three. A third power has been observed in the high temperature creep of a number of metal oxides (reviewed in Cannon & Sherby 1973 and Kirby & Raleigh 1973).

High temperature transient creep data of Goetze & Brace (1972) and of Goetze

Table 3   High temperature transient creep parameters[a]

Rock	m	n	Q (kcal/mole)	C_0[b]
Westerly Granite	0.49	3.5	75–80	2.9×10^4
San Marcos Gabbro	0.44	4.1	90–100	560
Maryland Diabase	0.35	5.1	80–90	63
Mt. Albert Peridotite	0.33	6	—	—

[a] Data of Goetze & Brace (1972) and Goetze (1971).
[b] In units of $s^{-m}\text{kbar}^{-mn}$.

(1971) are listed in Table 3. The parameters listed in this table are for the transient creep equation

$$\varepsilon = C_0 \{ t\sigma^n \exp(-Q/kT) \}^m \qquad\qquad 18.$$

where C_0 is a constant.

Hydrolytic Weakening

Griggs & Blacic (1965) discovered that a small amount of water in quartz greatly reduces the high temperature strength of this mineral. The existence of this water-weakening effect has been confirmed for quartz and a number of other silicate minerals (Griggs 1967, Blacic 1972, Riecker & Rooney 1969, Hobbs, McLaren & Paterson 1972, Balderman 1974, Baëta & Ashbee 1970a, Heard & Carter 1968). Only a few tenths of a percent or less by weight of water is necessary to cause weakening. The effect is observed in olivine or dunite as an increase in the high temperature creep rate and a decrease in the value of the activation energy (see Figure 6 and Table 2; Blacic 1972, Griggs 1967, Post & Griggs 1973, Carter & Ave'Lallemant 1970).

´ F. C. Frank and Griggs (see Griggs 1967) proposed a dislocation double kink catalysis mechanism to explain the water weakening. Their mechanism is similar to a double kink catalysis mechanism that has been used to explain solid solution softening of certain metal alloys (Urakami & Fine 1972, Arsenault 1967, 1969, Weertman 1958). Water at a site on a dislocation line is assumed to hydrolyze a silicon-oxygen bridge by the reaction

$$\text{Si–O–Si} + H_2O \rightarrow \text{SiOH} \cdot \text{HO–Si}$$

The bond between the two hydrogen atoms is weaker by an order of magnitude than the Si–O or Si–OH bonds. A dislocation kink can be created which moves where Si–O–Si bridges have been hydrolyzed. In the absence of the water the Peierls stress is so high that dislocations cannot move easily on their slip planes. This theory has been extended and fitted into the Alexander & Haasen (1968) theory of accelerating transient creep and upper yield point effect (Griggs 1974, Hobbs, McLaren & Paterson 1972, Balderman 1974).

No studies, experimental or theoretical, seem to have been made on the effect the presence of water might have on the diffusion constant D. If the value of D is changed, the creep rate will be changed. Insufficient experimental evidence exists according to Clauer, Seltzer & Wilcox (1971) and Pascoe & Hay (1973) to assess clearly whether deviations from stoichiometry cause large changes in diffusional (Nabarro-Herring) creep rates in oxides. If large changes are produced in diffusional creep, they are to be expected in dislocation climb creep. In lead sulfide, deviations from stoichiometry (Seltzer & Duga 1966, Seltzer 1967) do produce changes in the creep rate. Recent results of Burton & Reynolds (1973) indicate that a large degree of nonstoichiometry can increase the creep rate of uranium dioxide by about a factor of 50. The theory of Pascoe & Hay (1973) indicates that significant changes in the creep rates of oxides can be produced by nonstoichiometry. Perhaps the effect of water on the creep rate of silicates can be explained through a non-stoichiometry effect.

The existence of water within a silicate lattice certainly lowers the melting point of the material. An increase in creep rate can be explained, at least qualitatively, through this decrease of T_m.

Balderman (1974) determined that (wet) quartz follows Equation 17, the power law creep equation, where n and Q have the values $n = 3.64 \pm 0.18$ and $Q = 31.6 \pm 3.5$ kcal/mole, in the stress range of 0.5 to 14 kbar and the temperature region of 250 to 500°C. He points out that the measured value of Q agrees with the value of $Q = 31.9$ kcal/mole for the self-diffusion of oxygen in quartz measured at temperatures below 900°C (unpublished data of R. Haul cited by Baëta & Ashbee 1970b). (Above 900°C Haul measured a value of $Q = 51.6$ kcal/mole.) Because the value of D_0 for the self-diffusion of oxygen has not been published, it is impossible to compare the measured creep rate with values predicted from dislocation climb creep theories.

MANTLE VISCOSITY

The main purpose for carrying out creep experiments on rocks is to obtain data that can be applied to the deformation of the earth's mantle and crust. We would like to know what the creep rate of mantle rock would be at a given depth within the earth if the deviator stress is specified to have some particular value. Estimates of the creep rate as a function of depth are given in Kirby & Raleigh (1973), Stocker & Ashby (1973), and Weertman (1970).

Ashby (1969) has suggested a useful method for plotting theoretical estimates of steady-state creep rates. On a diagram of homologous temperature T/T_m versus normalized stress σ/μ, he plotted contours of constant strain rate $\dot{\varepsilon}_s$ calculated from the rate-controlling mechanism and called these plots deformation-mechanism maps. Stocker & Ashby (1973) used them to describe the estimated creep properties of the upper mantle.

In Figure 7 we have made a deformation-mechanism map for olivine (dunite) using the values (taken from the data of Kohlstedt & Goetze) of $g = 29$, $n = 3$, $C = 4.2 \times 10^{11}$ $s^{-1}kbar^{-3}$, $D_0 = 0.15$ m^2 s^{-1}. (This value of D_0 gives the same value of D at $T = 1673°$K as the diffusion constant given in Table 1 if $g = 29$

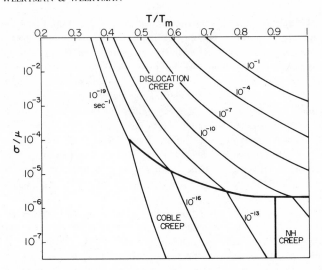

Figure 7   Deformation map for olivine showing contours of constant strain rate $\dot{\varepsilon}_s$. The grain size is taken to be 1 mm.

and $Q = 125$ kcal/mole.) For Coble creep the value of D_B was estimated by reducing the value of g by a factor of 2/3. The grain size is taken to be 1 mm. If the grain size is increased in value to 0.1 m, the creep rate predicted by the Nabarro-Herring mechanism and the Coble mechanism is decreased by a factor of 10^4.

The effective viscosity η of a material always can be defined by the equation

$$\sigma = 2\eta\dot{\varepsilon}_s \qquad\qquad\qquad\qquad 19.$$

For a Newtonian solid the viscosity η is independent of stress and strain rate. For a material that obeys a power law creep equation, the value of the effective viscosity η depends on the stress or the strain rate.

Figure 8 plots the effective viscosity of the earth's mantle as a function of depth calculated from Figure 7 for a constant strain rate of $\dot{\varepsilon}_s = 10^{-14}\,\text{s}^{-1}$. Figure 9 contains a similar plot, but the effective viscosity is calculated for a constant stress of $\sigma = 10$ bars (1 MN m^{-2}). Effective viscosities of the Moon, Mars, and Venus are also shown in Figures 8 and 9. The effective viscosity curves are calculated with the assumption that the grain size is so large that Nabarro-Herring or Coble creep can be ignored. Dislocation climb creep is assumed to be the dominant creep mechanism, and the creep parameters used to calculate the creep rates in Figure 7 are used again to calculate the effective viscosity curves. The estimated values of T_m/T as a function of depth that were used in Weertman (1970) are used to obtain the curves in Figures 8 and 9.

The effective viscosities given in Figures 8 and 9 are uncertain because the melting temperature profile and the actual temperature profile of these terrestrial planets are unknown. It also is not known how well the creep properties of olivine or dunite approximate the creep properties of the earth's mantle or of the rock in the interior of the moon, Mars, and Venus. Nevertheless, we believe one can conclude from the plots of Figures 7–9 that both Mars and the moon have thick, rigid outer crusts. (Results from the Apollo program have, of course, given direct evidence that the moon has a thick, rigid outer crust.) For the earth, there seems to be no doubt that the upper mantle can convect, and probably the lower mantle as well (as can both the interiors of the moon and of Mars). The rock of even the outer crust of Venus is so hot that appreciable high temperature creep should occur. An estimate of the height of craters or mountains on Venus would be extremely useful for calibrating the curves of Figures 7–9. (That is, an estimate of the stress in the crust can be made from the height of craters or mountains. The surface temperature of Venus is known. A lower bound on the creep rate of the outer crust can be made by assuming that a crater has persisted for 4.6 eons.)

The effective viscosity curves for the earth of Figures 8 and 9 show a minimum at a depth of the order of 200 km. [Gordon (1965, 1967) first predicted this minimum through use of the Nabarro-Herring creep mechanism.] This result is in

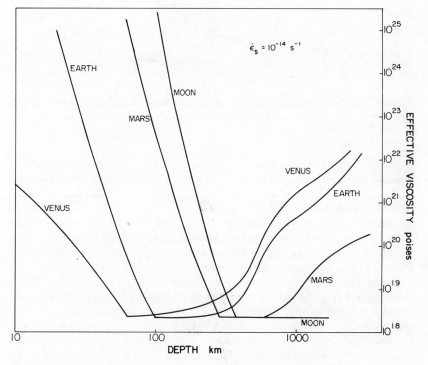

Figure 8   Effective viscosity vs depth for constant strain rate of $\dot{\varepsilon}_s = 10^{-14}\ s^{-1}$. (Note: 1 poise = 0.1 $N\ s^{-1}\ m^{-2}$.) Calculated from Equation 17 assuming that $g = 29$, $n = 3$, and $C = 4.2 \times 10^{11}\ s^{-1}\ kbar^{-3}$.

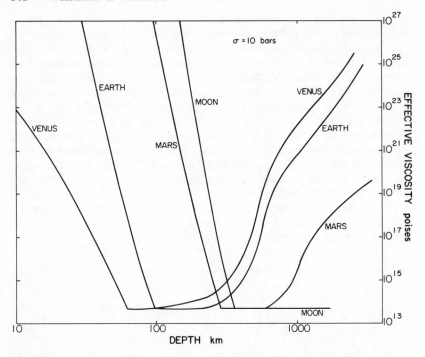

Figure 9 Effective viscosity vs depth for constant stress $\sigma = 10$ bars (1 $MN\ m^{-2}$). (Note: 1 poise $= 0.1\ N\ s^{-1}\ m^{-2}$.) Calculated from Equation 17 assuming that $g = 29$, $n = 3$, and $C = 4.2 \times 10^{11}\ s^{-1}\ kbar^{-3}$.

qualitative agreement with the estimates of the earth's viscosity based on isostatic rebound after the removal of ice age ice sheets and ice age lakes (O'Connell 1971, Walcott 1972, 1973, McConnell 1965, 1968, Chappell 1974, Christoffel & Calhaem 1973). The minimum value of the viscosity given in Figures 8 and 9 is considerably smaller than the usual value estimated from rebound data ($\eta \sim 10^{20}$ poises). Recent analyses by Post & Griggs (1973) and Brennen (1974) have led them to conclude that the postglacial rebound data are consistent with the rock of the upper mantle obeying a power law creep equation with $n \approx 3$.

An argument has been made (MacDonald 1963, 1966) that the viscosity of the lower mantle must be high ($\eta > 10^{25}$ poises) because the shape of the present figure of the earth is appropriate for an earth in hydrostatic equilibrium, spinning with the faster rotational velocity the earth had 10 million years ago. Dicke (1969), however has concluded from the time required for relaxation of second-order harmonic distortions of the earth that the lower mantle has a viscosity of the order of $\eta \sim 10^{22}$ poises. Goldreich & Toomre (1969) have concluded that in order to explain polar wandering, the value of the viscosity of the lower mantle must be much smaller than 10^{25} poises.

ACKNOWLEDGMENT

This work was supported by the National Science Foundation through the Northwestern University Materials Research Center.

Literature Cited

Alexander, H., Haasen, P. 1968. *Solid State Phys.* 22:28–158

Arsenault, R. J. 1967. The double-kink model for low-temperature deformation of b.c.c. metals and solid solutions. *Acta Met.* 15:501–11

Arsenault, R. J. 1969. Solid solution strengthening and weakening of b.c.c. solid solutions. *Acta Met.* 17:1291–97

Ashby, M. F. 1969. A first report on deformation-mechanism maps. *Acta Met.* 20:887–97

Baëta, R. D., Ashbee, K. H. G. 1970a. Mechanical deformation of quartz, 1, constant strain-rate compression experiments. *Phil. Mag.* 22:601–23

Baëta, R. D., Ashbee, K. H. G. 1970b. Mechanical deformation of quartz II. stress relaxation and thermal activation parameters. *Phil. Mag.* 22:624–35

Balderman, M. A. 1974. The effect of strain-rate and temperature on the yield point of hydrolytically weakened synthetic quartz. *J. Geophys. Res.* 79:1647–52

Blacic, J. D. 1972. *Flow and Fracture of Rocks*, ed. H. C. Heard, I. Y. Borg, N. L. Carter, C. B. Raleigh, 109–15. Washington DC: Am. Geophys. Union. 352 pp.

Brennan, C. 1974. Isostatic recovery and the strain rate dependent viscosity of the earth's mantle. *J. Geophys. Res.* 79:3993–4001

Buening, D. K., Buseck, P. R. 1973. Fe-Mg lattice diffusion in olivine. *J. Geophys. Res.* 78:6852–62

Burton, B., Reynolds, G. L. 1973. The influence of deviations from stoichiometric composition on the diffusional creep of uranium dioxide. *Acta Met.* 21:1641–47

Cannon, W. R., Sherby, O. D. 1973. Third-power stress dependence in creep of polycrystalline nonmetals. *J. Am. Ceram. Soc.* 56:157–60

Carter, N. L., Ave'Lallemant, H. G. 1970. High temperature flow in dunite and peridotite. *Geol. Soc. Am. Bull.* 81:2181–2202

Carter, N. L., Heard, H. C. 1970. Temperature and rate dependent deformation of halite. *Am. J. Sci.* 269:193–249

Chappell, J. 1974. Upper mantle rheology in a tectonic region: evidence from New Guinea. *J. Geophys. Res.* 79:390–98

Christoffel, D. A., Calhaem, I. M. 1973.

Upper mantle viscosity determined from Stoke's law. *Nature* 243:51–52

Clauer, A. H., Seltzer, M. S., Wilcox, B. A. 1971. Ceramics in Severe Environments, *Mater. Sci. Res.* 5:361–84

Coble, R. L. 1963. A model for boundary diffusion controlled creep in polycrystalline materials. *J. Appl. Phys.* 34:1679–82

Dicke, R. H. 1969. Average acceleration of the earth's rotation and the viscosity of the deep mantle. *J. Geophys. Res.* 74:5895–5902

Elliott, D. 1973. Diffusion flow laws in metamorphic rocks. *Geol. Soc. Am. Bull.* 84:2645–64

Garofalo, F. 1965. *Fundamentals of Creep and Creep-Rupture in Metals.* New York: Macmillan, 258 pp.

Goetze, C. 1971. High temperature rheology of Westerly granite. *J. Geophys. Res.* 76:1223–30

Goetze, C., Brace, W. F. 1972. Laboratory observations of high-temperature rheology of rocks. *Tectonophysics.* 13:583–600

Goetze, C., Kohlstedt, D. L. 1973. Laboratory study of dislocation climb and diffusion in olivine. *J. Geophys. Res.* 78:5961–71

Goldreich, P., Toomre, A. 1969. Some remarks on polar wandering. *J. Geophys. Res.* 74:2555–67

Gordon, R. B. 1965. Diffusion creep in the earth's mantle. *J. Geophys. Res.* 70:2413–18

Gordon, R. B. 1967. Thermally activated processes in the earth: creep and seismic attention. *Geophys. J.* 14:33–43

Griggs, D. T. 1967. Hydrolytic weakening of quartz and other silicates. *Geophys. J.* 14:19–31

Griggs, D. 1974. A model of hydrolytic weakening in quartz. *J. Geophys. Res.* 79:1653–61

Griggs, D. T., Blacic, J. D. 1965. Quartz: anomalous weakness of synthetic crystals. *Science* 147:292–95

Griggs, D. T., Handin, J., Eds. 1960. Rock Deformation. *Geol. Soc. Am. Mem.* 79. Boulder, Colo.: Geol. Soc. Am. 382 pp.

Heard, H. C. 1972. *Flow and Fracture of Rocks*, ed. H. C. Heard, I. Y. Borg, N. L. Carter, C. B. Raleigh, 191–209. Washington DC: Am. Geophys. Union. 352 pp.

Heard, H. C., Borg, L. Y., Carter, N. L., Raleigh, C. B., Eds. 1972. *Flow and Fracture of Rocks.* Washington DC: Am. Geophys. Union. 352 pp.

Heard, H. C., Carter, N. L. 1968. Experimentally induced "natural" intragranular flow in quartz and quartzite. *Am. J. Sci.* 266:1–42

Heard, H. C., Raleigh, C. B. 1972. Steady-state flow in marble at 500°C to 800°C. *Geol. Soc. Am. Bull.* 83:935–56

Herring, C. 1950. Diffusional viscosity of a polycrystalline solid. *J. Appl. Phys.* 21:437–45

Hobbs, B. E., McLaren, A. C., Paterson, M. S. 1972. *Flow and Fracture of Rocks,* ed. H. C. Heard, I. Y. Borg, N. L. Carter, C. B. Raleigh, 29–53. Washington DC: Am. Geophys. Union. 352 pp.

Jackson, D. D., Anderson, D. L. 1970. Physical mechanisms of seismic wave attenuation. *Rev. Geophys. Space Phys.* 8:1–63

Jeffreys, H. 1972. Creep in the earth and planets. *Tectonophysics* 13:569–81

Kennedy, A. J. 1963. *Processes of Creep and Fatigue in Metals.* New York: Wiley. 480 pp.

Kirby, S. H., Raleigh, C. B. 1973. Mechanisms of high-temperature, solid-state flow in minerals and ceramics and their bearing on the creep behavior of the mantle. *Tectonophysics* 19:165–94

Kohlstedt, D. L., Goetze, C. 1974. Low stress, high temperature creep in olivine single crystals. *J. Geophys. Res.* 79:2045–51

Lliboutry, L. 1965. *Traite Glaciologie,* Tome 2. Paris: Masson. 611 pp.

Lomnitz, C. 1956. Creep measurements in igneous rocks. *J. Geol.* 64:473–79

MacDonald, G. J. F. 1963. The internal constitutions of the inner planets and the moon. *Space Sci. Rev.* 2:473–557

MacDonald, G. J. F. 1966. *Advances in Earth Science,* ed. P. M. Hurley, 199–245. Cambridge, Mass.: MIT Press. 502 pp.

McConnell, R. K. Jr. 1965. Isostatic adjustment in a layered earth. *J. Geophys. Res.* 70:5171–88

McConnell, R. K. Jr. 1968. Viscosity of the mantle from relaxation time spectra of isostatic adjustment. *J. Geophys. Res.* 73:7089–7105

McKenzie, D. P. 1968. *The History of the Earth's Crust,* ed. R. A. Phinney, 28–44. Princeton: Princeton Univ. Press. 244 pp.

Misener, D. J. 1972. Interdiffusion studies in the system Fe_2SiO_4–Mg_2SiO_4. *Ann. Rep. Carnegie Inst. 1971–1972,* 516–20

Misra, A. K., Murrell, S. A. F. 1965. An experimental study of the effect of temperature and stress on the creep of rock. *Geophys. J.* 9:509–35

Nabarro, F. R. N. 1948. *Strength of Solids,* 75–90. London: Phys. Soc. 162 pp.

O'Connell, R. J. 1971. Pleistocene glaciation and the viscosity of the lower mantle. *Geophys. J.* 23:299–327

Pascoe, R. T., Hay, K. A. 1973. Theory of diffusion creep in pure, impure and non-stoichiometric ionic materials. *Phil. Mag.* 27:897–914

Paterson, W. S. B. 1969. *The Physics of Glaciers.* Oxford: Pergamon. 250 pp.

Post, R. L. Jr. 1973. The flow laws of Mt. Burnett dunite. PhD thesis. Univ. Calif. Los Angeles

Post, R. L. Jr., Griggs, D. T. 1973. The earth's mantle: evidence of non-Newtonian flow. *Science* 181:1242–44

Raleigh, C. B., Kirby, S. H. 1970. Creep in the upper mantle. *Mineral. Soc. Am. Spec. Pap.* 3:113–21

Riecker, R. E., Rooney, T. P. 1969. Water-induced weakening of hornblende and amphibolite. *Nature* 224:1299

Runcorn, S. K. 1972. Dynamical processes in the deeper mantle. *Tectonophysics* 13:623–37

Seltzer, M. S. 1967. Investigation of Steady-State Creep in Nonstoichiometric Compounds, 2nd Ann. Rep. Columbus, Ohio: Battelle Mem. Inst. 16 pp.

Seltzer, M. S., Duga, J. J. 1966. Investigation of Steady-State Creep in Nonstoichiometric Compounds, 1st Ann. Rep. Columbus, Ohio: Battelle Mem. Inst. 48 pp.

Sherby, O. D., Burke, P. M. 1967. *Progr. Mater. Sci.* 13:325–90

Sherby, O. D., Simnad, M. T. 1961. Prediction of atomic mobility in metallic systems. *Trans. Am. Soc. Metals* 54:227–40

Shewmon, P. G. 1963. *Diffusion in Solids.* New York: McGraw-Hill. 203 pp.

Stocker, R. L., Ashby, M. F. 1973. On the rheology of the upper mantle. *Rev. Geophys. Space Phys.* 11:391–426

Sully, A. H. 1949. *Metallic Creep,* 37–57. London: Butterworths. 278 pp.

Urakami, A., Fine, M. E. 1972. Solid solution softening by double kink catalysis. *Mech. Behav. Mater. Proc. Int. Conf., 1st, 1971.* I:87–96

Walcott, R. I. 1972. Late quaternary vertical movements in eastern North America: quantitative evidence of glacio-isotatic rebound. *Rev. Geophys. Space Phys.* 10:849–84

Walcott, R. I. 1973. Structure of the earth from glacio-isostatic rebound. *Ann. Rev. Earth Planet. Sci.* 1:15–37

Weertman, J. 1958. Dislocation model of

low temperature creep. *J. Appl. Phys.* 29:1685–89

Weertman, J. 1968. Dislocation climb theory of steady-state creep. *Trans. Am. Soc. Metals* 61:681–94

Weertman, J. 1970. The creep strength of the earth's mantle. *Rev. Geophys. Space Phys.* 8:145–68

Weertman, J. 1973. *Physics and Chemistry of Ice,* ed. E. Whalley, S. J. Jones, L. W. Gold, 320–37. Ottawa: Roy. Soc. Can. 403 pp.

Weertman, J. 1974. High temperature creep produced by dislocation motion. *Dorn Mem. Symp.,* ed. J. C. M. Li, A. K. Mukherjee, New York: Plenum. In press

Weertman, J., Weertman, J. R. 1965. *Physical Metallurgy,* ed. R. W. Cahn, 793–819. Amsterdam: North-Holland. 1100 pp.

Wesson, P. S. 1972. Mantle creep: elasticoviscous versus modified Lomnitz law, and problems of "The New Global Tectonics." *Am. Assoc. Petrol. Geol. Bull.* 56:2127–49

THE MECHANICAL BEHAVIOR OF PACK ICE ✖10042

D. A. Rothrock
Department of Atmospheric Sciences, University of Washington,
Seattle, Washington 98195

If one were camping on pack ice and took an occasional walk, one would encounter at times some very noticeable local activity. Floes hundred of meters wide would separate, exposing the ocean to the atmosphere. This open water would soon freeze over, leaving bands of thin ice. As the thicker floes moved again, the thin ice might be broken and pushed up into piles of rubble known as pressure ridges. One would have no way of detecting, however, that he was slowly moving under the force of the wind, so that in a period of many years he would drift thousands of kilometers. Essential to the occurrence of this drift is the continual opening and ridging which allow the floes to rearrange themselves as the ice cover deforms. A fascinating description of these phenomena can be found in Nansen's (1897) narrative, *Farthest North*. Other descriptions are those by Zubov (1943), Kovacs (1972), and Weeks, Kovacs & Hibler (1971).

Here, we review the formulation of continuum models of this drift. A continuum element of pack ice must contain many floes (Figure 1), and assumptions about its mechanical properties should be based as far as possible on an understanding of those local mechanical processes of floe interaction one would have witnessed near his camp. The model must predict the horizontal velocity, the mass balance, and any other properties peculiar to the model, as functions of the two horizontal coordinates. We review the elements and basic assumptions of such models. These models have been applied primarily to the ice cover on the Arctic Ocean, but are equally applicable to any ice cover in which the dominant mechanisms of large-scale deformation are ridging and opening.

In the first section we consider the kinematics of the large-scale drift. Evidence suggests that we can treat pack ice as a continuum over distances of about 100 km, and that a useful and reproducible measure of strain rate can be obtained by tracking points spaced at this distance. In the next two sections, we review the governing equations for a pack ice model: those determining the properties or the state variables of the ice cover, and the momentum equation. A constitutive equation to relate the contact stress between floes to kinematic variables is an essential element of a model, because the divergence of this stress is an important

318 ROTHROCK

term in the momentum balance. In the following section, we describe models of the mechanisms of floe interaction, rafting, and ridging, that have provided a new basis for formulating constitutive laws. Finally, we review the various constitutive equations that have been proposed for the ice cover. Because of the energetics of these mechanisms, the most promising constitutive equations are the plastic laws rather than the viscous laws.

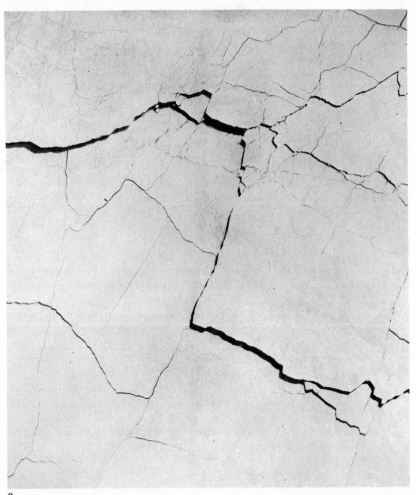

a

Figure 1   a (above): Aerial photograph of arctic pack ice in winter. In this photograph, the dark areas are thin ice or open water; the ice in light areas is several meters thick. Full scale is about 20 km. *b (opposite)*: Summer ice conditions in the central Arctic. In this negative format, the dark areas are thick ice; the light areas are open water. The frame is about 50 × 100 km (from Coon et al 1974).

b

KINEMATICS

A large portion of the recent activity in ice dynamics has been motivated by the realization that velocities alone are poor indicators of ice behavior. The mechanical behavior of the ice is related to velocity differences between floes or to velocity gradients in a continuum model. A sensitive test of the validity of a proposed ice model would therefore be to check those quantities involving velocity gradients, that is, vorticity and strain rate.

Although velocity gradients are well defined in a continuum model, they are not well defined in a real ice cover, where the velocity field exhibits spatial discontinuities associated with the fractured nature of pack ice itself (Nye 1973a). Velocities are continuous across a floe, but can change abruptly at floe boundaries. There has been a need, unanswered until recently, to measure velocity gradients in a real ice cover by spatial or temporal averaging. Temporal averages are not of much use. We want to measure the response of the ice to real weather systems, which vary over periods of about a day. To do this, our measurements must be averaged over periods of less than a day. Such averaging would filter out various high frequencies associated with, say, elastic waves or surface gravity waves (which most measurement techniques would not resolve anyway), but it would not smooth the discontinuities associated with the existence of discrete floes. If spatial averages are to be useful, there must be a continuum length scale l_c which is both large compared to the typical floe size l_i and small compared to the smallest space scales l_f of the driving forces (Maykut, Thorndike & Untersteiner 1972). That is, we require l_c to exist such that

$$l_i \ll l_c \ll l_f \qquad\qquad 1.$$

Although neither l_i nor l_f is well defined, we might say that rigid pieces of ice larger than 50 km across are rare, and atmospheric pressure systems in the Arctic are typically 1000 to 2000 km across.

Nye (1973a) has proposed spatially smoothing the velocity field to remove variations shorter than l_c, and defining the strain rate from the smoothed velocity field by the usual definition. This is a pleasing idea in principle, but it is not practical to collect this much information about the velocity field. Even with photographic or radar images (Tabata 1972a), velocities can be obtained only for those few points at which some ice feature can be identified in two sequential pictures. So, in practice, all measurements of strain rate have been accomplished by tracking a few (three or more) Lagrangian points. One method of calculating the strain rate tensor from these data is discussed by Thorndike (1974). The distance between these points—that is, the gauge length l_g of the measurement—should equal l_c. If l_c turns out to be a range of values satisfying inequality 1, strain rates measured on the same range of gauge lengths should equal each other.

There are now two pieces of empirical evidence about the correct size of l_g obtained by using completely different techniques. In the first, strain rates were simultaneously measured on a gauge of 100 km (Thorndike 1974) and on gauge lengths of 20 km and less (Hibler et al 1974b). If both these gauge lengths were within the range of continuum scales, the measured strain rates would correlate perfectly. The actual correlation is about 0.5. These measurements also showed that the net dilatation over a 25-day period (day 88 to day 113 in Hibler et al 1974b) was strongly dependent on the gauge length: 0.005 for $l_g = 100$ km, -0.019 for $l_g = 16$ km, -0.041 for $l_g = 8$ km, and -0.084 for $l_g = 5$ km. It is clear that one does not obtain the same estimate of strain rate and, more importantly, of strain by measuring over these different gauge lengths. No two or these lengths, then, lie within a possible range of continuum lengths l_c.

In the second case, a recent analysis of ERTS photographs (Nye & Thomas 1974) suggests that the variance of strain rate measurements decreases with increasing l_g, up to the largest lengths used, about 100 km. With strain rates of about 0.025 day^{-1}, the variance in the measured value was about 0.01 day^{-1} with $l_g = 50$ km, and decreased to about 0.004 day^{-1} with $l_g = 100$ km. Thus, it appears that the continuum length scale is 100 km or larger.

From the above measurements and from other data (Colony & Rasch 1975, Dunbar & Wittman 1963), we now have a good idea of typical velocities, vorticities, and strain rates in the central Arctic. Pack ice moves at speeds of 10 km day^{-1}, occasionally reaching twice that speed. The actual path followed by a point in the ice is several times the net distance it travels over a period of months (Dunbar & Wittman 1963). The wiggles in the paths of points spaced hundreds of kilometers apart correlate closely in time (Bushuyev et al 1967, Thorndike 1974). Instantaneous strain rates can be as large as several percent per day, but their long-term averages can be considerably smaller. For example, during the 40-day period documented by Thorndike (1974), the mean rates of divergence and shear (defined as the sum and the difference of the principal strain rates) were 0.001 and 0.002 day^{-1}. These numbers reflect the tendency of the ice to deform more in shear than in divergence or convergence. The vorticity behaves much the same as the rate of shear.

Strain rates are much larger in the shear zone. This is a band of intensely ridged ice near shore, whose width and distance from shore can vary. In an example taken from sequential ERTS photographs (Hibler et al 1974a), the shear zone exhibited a divergence of 1.6 day^{-1} and a vorticity and a rate of shear both of 0.5 day^{-1}. Similarly large vorticities have been deduced from rotations of a ship in close ice several miles off the coast in the Chukchi Sea (Lappo 1958). The ice must be pressed against the shore to create a shear zone; the zone does not exist when offshore winds detach the ice from the coast.

Tabata (1972b), in measuring the strains within a single piece of ice at the same time and location as the measurements discussed above (Thorndike 1974, Hibler et al 1974b), found strains of no more than about 3×10^{-4} over gauge lengths of several hundred meters. His measurements were made while the ice pack was undergoing strains of several percent on a gauge length of 100 km. The great difference between intra- and interfloe strain is evidence of a fact which appears obvious by intuition— that the strain which contributes to the large-scale circulation of pack ice is caused by the relative movement of pieces of ice and not by the continuous deformation of the pieces themelves.

THE STATE OF THE ICE COVER

Some properties of the ice cover, such as the rate of accretion and the resistance to deformation, vary by more than an order of magnitude depending on the ice thickness. The bulk of this variation is in the thickness range 0 to 50 cm, in what is usually called thin ice. These properties may also be sensitive to the size of floes. Only a very crude model can ignore all dependence on the state of the ice cover.

One approach to parameterizing the ice state has been to define the compactness S as the fraction of area covered by ice. The remaining fraction $1 - S$ is covered by open water. In the simplest development of a governing equation for compactness (Drogaitsev 1956, Nikiforov 1957), one assumes that the ice-covered area A within a total area B is conserved ($dA/dt = 0$), so that only the area covered by open water, $B - A$, can change. Then, we have

$$\frac{dS}{dt} = \frac{d}{dt}\frac{A}{B} = \frac{1}{B}\frac{dA}{dt} - \frac{A}{B^2}\frac{dB}{dt} = -S\,\mathrm{div}\,\mathbf{v} \qquad\qquad 2.$$

where $\mathbf{v}$ is the ice velocity in the horizontal. This follows because $(1/B)\,(dB/dt)$ is div $\mathbf{v}$ by definition. Replacing the material derivative d/dt by $\partial/\partial t + v_i\,\partial/\partial x_i$, we can write

$$\frac{\partial S}{\partial t} = -\mathrm{div}\,(S\mathbf{v}) \qquad\qquad 3.$$

This equation can be modified by a source term on the right-hand side to allow for thermodynamic changes in S, as done by Doronin (1970). A sink for S caused by ridging could also be included in the compactness equation. A complete mass balance requires either the assumption that the thickness of the ice-covered area is constant, or a balance equation for this thickness (Doronin 1970, Rothrock 1970). Soviet scientists have developed models which we will discuss later, in which the mechanical behavior of the ice depends on the compactness.

A more complete theory accounts for all the ice thicknesses present at one time in a continuum element. We can define a thickness distribution $g(h)$ such that the fraction of area covered by ice in the thickness band h_1 to h_2 is $\int_{h_1}^{h_2} g(h)dh$ (Maykut & Thorndike 1973). This distribution is governed by an equation analogous to the compactness equation, with the thermodynamic and mechanical sources explicitly shown (Thorndike & Maykut 1973, Coon et al 1974, Rothrock 1974a):

$$\frac{\partial g}{\partial t} = -\mathrm{div}\,(g\mathbf{v}) - \frac{\partial}{\partial h}(gf) + \psi \qquad\qquad 4.$$

where f is the thermodynamic growth rate and the redistribution function ψ describes the creation of open water and the transfer of ice of one thickness to another by rafting and ridging. To construct the redistribution function, it has been necessary to make assumptions about the distribution of the ice that fails in ridging, about the thickness of the ridged ice produced, and about the dependence of ψ on the strain rate. These assumptions have been guided by theoretical ridging models (Parmerter

& Coon 1972, 1973) which are discussed in a later section. More information about the sensitivity of the g distribution to these assumptions would be a guide to field measurements to check the assumed form of the function ψ. The direct measurement of the redistribution function requires the ability to measure $g(h)$ at two times and calculate the differences dg—a feat which presents a distinct challenge to our present technology.

Equation 4 contains the mass balance, since the mass per unit area m equals $\int_0^\infty \rho_{ice} \, hg(h) \, dh$; where, ρ_{ice} is the mass density of ice, which is approximately constant. In fact, knowing $g(h)$, one may calculate the mean of any quantity that depends on ice thickness in a known way; for example, the mean rate of ice production is $\int_0^\infty f(h)g(h) \, dh$.

Solving for $g(h)$ certainly complicates any model of pack ice; for practical purposes, it would be desirable to simplify the theory. For example, we can regard the variables S, defined as $\int_{0+}^\infty g(h) \, dh$, and H, defined as $\int_{0+}^\infty hg(h) \, dh$, as a two-parameter approximation of the full thickness distribution. However, the optimal way to reduce the complexity of $g(h)$ to a few parameters and still retain sufficient information has yet to be given much attention. A prerequisite is obviously an understanding of just what information in $g(h)$ is most crucial for modeling ice.

THE MOMENTUM EQUATION

Along with equations to determine the state variables, a model of pack ice must include the momentum equation

$$m\left(\frac{\partial \mathbf{v}}{\partial t} + (\mathbf{v} \cdot \nabla)\mathbf{v}\right) = -m\hat{f}\mathbf{k} \times \mathbf{v} + \nabla \cdot \boldsymbol{\sigma} + \boldsymbol{\tau}^w + \boldsymbol{\tau}^a - m\hat{g}\nabla\hat{H} - \bar{h}\nabla P \qquad 5.$$

This is a vector equation with two (horizontal) components. The Coriolis parameter $\hat{f}$ is twice the product of the earth's rate of rotation and the sine of the latitude; $\mathbf{k}$ is the unit vector normal to the $\mathbf{x}$ plane. The stress $\boldsymbol{\sigma}$ (a second rank tensor) is actually the difference between two terms (Nye 1973b). The first is the vertically integrated stress $\boldsymbol{\sigma}^{tot}$ in the ice. The second is the vertically integrated hydrostatic load $\boldsymbol{\sigma}^h$ and equals $-\frac{1}{2}\rho_w \hat{g}\bar{h'}^2 \mathbf{I}$, where ρ_w is the density of water, $\hat{g}$ is the acceleration of gravity, $\bar{h'}$ is the mean draft of the floating ice, and $\mathbf{I}$ is the identity tensor. Thus, $\boldsymbol{\sigma}$ is the contact stress transmitted between floes. Although for brevity we call $\boldsymbol{\sigma}$ a stress, it is actually a stress integrated through the ice thickness, and has dimensions of force/length. The constitutive equation which relates stress to kinematic variables is a major element of any model, and the subject of a later section. The air stress $\tau^a(\mathbf{x}, t)$ and water stress $\tau^w(\mathbf{v}, \mathbf{x}, t)$ are caused, respectively, by the surface winds and by the relative movement between the ice and the underlying ocean. Various treatments of these terms have been described by Campbell (1965), Karelin & Timokhov (1971), Smith (1971), and Brown (1973). The term $-m\hat{g}\nabla\hat{H}$ is the acceleration caused by the tilting sea surface; $\hat{H}(\mathbf{x}, t)$ is the height of the sea surface. The term $-\bar{h}\nabla P$ is due to the horizontal gradient of sea surface atmospheric pressure $P(\mathbf{x}, t)$; $\bar{h}$ is the mean ice thickness. This term has not been included previously in the momentum equation (e.g. Campbell 1965, Doronin 1970, Coon et al 1974), and is small.

A typical balance of terms in equation 5 as estimated by Hunkins (1974) is shown in Figure 2. That the temporal acceleration is negligible in Hunkins's force balance can be confirmed from station position data (AIDJEX Staff, 1972). The advective acceleration is always three orders of magnitude less than significant terms; the temporal acceleration can be significant but usually is not. Hunkins neglected the term $-\bar{h}\nabla P$. From the pressure data (Brown, Maier & Fox 1974) during the period covered by Figure 2a, we can see that this term had a magnitude of about 0.08 dyn cm^{-2} (taking $\bar{h} = 300$ cm). During a storm, this term is even less important relative to τ^a because of the nonlinear relation between τ^a and ∇P. The stress divergence $\nabla \cdot \sigma$ can be evaluated only as a residual in the momentum equation; all other terms have been measured directly.

We can characterize the floating ice, then, as highly damped. The wind drags the ice over a relatively stationary ocean, opposed mostly by water drag and resistance of the ice cover to deformation. The wind usually changes so slowly that the ice offers little inertial resistance to following it. Translated into a statement of kinetic energy balance, this means that the work of the wind is dissipated partly by work against internal stress and partly by work on the ocean. The kinetic energy of the ice cover and its changes are comparatively unimportant.

The simplest class of theories is that in which accelerations and stress divergence are assumed to be negligible. We might call this case equilibrium stress-free drift. Nansen (1902), Shuleikin (1938), and Reed & Campbell (1962) have compared such theories to observed drift and winds; Fel'zenbaum (1958) has predicted the mean drift in the Arctic Basin; Kagan (1967) has modeled response to tides. If the water

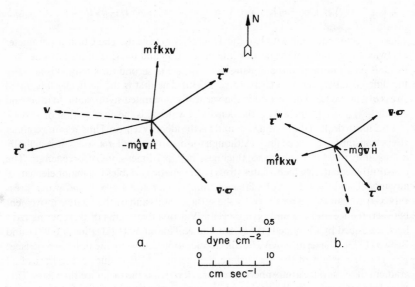

Figure 2 The balance of terms (solid vectors) in the momentum equation (5.) during two 12-hr periods (a: 1200–2400, 13 April 1972, b: 1200–2400, 19 April 1972) from Hunkins (1974). The velocity (dashed) is also shown.

stress varies linearly with ice velocity, the theory predicts that the ice drifts at a fixed angle of about 45° to the right of the wind stress vector, at about 2% of the speed of the wind 2 m above the ice. If the water stress is nonlinear in the ice velocity (Shuleikin 1938, Reed & Campbell 1962) this angle will increase to 90° and the ratio of ice speed to wind speed will decrease to zero as the wind speed decreases to zero. (The terms in $V\hat{H}$ and VP are usually neglected, but they can be added to τ^a if desired.)

MECHANISMS OF FLOE INTERACTION

If we knew what the constitutive equation for pack ice should be, we would not need to pay attention to the mechanisms of floe interaction. But the simple fact is that we are not at all sure about the constitutive equation. Because it has been too difficult to perform a field experiment that would conclusively determine the constitutive equation, we have turned to the study of these mechanisms—rafting, ridging, shearing, and opening—to deduce what we can about the large-scale mechanical behavior of pack ice.

Models of these mechanisms treat the ice as a plate (or beam) on an elastic foundation, this being the water in which the ice floats. Vertical loads can be applied in several ways to break the ice sheet in bending, by deflecting it up or down. These phenomena have horizontal length scales of tens of meters, determined by the plate's properties and the foundation stiffness (Parmerter 1974a). [Mohaghegh (1973a, b) has provided useful summaries of data on the strength of sea ice.]

Pack ice exists in a highly fractured state, although it is not yet clear how this state is maintained. Several mechanisms have been proposed. Thermal stresses induced by a rapid drop in air temperature cause cracks to be initiated in lead ice (Matsuoka 1972) and at a spacing of about 200 m in thick ice (Evans & Untersteiner 1971, Evans 1971, Milne 1972). Isostatic imbalance caused by variations in ice thickness such as those typically found can also initiate cracks (Schwaegler 1974). The question remains whether these cracks, having been initiated, will propagate vertically through the whole ice sheet, effectively separating it into two pieces. Work in progress (R. R. Parmerter, personal communication) suggests that they will.

Considerations of the cracking mechanisms have not taken into account the visco-elastic behavior of sea ice, even though relaxation times ranging from minutes to days have been observed (Tabata 1958). Thermal stresses and stresses caused by isostatic imbalance are applied over periods of days to months. It seems possible that ice could relieve these stresses by creep rather than by fracture; this possibility deserves further study. Whatever maintains the fractured state must act continuously or else thermodynamic processes would weld all the floes back together.

Models of rafting agree with observations (Kovacs 1972) that ice of typical strength cannot override without breaking if it is thicker than about 17 cm (Matsuoka 1972, Parmerter 1974b). The conditions necessary for rafting are not clearly understood. Matsuoka (1972) alludes to friction as a controlling factor in the initial overriding at the boundary between two floes. He shows that overriding requires a crack with a critical slant from vertical of 9° to 14°, depending on the coefficient of friction. Parmerter (1974b) ignores friction and calculates a horizontal load necessary for overriding to occur based on the required vertical displacements of the floating ice sheets near the crack. Ice 10 cm thick, for example, with a vertical crack, requires about 1.2×10^7 dyn cm^{-1} to raft. This is only 40% of the load required to cause the same ice sheet, treated as a beam on an elastic foundation, to buckle (Kheisin 1971), and an order of magnitude less than the load required to crush the ice (assuming a compressive strength of 2×10^7 dyn cm^{-2}). Once overriding has occurred, the horizontal load needs only to overcome friction between the overlying floes and produce some potential energy for rafting to continue. This process has not been modeled.

The mechanism of ridging (described by Weeks, Kovacs & Hibler 1971), reduces even further the capacity of the ice cover to carry horizontal loads. Like the breaking of an overriding thick flow just discussed, the ridging process breaks ice in bending: the weight of the sail is not distributed over the buoyancy of the keel, and the resulting moment can break the ice. The process has been modeled in detail by Parmerter & Coon (1972, 1973). Horizontal loads need only overcome the gravitational potential energy and frictional forces to build the pile, which then breaks off more ice to be incorporated in the ridge. In 10 cm ice, for example, a horizontal load of only 2×10^5 dyn cm^{-1}—two orders of magnitude less than the load needed to raft the same ice—can keep the ridge building once the original rubble pile exists. But the rubble pile is a catalyst for the process, and it must be started by some other mechanism such as the breaking of overriding ice.

This knowledge about the mechanisms of floe interaction has drastically altered our view of how much stress is necessary to cause the ice cover to deform. Where it was once assumed that loads had to be large enough to crush or buckle ice sheets, we now know that much smaller loads can raft and ridge the ice.

But we still do not know how the effects of many processes acting together in a continuum element should be averaged to deduce the large-scale properties and response of pack ice. This subject has received only little and qualitative attention (Hartwell 1972, Hibler et al 1973). Looking inside the continuum element, we see that where ridges form in thin ice, the stress is small. Elsewhere, in cracked but unridged thin ice, larger loads must be present to cause the ice to override; and in still other places, where thick ice bears on thick ice, quite large stresses can be transmitted with little noticeable local failure. How are the large stresses generated to activate a "strong" process such as overriding if the "weak" process of ridging is occurring in the vicinity?

Work is also needed in parameterizing the friction between rough, jagged blocks of ice and floes. This friction is undoubtedly a significant sink of mechanical energy in ridge building (Parmerter & Coon 1973, Rothrock 1974c) and may also be important in rafting and in shearing at floe boundaries.

CONSTITUTIVE EQUATIONS

Requirements of the Constitutive Equation

During a period when the motivation for modeling pack ice exceeded our understanding of the mechanisms of large-scale deformation, constitutive laws were hypothesized that attributed a viscous behavior to the ice. Viscosity, however, is not compatible with what we now know about ice mechanisms: that their load-displacement relations are independent of rate. To accommodate this rate-independence, a plastic constitutive law has been formulated.

The plastic constitutive law is supported by the energetics of floe interaction mechanisms. The rates of energy dissipation by potential energy and friction in ridge building (Parmerter & Coon 1973) and by frictional losses in shear between floes are proportional to the rate at which these processes proceed. And it seems natural to assume that these processes proceed at a rate proportional to the large-scale strain rate. This linear relation between the dissipation rate and the strain rate is characteristic of plastic flow, not of viscous flow. (In viscous materials, the rate of energy dissipation is proportional to the square of the strain rate.)

All that we have said about formulating the constitutive equation from a knowledge of mechanisms concerns the real-time response of pack ice—that is, the response to the highest frequencies of the large-scale forcing by winds and currents, say, several cycles per day. Modeling the time-averaged response to time-averaged forcing is, however, a different problem to which ad hoc laws may be applicable. Of course, the best way to deduce the time-averaged response is from a good model of the real-time response.

It will be helpful to think of two scalar invariants of the stress tensor: σ_I, which equals half the sum of the principal stresses (that is, the isotropic part of the stress or the negative pressure); and σ_{II}, equal to half the difference of the principal stresses, a measure of the shear stress or of the size of the stress deviator:

$$\sigma' = \sigma - \tfrac{1}{2}(\mathrm{tr}\,\sigma)\mathbf{I} \qquad\qquad 6.$$

The stress tensor can be written in terms of σ_I, σ_{II}, and a tensor which determines the direction of the principal axes and has a determinant equal to -1. Each constitutive equation implies that the stress state in pack ice can occupy a certain region in the σ_I, σ_{II} plane. Comparing these regions illuminates differences between the assumed equations. For example, the stress-free theories assume that σ_I and σ_{II} are both zero.

The behavior of pack ice which we expect to see reflected in a constitutive law should have two characteristics. First, the material might support tensile stress if it is not fully fractured; but even so, it is brittle in tension and can support no tensile stress as soon as it begins to extend. This means that the principal stresses σ_1 and σ_2 shown in Figure 3 cannot be positive. Second, resistance to pure convergence or to shear should be the same size since they are both due to the same processes: rafting, ridging, and shearing between floes.

For the stress divergence over distances of about 500 km to be about equal to the wind stress (1 dyn cm^{-2}), stresses must be of the order of 5×10^7 dyn cm^{-1}. The allowable stresses then should occupy the part of the σ_I, σ_{II} plane shown hatched in Figure 3.

One source of misunderstanding has been to treat the stress divergence $\nabla \cdot \sigma$ in the momentum equation as a term $\mathbf{R}$ which can be written down without ever defining stress. For example, the assumption $\mathbf{R} = -k\mathbf{v}$ (Sverdrup 1928, Rossby & Montgomery 1936) is appropriate for the friction between a moving object and a stationary object,

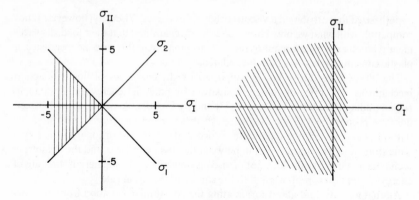

Figure 3   (*left*) The (σ_I, σ_{II}) plane. The units are 10^7 dyn cm^{-1}.

Figure 4   (*right*) The stress states (hatched area) implied by the viscous law of equations (12) and (13).

but it has no physical basis in the present problem where stress is related to relative movement of the material—that is, to gradients of displacement or of velocity. To avoid inconsistencies and to enable the formulation of boundary conditions on stress, it is desirable to write σ explicitly before forming $\nabla \cdot \sigma$.

Viscous Behavior as the Time-Averaged Response

Although there is at present no explanation for viscous behavior of pack ice on real time scales, viscous laws may be appropriate for time-averaged response (Nye 1973c). But the viscosity would then be an inverse measure of the mobility induced by averaged high frequency motion caused by storms which loosen the ice temporarily and allow it to respond to the time-averaged driving forces. We must distinguish, however, just what it is that viscosity is being assumed to parameterize. In Nye's argument, it would account for the contact stresses between floes; viscosity in this sense may turn out to be a useful concept in modeling the time-averaged response.

There is no basis, however, for associating viscosity with Reynolds stresses—that is, the net advective transport of momentum by high-frequency motions which make no contribution to the mean velocity field. The Reynolds stress σ_{ij}^R can be defined (Kheisin 1971) as $-m\overline{v_i' v_j'}$ where $v_i'(t)$ denotes the difference between the mean value $\bar{v}_i$ and the instantaneous value v_i. The Reynolds stress σ_{ij}^R is the net transport by these velocity fluctuations of momentum mv_i' in the j direction. Thorndike (personal communication) has calculated σ^R from the records of drifting station positions in the central Arctic Ocean (Transpolar Drift Stream) and finds it to be of the order of -10^4 dyn cm^{-1}. We have asserted above that stresses need to be of the order 10^7 dyn cm^{-1} to have significant gradients. Saying that Reynolds stresses are not significant is only to restate that the advective accelerations $mv_j \partial v_i / \partial x_j$ are insignificant.

An idea borrowed from fluid dynamics is that an eddy viscosity κ exists in analogy with molecular viscosity. In a parallel flow $\bar{v}_1(x_2)$, for example, the Reynolds stress would be proportional to the rate of shear of the mean flow:

$$\sigma^R_{12} \equiv -m\overline{v'_1 v'_2} = m\kappa \frac{\partial \bar{v}_1}{\partial x_2} \qquad\qquad 7.$$

When Thorndike correlated σ^R_{12} and $\partial\bar{v}_1/\partial x_2$ he found, to a high level of confidence, an effective eddy viscosity of -10^7 cm^2 sec^{-1} or a dynamic viscosity $m\kappa$ of -3×10^9 g sec^{-1}. The negative sign of κ is striking; it indicates that the fluctuating velocities v'_i tend to concentrate rather than diffuse momentum. The magnitude of the eddy viscosity has been estimated by mixing length theory (which is a scaling argument and cannot determine signs) to be 10^3 to 10^5 cm^2 sec^{-1} (Gorbunov & Timokhov 1968, Volkov, Gudkovich & Uglev 1971). These values of viscosity are several orders of magnitude too small to influence the momentum equation. A viscosity of the order of 10^{14} g sec^{-1} is required to create stresses of 5×10^7 dyn cm^{-1}, given typical strain rates of 0.02 day^{-1}. Thus, Reynolds stress and eddy viscosity cannot be invoked as the basis for a viscous model of time-averaged response of pack ice.

Viscous Constitutive Equations

Several viscous constitutive equations have been proposed. One of the first was

$$\sigma = 2\eta\dot{\varepsilon}' \qquad\qquad 8.$$

which states that the ice is viscous in shear (Ruzin 1959, Campbell 1965, Yegorov 1970). The strain rate tensor $\dot{\varepsilon}_{ij}$ is

$$\frac{1}{2}\left(\frac{\partial v_i}{\partial x_j} + \frac{\partial v_j}{\partial x_i}\right) \qquad\qquad 9.$$

The prime denotes the deviator as in equation (6), and η is the shear viscosity. This law (8) gives rise to the term $\eta\nabla^2\mathbf{v}$ in the momentum equation. Only stresses on the σ_{II} axis in Figure 3 can exist. Although the ice resists shearing deformation, it can converge without resistance, an ability for which there is no physical explanation. Campbell & Rasmussen (1972, Rheology II) have proposed making the shear viscosity depend on the divergence of velocity

$$\sigma = 2 \left\{ \begin{array}{l} \eta_D\,(\text{small}), \text{div } \mathbf{v} > 0 \\ \eta_C\,(\text{large}), \text{div } \mathbf{v} < 0 \end{array} \right\} \dot{\varepsilon}' \qquad\qquad 10.$$

but the stress is still limited to the σ_{II} axis and convergence is still unopposed.

The viscous assumption becomes more acceptable when an isotropic stress is added. The most general form of a viscous law including an isotropic term is

$$\sigma = f_1(\dot{\varepsilon}_I, \dot{\varepsilon}_{II})\mathbf{I} + f_2(\dot{\varepsilon}_I, \dot{\varepsilon}_{II})\dot{\varepsilon}' \qquad\qquad 11.$$

where $\dot{\varepsilon}_I$ and $\dot{\varepsilon}_{II}$ are two invariants of the strain rate tensor, say, the sum and the difference of the principal strain rates (Glen 1958, 1970). The functions f_1 and f_2 can also depend on other scalars such as compactness. Campbell & Rasmussen (1972, Rheology I) and Hibler (1974) have assumed the linear form of (11)

$$\boldsymbol{\sigma} = \zeta \cdot \text{tr}\,\dot{\boldsymbol{\varepsilon}}\mathbf{I} + 2\eta\dot{\boldsymbol{\varepsilon}}' \qquad\qquad 12.$$

where ζ is the bulk viscosity. But then the right- and left-hand sides of the σ_I, σ_II plane are equally accessible to the material, since $\text{tr}\,\dot{\boldsymbol{\varepsilon}}\,(\equiv\text{div}\,\mathbf{v})$ is typically the same size regardless of its sign.

We have suggested, however, that tensile stresses should be negligible compared to compressive stresses. This behavior would require a nonlinear form of (11), with a bulk viscosity in (12) given by

$$\zeta = \left\{ \begin{array}{l} \zeta_\text{D}\,(\text{small}),\,\text{div}\,\mathbf{v} > 0 \\ \zeta_\text{C}\,(\text{large}),\,\text{div}\,\mathbf{v} < 0 \end{array} \right\} \qquad\qquad 13.$$

The shear viscosity could be a constant, and both the bulk and the shear viscosity might be assumed to depend on some material property such as compactness. This law would produce stress states in the hatched area in Figure 4. It is the most realistic viscous law which could be assumed for either the real time or the time-averaged response.

Rheology III proposed by Campbell & Rasmussen (1972) is not a permissible law, being of the form

$$\sigma_{ij} = 2m\kappa_i\dot{\varepsilon}'_{ij} \qquad\qquad 14.$$

where no summation is implied on the right-hand side. The coefficient of $\dot{\varepsilon}'_{ij}$ should be a scalar f_2, but is defined as a quasi-vector. Furthermore, f_2 must be a function of invariants; but κ_1, for example, is defined as a function of $\partial v_1/\partial x_1$, which is not an invariant. Another example of an improper constitutive law is the viscous law used by Doronin (1970),

$$\sigma_{ij} = 2\alpha S\,\frac{\partial v_i}{\partial x_j} \qquad\qquad 15.$$

where α is a constant and $S(\mathbf{x},\,t)$ is the compactness. This stress tensor is not symmetric; $\partial v_i/\partial x_j$ should be replaced by $\dot{\varepsilon}'_{ij}$.

The Magnitude of Viscosity

Just as there is no agreement on just what form of equation (11) is appropriate, there is also no agreement on how large the coefficients of viscosity should be. The basis for choosing a value is purely empirical. We will refer primarily to dynamic viscosity (η and ζ), which has dimensions[1] of mass/time and is approximately 300 g cm^{-2} times the kinematic viscosity, with dimensions of length^2/time. We have already stated that the dynamic viscosity must be about 10^{14} g sec^{-1} for the stress divergence to be of a magnitude comparable with the wind stress.

In calculating the mean drift in the Arctic Basin, Campbell (1965) found that

[1] In a two-dimensional theory, density becomes mass/length^2 and these dimensions follow.

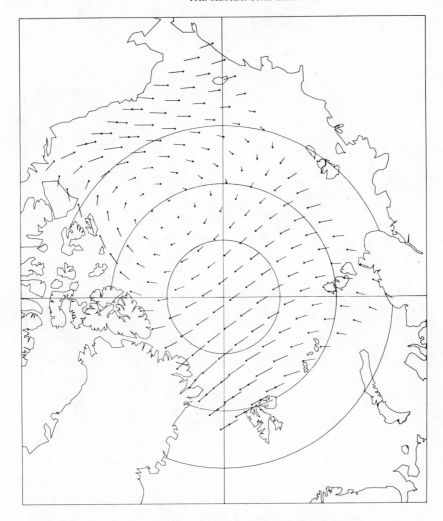

Figure 5   The mean drift in the Arctic Ocean, similar to that calculated by Campbell (1965). The results shown here and in Figures 6, 8, and 9 were calculated by R. Colony. The terms were treated as in Rothrock (1975, standard case) except that (1) the water drag coefficient was increased to 0.15 g cm^{-2} sec^{-1} to allow for the form drag of pressure ridge keels; (2) the stress divergence ($-\nabla p$ in Rothrock 1975) was either omitted or replaced, depending on the case; (3) the boundary conditions were altered appropriately; and (4) div **v** is specified only in Figure 9. In this figure, div $\boldsymbol{\sigma}$ is 10^{15} g sec^{-1} × ∇^2**v**, and the velocity is taken to be the stress-free value everywhere on the boundary. A vector one grid space long represents a velocity of 5 cm sec^{-1}; an x through a vector indicates a velocity greater than 5 cm sec^{-1}.

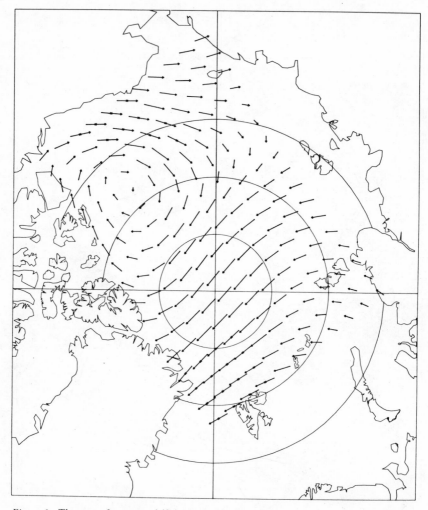

Figure 6 The stress-free mean drift in the Arctic Ocean. See the legend in Figure 5. Here, div $\boldsymbol{\sigma} \equiv 0$, and there are no boundary conditions since the problem is algebraic.

$\eta = 10^{15}$ g sec^{-1} gave the most realistic result,[2] reproduced in Figure 5. But this calculation assumed stress-free (purely wind-driven) drift at the boundaries, the least likely place. Because of this boundary condition, the whole flow is only slightly different from the stress-free drift shown in Figure 6. The observed drift is shown in Figure 7 for comparison.

[2] An error in Campbell's (1965) paper has led to some confusion. The units of K_h in his table on page 3292 should be cm^2 sec^{-1} rather than g sec^{-1}.

Hibler (1974) reports a similar result. Using field observations of real time response, he concluded that shear and bulk viscosities of the order of 10^{15} g sec^{-1} gave the best correlation between divergence of velocity and vorticity, on the one hand, and, on the other, the curvature of the surface atmospheric presure (theoretically proportional to the curl of the wind stress). Like Campbell, he neglected the effects of boundaries.

The results of Campbell's time-averaged problems and of Hibler's real-time problems are compatible, but their neglect of what are surely considerable effects of

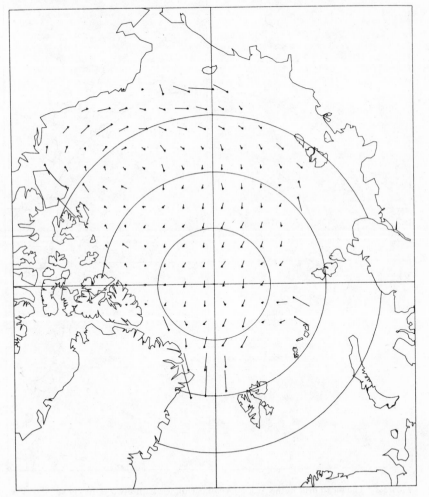

Figure 7 The observed mean drift in the Arctic Ocean, from Colony & Rasch (1975). A velocity vector one grid space long represents 5 cm sec^{-1}.

the boundary throws both studies into question. The boundary condition usually applied to viscous flows is that both velocity components be zero (no slip), and this condition seems applicable in the present problem. With no slip and a shear viscosity of 10^{15} g sec^{-1}, the mean drift is very unrealistic, as shown in Figure 8. The viscous drag of the boundaries has nearly stopped the Beaufort Sea gyre.

A much smaller shear viscosity of 6×10^{12} g sec^{-1} to produce shear zones about

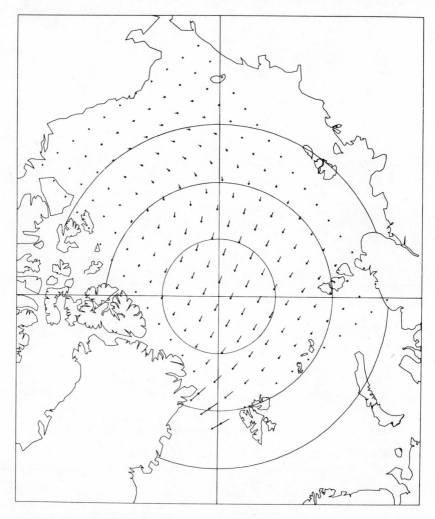

Figure 8 The mean drift in the Arctic Ocean calculated with div $\sigma = 10^{15}$ g sec$^{-1} \times \nabla^2 \mathbf{v}$. The no-slip boundary condition applies everywhere except between Greenland and Spitzbergen where the stress-free velocity is specified. See the legend in Figure 5.

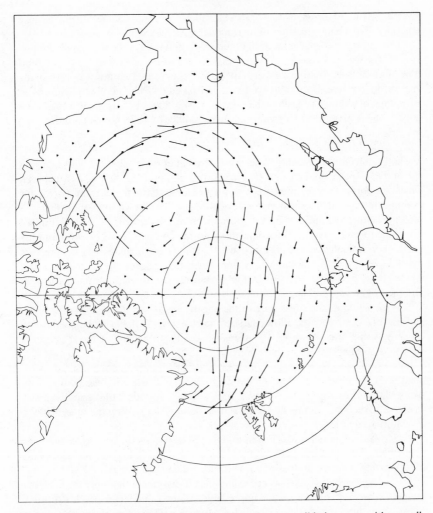

Figure 9 The mean drift in the Arctic Ocean for an incompressible ice cover with a small shear viscosity. Here, div $\sigma = -\nabla p + (6 \times 10^{12}$ g sec$^{-1})\nabla^2\mathbf{v}$. The no-slip boundary condition applies everywhere except between Greenland and Spitzbergen where an outward velocity of 11.4 cm sec^{-1} is specified. This export is produced by specifying div $\mathbf{v} = 0.67 \times 10^{-8}$ sec^{-1}. See the legend in Figure 5.

100 km wide (Rothrock 1975), combined with the assumption of incompressibility (comparable to a very large bulk viscosity), gives a more realistic mean drift, as Figure 9 shows. Campbell & Rasmussen (1972) have made calculations of real time response far from any boundaries. They find that shear and bulk dynamic viscosities ranging from 3×10^{11} to 3×10^{13} g sec^{-1} permit convergence and

divergence on the order of the observed values, 10^{-7} sec. Soviet scientists have generally used small viscosities in their models; 10^{11} g sec^{-1} is typical.

There is an irreconcilable conflict between choosing viscosity small enough (6×10^{12} g sec^{-1}) to model shear zones, where strain rates reach 0.5 day^{-1}, and choosing a sufficiently large viscosity (10^{14} g sec^{-1}) to cause significant stresses in the interior flow, where strain rates are typically 0.02 day^{-1}. This conflict illustrates an advantage of assuming plastic behavior. In plastic flow, the stress can remain the same order of magnitude for strain rates which differ by orders of magnitude.

Laws Based on Incompressibility

A wholly different approach has been used by Nikiforov et al (1967) and Doronin (1970). They appear to have modeled ice as a cavitating fluid, that is, as a material which supports isotropic pressure but not isotropic tension. [Doronin (1970), for example, has included a shear viscosity of 10^{11} g sec^{-1}, but we will regard this as a separate feature and ignore the stress deviator in this discussion.] The cavitating ice is assumed to have two regimes depending on the compactness S, which naturally fits this conceptualization of the ice cover. When the ice is open ($S < 1$), the pressure must be zero, and the unknowns S and v are found from the momentum and compactness balance equations. In the second regime, the ice is compact ($S \equiv 1$) and the unknowns, pressure ($\geqq 0$) and v, are found from the same equations.[3]

This model limits the stress state to the negative σ_{I} axis in the (σ_{I}, σ_{II}) plane. The addition of a shear viscosity allows the stress to be anywhere in the left half plane. For observed strain rates, however, and a small viscosity (10^{11} g sec^{-1}), the shear stress σ_{II} would be small, except in shear zones.

Initial results from cavitating models look realistic. Doronin obtains good agreement with the observed retreat of the ice boundary in the Kara Sea during late spring. This retreat, however, is shown to be as much a product of thermodynamics as one of advection, and this should be kept in mind when judging the success of the constitutive equation.

The cavitating model is pleasing in that its physical basis is clear; it accounts for opening of cracks to form leads and the closure of leads to form compact ice capable of supporting pressure. It does not account for ridging or rafting.

The cavitating fluid model has been simplified by assuming that only the compact regime exists so that the ice is incompressible (Witting & Piacsck 1972, Rothrock 1975). With this assumption, a realistic pattern of long-term mean drift and a reasonable pressure field are obtained (Figure 9). The calculated pressure is large (4 to 12×10^7 dyn cm^{-1}) north of Greenland, where ridging is known to be intense.

Two Other Models

For calculating the tidal motions of an ice-covered ocean, Kheisin & Ivchenko (1973) have used the viscoelastic constitutive equation

$$\sigma = -p_s \mathbf{I} + 2\eta \dot{\varepsilon}' \qquad\qquad 16.$$

---

[3] I have not been able to find the rationale for the device Doronin (1970) uses to "correct" velocities in compact regions. In the simplest case, the compactness equation would require div $\mathbf{v} = 0$, but I cannot make Doronin's equation (25) compatible with this equation.

where

$$p_s = \begin{cases} k\ \delta S, & \delta S > 0 \\ 0, & \delta S < 0 \end{cases}$$

17.

Here, k is the bulk modulus of elasticity; δS is a small variation in compactness about an equilibrium value S_0 and satisfies a linearized compactness equation

$$\frac{1}{S_0}\frac{\partial}{\partial t}(\delta S) = -\operatorname{div} \mathbf{v}$$

18.

For the parameters they chose, $k = 10^8$ dyn/cm and viscosities between 10^{12} and 6×10^{12} g sec^{-1}, the solutions of ice velocity were affected only slightly by the stress divergence, and primarily near the coasts.

A one-dimensional model based on momentum transfer by inelastic collisions between floes has been developed by Solomon (1973). It predicts a nonlinear resistance to compression by compact ice that is qualitatively reasonable. However, the model seems inapplicable for several reasons. First, it describes conditions which do not seem very realistic: namely, that the compactness is near unity, but the floes are generally not contiguous. In actuality, floes in compact ice appear to be in almost continuous contact. Second, the fact that accelerations as large as 1 cm sec^{-2} have been measured at first appears to support the notion of a series of impulses caused by collisions. But these accelerations are so infrequent that they do not add up to a significant net acceleration over a period such as an hour. The presence of such high frequency accelerations can be attributed to effects other than collisions, such as sporadic building of ridges and gustiness of winds (Craig, 1972). Third, the term Solomon suggests adding to the x_1 component of the momentum equation 5 in place of $\mathbf{V} \cdot \boldsymbol{\sigma}$ is, in our notation,

$$-m\frac{\partial}{\partial x_1}\left[\frac{l_i^2 S}{2(1-S)}\left(\frac{\partial v_1}{\partial x_1}\right)^2 \right], \qquad \text{for } \frac{\partial v_1}{\partial x_1} < 0$$

19.

With a length scale of 100 km for significant spatial gradients, and $l_i \sim 10$ km, $S \sim 0.99$, and $\partial v_1/\partial x_1 \sim 10^{-7}$ sec^{-1}, this term is on the order of 10^{-9} dyn cm^{-2}, which is extremely small compared with a typical wind stress τ^a of 1 dyn cm^{-2}.

Plastic Models

Other models which are rather clearly formulated in terms of the controlling physical mechanisms are plastic models (Coon 1974, Coon & Pritchard 1974, Coon et al 1974, Pritchard & Colony 1974, Rothrock 1974a, b, c). In these models, the stress can lie only within a curve in the left half of the σ_I, σ_{II} plane. This curve, called the yield curve $F(\sigma_I, \sigma_{II}) = 0$, is shown in Figure 10. The similarity to the shaded area in Figure 3 is not coincidental; the reasoning behind both figures is the same.

The ice can display two distinct types of behavior. If the stress is on the yield curve, $F = 0$, the ice flows plastically subject to the flow rule

$$\dot{\varepsilon}_{ij} = \lambda \frac{\partial F}{\partial \sigma_{ij}}$$

20.

which specifies the ratio of divergence to shear, but not the rate (λ) at which the strain occurs. Plastic flow provides all of the deformation significant to the large-scale circulation of the ice cover and is identified with the mechanisms of pack ice movement: ridging alone during pure convergence, opening alone during pure divergence, and both during a general rate of strain.

When the stress is strictly within the yield curve, the ice responds as a stiff elastic material deforming slightly, but not flowing. This represents the elastic strain of solid pieces of ice, which we have stated earlier contributes little to the movement of the pack. [Coon (1974) and Rothrock (1974a) describe this behavior as rigid.]

The shape and size of the yield curve are determined by the principle that the rate of plastic work equals the rate of energy dissipation by processes such as ridging and shearing (Rothrock 1974c). The shape is constant, and we can only guess that it resembles the shape in Figure 10. The size of the yield curve, however, which is determined by the (variable) yield strength p^*, depends on the thickness of the ice that will be ridged during deformation, and this in turn depends on the thickness distribution. Thus, as either thermodynamics or deformation alter the thickness distribution, p^* and the size of the yield curve change, increasing in some circumstances and decreasing in others.

We have said earlier that in order for the stress divergence to be significant in the momentum equation, the stress must be about 5×10^7 dyn cm^{-1}. Coon et al (1974) find that the yield strength fluctuates near 10^6 dyn cm^{-1}. But the value of p^* cannot be regarded as well known. In estimating the rate of energy dissipation, Coon et al consider only the rate of production of potential energy by ridging. Including other energy sinks, such as frictional losses in ridging and in shearing at floe boundaries, will raise the theoretical value of p^*, perhaps by a factor of two or three. In addition, the value of p^* depends on some assumptions that were rather arbitrarily made about the thicknesses of ice being ridged. The predictive capacity of plastic models has not been tested. But they have been formulated from a complete, although somewhat idealized, mechanistic picture of the interaction of floes, and hold considerable promise.

It is clear that internal ice stress σ is important in the momentum balance of pack ice; a great deal of attention has been directed toward modeling this stress. We suffer now from too many hypotheses and too few data of sufficient accuracy to test them critically.

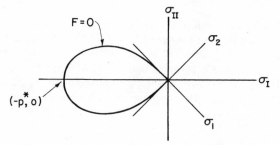

Figure 10   The yield curve $F(\sigma_I, \sigma_{II}) = 0$. The stress can be within or on the curve.

CONCLUSION

The nature of research in pack ice dynamics has changed beyond recognition in the past decade. The subject is now at a stage of rapid and fruitful development, thanks in part to the advent of remote sensing as a scientific tool. Although the most encouraging advances have been realized in the investigation of mechanisms responsible for large-scale deformation, important problems remain to be solved in that area. What, for example, is the role of friction in the rubble pile of a pressure ridge and in the grinding of floe against floe? Under what conditions can pack ice support tension; and does the formation of shorefast ice depend on tensile strength? How do the processes work (or what are the processes) that maintain a fractured ice cover? How well can we actually deduce the behavior of a continuum element of the ice cover from models of ridging and other processes?

As quickly as we design more complex models, such as those based on the thickness distribution, the need will grow to simplify them systematically to provide more efficient yet still effective models for extended use, both in ice forecasting and in climate modeling.

ACKNOWLEDGMENTS

I am grateful to M. D. Coon, R. J. Evans, G. A. Maykut, R. S. Pritchard, and A. S. Thorndike for their review of the manuscript and their helpful comments, and to A. Johnson for her editorial assistance. It was a great help to have access to A. S. Thorndike's analysis of drifting station data, and to have the results of R. Colony's calculations shown in Figures 5, 6, 8, and 9. J. Fitzgerald, N. Fukuyama, and C. Barnard prepared the figures and the manuscript. I am very much indebted for all of the assistance.

Literature Cited[4]

AIDJEX Staff. 1972. Station positions, azimuths, weather. 1972 AIDJEX pilot study preliminary data. *AIDJEX Bull.* 14: 63 ff. NTIS No. PB 220 859

Brown, R. A. 1973. On the atmospheric boundary layer: theory and methods. *AIDJEX Bull.* 20:1–141. NTIS No. PB 222 057

Brown, R. A., Maier, P., Fox, T. 1974. Surface atmospheric pressure fields and derived geostrophic winds. AIDJEX 1972. *AIDJEX Bull.* 26:173–203 NTIS No. PB 236 665/AS

Bushuyev, A. V., Volkov, N. A., Gudkovich, Z. M., Loshchilov, V. S. 1967. Rezul'taty ekspeditsionnykh issledovanii dreifa i dinamiki ledianogo pokrova arkticheskogo basseina vesnoi, 1967g. (Results of a Soviet experiment investigating the drift and dynamics of the Arctic Basin ice cover.) *Tr. Arkt. Antarkt. Nauch. Issled. Inst.* 257:26–44 (English transl. NTIS No. PB 196 063)

Campbell, W. J. 1965. The wind-driven circulation of ice and water in a polar ocean. *J. Geophys. Res.* 70:3279–3301

Campbell, W. J., Rasmussen, L. A. 1972. A numerical model for sea ice dynamics

[4] Many references are to the *AIDJEX Bulletin*, a series of progress and technical reports published by the Arctic Ice Dynamics Joint Experiment, University of Washington. The *Bulletin* may be obtained from the National Technical Information Service, 5285 Port Royal Road, Springfield, Virginia 22151. The classification numbers of this Service (NTIS No.) are supplied for each referenced *Bulletin*.

incorporating three alternative ice constitutive laws. *Sea Ice: Proc. Int. Conf., Reykjavik, Iceland,* 176–87. Reykjavik: Nat. Res. Counc.

Colony, R., Rasch, P. 1975. Drift paths in the Arctic Ocean. *AIDJEX Bull.* In press

Coon, M. D. 1974. Mechanical behavior of compacted arctic ice floes. *J. Petrol. Technol.* 26:466–70

Coon, M. D., Maykut, G. A., Pritchard, R. S., Rothrock, D. A., Thorndike, A. S. 1974. Modeling the pack ice as an elastic-plastic material. *AIDJEX Bull.* 24:1–105. NTIS No. PB 232 231/AS

Coon, M. D., Pritchard, R. S. 1974. Application of an elastic-plastic model of arctic pack ice. *The Coast and Shelf of the Beaufort Sea,* ed. J. C. Reed, J. E. Sater, 173–93. Arlington, Va: Arctic Inst. N. Am.

Craig, L. W. 1972. *High frequency accelerations of arctic pack ice.* MS thesis. Univ. Washington, Seattle.

Doronin, Yu. P. 1970. K metodike rascheta splochennosti i dreifa l'dov. (On a method of calculating the compactness and drift of ice floes.) *Tr. Arkt. Antarkt. Nauch. Issled. Inst.* 291:5–17 (English transl. *AIDJEX Bull.* 3:22–39. NTIS No. PB 196 063)

Drogaitsev, D. A. 1956. Zones of compression and rarefaction of ice in the atmospheric pressure field. *Izv. Akad. Nauk. SSSR Ser. Geophys.* 11:1332–37 (Transl. by Defense Res. Board Can., T 264 R)

Dunbar, M., Wittman, W. 1963. Some features of ice movement in the Arctic Basin. *Proc. Arct. Basin Symp.,* 90–104. Washington DC: Arctic Inst. N. Am.

Evans, R. J. 1971. Cracks in perennial sea ice due to thermally induced stress. *J. Geophys. Res.* 76:8153–55

Evans, R. J., Untersteiner, N. 1971. Thermal cracks in floating ice sheets. *J. Geophys. Res.* 76:694–703

Fel'zenbaum, A. I. 1958. Teoriya ustanovivshegosya dreifa l'dov i raschet srednego mnogoletnego dreifa v tsentral'noi chasti Arkticheskogo basseina. (The theory of the steady drift of ice and the calculation of the long period mean drift in the central part of the Arctic Basin.) *Probl. Sev. Akad. Nauk. SSSR* 2:16–46 (English transl. *Problems of the North* 2:13–44)

Glen, J. W. 1958. The flow law of ice. *Symp. Chamonix,* 171–83. Gentbrugge

Glen, J. W. 1970. Thoughts on a viscous model for sea ice. *AIDJEX Bull.* 2:18–27. NTIS No. PB 195 636

Gorbunov, Yu. A., Timokhov, L. A. 1968. Investigation of ice dynamics. *Izv. Atmos. Oceanic Phys.* 4(10):623–26

Hartwell, A. D. 1972. Airphoto analysis of ice

deformation in the Beaufort Sea. *AIDJEX Bull.* 13:1–33. NTIS No. PB 220 577

Hibler, W. D. 1974. Differential sea ice drift II: Comparison of mesoscale strain measurements to linear drift theory predictions. *J. Glaciol.* In press

Hibler, W. D., Ackley, S. F., Crowder, W. K., McKim, H. L., Anderson, D. M. 1974a. Analysis of shear zone ice deformation in the Beaufort Sea using satellite imagery. *The Coast and Shelf of the Beaufort Sea,* ed. J. C. Reed, J. E. Sater, 285–96. Arlington, Va: Arctic Inst. N. Am.

Hibler, W. D., Weeks, W. F., Ackley, S., Kovacs, A., Campbell, W. J. 1973. Mesoscale strain measurements on the Beaufort Sea pack ice (AIDJEX 1971). *J. Glaciol.* 12:187–205

Hibler, W. D., Weeks, W. F., Kovacs, A., Ackley, S. F. 1974b. Differential sea ice drift I: Spatial and temporal variations in sea ice deformation. *J. Glaciol.* In press

Hunkins, K. 1974. The oceanic boundary layer and ice-water stress during AIDJEX '72. *AIDJEX Bull.* 26:109–27 NTIS No. PB 236 665/AS

Kagan, B. A. 1967. Tidal ice drift. *Izv. Atmos. Oceanic Phys.* 3(8):512–16

Karelin, I. D., Timokhov, L. A. 1971. Experimental determination of the wind drag on an ice sheet. *Tr. Arkt. Antarkt. Nauch. Issled. Inst.* 303:155–65 (English transl. *AIDJEX Bull.* 17:41–53. NTIS No. PB 214 164 T)

Kheisin, D. Ye. 1971. Excitation of compressive stresses in ice during the hydrodynamic stage of compact ice drift. *Tr. Arkt. Antarkt. Nauch. Issled. Inst.* 303:89–97 (English transl. *AIDJEX Bull.* 16:97–107. NTIS No. PB 214 164 T)

Kheisin, D. Ye., Ivchenko, V. O. 1973. A numerical model of tidal ice drift with allowance for the interaction between floes. *Izv. Atmos. Oceanic Phys.* 9(4):420–29

Kovacs, A. 1972. On pressured sea ice. *Sea Ice: Proc. Int. Conf., Reykjavik, Iceland,* 276–95. Reykjavik: Nat. Res. Counc.

Lappo, S. D. 1958. O vrashchatel'nom dvizhenii dreyfushchikh ledyanykh polei. (On the rotational motion of drifting ice fields.) *Probl. Sev. Akad. Nauk. SSSR* 1:30–41 (English transl. *Problems of the North* 1:31–43)

Matsuoka, K. 1972. *The mechanics of fracture of sea ice in leads.* Master's thesis. Univ. Wash., Seattle, Wash.

Maykut, G. A., Thorndike, A. S. 1973. An approach to coupling the dynamics and thermodynamics of arctic sea ice. *AIDJEX Bull.* 21:23–29. NTIS No. PB 223 387

Maykut, G. A., Thorndike, A. S., Untersteiner, N. 1972. AIDJEX scientific plan. *AIDJEX Bull.* 15. NTIS No. 220 744. 67 pp.

Milne, A. R. 1972. Thermal tension cracking in sea ice; a source of under-ice noise. *J. Geophys. Res.* 77:2177–92

Mohaghegh, M. M. 1973a. Determining the strength of sea ice sheets. *AIDJEX Bull.* 18:96–109. NTIS No. PB 221 714

Mohaghegh, M. M. 1973b. Analysis of the failure of sea ice beams. *Preprints, Offshore Technology Conference, Houston, Texas, 1–3 May 1972.* Paper No. OTC 1809. Am. Inst. Mining Met. Eng.

Nansen, F. 1897. *Farthest North.* Westminster: Archibald Constable

Nansen, F. 1902. The oceanography of the north polar basin. *The Norwegian Polar Expedition, 1893–1896, Scientific Results,* 3:357–86. Christiana (Oslo): A. W. Brogger

Nikiforov, Ye. G. 1957. Ob izmenenii splochennosti ledianogo pokrova v sviazi s ego dinamikoi. (A change in the concentration of the ice cover in connection with its dynamics.) *Probl. Arkt.* 2: 59–71 (Transl. by Am. Met. Soc. NTIS No. AD 232204)

Nikiforov, Ye. G., Gudkovich, Z. M., Yefimov, Yu. I., Romanov, M. A. 1967. Osnovy metodiki rascheta peraspredeleniya l'da b arkticheskikh moryakh v navigatsionnyy period pod vozdeystviem vetra. (Principles of a method for computing ice redistribution in arctic seas under the influence of wind during the navigation system.) *Tr. Arkt. Antarkt Nauch. Issled. Inst.* 257:5–25. (English transl. *AIDJEX Bull.* 3:40–64. NTIS No. PB 196 063)

Nye, J. F. 1973a. The meaning of a two-dimensional strain-rate in a floating ice cover. *AIDJEX Bull.* 21:9–17. NTIS No. PB 223 387

Nye, J. F. 1973b. The physical meaning of two-dimensional stresses in a floating ice cover. *AIDJEX Bull.* 21:1–8. NTIS No. PB 223 387

Nye, J. F. 1973c. Is there any physical basis for assuming linear viscous behavior for sea ice? *AIDJEX Bull.* 21:18–19. NTIS No. PB 223 387

Nye, J. F., Thomas, D. R. 1974. The use of satellite photographs to give the movement and deformation of sea ice. *AIDJEX Bull.* 27:1–21

Parmerter, R. R. 1974a. Dimensionless strength parameters for floating ice sheets. *AIDJEX Bull.* 23:83–95. NTIS No. PB 230 378/AS

Parmerter, R. R. 1974b. A mechanical model of rafting. *AIDJEX Bull.* 23:97–115.

NTIS No. PB 230 378/AS

Parmerter, R. R., Coon, M. D. 1972. Model of pressure ridge formation in sea ice. *J. Geophys. Res.* 77:6565–75

Parmerter, R. R., Coon, M. D. 1973. Mechanical models of ridging in the arctic sea ice cover. *AIDJEX Bull.* 19:59–112. NTIS No. PB 221 039

Pritchard, R. S., Colony, R. 1974. One-dimensional difference scheme for an elastic-plastic sea ice model. *Proc. Int. Conf. Computational Methods in Nonlinear Mechanics, 1st.* In press

Reed, R. J., Campbell, W. J. 1962. The equilibrium drift of ice station Alpha. *J. Geophys. Res.* 67:281–97

Rossby, C.-G., Montgomery, R. B. 1936. On the momentum transfer at the sea surface. *Papers in Physical Oceanography and Meteorology,* MIT and Woods Hole Ocean. Inst. 4: No. 3. 30 pp.

Rothrock, D. A. 1970. The kinematics and mechanical behavior of pack ice : the state of the subject. *AIDJEX Bull.* 2:1–10. NTIS No. PB 195 636

Rothrock, D. A. 1974a. Redistribution functions and their yield surfaces in a plastic theory of pack ice deformation. *AIDJEX Bull.* 23:53–81. NTIS No. PB 230 378/AS

Rothrock, D. A. 1974b. A relation between the potential energy produced by ridging and the mechanical work required to deform pack ice. *AIDJEX Bull.* 23:45–51. NTIS No. PB 230 378/AS

Rothrock, D. A. 1974c. The energetics of plastic deformation in pack ice. *AIDJEX Bull.* 27:63–83

Rothrock, D. A. 1975. The steady drift of an incompressible arctic ice cover. *J. Geophys. Res.* 80:387–97

Ruzin, M. I. 1959. O vetrovom dreife l'dov v neodnorodnom pole davleniya. (The wind-induced ice drift in an inhomogeneous pressure field.) *Tr. Arkt. Antarkt. Nauch. Issled. Inst.* 226:123–35

Schwaegler, R. T. 1974. Fracture of ice sheets due to isostatic imbalance. *AIDJEX Bull.* 24:131–46. NTIS No. PB 232 231/AS

Shuleikin, V. V. 1938. The drift of ice fields. *C. R. Dokl. Akad. Sci. SSSR* 19:589–94

Smith, J. D. 1971. AIDJEX oceanographic investigations. *AIDJEX Bull.* 4:1–7. NTIS No. PB 220 357

Solomon, H. 1973. A one-dimensional collision model for the drift of compact pack ice. *Geophys. Fluid Dyn.* 5:1–22

Sverdrup, H. U. 1928. The wind-drift of the ice on the North Siberian Shelf. *The Norwegian North Polar Expedition with the "Maud", 1918–1925, Scientific Results,* 4:1–46

Tabata, T. 1958. Studies of viscoelastic properties of sea ice. *Arctic Sea Ice*. Nat. Acad. Sci.-Nat. Res. Counc. Publ. 598: 139–47

Tabata, T. 1972a. Observations of deformation and movement of ice field with the sea ice radar network. *Sea Ice: Proc. Int. Conf., Reykjavik, Iceland*, 72–79. Reykjavik: Nat. Res. Counc.

Tabata, T. 1972b. Microscale strain experiment. *AIDJEX Bull.* 14: 26–27. NTIS No. PB 220 859

Thorndike, A. S. 1974. Strain calculations using AIDJEX 1972 position data. *AIDJEX Bull.* 24: 107–29. NTIS No. PB 232 231/AS

Thorndike, A. S., Maykut, G. A. 1973. On the thickness distribution of sea ice. *AIDJEX Bull.* 21: 31–47. NTIS No. PB 223 387

Volkov, N. A., Gudkovich, Z. M., Uglev, V. D. 1971. Results of the study of non-uniform ice drift in the Arctic Basin. *Tr. Arkt. Antarkt. Nauch. Issled. Inst.* 303: 76–88. (English transl. *AIDJEX Bull.* 16: 82–96. NTIS No. PB 214 164 T)

Weeks, W. F., Kovacs, A., Hibler, W. D. 1971. Pressure ridge characteristics in the arctic coastal environment. *Proc. Int. Conf. Port Ocean Eng. Arct. Cond., 1st.* Tech. Univ. Norway. 1: 152–83

Witting, J., Piacsek, S. A. 1972. Arctic ice circulation model. *EOS, Trans. Am. Geophys. Union* 53: 1016 (Abstr.)

Yegorov, K. L. 1970. K teorii dreyfa ledyanykh poley v gorizontal'no neodnorodnom pole vetra. (The drift theory of floes in a horizontally heterogeneous wind field.) *Probl. Arkt. Antarkt.* 34: 71–78 (English transl. *Problems of the Arctic and Antarctic* 33–35: 231–38. NTIS No. TT 72-50006)

Zubov, N. N. 1943. *Arctic Ice* (English transl. NTIS No. AD 426 972)

MECHANICAL PROPERTIES OF GRANULAR MEDIA

×10043

J. Rimas Vaišnys and Carol C. Pilbeam
Kline Geology Laboratory, Yale University, New Haven, Connecticut 06520

INTRODUCTION

Were it possible to endow the phrase "in principle" with even a small fraction of the operational significance usually attributed to it, most problems encountered in describing and understanding granular systems would vanish. All the theoretical and experimental principles required for dealing with granular systems fall clearly in the province of very classical physics. Each individual step of any envisioned grand algorithm for the complete description of the mechanical aspects of granular systems can be carried out not only in principle, but also in practice. Nevertheless, the overall problem of describing and understanding granular media is such that all approaches known to us do not even approximately account for the properties of an arbitrary granular system. Given this situation of prejudice (with regard to what is understood) and frustration (with regard to what can be demonstrably accomplished), it is probable that, were it not for the ubiquity of granular media in various natural and technological settings, the problems met in dealing with them would be solved simply by declaring them to be nonproblems.

A granular system may be defined as a system made up of discrete material particles arranged so that direct, but noncohesive, forces between the particles are significant in determining the properties and behavior of the system. The set of granular media is thus a fuzzy set, reflecting the natural variety. We encounter aggregates in the form of beach sands, fault gouge, sediments, partially melted mantle rock, accreting planets, and lunar surfaces. In a technological setting, they are met in the processing of aspirin, catalysts, ores, and zeolites. Less specifically granular, but sharing properties in common, are systems such as sintered compacts and fractured rocks.

As may be imagined, the variety of guises under which granular media are found poses problems in reviewing the subject. Even if the topic is restricted to the mechanical properties of granular media, as it is here, one must deal with hundreds of works. Sheer bulk, variety in viewpoints, nonconformity of calibration, terminology, and procedures, and finally, dissimilarity in standards of measurement and reasoning make even a partial synthesis essentially impossible. Nevertheless, certain ideas and certain factual information can be transferred between areas of research which are very different in orientation and aims. Therefore we have

343

organized the review around two major themes. The first section considers aspects of methodology common to fields which study granular media. The second section discusses specific results from a particular theoretical viewpoint. Because of space limitations, background material is only briefly presented, and the article by Deresiewicz (1958a) is recommended as collateral reading.

GENERAL APPROACHES

Most problems involving the mechanical properties and behavior of granular systems may be reduced to the following form: we are given some materials, for which the constitutive equations are known either from calculation or measurement, in a certain known state at time t_0. A process is carried out to give a granular system in a certain state at time t_1. At time t_1 the granular system is subject to some boundary and body forces that have a known time development, until time t_2. We wish to know the state of the granular system at time t_2. The above procedure, with due care in the choice of process and force histories, presents a well-defined problem, in both the physical and mathematical senses, which is solvable in the best "in principle" tradition. Preparatory to discussing the difficulties which arise in practice, the above paradigm may be divided into the following stages: 1. specification of the constitutive equations, 2. specification of the mass distributions at t_1, 3. specification of the dynamical variables at t_1, 4. specifications of the applicable equations of motion and any other constraints, 5. specification of the forces acting on the system, possibly as functions of time, 6. solution of the equations of motion and constraint. In the above, it will be noted that steps 2 and 3 implicitly specify the process which is used to generate the granular system. The other steps schematize the usual approach used in problems of classical physics.

To gain an appreciation of two broad categories of problems encountered in dealing with granular systems, let us consider three specific "simple" systems to which the above paradigm may be applied: (a) A number of elastic spheres of equal size and of a known material are packed in a cubic close-packed array and subjected to a hydrostatic pressure. (b) A number of spheres as in a are packed in an irregular but well-specified packing and subjected to the same stress history as system a. (c) A situation identical to b exists, except that the spheres are packed randomly. In describing system a, we encounter no difficulties in carrying out the program directed by the paradigm, provided, first, the system remains within the regime of linear infinitesimal elasticity; second, we are a bit more precise in specifying the packing process so that the initial conditions are well-defined; and third, pressure is applied adiabatically. Indeed, even such real complications as frictional effects may be accounted for to a good approximation.

In attempting to describe system b, we have a situation which in principle is no different from that encountered with system a; in practice, however, serious difficulties arise when we attempt to carry out point 6 of the paradigm. It should be noted that system b is no less well-defined or determined than system a. The lack of symmetry in system b, however, makes an analytical solution impossible; we must either turn to numerical computer solutions or devise some sort of acceptable analytical approximations. The extent to which either approach is possible or satisfactory will be discussed below. At present we simply wish to note that consideration of system b has introduced us to one category of problems encountered in analyzing granular systems, a category we will refer to as computation-limited problems.

We now consider system c in light of the paradigm. The specification of random packing is shorthand for an incompletely specified or described packing process which leads to incomplete knowledge about the configuration of the spheres in system c. A problem arises because the random packing process forces us to consider a number of different possible packings, yet provides us with insufficient information to choose a unique structure for consideration under the paradigm. There is no doubt (within the underlying classical physics outlook) that system c involves a well-defined and definite configuration of spheres; the problem arises because we are unable to decide which of the possible configurations actually obtains. We are thus confronted with an example from the second category of problems encountered in dealing with granular media, a category we refer to as information-limited problems.

Let us now examine a fourth and perhaps a more interesting system in light of the above discussion. As an example of a geologically interesting granular assemblage, consider a region of partially melted rock in the mantle, with a view towards accounting for its seismic properties. It is probably clear that we shall encounter both computation- and information-limited problems at each step of our paradigm in attempting to describe this system. The processes leading to the formation of this assemblage are known only in general terms, so that neither the constitutive equations, the initial configuration, nor the existing stress conditions are uniquely defined. Even if the information were available to define a unique physical system, it is probable that complexity of geometry, pertinent constitutive equations, and boundary conditions would make the problem intractable without serious approximations.

With real systems, problems of both categories arise simultaneously and may indeed so interact that it becomes inappropiate to differentiate sharply between them. We have taken pains to distinguish between the two types of difficulties because fundamentally different approaches are involved in solving the two kinds of problems. For mathematical problems, essentially mathematical cures are required. These may involve theoretical developments, numerical analysis, or even experiments designed either to bridge a mathematical hiatus or to check the validity of an approximation. In the face of an information-limited problem, no amount of mathematizing in itself can be of much help. Additional information is required, which may be gained by further observations and experiments or by theoretical analysis, but on a different epistemological level from the problem level (e.g. the use of an atomic level theory to predict a constitutive relation).

Computation-Limited Problems

We now turn to an examination of ways in which the computation-limited problems (as classified above) may be approached.

REGULAR SYSTEMS   One of the major complications presented by real granular systems is the complexity of grain shapes, size distributions, and arrangements. It is therefore natural to examine simpler systems, some of which are realizable experimentally, involving more symmetrical, and hence analytically tractable, geometries. The simplest system representing a granular system is that of two elastic spheres in contact, investigated by Hertz (1881). These solutions, and their extensions to more general boundary surfaces, remain basic to much of the theoretical work on the mechanics of granular media. [Some of the solutions seem to be implicit in the earlier work of Craig (1880) on the distortion of the earth by polar ice caps.] Also relevant, though more directly applicable to composite rather than to granular media, is the work of Rayleigh (1892) on the properties of regular arrays of spheres and cylinders embedded in a matrix. This paper, which seems to be more often cited than read, is interesting for its combination of mathematical and physical reasoning. (At this point it is appropriate to point out the existence of helpful formal similarities and relations, realized by Rayleigh, between the descriptions of mechanical properties and the descriptions of electrical, magnetic, thermal, diffusive, and even viscous flow processes. In these regards, granular media are also closely related to heterogeneous and composite media.) Since then, the study of granular systems, chosen for the tractability of the corresponding mathematical analyses, has mostly concentrated on the study of various regular packings of uniform, elastic spheres. This work is referred to in the next section. At this point we wish to note two general conclusions: first, there is good agreement between the theoretical calculations and experiments for such systems, so that it is fair to say that a number of aspects of such systems are well-understood. Second, it has been found that the accurate experimental realization of these simple systems is harder than expected at first sight, so that conclusive results appear at a slow rate. Finally, we wish to mention directions in which further work, both theoretical and experimental, with these simple systems would be desirable: further study of energy-loss mechanisms under small amplitude mechanical excitation, the stress-strain response for spheres having nonlinear elastic and plastic constitutive relations, and the onset of particle fracture in regular arrays. Work with different contact geometries would also be desirable.

EFFECTIVE COMPONENTS   When systems have geometrical irregularities, whether in particle size, shape, or packing, analytical results are almost impossible to obtain without drastic simplifications. One frequently used approximation is that of an effective or representative particle, in which the parameters describing spheres are adjusted to represent the much more complex real particle interactions. Because it is usually the numbers of contacts and the contact geometries that are varied by the investigator, this approach may be called a contact theory approximation. The approximation, often justified on elementary probabilistic grounds in terms of average numbers of contacts, etc, is discussed in the next section.

   Both the exact and the approximate analyses mentioned so far attempt to arrive at a description of a granular system which both retains the maximum detail possible and also reflects the physical mechanisms as faithfully as possible. Thus, to the extent that the analyses are successful, one can delay formulation of specific questions until after the analysis is performed. If one is willing at the start to limit the sort of phenomena that will be dealt with, it becomes possible to generate other kinds of mathematical descriptions. An example of a more limited orientation is provided by the notion of effective-constitutive relations (e.g. effective elastic moduli), used in studies of polycrystalline and composite materials (Polder & Van Santen 1946, Landauer 1952, Kerner 1956, Hill 1963). The approach is included here because it is felt that using it more extensively in problems with granular media is likely to be productive.

Consider a granular system with some complicated internal geometry, where the largest particle size is much smaller than the overall size of the system, and subject to boundary stresses. If it is known in advance that any additional applied stresses will not change the particle geometry of the system appreciably, or that the gradient of the scaled additional stresses is small compared to the largest particle size in the system, one would probably expect, on an intuitive basis, to write relatively simple effective-constitutive relations between macrostrains and macrostresses (those measured over subsystem volumes that are large compared to the particle sizes, but still small compared to the system volume). Thus, for certain granular media (see below), one finds that an effective bulk modulus may be defined which is proportional to the one-third power of the hydrostatic pressure. The introduction of such an effective modulus, when appropriate, has the advantage that the sets of differential equations subject to the usually very complex particle boundary conditions are replaced by a smaller set of differential equations which involve only the much simpler system boundary conditions. Some disadvantages of such a simplification are that often the new differential equations are no longer linear, even when the underlying description itself is linear (as in the example above) and that information which indicates the limits of valid approximation is eradicated from the explicit formalism. Examples of analysis depending on effective-constitutive relations, and also involving systems containing fluid may be found in Frank (1965) and Biot (1973). Only rarely can effective-constitutive relations be derived theoretically. When one must rely almost entirely on experiments and observations for constitutive relations, it is well to recognize that a classical description imposes some basic restrictions, such as invariance requirements, which can be profitably incorporated in the form of the proposed relationships.

The mathematical description of granular systems in which large-scale deformations occur presents special problems. Even when the deformations involve only particle rearrangements and the geometry changes can be considered to occur quasi-statically, the deformation will usually depend on the stress history (Thurston & Deresiewicz 1959). Analyses of motions in which the kinetic energy of the particles is a significant factor have primarily been restricted to kinematic considerations (Brown & Richards 1965). Deformations involving changes in particle shape are discussed below.

VARIATIONAL APPROXIMATIONS   The Ritz (1908) variational method is a procedure for generating approximations that deserves to be used more in analyzing granular systems. This method may be employed whenever a physical system can be described by an extremum principle. It is attractive because auxiliary information may be incorporated in choosing the approximate or trial solutions, because it may be used with descriptions both on the particle and on the effective-constitutive relation levels, and because it is naturally related to the finite element techniques used in computational work. In the case of both elastic and plastic constitutive relations, both upper and lower bounds may be derived from the extremum principles (e.g. principles of minimum potential energy and complementary energy in elasticity) so that an indication of the accuracy of a proposed approximation may be obtained. Because the bounds are integral in nature, the approach is concordant with the intuitive description of a system, in terms of the macroscopic effective-constitutive parameters referred to earlier. For systems which can be described in terms of contacts, the natural trial functions are the analytically known contact stress-or-strain functions centered at the corresponding contacts, and the natural variational parameters are then the corresponding contact forces or contact displacements. It seems that for such systems, computer numerical experimentation with granular systems consisting even of irregular particles in irregular arrangements ought to be possible. A summary of variational approaches is provided, for example, by Pearson (1959) and Hill (1967), and a systematic and readable discussion of approximations in terms of function space methods is given by Synge (1957).

Information-Limited Problems

We now consider the problems that arise in describing and understanding granular systems when the available information is insufficient to define a unique structure or process. Under these circumstances, conventional wisdom indicates that a statistical approach is appropriate. It will be recalled from the discussion following the description paradigm that difficulties caused by the lack of necessary information can arise at any step of the paradigm. The statistical formalism may be used at any, or all, stages of the process, but to simplify the discussion both conceptually and notationally, it will be presented in terms of a simple, concrete example. In the following we will consider a granular system for which the only source of uncertainty is an incompletely specified elastic modulus which varies with position and is denoted by $m(r)$. Such a system might arise from a process in which one or more materials, characterized by given elastic moduli, are divided into particles. The particles are combined in known proportions and assembled into a specific granular system of volume V. Then the system is presented to us for description of its effective modulus. In the following, we outline a statistical approach with reference to the incompletely specified $m(r)$ and then examine some specific questions that may arise with the practical use of such descriptions.

STATISTICAL DESCRIPTION   In the case under discussion, the given information about the component volume fractions is insufficient to determine a unique $m(r)$ to which the paradigm could be applied. To deal with this underdeterminacy, we introduce the notion of an ensemble of systems, i.e. a collection of all possible systems which have different modulus functions $m(r)$ but the same volume and composition.

Let us distinguish the different ensemble members by adjoining an index v to $m(r)$, so that we write $m(r, v)$. If we fix our attention on any modulus function in the ensemble, $m(r, v)$, the corresponding physical system is uniquely specified, and there is a correspondingly well-defined mathematical problem whose solution would define the corresponding effective modulus, $m_e(v)$. In principle then, we could calculate the effective modulus for each member of the ensemble, thereby obtaining an ensemble of effective moduli.

Because in most cases an ensemble contains an astronomical number of systems (the membership may not even be countable), an alternative procedure is used. We describe the ensembles in probabilistic terms. For example, we introduce the probability density functional $P[m(r)]$ for the modulus function (this is usually interpreted loosely as giving the probability of finding a system described by the given modulus function) and the probability density function $P(m_e)$ for the effective modulus. By use of the applicable equations of the paradigm discussed earlier, it is possible to derive a differential equation relating $P(m_e)$ to $P[m(r)]$. Because solutions to such functional equations are not generally known, and because the available information may not specify the initial functional—here $P[m(r)]$—completely, the above equations are usually integrated (in effect) to give simpler, although less informative, differential equations involving the moments of the various distributions. Unfortunately, the moment equations are usually not closed, so that approximations must be made and sets of equations must be solved, even to calculate the expected value of some quantity. The reader is referred to Batchelor (1956) and Blanc-Lapierre & Fortet (1968) for examples of applications of statistical descriptions to physical systems.

PROBLEMS AND APPLICATIONS   We have already indicated that the use of complete statistical descriptions is beset by mathematical problems. In addition, there are conceptual difficulties involving the interpretation of the basic concepts in probability (Savage 1972). (For example, a frequency interpretation of probability can introduce difficulties in using the ensemble concept when dealing with a unique object such as the earth.) With ambiguities on the conceptual level, difficulties can also arise in setting up clear correspondences between physically measured quantities and the entities of the mathematical representation.

It may be pointed out that under certain circumstances the direct use of a statistical approach may be unnecessary or simply inefficient. In the case of the physical example discussed above, $P[m(r)]$ in principle represents all the information available about the system. However, suppose we are interested only in assessing the effective modulus of the most and least rigid systems, compatible with the preparation procedure. In this case, for example, it would probably be more efficient to use a dynamical programming approach to generate the two systems of interest than to attempt calculations of $P(m_e)$. Also, in this case, a statistical calculation in terms of the moments, unless done quite exactly, would do little to provide the desired answer.

In the description of granular systems, statistical approaches have been almost completely confined to describing particle-packing geometries, discussed in the next section. Methods used by Frisch & Stillinger (1963) and Miller (1969) to describe liquids and heterogeneous media, respectively, may be applicable to the description of more complex granular systems as well.

As stated above, a statistical approach may be used to cope with what we have termed an information-limited problem. The above discussion should make it evident that introducing a probability density functional and a statistical formalism does not per se correct the lack of information; instead, what is achieved is a consistent codification of the lack of knowledge.

CONTACT DESCRIPTION OF AGGREGATES

Packing Properties

Because of their analytical simplicity, regular arrays of equal spheres have frequently been employed to model granular aggregates. The most popular candidates have been the close-packed structures, fcc and hcp packings, with coordination number 12 and packing density 0.74. Although many real aggregates do consist largely of spherical particles, their packing is much more likely to be random than regular. Indeed it was the similar divergence of both liquid molecular structure and the irregular packings of hard spheres from regular crystalline structures that led to extensive work on the random arrangements of spheres (Bernal 1964). Direct measurements on thousands of steel balls (Bernal & Mason 1960, Scott 1962, Scott & Kilgour 1969) have determined the maximum packing density and the average coordination number for randomly packed equal spheres to be 0.64 and 6+, respectively. There is also a random loose packing with a packing density of about 0.60, but it is less reproducible than the random close packing (Scott & Kilgour 1969). If a regular array is considered necessary to model aggregate properties that depend strongly on contact behavior, a simple cubic array is probably more representative than the close-packed arrays.

A major difficulty in applying contact theory to random packings of spheres is that for any sphere in the packing there is, instead of a simple coordination number, only an average or expected number of neighbors at a particular radial separation. This distribution of neighbors can be expressed mathematically as a radial distribution function which gives the average number of sphere centers as a function of increasing radial distance from a central sphere. A discussion of some characteristic features of the radial distribution functions derived from the model experiments on steel balls is given by Finney (1970).

An example of one of the problems with which a contact theory description of aggregates must cope is found in Bernal & Mason (1960). Using paint to mark the contacts in a random packing of ball bearings, they measured an average of 8.5 "contacts" per ball in the maximum density packing. Of these average contacts, 2.1 were near contacts, i.e. separated from the central sphere by gaps up to 5% of the sphere radius. The remaining 6.4 contacts were called close contacts. These results suggest that variation in the average number of close contacts under various measurement conditions will be an important feature of aggregate behavior. This discussion also points out the basis for ambiguity in the literature regarding the definition of contact. Some authors, such as Bernal & Mason, implicitly define the average number of contacts per sphere as the average number of first or nearest neighbors (the first peak in the radial distribution function), a definition which reflects their interest in the structure of liquids and perhaps the experimental difficulties in distinguishing between close and near contacts as well. For contact theory purposes, however, the distinction is important because a near contact does not transmit forces.

A recent approach to the statistics of random packings of spheres is computer simulation. Examples are the studies by Norman et al (1971), Adams & Matheson (1972), Bennett (1972), and Visscher & Bolsterli (1972). The basic procedure for generating the models in these studies was to select a sphere with a particular radius from some prescribed probability distribution of radii, and to place the sphere in the growing assembly according to a deterministic algorithm. In the study by Visscher & Bolsterli, the packing rule was chosen to reproduce the effects of a vertical gravitational force, whereas in the other three studies, the algorithm imitated a central, long-range, attractive force. To date, in addition to elucidating the problems involved, a major achievement of such simulations has been to generate equal sphere models that almost reproduce the packing density and average coordination values obtained in the experimental work. Norman et al (1971) attempted to use computer models, which had been constructed to reproduce experimental results, to predict additional statistics not available from experiments. The models studied were relatively small, however, and it is not clear whether the resulting statistics warranted such an extension. What does seem clear is the observation by Finney (1970) that to make real progress in the statistics of irregular sphere packings, it is necessary to formulate the statistical geometry in mathematical terms.

Looking for a relationship between density and average coordination number, Norman et al (1971) graphed data from regular array calculations, from experimental studies on random packings, and from their simulated packings involving several radii distributions. The empirical fit to the data was given by

$$\bar{N} = 1.126 \exp(3.196\phi) + 0.860 \exp(-3.50\phi)$$

where $\bar{N}$ is the average number of contacts per particle and ϕ is the packing density. A somewhat similar expression has been suggested by Ridgway & Tarbuck (1967), who pointed out that there is no simple relationship between $\bar{N}$ and ϕ, and that their expression represented a trend rather than an accurate relationship. In both of these studies the definition of $\bar{N}$ should really be the average nearest neighbor coordination, inasmuch as the experimental $\bar{N}$ values used included both close and near contacts.

Very little is known about the statistics of random packings of irregularly shaped particles. One effect of decreasing the average particle sphericity is to increase the aggregate porosity, as observed early by White & Walton (1937). In recent work by Koerner (1970) on crushed quartz powders, both maximum (vibrated powder) and minimum (poured powder) densities decreased with decreasing sphericity. Similar results were found by Riley & Mann (1972) for differently shaped glass particles, with the exception of flaky particles. The more open structure of such packings can be qualitatively explained by the occurrence of contacts at projecting points, but whether the lower densities indicate fewer contacts per grain is not known.

Contrary to geometrical arguments, increasing porosity usually accompanies decreasing average grain size for the finer sizes of both spherical and irregular particles. Westman & Hugill (1930) observed this size effect in sands having average grain diameters from about 0.45 to 0.009 cm. In Koerner's work (1970) on quartz powders, both maximum and minimum densities decreased with decreasing particle size, at essentially equal rates, for average grain sizes from about 0.3 to 0.004 cm. The results of a study made by Arakawa & Suito (1969) of loosely packed powders covering a range of average particle diameter from about 10^{-2} to 10^{-4} cm could be described by the relationship $v \propto (1/d)^n$, where v is void volume and d is average grain diameter. The exponent n, constant for any particular powder, was found to vary approximately linearly with the reciprocal of the material density. In contrast with the mixing results obtained by Westman & Hugill and Koerner, the results of Arakawa & Suito show that the porosity increased with the addition of finer sizes. The standard coarse powder used for their mixing experiments was about 12 μm in diameter.

The origin of this size effect must be related to the increasing importance of surface area and, consequently, of surface cohesive and frictional forces and of surface relief with decreasing particle size. If there are short-range attractive forces between particles, fewer contacts per particle are necessary for a stable packing. For example, a computer analysis by Vold (1959) has shown that cohesive forces between particles can give rise to very open packings with a mean coordination number of two. Also, increasing the relative importance of surface roughness should be analogous to decreasing the sphericity.

The decrease of porosity with the mixing of particle sizes (except for the very small sizes) has long been a subject of empirical studies, the results of which have generally been limited to very special cases. References to much of this work may be found in Deresiewicz (1958a) and Norman et al (1971). Some of the complexities encountered in trying to describe ternary mixtures of spheres are illustrated by the work of Dexter & Tanner (1971).

Small Deformations

Given any assembly of particles that is being loaded, the distribution of contacts will determine the distribution of forces within the aggregate. The results of random packing studies suggest that the forces will not be distributed uniformly. Direct observations of stress distribution have been made by Dantu (1957) and by Drescher & de Josselin de Jong (1972) in photoelastic studies of a two-dimensional model of granular media constructed of plastic discs stacked between two glass plates and subjected to various applied loads. They found that the load was largely transmitted by relatively rigid, heavily stressed lines of discs, forming a network of contact forces. Groups of discs separating the lines of stress were only lightly loaded. In preliminary work by Chen (1974) on a similar model, lines of stressed particles were found to carry 50–90% of the load while occupying only 25–35% of the total sample area. The load on any one contact and the total number of loaded contacts increased with increasing applied load in such a manner than the average load per contact increased relatively slowly. A similar interrelation between asperity load and the number of loaded asperities is indicated for rough surfaces in contact (e.g. Archard 1953, Greenwood & Williamson 1966).

In view of the complicated nature of the distribution of contacts and contact forces in aggregates, solving the general problem of aggregate dynamical behavior appears unlikely, given the present state of contact theory. One approach is to isolate a part of the problem by asking a specific question, then trying to answer the question with a relatively simple model—one that can be treated analytically, although it may have limited application. The question of the pressure dependence of velocities in aggregates has been approached in this manner (e.g. Brandt 1955, 1967, Pilbeam & Vaišnys 1973) and some of the results are discussed below. The discussion first considers aggregates at low confining pressures, where large-scale changes in geometry are not important, then continues to the higher compaction pressure regime, where they are.

The discrete analysis of wave propagation in aggregates begins with the elastic response of single contacts to applied forces. The basic relationships, which have been empirically verified (e.g. Mindlin et al 1951, Johnson 1955), are the normal force-displacement relations derived by Hertz (Timoshenko & Goodier 1951) and the tangential force-displacement relations derived by Mindlin (1949). These relations have been applied by Duffy & Mindlin (1957), Deresiewicz (1958b), Duffy (1959), and Thurston & Deresiewicz (1959) to ordered arrays of uniform spheres to obtain incremental stress-strain relations for the arrays.

Both the tangential and normal load-displacement relations are nonlinear, and the tangential relation is also inelastic; consequently, the incremental stress-strain relations are dependent on loading history. However, for variations in stress that are small compared to the initial state, such as those accompanying acoustic wave propagation, the incremental form of the stress-strain relations can be used directly, and the change in contact stiffnesses with the stress variation can be neglected. Following this approach, velocities have been computed for close-packed regular arrays of equal spheres under isotropic confining stress and compared with experiments on bars constructed of such arrays (Duffy & Mindlin 1957, Duffy 1959). For all modes of propagation, the theory predicts regular array velocities of the form $\bar{\rho}V^2 = Kp^{1/3}$, where $\bar{\rho}$ is the array density and p is the confining pressure. The coefficient K is dependent on the elastic constants of the solid material and on the array geometry and direction of wave propagation. The experimental regular array results, obtained at confining pressures of less than 1 bar, substantiate the basic theoretical formulation but indicate that even under such ideal conditions, the making and breaking of contacts with pressure variations cannot be ignored. A comprehensive review of contact theory and related research is given by Deresiewicz (1958a, 1966).

The predicted dependence of the velocity modulus, $M = \bar{\rho}V^2$, on the third root of the confining pressure is inherent in the definition of Hertzian contact, and is independent of grain size and shape and of packing geometry, as long as the geometry is unchanging with pressure. This pressure dependence is, then, the dependence that would be expected for randomly packed, elastically deforming aggregates if it were not for the variation of mean coordination with varying pressure.

To study the applicability of contact theory to real systems, Pilbeam & Vaišnys (1973) made laboratory measurements of extensional and torsional velocities of irregular packings of several types of particles at confining pressures of less than one bar. Changes in $\bar{\rho}$ over the pressure range studied were immeasurable. Velocities of aggregates of spherical particles were found to vary as the 1/3 to 1/3.3 power of the pressure. To interpret these results, the aggregate was modeled by a section of a simple cubic array of like spheres modified to include missing contacts. From the model, a relationship of the form $\bar{\rho}V^2 \propto (n_c/n)^{2/3}p^{1/3}$ was derived, where n_c/n is the ratio of close contacts to total possible contacts at confining pressure p (Pilbeam 1971). If the number of possible contacts is assumed to vary linearly in the direction of relative displacement of sphere centers, the relationship $\bar{\rho}V^2 \propto p^{1/3.3}$ can be derived (Pilbeam & Vaišnys 1973), in good agreement with the observations. Other aspects of the study, such as the influence of particle shape and size on the velocity, were qualitatively explained in terms of contact theory.

There is a large literature on velocities measured in rocks and sediments and model aggregates, mostly at low pressures. References to much of this work may be found in Deresiewicz (1958a, 1966), Hardin & Richart (1963), Krizek (1971), and Hamilton (1972). With the exception of the work of Brandt (1955, 1967), White (1965), and Pilbeam & Vaišnys (1973), little attempt has been made to interpret the measurements within the framework of contact theory. Because a large number of factors, such as particle shape, size distribution, and surface condition, can affect the predictions of contact theory in many complicated ways, it would be an inconclusive exercise to try to analyze large volumes of data that were obtained without contact theory relations in mind. It can be said, however, that the general features of aggregate velocity data are consistent with contact theory principles. We therefore confine the following brief discussion to some general comments regarding contact theory expectations for velocities.

For wavelengths much larger than the grain size, the velocity is expected to be independent of frequency, and this expectation is borne out by empirical data. As the amplitude of vibration increases, the nonlinearity of the contact stiffnesses will be reflected in the velocities. Both normal and tangential stiffnesses vary directly with the contact area, and the contact area increases with increasing normal force. Mindlin (1949) has shown that under tangential loading, slipping occurs at the contact periphery and progresses inward as the tangential force increases. Consequently, the contact area decreases as the ratio of tangential to normal force increases. In general, for a constant vibration amplitude, the effects of non-linearity will become more obvious as the confining pressure decreases. The presence of a network of highly stressed regions, as discussed above in connection with the photoelastic studies, may act to enhance the basic nonlinearity. Under some conditions, the network may give rise to an additional channel for wave propagation, the more rapid path being through the lines of heavily stressed contacts.

The inclusion of tangential contact stiffnesses in calculating velocities for regular arrays under isotropic stress (Duffy & Mindlin 1957, Duffy 1959) was an important step because the tangential stiffness is of the same order of magnitude as the normal stiffness. In real aggregates, considerable tangential forces arise at contacts on the application of external loading, and the inclusion of tangential stiffnesses derived from a small stress approximation is probably unrealistic. The major effect of neglecting the tangential stiffnesses in calculations is to predict lower velocities than would be predicted otherwise.

The dependence of velocity on porosity is a long-debated topic. In contact theory the porosity affects the velocity directly only through the term for bulk density. The velocity modulus is some function of the average number of close contacts per particle, and the porosity enters here only insomuch as the porosity and the number of close contacts are related. Generally, the velocity is observed to decrease as the porosity increases, with the modulus decreasing more rapidly than the bulk density (e.g. Hardin & Richart 1963, Watkins et al 1972). This trend recalls the relationship between bulk density and average number of contacts per grain discussed above. As was noted there, however, the term contacts included both close and near contacts; and as the discussion of the velocity-pressure results has shown, the conversion of near to close contacts can increase the velocity modulus significantly with immeasurable change in bulk density. On the other hand, examples can be constructed in which the density changes significantly but the modulus does not. Consider initially a simple cubic array of like spheres. Imagine each sphere to be replaced by a cube that falls slightly short of circumscribing the sphere, leaving the projecting sphere as bumps on each face at the original contact points. Although the bulk density is greater in the modified array, the number of contacts per particle, the effective particle radius, and the radius of curvature at contacts are all the same for both cases, and thus the velocity modulus is the same.

One last point, elicited by some confusion in the literature, is that contact theory predicts the velocity modulus will increase with increasing contacts per particle, not simply with increasing contacts. The predicted velocities of contact theory are independent of grain size. For example, the same velocity is predicted for two simple cubic arrays of equal volume, one with spheres twice the size of the other and hence with one eighth the total number of contacts. If the geometry of either array is altered in the direction of closer packing, for example by sliding every other layer of spheres over by a sphere radius, then the modulus is increased.

The inclusion of the effects of slipping at contacts under small oscillating tangential or oblique forces gives rise to energy dissipation (Mindlin et al 1951, Mindlin & Deresiewicz 1953). For single contacts subject to tangential oscillations, the fractional energy loss per cycle, $2\pi Q^{-1}$, is proportional to T^*/N_0, for $T^*/N_0 \ll \mu$. T^* is the maximum amplitude of the tangential component of force, N_0 is the normal preloading force, and μ is the coefficient of friction. As T^*/N_0

approaches μ, Q^{-1} increases with higher powers of T^*. Extension of the contact relations to energy losses in regular arrays during wave propagation indicates that Q^{-1} should increase with the first power of the wave amplitude (Duffy & Mindlin 1957). Experiments on single contacts by Mindlin et al (1951), Johnson (1955, 1961), and Goodman & Brown (1962) show that, although the force-displacement relations are verified, the theoretical energy dissipation relations correlate with observations only at the higher values of T^*/N_0. At the lower values characteristic of acoustic wave propagation, the single contact Q^{-1} is constant, independent of T^* and N_0, and is considerably larger than the predicted Q^{-1}. It is not surprising, therefore, that the data from regular arrays (Duffy & Mindlin 1957) and from irregular packings (Pilbeam & Vaišnys 1973) indicate that Q^{-1} for acoustic propagation is independent of wave amplitude and frequency, and is higher than predicted. Some experimenters do find some dependence of Q^{-1} on amplitude in vibration testing (e.g. Hall & Richart 1963), but the dependence is less than predicted. As the vibration amplitude increases, and T^*/N_0 at contacts increases, the single contact data suggest that Q^{-1} should begin to increase with amplitude. As was true for the velocities, the point at which this nonlinearity becomes important will vary with the packing properties and the confirming pressure.

To explain the discrepancy between theory and experiment, White (1966) proposed that energy dissipated during small amplitude displacements at contacts primarily reflects the work required to overcome the difference between static and kinetic friction coefficients before slipping can occur. His expression for Q^{-1}, for both single contacts and regular arrays, depends only on the difference between the coefficients and has never been experimentally tested. The problem arises, however, that Q^{-1}, both for regular arrays and irregular packings at low confining pressures, decreases with increasing confining pressure (Duffy & Mindlin 1957, Pilbeam & Vaišnys 1973). And this pressure dependence cannot be explained simply by missing contacts, because for White's mechanism both energy loss and energy storage change proportionally with changing numbers of contacts. At higher pressures there is some evidence that Q^{-1} may become independent of confining pressure (Gardner et al 1964).

There is, then, no satisfactory interpretation of energy dissipation in aggregates during wave propagation. About the strongest statement that can be made is that the losses are associated with movement at contacts, as indicated by the pressure dependence of Q^{-1} and by the increase of Q^{-1} with decreasing contact friction (Pilbeam & Vaišnys 1973).

Compaction

At the higher pressures characteristic of compaction tests on aggregates, major changes in the packing structure occur, changes which depend on the mode of deformation of the aggregate material. The microscopic mechanisms of the compaction of powdered materials have been recently studied by Hardman & Lilley (1973). They measured changes of dominant pore radius and particle surface area during the compaction of powdered sodium chloride, sucrose, and coal, and compared these measurements with observations of sections of the compacts made by scanning electron microscope. Although the compaction curves were similar for all three, the sodium chloride particles deformed plastically, whereas the compaction of sucrose and coal was dominated by particle fracture. The filling of pores with fragments was found to be an important process in the compaction of sucrose. Fracture fragments in the coal compacts tended to remain as chains or bridges of small particles wedged between larger particles, rather than filling in pores. This last observation suggests the descriptions of fracturing in model aggregates reported by Gallagher (1971). Looking at sections of sand aggregates that had been raised to confining pressures of 350 and 1000 bars, he found chains of fractured particles oriented at small angles to the direction of greatest load. The chains were more numerous in samples that had been raised to the higher pressure.

It would be surprising, perhaps, to find any aspect of such a complex process as compaction that could be consistently interpreted in terms of contact theory. Nevertheless, the dependence of velocity on compacting pressure is again suggestive of contact theory relations.

The variation of compressional velocity during the compaction of several types of powders under uniaxial loading from 10 to 10^3 bars was studied by Vaišnys & Gordon (1965). Although some of the powders were compacted primarily by plastic deformation at contacts and others by fracture, velocities on the initial loading of all powders increased with the 1/3 power of pressure until a pressure comparable to the characteristic strength of the material was reached, at which point the velocity exponent began to decrease. In all cases, no more than 10–15% of the power dependence of velocity on pressure was accounted for by the decrease of bulk density with increasing pressure. On cycling the pressure below the maximum value reached on initial loading, the velocity-pressure exponent was 1/4 to 1/5 for the fractured powders and 1/5 to almost zero for those in which the contacts had deformed plastically. For all powders, the void ratio (the ratio of void volume to solid volume) was permanently decreased on initial loading and showed very little recovery during cycling. Comparison of void ratio and velocity changes during loading and cycling is another reminder that velocity and porosity are not uniquely related. Data reported by Warren & Anderson (1973) on sands do not appear to be in disagreement with the above results.

A qualitative explanation of these results can be given as follows. For powders of brittle materials, the process of fracturing provides a reservoir of new contacts and maintains a supply of near contacts to be converted to close contacts with increasing pressure. The velocity-pressure exponent of 1/3 on initial loading can therefore be explained in terms of increasing average contacts per particle, as discussed above for the velocity measurements at low confining stresses. The smaller exponent on pressure cycling is explained by a decrease in the extent of fracturing. Plastically deforming contacts are assumed to respond elastically to the small stress oscillations associated with wave propagation. The velocity-pressure exponent is greater than the Hertzian exponent of 1/6 for plastically deforming powders on initial loading because of the more rapid increase of contact area with plastic deformation than with elastic deformation. Once a plastically deforming contact has been loaded to some maximum, the contact will deform elastically at all loads less than that maximum if there has been no welding at the contact. For those plastically deforming powders without welding, the velocity-pressure exponent approaches the Hertzian exponent on pressure cycling; and for those powders with welding, the contact area can not change during pressure cycling and the velocity is observed to be essentially independent of pressure.

In this same study (Vaišnys & Gordon 1965), the velocity at a given stress was found to decrease with decreasing particle size, a result similar to the findings of Pilbeam & Vaišnys (1973) at low pressure. This trend is thought to be related to a decrease in average particle coordination with decreasing particle size for the finer sizes of particles, a possibility discussed in the section on packing properties. The rate of volume decrease with compacting stress is smaller for smaller particle sizes as well. Leiser & Whittemore (1970) attribute this to a decrease in the number of flaws with decreasing particle size, so that smaller particles are more resistant to fracture. Such a proposal may be related to an empirical generalization, known in fracture mechanics as Auerbach's law, that the critical load for Hertzian fracture is proportional to the radius of the indenter (Nadeau 1973). The rate of compaction has been found to be correlated with particle hardness (Cooper & Eaton 1962). In the study of Vaišnys & Gordon (1965), where compaction was described by the equation $e = k \log [P_0/(P + P_1)]$, P_0 was found to vary directly as the hardness of the material. In the above equation, e is the void ratio, P the compacting stress, and k, P_0, P_1 are constants. P_0 may be interpreted as a stress extrapolated to a hypothetical zero void volume state.

At the relatively high stress levels found in compaction experiments, time-dependent aspects of deformation processes are accentuated. For example, Lawrence (1973) has reported some creep measurements in powders at room temperature. Progress in understanding the inherently complex time-dependent effects is likely to require a return to the geometrically simple systems which have contributed so much to elucidating the structural and quasi-static aspects of granular systems. It seems probable that the description of energy-loss and time-dependent phenomena will involve an examination of contact micromechanics (Yip & Venart 1971, Liebensperger & Brittain 1973), as well as closer attention to interface properties and to material deformation mechanisms.

Literature Cited

Adams, D. J., Matheson, A. J. 1972. Computation of dense random packings of hard spheres. *J. Chem. Phys.* 56:1989–94

Arakawa, M., Suito, E. 1969. Packing structure of fine powder. *Bull. Inst. Chem. Res. Kyoto Univ.* 47:412–25

Archard, J. F. 1953. Contact and rubbing of flat surfaces. *J. Appl. Phys.* 24:981–88

Batchelor, G. K. 1956. *The Theory of Homogeneous Turbulence.* Cambridge: Univ.

Press. 197 pp.

Bennett, C. H. 1972. Serially deposited amorphous aggregates of hard spheres. *J. Appl. Phys.* 43:2727–34

Bernal, J. D. 1964. The Bakerian lecture, 1962. *Proc. Roy. Soc. A* 280:299–322

Bernal, J. D., Mason, J. 1960. Co-ordination of randomly packed spheres. *Nature* 188:910–11

Biot, M. A. 1973. Non-linear and semi-linear

rheology of porous solids. *J. Geophys. Res.* 78:4924–37

Blanc-Lapierre, A., Fortet, R. 1968. *Theory of Random Functions* 2. New York: Gordon & Breach. 324 pp.

Brandt, H. 1955. A study of the speed of sound in porous granular media. *J. Appl. Mech.* 22:479–86

Brandt, H. 1967. Compressional wave velocity and compressibility of aggregates of particles of different materials. *J. Appl. Mech.* 34:866–72

Brown, R. L., Richards, J. C. 1965. Kinematics of the flow of dry powders and bulk solids. *Rheol. Acta* 4:153–65

Chen, T. C-T. 1974. Stress distribution in two-dimensional granular media. Senior Thesis. Yale Univ., New Haven, Conn. 26 pp.

Cooper, A. R., Eaton, L. E. 1962. Compaction behavior of several ceramic powders. *J. Am. Ceram. Soc.* 45:97–101

Craig, T. 1880. Distortion of an elastic sphere. *J. reine angew. Math.* 90: 253–66

Dantu, P. 1957. Contribution à l'étude mécanique et géometrique des milieux pulvérulents. *Proc. Int. Congr. Soil Mech. Found. Eng., 4th* 1:144–48

Deresiewicz, H. 1958a. Mechanics of granular matter. *Advan. Appl. Mech.* 5:233–306

Deresiewicz, H. 1958b. Stress-strain relations for a simple model of a granular medium. *J. Appl. Mech.* 25:402–6

Deresiewicz, H. 1966. On the mechanics of granular media. In *Applied Mechanics Surveys*, ed. H. N. Abramson, H. Liebowitz, J. M. Crowley, S. Juhasz, 277–84. Washington DC: Spartan

Dexter, A. R., Tanner, D. W. 1971. Packing density of ternary mixtures of spheres. *Nature Phys. Sci.* 230:177–79

Drescher, A., de Josselin de Jong, G. 1972. Photoelastic verification of a mechanical model for the flow of granular material. *J. Mech. Phys. Solids* 20:337–51

Duffy, J. 1959. A differential stress-strain relation for the hexagonal close-packed array of elastic spheres. *J. Appl. Mech.* 26:88–94

Duffy, J., Mindlin, R. D. 1957. Stress-strain relations and vibrations of a granular medium. *J. Appl. Mech.* 24:585–93

Finney, J. L. 1970. Random packings and the structure of simple liquids. I. The geometry of random close packing. *Proc. Roy. Soc. A* 319:479–93

Frank, F. C. 1965. On dilatancy in relation to seismic sources. *Rev. Geophys.* 3:485–503

Frisch, H. L., Stillinger, F. H. 1963. Contribution to the statistical geometric basis of radiation scattering. *J. Chem. Phys.* 38:2200–7

Gallagher, J. J. Jr. 1971. Photomechanical model studies relating to fracture and residual elastic strains in granular aggregates. PhD thesis. Tex. Agr. Mech. Univ., College Station, Texas. 141 pp.

Gardner, G. H. F., Wyllie, M. R. J., Droschak, D. M. 1964. Effects of pressure and fluid saturation on the attenuation of elastic waves in sand. *J. Petrol. Technol.* 16:189–98

Goodman, L. E., Brown, C. B. 1962. Energy dissipation in contact friction: Constant normal and tangential loading. *J. Appl. Mech.* 29:17–22

Greenwood, J. A., Williamson, J. B. 1966. Contact of nominally flat surfaces. *Proc. Roy. Soc. A* 295:300–19

Hall, J. R. Jr., Richart, F. E. Jr. 1963. Dissipation of elastic wave energy in granular soils. *J. Soil Mech. Found. Div. Amer. Soc. Civil Eng.* 89(SM6):27–56

Hamilton, E. L. 1972. Compressional-wave attenuation in marine sediments. *Geophysics* 37:620–46

Hardin, B. O., Richart, F. E. Jr. 1963. Elastic wave velocities in granular soils. *J. Soil Mech. Found. Div. Amer. Soc. Civil Eng.* 89(SM1):33–65

Hardman, J. S., Lilley, B. A. 1973. Mechanisms of compaction of powdered materials. *Proc. Roy. Soc. A* 333:183–99

Hertz, H. 1881. Ueber die Berührung fester elastischer Körper. *J. Reine Angew. Math.* 92:156–71

Hill, R. 1963. Elastic properties of reinforced solids: Some theoretical principles. *J. Mech. Phys. Solids* 11:357–72

Hill, R. 1967. *The Mathematical Theory of Plasticity.* Oxford: Clarendon. 355 pp.

Johnson, K. L. 1955. Surface interaction between elastically loaded bodies under tangential forces. *Proc. Roy. Soc. A* 230:531–49

Johnson, K. L. 1961. Energy dissipation at spherical surfaces in contact transmitting oscillating forces. *J. Mech. Eng. Sci.* 3:362–68

Kerner, E. H. 1956. The elastic and thermo-elastic properties of composite media. *Proc. Phys. Soc. London B* 69:808–13

Koerner, R. M. 1970. Limiting density behavior of quartz powders. *Powder Technol.* 3:208–12

Krizek, R. J. 1971. Rheologic behavior of cohesionless soils subjected to dynamic loads. *Trans. Soc. Rheol.* 15:491–540

Landauer, R. 1952. The electrical resistance

of binary metallic mixtures. *J. Appl. Phys.* 23:779–84

Lawrence, P. 1973. The cold compaction of several mineral powders. II. The time-dependent deformation. *J. Mater. Sci.* 8: 770–76

Leiser, D. B., Whittemore, O. J. Jr. 1970. Compaction behavior of ceramic particles. *Am. Ceram. Soc. Bull.* 49:714–17

Liebensperger, R. L., Brittain, T. M. 1973. Shear stresses below asperities in Hertzian contact as measured by photoelasticity. *J. Lubric. Technol.* 95:277–83

Miller, M. N. 1969. Bounds for effective electrical, thermal, and magnetic properties of heterogeneous materials. *J. Math. Phys.* 10:1988–2004

Mindlin, R. D. 1949. Compliance of elastic bodies in contact. *J. Appl. Mech.* 16:259–68

Mindlin, R. D., Deresiewicz, H. 1953. Elastic spheres in contact under varying oblique forces. *J. Appl. Mech.* 20:327–44

Mindlin, R. D., Mason, W. P., Osmer, T. F., Deresiewicz, H. 1951. Effects of an oscillating tangential force on the contact surfaces of elastic spheres. *Proc. Nat. Congr. Appl. Mech., 1st,* 203–8

Nadeau, J. S. 1973. Hertzian fracture of vitreous carbon. *J. Am. Ceram. Soc.* 56: 467–72

Norman, L. D., Maust, E. E., Skolnick, L. P. 1971. Computer simulation of particulate systems. *US Bur. Mines Bull.* 658. 55 pp.

Pearson, C. E. 1959. *Theoretical Elasticity.* Cambridge: Harvard Univ. Press. 218 pp.

Pilbeam, C. C. 1971. Acoustic velocities and energy losses in granular aggregates. PhD thesis. Yale Univ., New Haven, Conn. 145 pp.

Pilbeam, C. C., Vaišnys, J. R. 1973. Acoustic velocities and energy losses in granular aggregates. *J. Geophys. Res.* 78:810–24

Polder, D., Van Santen, J. H. 1946. The effective permeability of mixtures of solids. *Physica* 12:257–71

Rayleigh, L. 1892. On the influence of obstacles arranged in rectangular order upon the properties of a medium. *Phil. Mag.* 34:481–502

Ridgway, K., Tarbuck, K. J. 1967. Progress report on random packed beds. *Brit. Chem. Eng.* 12:384–87

Riley, G. S., Mann, G. R. 1972. Effects of particle shape on angles of repose and bulk densities of a granular soil. *Mater. Res. Bull.* 7:163–70

Ritz, W. 1908. Über eine neue Methode zur Lösung gewisser Variations probleme der mathematischen Physik. *J. Reine Angew. Math.* 135:1–61

Savage, L. J. 1972. *The Foundations of Statistics.* New York: Dover. 310 pp.

Scott, G. D. 1962. Radial distribution of the random close packing of equal spheres. *Nature* 194:956–57

Scott, G. D., Kilgour, D. M. 1969. The density of random close packing of spheres. *Brit. J. Appl. Phys. (J. Phys. D)* 2:863–66

Synge, J. L. 1957. *The Hypercircle in Mathematical Physics.* Cambridge: Cambridge Univ. Press. 424 pp.

Thurston, C. W., Deresiewicz, H. 1959. Analysis of a compression test of a model of a granular medium. *J. Appl. Mech.* 26:251–58

Timoshenko, S., Goodier, J. N. 1951. *Theory of Elasticity.* New York: McGraw-Hill. 506 pp.

Vaišnys, J. R., Gordon, R. B. 1965. High pressure effects in granular media: Sound velocity and compression. *Trans. Am. Geophys. Union* 46:162 (Abstr.)

Visscher, W. M., Bolsterli, M. 1972. Random packing of equal and unequal spheres in two and three dimensions. *Nature* 239: 504–7

Vold, M. J. 1959. A numerical approach to the problem of sediment volume. *J. Colloid. Sci.* 14:168–74

Warren, N., Anderson, O. L. 1973. Elastic properties of granular materials under uniaxial compaction cycles. *J. Geophys. Res.* 78:6911–25

Watkins, J. S., Lawrence, A. W., Godson, R. H. 1972. Dependence of in situ compressional-wave velocity on porosity in unsaturated rocks. *Geophysics* 37:29–35

Westman, A. E. R., Hugill, H. R. 1930. The packing of particles. *J. Am. Ceram. Soc.* 13:767–79

White, J. E. 1965. *Seismic Waves: Radiation, Transmission, and Attenuation.* New York: McGraw-Hill, 302 pp.

White, J. E. 1966. Static friction as a source of seismic attenuation. *Geophysics* 31:333–39

White, H. E., Walton, S. F. 1937. Particle packing and particle shape. *J. Am. Ceram. Soc.* 20:155–66

Yip, F. C., Venart, J. E. S. 1971. An elastic analysis of the deformation of rough spheres, rough cylinders and rough annuli in contact. *Brit. J. Appl. Phys. (J. Phys. D)* 4:1470–86

ADAPTIVE THEMES IN THE EVOLUTION OF THE BIVALVIA (MOLLUSCA)

×10044

Steven M. Stanley

Department of Earth and Planetary Sciences, The Johns Hopkins University,
Baltimore, Maryland 21218

INTRODUCTION

For many decades the evolutionary history of the Bivalvia remained in relative obscurity. Now, for the first time, it is being elucidated in some detail through a variety of scientific approaches. Ironically, the bivalve fossil record is yielding a better conceptual picture of evolution than are the records of other groups that have been studied much more intensively over the years. The reason for this paradox is that the great diversity of living bivalves offers an excellent opportunity for direct study of many adaptive features that have homologies and analogies among extinct taxa. Because of the nature of the fossil record, it is largely from shell form that the adaptations of extinct taxa are inferred. No effort is made here to review comprehensively the functional morphology of the bivalve shell. An earlier study, based on Western Atlantic bivalves (Stanley 1970), provided a start in this direction, and additional contributions have appeared since its publication. My present goal is to distill from the evolutionary record of the Bivalvia, via functional morphology, certain basic adaptive tendencies and patterns that seem, at our present state of knowledge, to be most fundamentally characteristic of the evolution of this important class of mollusks. For the sake of brevity, only a few examples are offered by way of illustration. Many of the adaptive themes discussed characterize, to varying degrees, the evolution of other animal taxa as well. Relevant concepts of functional morphology are introduced at the outset as background for later discussion.

FUNCTIONAL AND NONFUNCTIONAL INTERPRETATIONS

It is commonly assumed that the salient morphologic features of any taxon are functional. If selection for the presence of a structure ceases, the vestigial structure usually tends to disappear, either because its development is wasteful of metabolic energy or because it is somehow linked by pleiotropy to other functional features

361

upon which selection operates to the detriment of the vestigial structure. It is true that pleiotropy can also maintain nonfunctional structures through developmental linkage with functional structures, but it is difficult to imagine that such linkage can initiate growth of large or complex structures or modify preexisting structures to any appreciable degree. Growth by pleiotropic linkage has been invoked chiefly to account for striking examples of allometry. Its possible role has, however, been rejected in recent reevaluations of two classic examples of allometry: relative growth of the antlers of the Irish "Elk" (Gould 1973) and relative growth of the horn of titanotheres (Stanley 1974).

In the past it has been suggested at least informally that certain prominent features of the bivalve shell are functionless. Among these features are genetically produced spines and ridges of the shell exterior. They have been regarded as functionless because no likely function has been envisioned. Recent studies show that shell surface features do indeed have functions. In fact, it can never be proven that a structure is functionless, in that the possibility always remains that the true function has not been considered.

Another classical argument against the functional origin of morphologic features by natural selection is that certain features could not have been functional in their incipient stages. This view was rejected by Darwin, who argued that he knew of no valid examples. Darwin also pointed out that certain structures that now perform one type of function originated to perform a totally different type of function. We now term such adaptations preadaptations. The transition from swim bladder to lung in the evolution of the vertebrates formed one of Darwin's examples. An apparent example in the Bivalvia is the behavioral pattern of valve clapping, which

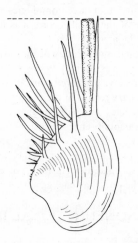

Figure 1   Probable disposition of siphons (stippled) and spines of *Hysteroconcha* in life position. Dashed line represents the sediment-water interface. At any stage of ontogeny, only the last-formed spines protect the siphons.

probably arose to perform a mantle-cleansing function, but was transformed into a jet propulsion mechanism for swimming in scallops (Yonge 1936b) and in less adept swimmers, like *Solemya* (Stanley 1970). Another apparent example is the elongate hinge line of pterioid bivalves, which seems to have served at first to strengthen articulation of the valves, but in so doing formed a posterior auricle that subsequently became an adaptation to permit effective ejection of the exhalent water current (Stanley 1972).

ADAPTIVE BARRIERS AND INCIPIENT FEATURES

What has perhaps not been fully appreciated is that the limited adaptive value of incipient evolutionary stages of many organs offers a solution to a general problem in paleontology. It is sometimes claimed that a given type of structure probably did not function in a postulated manner because numerous taxa lack the structure even though they have the same physiological, behavioral, or ecological traits that make the structure useful to taxa that have it. The main reason that structures of widespread potential utility are not universally present is that they arise only with difficulty, not just because of morphological or developmental constraints, but because of the limited selective value of their incipient stages. There is an adaptive barrier to the origin of many structures unless they have distinct preadaptive functions. Three intergrading conditions can be envisioned for the origin of morphologic features:

1. In some instances the incipient development of a feature is clearly of selective value, especially if the feature is connected to the preexisting condition by a *morphological continuum*. An example would be the appearance of ridges on the shell surface that function as aids to burrowing (Stanley 1969). Even as faint expressions, appropriately formed ridges can aid burrowing by gripping the sediment. There is no clear-cut division between the presence and absence of such ridges and, for simple ridge configurations at least, there is no barrier to be crossed at the outset.

2. Other structures or conditions are *discrete* entities in the sense of being clear-cut in their presence or absence. Meristic characteristics are of this type, but so are other functions. Fusion of the two halves of the bivalve mantle across the commissure is quite distinct from simple, part-time apposition of the mantle margins by muscular contraction. Mantle fusion was of immense importance in the evolution of the bivalves (Yonge 1957, Stanley 1968) but was not established in major taxa until the late Paleozoic, tens of millions of years after the class began its initial adaptive radiation. A definitive barrier must be crossed for a feature like mantle fusion to appear, and the barrier can be crossed only with difficulty.

3. Finally, there may be a morphological continuum between presence or absence, but a barrier still interferes because the structure must cross an *adaptive threshold* of size or complexity before it is sufficiently functional to be selected for. An example is provided by the long, sharp posterior spines of the living clam genus *Hysteroconcha* (Figure 1). Except for the presence of these remarkable structures, this genus has the morphology of a typical clam. It is likely that the spines serve to thwart predators

(Carter 1967b). In particular, the spines probably prevent fishes from nipping off the clam's siphons (Stanley 1970). The spines are grown on either side of the extended siphons, and fishes are known to feed on the siphons of many species of clams. Clams commonly regenerate amputated siphons, but only by the expenditure of considerable metabolic energy (Trevallion et al 1970). Despite the widespread problem of siphon loss for burrowing clams, spines like those of *Hysteroconcha* are exceedingly rare among Recent and fossil clams. The reason that nearly all species have failed to grow them is undoubtedly that, until they are several millimeters long, the spines cannot offer significant protection for the siphons. If, as seems probable, incipient spines are not likely to serve any preadaptive function, the adaptive threshold beyond which spine growth will be selected for will be crossed only rarely.

In summary, the Bivalvia illustrate quite effectively why certain morphologic traits that would be of widespread utility among a variety of similar taxa arise only rarely and sporadically.

MORPHOLOGICAL COMPROMISES IN EVOLUTION

Evolution in all but its simplest form represents compromise. Most often an adaptation that arises because of utility for one or more purposes does so at the expense of some alternative evolutionary change that would have offered other benefits. Nowhere in the animal kingdom does morphological compromise seem more evident than in the evolution of the bivalve shell, perhaps because of the basic simplicity and accretionary mode of growth of this calcareous organ. Three major sources of compromise are treated separately below.

Adverse Genetic Linkage

While pleiotropy seldom produces complex nonfunctional features, it plays a widespread role in modifying adaptive features by balancing their selective values against those of other genetically linked features that are subjected to different selection pressures. So little is known of the genetics of most taxa other than *Drosophila* that examples of this obviously important type of compromise have been little documented, and it seems likely that no example is yet known for the Bivalvia.

Mutual Exclusion of Alternatives

Commonly a given morphological feature has adaptive significance in more than one aspect of physiology, behavior, or ecology. Under such circumstances, even without genetic linkage to other features, the direction of natural selection will represent a compromise of all relevant factors. The compromise can be envisioned as being represented by a point in an n-dimensional space, the coordinates of which represent morphological parameters describing the potential geometry of the structure. The compromise point will not simply be that one in the hyperspace for which the selection coefficient is lowest (fitness is highest). Exactly where the point lies will be influenced by the ease with which different morphologies can evolve from the preexisting morphology, one factor in which is the inadaptive nature of incipient stages described above. In addition, stochastic factors play an important role. Early fixation of a mutation that happens to move morphology toward one point of high fitness may preclude evolution toward a point of even higher fitness that lies in a different direction.

Figure 2 Effects of the concentric ridges of *Chione cancellata*. *A, B*: anterior and left-lateral views of shell (both × 1). *C*: top view of animals leaving crawling trails while having difficulty burrowing after being placed on their native sediment in an aquarium. *D*: top view of two casts of shells placed in life orientation in medium-grained sand in a wave tank and subjected to current oscillations parallel to the long axis of the 5 cm scale. The upper cast in the picture was unmodified, and the lower one, from the same mold, was made smooth by removal of the concentric ridges on its surface. The casts were 3.1 cm long, and each was placed so that 1.1 cm of its posterior tip was exposed above the original flat surface of the sediment and its plane of commissure was oriented at 60° to the direction of flow. The smooth cast suffered much greater scour.

Ridges on the external surface of the bivalve shell can be viewed in the frame-work just outlined. As an example, let us consider crudely concentric ridges in two extant tropical species of the family Veneridae. In *Chione cancellata* (Figure 2) the ridges are sharp structures that stand well out from the shell surface. They are thin relative to their height, and serrated, apparently for strength. Even with serrations they are commonly found damaged in nature. They are also supported basally by less pronounced radial ribs. The concentric ridges cause animals of this species difficulty in burrowing. Individuals require fewer burrowing movements for burial after their ridges have been removed than before alteration (S. M. Stanley, in preparation). On the other hand, the ridges are helpful in reducing scour around the shell, which is commonly partly exposed above the sediment-water interface in nature (Figure 2D). Although often figured as crawling animals in textbooks, most suspension-feeding marine clams move about very little once buried in sedi-ment. Most species burrow chiefly to reassume normal positions of burial after being exposed by scour of surrounding sediment. In essence, there are two direc-tions that adaptations to the scour problem can take. One is for features to evolve to reduce scour, which is the path followed in the evolution of *C. cancellata,* as just described. The other is for improved burrowing efficiency to evolve, allowing a clam to reburrow rapidly, once exhumed. With respect to the morphology of the ridges, these two adaptive pathways tend to be mutually exclusive. The kinds of ridges that prevent scour differ in form from ridges that aid in burrowing. The second type of ridge is present in *Anomalocardia brasiliana,* another venerid clam (Figure 3). The ridges of this species are less pronounced and are well developed only in the anterior region, where they are asymmetrical in cross section (the posterior region has only weak, variable corrugations reflecting irregular growth). The dorsal slope of each anterior rib is steep, and the ventral slope is gentle. Thus, as the shell rotates back and forth in its normal burrowing movement, the ribs alternately grip and slide through the sediment in the way that a ratchet grips in one direction and slides freely in the other. It has been suggested that this gripping effect aids the animal to pry itself downward more effectively than it could in the absence of the ridges (Stanley 1969). This function has now been confirmed experimentally by measuring numbers of rotational movements required for burial

Figure 3   Anomalocardia, showing ridges that aid in burrowing in anterior (*A*) and left-lateral (*B*) views (both × 1). The dorsal slope of each rib is steep, and the ventral slope is gentle.

by unaltered animals and by the same animals with the valleys between the ridges filled with wax made neutrally buoyant by the addition of ferric oxide (S. M. Stanley, in preparation).

It is debatable why different alternatives for ridge growth have been followed in the evolution of *C. cancellata* and *A. brasiliana.* The selection of alternatives may, in part, have been accidental or stochastic, in the sense that the type of morphology for which mutations happened to arise first in a given lineage was adopted. It seems, however, that in this particular instance the differing adaptive pathways may not have arisen by chance. *Chione* has a very shallow life position, with the posterior tip of the shell often supporting the growth of epifaunal algae. The siphons are very short. Scour is a constant threat and would seem to be a very immediate problem for survival. In contrast, *Anomalocardia* lives slightly deeper and is seldom exposed at the sediment surface. Furthermore, the posterior region of its shell is rostrate. The elongate tip of the shell, from which the siphons emerge, is reduced in size both parallel and perpendicular to the sagittal plane. This reduction greatly reduces scour because the current velocity at which scour begins and the amount of scour at a given velocity decrease greatly with the dimensions of the obstructional object producing the scour. The rostrate posterior is common within the Bivalvia and seems generally to have evolved to reduce scour. With the appearance of this feature, it would seem that the shell surface ornamentation of *Anomalocardia* was freed to assume a separate function, that of improving burrowing efficiency. It is possible, of course, that habitat differences between the two species might also account for their differing ornamentation patterns. *A. brasiliana,* however, tends to live in very restricted habitats, whereas it is in areas of predictably strong water movement that a premium is nearly always placed on rapid burrowing because frequent exposure by scour is inevitable (Yonge 1950, Stanley 1970).

The example just offered considers alternative morphologies for a given type of structure. Another possibility is that two *different* classes of morphological features may be incompatible. Continuing with the example of shell surface ridges, such ridges could not perform the function of aiding burrowing, as in *Anomalocardia,* if pronounced radial ribs were secreted to perform some separate function. Not only would simultaneous secretion of the two types of ridges be difficult, but functional interference would permit only a single alternative to be followed.

Ontogenetic Preclusion

Rudwick (1968) pointed out that for brachiopods, which like bivalves grow by shell accretion, certain shell features may be useful during only a fraction of ontogeny. If they are functional only before the final stages of growth, they may limit the morphological options available for later ontogeny or, by their continued presence, reduce fitness when they are no longer adaptive. Torsion in gastropods, if useful only to larval stages as some have suggested (Garstang 1928), would be an extreme example. Taxa that grow by molting (arthropods) or that resorb and reform mesodermal skeletal elements easily (echinoderms and vertebrates) are not so strongly affected late in life by the course of early development. In the Bivalvia, although the juvenile shell becomes part of the adult, selection pressure for specific shell shapes tends to be rather weak early in postlarval ontogeny. For this reason, postlarval stages in the Bivalvia tend to look more alike than do adult bivalves. Recapitulation of phylogeny by ontogeny plays a role here, but it is not the only factor because species that are small as adults are also, on the average, less variable in shape than are larger species. Consequently, for small bivalves, whether juvenile or adult, it is difficult to infer life habits from shell form (Stanley 1972). The reason is that scaling problems of various types are relatively unimportant at small size. This explains why all free-living epifaunal clams that are adapted for crawling about on hard substrate are minute. Clams in general are not well adapted for this mode of life, and for large ones, it is a virtually impossible one to adopt. As another example, small bivalves that attach to hard substrata tend to do so by lodging in crevices, rather than having to attach firmly upon flat surfaces by special morphological adaptations. Tiny clams occupying soft substrata do not face the same burrowing problems as large clams because they are closer in size to the sedimentary particles that surround them. For them the substratum resembles that of an epifaunal, hard-substratum dweller that nestles in a crevice. All of these factors reduce tiny clams' need for specialized morphological features. Because the juvenile shell form of most bivalve species has tended to remain somewhat generalized, few constraints seem to have been placed on adult morphology. In a few fossil forms, for unknown reasons, major changes of shape have occurred during ontogeny. Examples are the genera *Slava* and *Butovicella* from the Silurian of Bohemia (Kříž 1969). Whether the conical umbones of *Slava* (Figure 4) were a hindrance in the adult habit is unknown. In the living genus *Mercenaria* and certain other venerid clams, sharp concentric ridges like those of *C. cancellata* are secreted during the first year of growth, apparently to prevent exposure by scour, but these are worn off in succeeding years, when such ridges are no longer secreted (Stanley 1972).

THE PRINCIPLE OF CORRELATION

What seems today an obvious aspect of functional morphology was to the early Nineteenth Century contemporaries of its popularizer, Baron Georges Cuvier, a revolutionary concept. Cuvier called this concept the Principle of Correlation. It describes what Darwin termed "coadaptation," or the unified nature of organs and

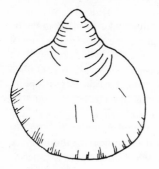

Figure 4 Lateral view of the genus *Slava* from the Silurian of Bohemia (× 1).

organ systems that permits them to interact harmoniously in fitting a given species to a particular mode of life. As obvious as the principle of correlation may now seem, its implications are easily overlooked, as illustrated by the following example.

The long-held view that the Bivalvia arose as burrowers has recently been confirmed by Pojeta (1971) in a detailed review of the sequence of appearance of Ordovician taxa. In an analysis of the evolution of byssal attachment within the class, I concluded that the fossil record bears out the hypothesis of Yonge (1962) that taxa that attach by a byssus as adults arose by neoteny, from the byssate postlarval stages of certain primitive burrowing clams (Stanley 1972). Most byssally attached bivalves are epifaunal, and it was predicted from Yonge's hypothesis that in the ancestry of epifaunal byssate groups, transitional forms should be found that possessed a byssus but that still lived infaunally. Such precursors were, in fact, found for most major *epibyssate* taxa, and their habits were termed *endobyssate*. Some endobyssate taxa, like the living mytilid genus *Modiolus,* survive today. Features of *Modiolus* that are usually associated with the endobyssate habit in other taxa as well are: an anterior lobe, which is really just the remnant of a circular anterior region of a burrowing ancestor; a broad ventral byssal sinus; an elongate shape; and moderate convexity of both valves without appreciable ventral flattening. One important byssate group for which it appeared that an endobyssate ancestry could not be demonstrated was the Pteriacea, although during the Middle and Late Paleozoic, many species of this superfamily were clearly endobyssate.

In a comprehensive review of Ordovician bivalves, Pojeta (1971) noted that *Ahtioconcha* from the Ordovician Kukruse Stage (early Caradoc) of the Baltic region is the oldest apparent pteriacean genus. In lateral view (Figure 5A,B) *Ahtioconcha* resembles living endobyssate species of the Mytilidae, many of which belong to the genus *Modiolus*; it is elongate, possesses a distinct anterior lobe that housed the anterior adductor muscle, and has a broad byssal sinus. The genus *Ahtioconcha* was erected by Öpik in 1930. Relying on a subsequent description of the genus as having "a flat, to concave right valve" (Pojeta 1971), I could not unequivocally assign it an endobyssate habit. A flat or concave right valve is characteristic of pteriaceans that live epifaunally, with the right valve resting against the substratum. It was also difficult to assign an epifaunal habit to *Ahtioconcha* because epifaunal pteriaceans generally possess a more circular and acline shape than *Ahtioconcha* and have a distinct byssal notch, for stabilization (Figure 6); the shell margin presses against the substratum at two positions distal to the point of byssal emergence (Stanley 1970, 1972). This shape also provides a broad region for drawing water into the mantle cavity (Yonge 1953), in contrast to the narrow inhalent region of infaunal and semi-infaunal forms, which is all that projects above the sediment. In other words, various features of the lateral-view outline of *Ahtioconcha* clearly pointed to an endobyssate habit, but descriptions of the outline viewed dorsally or ventrally seemed to portray characteristically epifaunal features. The Principle of Correlation seemed to be violated.

Further search into the literature has now revealed a more detailed morphological picture of *Ahtioconcha*. Eberzin (1960) provided a photograph displaying the shell in ventral view (Figure 5C). Although the commissure can be seen to be bent in this view, the right valve is clearly convex in any section taken perpendicular to the commissure. It turns out that Öpik's (1930) original description of *Ahtioconcha*, in German, accurately stated that its shell is almost equivalve and is simply twisted to the right, posterior to the beaks. These features are quite compatible with the endobyssate features of the lateral-view outline and in fact represent the incipient inequivalve condition that was predicted for transition to epifaunal habits within the Pteriacea (Stanley 1972). Curvature of the commissure, which apparently developed

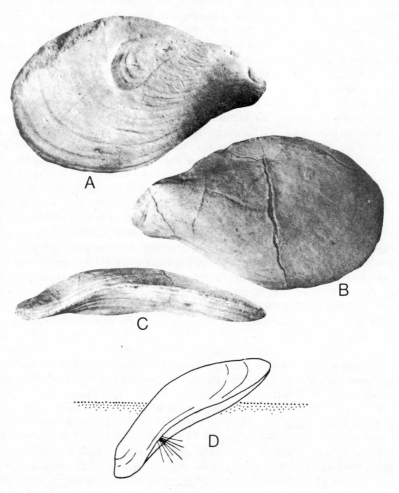

Figure 5   Ahtioconcha auris from the Middle Ordovician of the Baltic region. *A*, right-lateral view; *B*, left-lateral view; and *C*, ventral view of a composite mold (× 0.9). From Eberzin (1960). *D*: reconstruction of the inferred endobyssate life position (oblique ventral view).

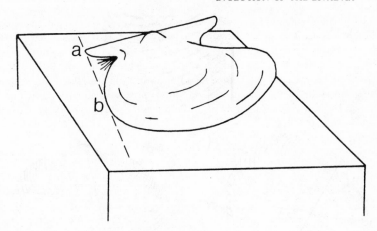

Figure 6   Life position of an idealized epifaunal pteriid, showing how the shell margin pressed against the substratum at two positions (*A, B*) distal to the site of byssal attachment. The shell can be overturned about the dashed line only by breaking the shell or byssus.

as the pteriaceans evolved from more deeply buried, equivalved ancestors, probably improved the stability of the shell by reducing the profile exposed to horizontal water movements. A broad, blade-like structure that happens to be oriented at right angles to current direction causes much greater scour of surrounding sediment than a low, conical structure (S. M. Stanley, in preparation). Pteriacea are also characterized by relatively weak byssal fixation, and assumption of a low angle life position would also have reduced the chance of dislodgement by strong water movements (Stanley 1972). Now that its morphology is understood, *Ahtioconcha* can be seen to display a degree of coadaptation equivalent to that found in other bivalves.

Numerous other illustrations of the Principle of Correlation are evident in the Bivalvia. For example, spines and ridges on the shell surface that prevent scour of sediment but hinder burrowing are found only in burrowing species with low ratios of shell length to shell height and shell height to shell width. Blade-like, disc-like, and cylindrical shells offer less resistance to sediment penetration and tend to belong to species that have evolved morphologies for rapid burrowing rather than for resistance to scour (Stanley 1970). Divaricate ornamentation that aids in burrowing (Figure 7) occurs only in species having circular shell outlines and using considerable shell rotation to burrow (Stanley 1969). Such ornamentation would be useless to an elongate razor clam, which pushes into the sediment with no rotation of the shell by virtue of its small cross-sectional area perpendicular to the direction of penetration.

The degree to which the coadaptation of morphological features is understood within a taxon is perhaps a measure of the resolution attained in study of the taxon's functional morphology. For the Bivalvia, at least in comparison to other invertebrate taxa with fossil records, we have attained a rather high degree of resolution, owing in large part to the diversity of living species.

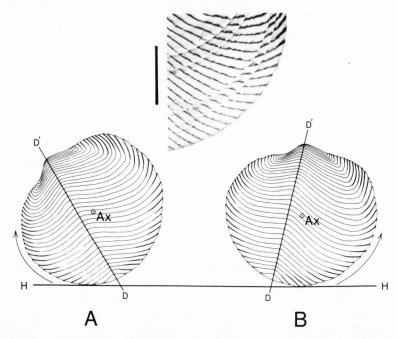

Figure 7 Divaricate ridges that aid *Divaricella* in burrowing. *A*: orientation of the shell before backward rotation (*Ax*, axis of rotation; *HH'*, horizontal line; *DD'*, line of demarcation between anterior and posterior ridges). As the shell rotates backward (arrow), the steep dorsal slopes of the anterior ribs will grip the sediment and the gentle ventral slopes of the posterior ribs will slide easily through it. *B*: orientation of the shell before forward rotation. With forward rotation (arrow) the posterior ribs will grip and the anterior ribs will slide easily. Photo inset: segment of a real shell showing the steep dorsal slopes of the divaricate ridges and their offset at growth rings, which represent periods of interrupted or very slow shell secretion (vertical bar 3 mm long).

CONVERGENCE, ITERATIVE EVOLUTION, AND INHERENT ADAPTIVE LIMITATIONS

It is possible to characterize any diverse but coherent taxon by a set of adaptive features that occur in a large percentage of its subgroups. Innumerable other features found in other groups will be absent. It is never possible to say that a taxon could not evolve a given feature, or at least that it could not evolve it without becoming a new taxon, but clearly many potentially useful features never arise in a given taxon because its potential to develop them is very small. Moreover, certain features that would seem to be easily formed are absent because they would be incompatible with sets of preexisting features. In this sense, diversification of any taxon is self-limiting.

In the Bivalvia, as in many other taxa, certain basic morphologies or character complexes have tended to recur widely in phylogeny through iterative evolution and convergence. The Principle of Correlation, discussed above, offers a partial explanation, but a remaining question is why only a few basic morphologies are especially prominent. The answer seems to be that fundamental characteristics of the class impose severe morphological constraints. For example, epibenthos can be either mobile or sessile. Because of the bivalves' basic specialization for infaunal existence, including their sac-like foot and laterally compressed shell form, they have never been able to develop efficient mechanisms for locomotion upon the surface of the substratum. Firm attachment has remained as their only viable option for large-scale radiation into epifaunal niches (Stanley 1972). The byssus has represented the only significant pathway by which the transition to epifaunal fixation has been made, and only two general sets of adaptive features have been of widespread importance in permitting firm byssal attachment to hard surfaces. One includes ventral flattening, whereby a broad base is provided to improve the mechanical advantage of the byssus against potential over-turning forces (Figure 8). Usually, in association with ventral flattening, byssal musculature is strengthened and shifted so as to pull the ventral shell surface directly against the substratum. The second important morphological complex for epibyssate life has arisen with the evolution of life habits in which the commissure plane is held at a low angle. Such a posture is usually attained by flattening of the lower valve. In addition a broad curvature of the ventral margin or byssal notch and auricle brings the shell margin into contact with the substratum at two points distal to the position of byssal attachment so that over-turning by water currents or predators is resisted (Figure 6). Clearly, given the constraints imposed by the set of characteristics found in ancestral burrowing clams, very few avenues to epifaunal life have been readily available within the Bivalvia.

Several shell forms have tended to evolve recurrently in the evolution of the Bivalvia. One is the epifaunal mussel shape just described. It is found in *Mytilus* (Figure 8) and also, for example, in Paleozoic ambonychiids (Stanley 1972), in limids of the Jurassic (Seilacher 1954) and Cenozoic (Stanley 1970), and in the living Dreissenacea (Yonge & Campbell 1968). Another morphology converged

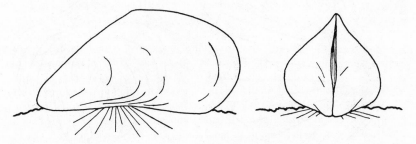

Figure 8   *Mytilus edulis* in life position on a hard substratum, showing the typical shape of an epifaunal mussel shell, in which the anterior is reduced and the ventral region is flattened.

upon by numerous taxa is the cockle shape of the extant Cardiidae (Figure 9). This compact shape, also found in shallow-burrowing arcids and trigoniids, is characterized by a flattened or low-pyramidal siphonal region, which reduces exposure of the shell by scour (S. M. Stanley, in preparation). The Mactracea exemplify several types of convergence with other taxa, some of which are shown in Figure 10. Discordant ornamentation that aids in burrowing (Stanley 1969) has also arisen polyphyletically. Its form varies in detail but tends to fall into several discrete categories (Seilacher 1972). The chevron pattern, for example, is strikingly similar in the circular genera *Divaricella* (Figure 7) and *Strigilla*. Such ornamentation is more difficult to secrete than radial and concentric structures. The point of secretion of a discordant ridge must migrate laterally while the shell grows radially, to produce a diagonal resultant direction of growth (Stanley 1969). The fact that diagonal ridges are offset where concentric interruption rings are formed during periods of minimal growth (Figure 7) demonstrates that lateral migration of a site of ridge secretion is independent of the growth rate of the shell.

As difficult as it might seem for mechanisms to evolve for production of well-patterned chevron-like ornamentation, other morphological features have originated far less frequently because of linkage to other rare attributes. Windows in the valves of *Corculum* for exposure of symbiotic zooxanthellae are an example (Seilacher 1972). Similarly, the distorted shape of *Tridacna* reflects an unusual association with zooxanthellae (Yonge 1936a, Stanley 1970). Taxa like these or, for example, the unusual and poorly understood Mesozoic rudists or the bizarre living Clavagellacea are thereforefore unpredictable members of the Bivalvia in the sense that an observer, having viewed the evolution of the class for millions of years before their origin, could not have forecast their appearance, or at least could not have forecast it with a high degree of probability. In contrast, the readily evolved and ecologically flexible forms of, say, mussels and cockles could, from our Recent vantage point, be predicted to reappear again and again in future evolution, assuming that the Bivalvia were to flourish for another two or three hundred million years.

In summary, once the basic body plan of the Bivalvia arose, it placed constraints

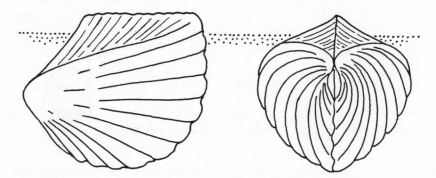

Figure 9 A shell of the typical cockle shape, in which the posterior-dorsal region is flattened or pyramidal and reduces scour of surrounding sediment.

Deep burrowing in firm, stable substrata	Shallow burrowing in stable substrata	Rapid burrowing in shifting sand	Rapid burrowing along shifting-sand beaches	Shallow burrowing in soft, muddy substrata
Shell morphology	Shell morphology	Shell morphology	Shell morphology	Shell morphology
Elongate and tubular (to minimize siphon length) Siphonal and pedal gapes (for rapid extension and withdrawal of siphons and foot parallel to shell's long axis; permitted by deep, protected life position) Valves thin (metabolically economical; permitted by deep, protected life position) Loss of hinge teeth (for rocking valves about dorsal-ventral axis during siphonal and pedal movements)	Posterior elongate (to minimize siphon length) Center of gravity near anterior (for stability) Valves thick (for stability)	Anterior triangular (for easy penetration) Shell thick – Tivela (for stability) Exterior smooth (for streamlining)	Anterior very elongate (to accommodate large foot) Posterior truncate (for pedal emergence opposite siphons) Slender; maximum width near posterior (for rapid burrowing) Valves thick (for stability) Exterior smooth (for streamlining)	Small (for flotation) Valves thin – Mulinia (for flotation)

Mya (Myacea)   Tresus   Mercenaria (Veneracea)   Rangia   Tivela (Veneracea)   Spisula   Donax (Tellinacea)   Mesodesma   Corbula (Myacea)   Mulinia

Mactracea

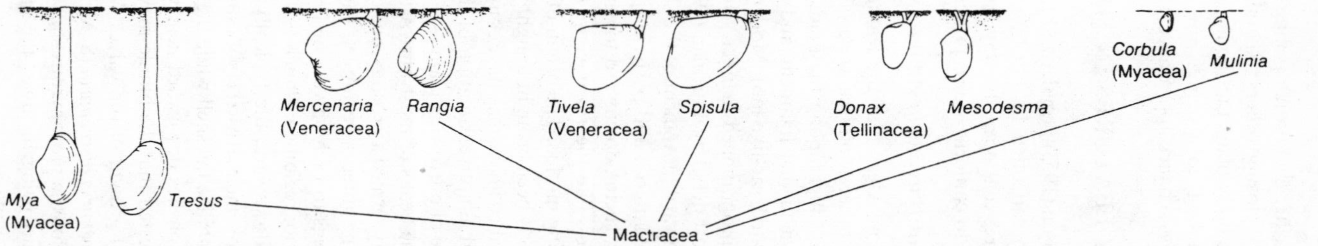

Figure 10 Convergence in form and habits between living bivalve genera. The right-hand member of each pair belongs to the Mactracea and the left-hand member, to another superfamily (From Stanley 1970).

on further evolution, opening up highly probable avenues for further evolution and nearly closing others. Readily evolved body plans have arisen polyphyletically, while body plans that could evolve less easily have appeared rarely or not at all, depending, to a degree, on stochastic factors such as infestation by symbiotic algae or chance appearance of necessary preadaptations.

KEY FEATURES IN THE ORIGIN OF THE BIVALVIA

Simpson (1953) popularized the idea that the initial adaptive radiation of a higher taxon can often be attributed to the appearance of one or more key adaptive features that open the way for the invasion of a new adaptive zone. The Bivalvia seem to conform to this principle.

Initial Burrowing Adaptations

It seems clear that the features by which the bivalve line of evolution diverged from other Mollusca were largely adaptations for exploitation of the burrowing habit. Chief among these adaptations were the laterally compressed shell, sprung with a medial ligament and closed by adductor muscles, and the hydraulically operated, sac-like foot. Modern bivalves of all types tend to burrow in a standard, well programmed manner. The typical sequence of movements has been described in most detail by Trueman (1966). Usually the final step that produces a single net downward movement of the shell involves shell rotation. By sequential contraction of the anterior and posterior pedal muscles, the shell rocks forward and then backward, slicing downward toward the temporarily anchored foot. In a few elongate species, like the razor clam *Ensis,* the shell is forced into the sediment without rocking motion. The origin of the Bivalvia from more primitive Mollusca has only recently been brought to light. Before considering how the key features and behavior described above may have come into being and what role they may have played in the initial adaptive radiation of the Bivalvia, it will be necessary to analyze the basic shape of burrowing clams.

The beaks of most burrowing clams are directed forward, forming what is commonly termed a prosogyrous condition. A flattened, heart-shaped lunule commonly fills the anterior region left vacant by the spiral growth of the two valves. Half of it belongs to each valve. The lunule and the adjacent umbones form a broad, blunt anterior region. Inasmuch as the net direction of burrowing for most clams is forward and downward, the typically blunt anterior might seem to represent a hindrance to movement. Ansell (1962) suggested that the lunule might serve as a pressure plate to stabilize the shell while the foot probes downward. The problem here is that the foot pushes downward nearly parallel to the plane of the lunule. Carter (1967a) suggested that the lunule may have no function other than to fill the void left by spiral growth of the valves. Work just completed on the typical burrowing clam *Mercenaria mercenaria* (S. M. Stanley, in preparation) demonstrates a special function for the prosogyrous shape and flattened lunule. As for many other adaptations of burrowing bivalves, the critical relation is with the rotational movements of burrowing.

The kinematics of burrowing in *M. mercenaria,* determined by analysis of movies, are quite simple (Figure 11). Movement of the shell is almost perfectly rotational. There is no significant translational component. Forward rotation follows probing of the foot and proceeds from a standard orientation of the shell, which has been termed the erect probing orientation (Stanley 1970). The axis about which forward rotation takes place lies in the middle of the shell, near the center of the circle of which the anterior-ventral shell margin forms a segment. The location of this axis seems to be determined chiefly by resistence of the sediment against the rounded portion of the shell, rather than by the distribution of the shell's weight, because the shell's weight is insignificant relative to the forces of muscular contraction and sediment resistence. Backward rotation restores the shell to its erect probing orientation. If backward rotation occurred about the same axis as forward rotation, the anterior would be raised to its previous position and no downward progress would be made. One can predict that the axis of rotation must shift, and movies show that it does in fact shift to a position above and in front of its position for forward rotation. Thus, backward rotation leads to a new erect probing orientation in which the shell has moved forward and downward. The shift of the axis must be accomplished partly by the differing position of the anterior and posterior pedal retraction muscles and partly by the distribution of resistant forces around the anterior end of the clam. It would seem that the blunt anterior formed by the

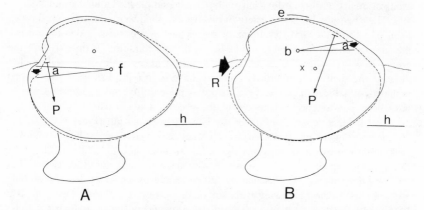

A   B

Figure 11 Burrowing mechanism of *Mercenaria mercenaria,* as shown by one forward-and-back rocking movement documented by a movie. Left-hand diagram: partly buried shell rocking forward from the erect probing position (solid outline) to the position of maximum forward rotation (dashed outline). The dotted line represents the sediment surface. *P,* direction of contraction of the pedal retractor causing rotation; *f,* axis of forward rotation; *a,* angle of rotation of the shell; *h,* horizontal line. Right-hand diagram: backward rotation from the position of maximum forward rotation (shown here by solid outline) to the next erect probing position (dashed outline). *b,* axis of backward rotation; *R,* resistant force of sediment against the blunt anterior region of the shell; *e,* position of the dorsal shell margin in the previous erect probing position (solid outline of left-hand diagram), showing downward progress; other symbols as in the upper diagram.

prosogyre beaks and lunule represents an adaptation providing such resistance. In effect, the animal seems to pry downward into the sediment with backward rotation because its anterior region tends to lock itself into the sediment instead of slipping upward.

To test this idea robot clams were made from epoxy casts of real clam valves. The robots were designed to be pushed downward by a simple machine applying forces that stimulated the pull of pedal retractor muscles. An unaltered robot was found to penetrate sand in very much the same fashion as a real clam. Using standard conditions, its downward progress was compared with that of a robot which was identical except for modification of its anterior region. Epoxy was added to this robot to fill in the area between the umbones and the anterior margin. In effect, the lunule was eliminated and the anterior end of the shell was made circular, like the segment of a discus. As predicted, the modified robot, having a shell that offered less resistance to forward and upward movement, made less downward progress in a given number of forward-and-back rocking movements than the unaltered robot (S. M. Stanley, in preparation).

It seems evident that the typical prosogyre shape of clams is an adaptation that increases burrowing efficiency. If the Bivalvia were primitively burrowers, a critical question is when the prosogyre shape arose. Pojeta et al (1972) have recently offered convincing evidence that the Bivalvia descended from the ribeiriids. This group of small Cambro-Ordovician mollusks had a simple shell folded about a dorsal axis. The ribeiriids display a variety of bivalve-like shell shapes but are not known to have had adductor muscles. In many species the beaks lie near, and point toward, the anterior end of the shell (Kobayashi 1933). This condition, if primitive for the group, may have been inherited from univalved ancestors because coiling that directed the shell apex anteriorly seems to have been typical of early cap-shaped mollusks. Whatever may have been its preadaptive utility, this condition seems to have been put to good advantage in the evolution of the Bivalvia. The oldest apparent bivalve genus, *Fordilla* of the late Early Cambrian (Pojeta et al 1973), is clearly prosogyre and exhibits a circular anterior, which suggests that it employed rocking movements in burrowing (Figure 12). Such motion cannot be ruled out for the ancestral ribeiriids, some of which also have circular anterior regions, but lacking knowledge of the musculature and pedal anatomy of ribeiriids, we can only speculate as to the burrowing behavior of this group. *Fordilla* is regarded as a true bivalve because of its possession of anterior and posterior adductor muscles (Pojeta et al 1973).

What is puzzling about the seemingly correct assignment of *Fordilla* to the Bivalvia is that there is at present no unequivocal Middle and Late Cambrian record for the Bivalvia linking *Fordilla* to Ordovician taxa. Obviously, the Bivalvia subsisted at quite a low level of diversity for several tens of millions of years before becoming a significant class of mollusks. As a possible explanation for the delayed adaptive radiation, I suggested that *Fordilla* may somehow have lacked a fully developed burrowing system (Stanley 1973). This is conceivable because the genus, though having a circular anterior ventral region, lacks the blunt anterior of modern clams and displays unusual and problematical muscle scars in the region

of pedal emergence. In contrast, a *Mercenaria*-like anterior is displayed by Early and Middle Ordovician genera like *Redonia*. I now favor a different explanation for the delayed adaptive radiation of the Bivalvia and will outline it in the following section.

The Postlarval Byssus as a Final Adaptive Breakthrough

Among living mollusks, the byssus is unique to the Bivalvia. As mentioned earlier, it apparently arose as a postlarval organ of attachment. It appeared at some time in the ancestry of the bivalves or ribeiriids. It seems likely that this time was the Early Ordovician, long after divergence from the ribeiriids, and that the byssus was what, literally and figuratively, first gave the Bivalvia a foothold for success on the sea floor. While we tend to think of the byssus chiefly as an organ for attachment of certain adult taxa, even today it is also of enormous importance in performing its original role for burrowing clams. A postlarval byssus is apparently present in virtually all living species of burrowing bivalves. Attachment following settlement is critical because a tiny clam, being the same general size as a grain of sand and even less dense, is highly vulnerable to transport and destruction by bottom currents. It seems likely that the juvenile byssus greatly increased the survival rate of ancestral burrowing bivalve species and triggered their diversification. If the postlarval byssus were somehow eliminated from modern clams, it seems likely that mass extinction would ensue. It is quite easy to believe that clams survived only marginally in the Cambrian because their postlarval stages lacked a byssus. Even as an adult, *Fordilla* (Figure 12) was quite small.

The fact that byssally attached adult taxa also arose for the first time and radiated rapidly during the first half of the Ordovician is suggestive of an Early

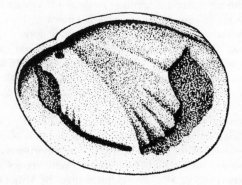

Figure 12 Composite reconstruction of the interior of the right valve of *Fordilla troyensis,* the oldest fossil bivalve now recognised (from the Lower Cambrian of New York State.) The shell is 4 mm long. It is not known whether this species had cardinal teeth, but there are no anterior or posterior lateral teeth. The scars extending ventrally from the anterior and posterior adductor scars represent muscles of uncertain morphology and function (From Pojeta et al 1973).

Ordovician origin for the byssus. Thus, it seems likely that byssal attachment was a key feature both for the diversification of burrowing taxa and for the origin and elaboration of sessile modes of life.

LATER ADAPTIVE BREAKTHROUGHS AND NET ADVANCES

It now seems likely that the earliest bivalves, including *Fordilla,* were suspension feeders. The nuculoids, which apparently arose in the Early Ordovician and persist today as a predominantly deposit-feeding group, appear to have descended from suspension feeders (Pojeta et al 1973), and their origin represented an important expansion of the bivalves' adaptive zone. In terms of habitat exploitation, two major types of adaptive breakthrough were especially important in the Paleozoic and Mesozoic evolution of the class. Much to his credit, Yonge (1957, 1962) anticipated the general significance of both without study of the fossil record. The first was the development of adult byssal attachment. As discussed above, this occurred by neotenous transfer of the postlarval byssus to the adult stage. It formed a polyphyletic pathway to epifaunal modes of life within the Bivalvia (Stanley 1972). The second major type of adaptive breakthrough was mantle fusion, including siphon formation, which led the way to more successful burrowing habits, again polyphyletically (Stanley 1968). Siphons permitted deeper burial, for protection against environmental fluctuations in the intertidal zone and against predation and current scour in a wider range of habitats. Mantle fusion also improved the hydraulic efficiency of the burrowing mechanism by preventing leakage of water from the mantle cavity during adduction of the valves. Except for the unique protobranchs, which burrow in muddy sediments with a muscular, flanged foot (Stanley 1970), all rapid-burrowing bivalves today have fused mantle margins. Adult byssal attachment became a common mode of life during the Ordovician, but only one pre-Carboniferous bivalve genus is known to have a pallial sinus, indicating the presence of a siphon. This is the aberrant form *Lyrodesma,* which is tentatively placed in a family by itself but may be related to the trigoniids (Newell & LaRocque 1969). The sinus of *Lyrodesma* is extremely shallow, and the genus could not have burrowed deeply. Furthermore, it seems to have become extinct without leaving siphonate descendents. It was not until late in the Paleozoic that siphons truly became established in the Bivalvia.

As mentioned above, Pojeta (1971) noted that infaunal bivalves tended to dominate Early Ordovician faunas. In fact, based on recognition of the endobyssate habit, which undoubtedly characterized nearly all modiomorphids (Stanley 1972), I see no evidence that any known Early Ordovician bivalve species was adapted for epifaunal life. Pojeta (1971, p. 30) also concluded that "By Late Ordovician time, pelecypods [bivalves] . . . had already explored most of the major modes of life utilized by younger forms except for the cementing of the shell to the substrate and swimming, although the degree of exploration of some of the modes of life was not as advanced as it was to become." To elaborate, it should be pointed out that by the Late Ordovician, there were still no rapid-burrowing, suspension-feeding clams, nor were there any deep burrowers. The first apparent deep burrowers were the early lucinids of the Devonian (Stanley 1968), and this highly specialized living group, which employs a unique anterior mu ?ous tube and single posterior siphon (Allen 1958), has never given rise to other deep-burrowing taxa. It was not until the Permo-Carboniferous that the siphonate anomalodesmatids produced a variety of deep burrowers and a few forms that may have been rapid burrowers. Scallops with shell characteristics indicative of a potential for adept swimming did not arise until the Carboniferous (Stanley 1972), and cementation did not evolve until the Permian, among oyster-like bivalves (New ?il & Boyd 1970).

Not only did the life-habit spectrum of the Bivalvia expand markedly during the Paleozoic, but diversities of different life-habit groups underwent major shifts (Stanley 1972). With the appearance of improved burrowing adaptations, burrowing taxa increased in diversity during the middle and late Paleozoic. Endobyssate taxa, though arising from burrowers, had immediately overshadowed them in the Middle and Late Ordovician. Most of the early endobyssate taxa had been cyrtodontids and modiomorphids. Their success relative to burrowers in the Ordovician probably stemmed from the advantage they had in maintaining stable life positions. Before adaptations for rapid burrowing and deep burrowing arose, free-living burrowers must have been highly vulnerable to disruption by scour. They were restricted to life at the sediment-water interface and could not reburrow efficiently if exhumed. It seems likely that endobyssate taxa then declined during the Paleozoic because although their usual life positions of partial burial were more stable than those of unattached primitive burrowers, these taxa still could not reburrow as efficiently as modern clams. While they declined, various epifaunal byssate groups evolved from them and many of these invaded hard substrata. The most diverse new epifaunal group was the Pectinacea. By the end of the Paleozoic the adaptive zone of the Bivalvia had been greatly expanded by the appearance of epifaunal taxa and infaunal taxa that burrowed deeply and efficiently. In light of this earlier analysis (Stanley 1972), I find no justification for the conclusion of Bretsky (1973) that "changes in the distribution and diversity among the organisms chosen [bivalves of Ordovician to Jurassic age] should reflect no drastic adaptive shifts, only the expression of a limited number of adaptive responses to environmental change."

Other adaptive breakthroughs of importance have occurred since the Paleozoic. The remarkable muscular foot of the trigoniids clearly endowed them with the capacity for more rapid burrowing than had characterized perhaps any of the suspension-feeding clams of the Paleozoic (Stanley 1972). It was this adaptation that permitted the trigoniids to diversify greatly in the Jurassic and Cretaceous, just as a similar adaptation has led to the more recent success of the Cardiacea. These and other post-Paleozoic taxa have also employed shell surface ornamentation as an aid to burrowing. Such ornamentation, varieties of which were discussed earlier in this paper, is virtually unknown among Paleozoic clams. Types of ornamentation that prevent scour were also quite uncommon before the Mesozoic. Among epifaunal taxa, many Cenozoic scallops display more advanced swimming adaptations than scallops of the Paleozoic (Stanley 1972). In addition, Waller (1972) has suggested that increased use of foliated calcite in their shell structure since the Paleozoic has given scallops greater success in shallow-water habitats.

In short, there is no question that Mesozoic bivalve faunas, in general, display net adaptive advances over Paleozoic bivalve faunas and that Cenozoic faunas show further net advances. The following section, however, will document and analyze the tendency of many archaic groups to survive or even arise and diversify along with advanced taxa of the types just described.

EVOLUTIONARY REVERSIONS AND PERSISTENCE OF ARCHAIC FORMS

A remarkable aspect of the phylogeny of the Bivalvia is that the fossil record shows that several extant bivalve groups judged by biologists to be primitive actually evolved quite recently, long after the class had attained great diversity and long after many advanced families were in existence. Yonge (1953), for example, suggested that the living arcoid genus *Glycymeris* seems to embody the features one would expect to find in the ancestral group of burrowing, suspension-feeding clams. The fossil record shows that the early Paleozoic cyrtodontids, the ancestral group of arcoids, actually did resemble the glycymerids in certain ways and apparently played an important evolutionary role in giving rise to various other nonarcoid taxa. Primitive cyrtodontids resembled glycymerids in having simple, circular shells, and probably also resembled them in being sluggish, shallow-burrowing, filibranch animals. The glycymerids, however, did not arise until the Cretaceous, and they diversified during the expansion of modern heterodont clams with advanced burrowing behavior and siphons. A similar example is found in the Lucinacea. From detailed anatomical study of living species, Allen (1958) concluded that the Lucinidae are the most modern of the three extant lucinacean families and that the Thyasiridae and Unigulinidae (Diplodontidae) were ancestral to it. In fact, the latter groups, which lack a posterior siphon, evolved in the Mesozoic, whereas the Lucinidae arose in the mid-Paleozoic. Allen (1968) also noted many seemingly primitive features in the living genus *Crassinella* and suggested that it might represent an ancestral group within the Heterodonta. In fact, it originated in the Eocene, while the Heterodonta range back into the Paleozoic, and its seemingly primitive features apparently represent neotenous reversions (Stanley 1972). Still another diverse modern arcoid group, the Anadarinae, gives the appearance of being primitive, in having filibranch gills, sluggish burrowing behavior, and unfused mantle margins, yet this superfamily arose in the Late Cretaceous and diversified during the Cenozoic.

A second striking aspect of bivalve phylogeny is that many genuinely primitive groups have persisted for hundreds of millions of years. The *Treatise on Invertebrate Paleontology* lists over 35 living genera that arose before the Cretaceous Period. These groups are not confined to geographic refugia, nor are they represented by only a few living species, yet they survive side-by-side with advanced modern taxa in numerous habitats throughout the world.

The tendency of modern Bivalvia to foster reversion to primitive morphologies and to harbor truly primitive taxa suggests that competition between species is relatively weak in this class (Stanley 1973). By way of comparison, it can be noted that comparable traits are lacking in the phylogeny of the Mammalia, where displacement of primitive taxa has been a dominant feature. Relict groups of primitive mammal taxa, like the lemuroids, tend to be confined to geographical refugia. Similarly, convergence among major groups of mammals has arisen largely through geographic separation, as between the marsupials of Australia and the placental mammals of other continents (Kurtén 1969). Also through competitive interactions, mammals have tended to maintain distinct adaptive zones (Van Valen 1971). In contrast, convergence among bivalve groups, like those of Figure 10, show no significant relation to geographic barriers. Also orders of bivalves do not occupy distinct adaptive zones, but overlap to a considerable degree in form and habits. On a finer scale, abutments of geographic range are extremely common among congeneric mammal species, whereas they are virtually unknown in the Bivalvia, except where coinciding with physiographic barriers (Stanley 1973).

Predation upon marine clams, especially juveniles, is extraordinarily heavy (Thorson 1966, Muus 1973). Crabs, snails, starfish, and certain fish groups take very heavy tolls. Because population densities are limited chiefly by predation and physical disturbance, space and food are generally superabundant. A typical suspension-feeding marine clam, with primitive sensory devices, sedentary habits, and abundant food supply, has little interaction with its neighbors. For all of these reasons, competition is relatively weak, except perhaps among deposit feeders, which rely upon local food reservoirs. Population densities can be greatly increased in nature by screening out predators. Thus there seems to be no real evidence that extensive niche partitioning has arisen among suspension-feeding clams from competitive interactions. Even among epifauna, for which surface space can potentially be in limited supply, predation and physical disturbances seem usually to permit the coexistence of similar species. A good example discussed in great detail by Harger (1972) is the co-occurrence of *Mytilus edulis* and *M. californianus* on rocky shores in California.

The Bivalvia can be contrasted with the mammals and other vertebrate groups, in which 1. populations tend to be less heavily preyed upon and more often limited by food supply, and 2. species tend to display well-developed mobility, good sensory perception, and territoriality. Certain invertebrate groups, like arthropods and cephalopods, share these features with vertebrates. It is not hard to see how competition has eliminated archaic groups of mammals during the emergence of the late Cenozoic faunas and how it has had similar effects within other taxa characterized by intense interspecific competition, but has had little impact in the evolution of the Bivalvia. It seems evident that weak competition among bivalve taxa has permitted stochastic factors to play a larger role in evolution than it has for more competitive groups like the mammals. The phylogenies of many bivalve taxa contain back-and-forth trends from one mode-of-life to another: for example, between burrowing and endobyssate habits and between endobyssate and epibyssate habits (Stanley 1972). The history of the Bivalvia shows net trends, as described in the previous section, but the trends have occurred gradually and with dilution by numerous evolutionary reversions. They represent, for the most part, simply the statistically greater success of certain taxa than others in the face of potential agents of extinction.

Thus, the adaptive zones of highly competitive groups, like mammals, are formed of subzones of member taxa that abut against one another, except where the subtaxa are geographically separate. Diversification of an order or family in a given geographic region is accomplished largely through 1. expansion of the original adaptive zone or 2. narrowing of subzones. Within the Bivalvia, however, evolution is quite different. The diversity of morphologic features of this class does not seem to have arisen through character displacement. The set of characters that we see today is the result of relatively weak selection pressures acting separately upon individual lines of descent. Many adaptive features have appeared iteratively or convergently, as adaptive themes that are readily evolved within the bivalves' basic body plan are played out again and again. The sets of seemingly archaic features are less abundant today than advanced morphologies, but are still accommodated within the total adaptive zone because subzones and niches of subtaxa can overlap even within single geographic regions. Taxa of primitive morphology suffer higher extinction rates than taxa of advanced morphology, but are free to reappear again and again. Peak diversity of families in the Mammalia was attained by the Oligocene, shortly after the class began its major adaptive radiation. In contrast, the Bivalvia continue to diversify at the family level almost five hundred million years after the onset of their initial adaptive radiation (Stanley 1973).

Literature Cited

Allen, J. A. 1958. On the basic form and adaptations to habitat in the Lucinacea (Eulamellibranchia). *Phil. Trans. Roy. Soc. London B* 241:421–84

Allen, J. A. 1968. The functional morphology of *Crassinella mactracea* (Linsley) (Bivalvia: Astartacea). *Malacol. Soc. London, Proc.* 38:27–40

Ansell, A. D. 1962. Observations on burrowing in the Veneridase (Eulamellibranchia). *Biol. Bull.* 123:521–30

Bretsky, P. W. 1973. Evolutionary patterns in the Paleozoic Bivalvia: documentation and theoretical considerations. *Geol. Soc. Am. Bull.* 84:2079–96

Carter, R. M. 1967a. On the nature and definition of the lunule, escutcheon and corcelet in the Bivalvia. *Proc. Malacol. Soc. London* 37:243–63

Carter, R. M. 1967b. The shell ornament of *Hysteroconcha* and *Hecuba* (Bivalvia): A test case for inferential functional morphology. *Veliger* 10:59–71

Eberzin, A. G., Ed. 1960. *Osnovy Paleontologii, Spravochnik diya Paleontologov i Geologov SSSR*. Vol. 3. *Mollyuski; Pantsirnye, Dvustvorchatye, Lopatonogie*. Moscow: Akad. Nauk SSSR. 300 pp.

Garstang, W. 1928. Origin and evolution of larval forms. *Rep. Brit. Assoc. Sect. D*, 77

Gould, S. J. 1973. Positive allometry of antlers in the "Irish elk," *Megaloceras giganteus*. *Nature* 244:375–76

Harger, J. R. 1972. Competitive co-existence: maintenance of interacting associations of the sea mussels *Mytilus edulis* and *Mytilus californianus*. *Veliger* 14:387–410

Kobayashi, T. 1933. Faunal study of the Wanwanian (basal Ordovician) series with special notes on the Ribeiridae and the ellesmeroceroids. *Tokyo Imp. Univ. Fac. Sci. J.* (2) 3:249–328

Kříž, J. 1969. Genus *Butovicella* Kříž, 1965 in the Silurian of Bohemia (Bivalvia). *Sb. Geol. Věd. Paleontol.* 10:105–39

Kurtén, B. 1969. Continental drift and evolution. *Sci. Am.* 220(3):54–64

Muus, K. 1973. Settling, growth and mortality of young bivalves in the Øresund. *Ophelia* 12:79–116

Newell, N. D., Boyd, D. W. 1970. Oyster-like Permian Bivalvia. *Am. Mus. Nat. Hist. Bull.* 143:217–82

Newell, N. D., LaRocque, A. 1969. Family Lyrodesmatidae Ulrich 1894. *Treatise on Invertebrate Paleontology*, ed. R. C. Moore, 471. Lawrence, Kansas: Geol. Soc. Am. and Univ. Kansas

Öpik, A. 1930. Beitrage zur kenntnis der Kukruse $(C_2 - C_3)$ Stufe in Eesti: *Acta Commentat. Univ. Tartu. Dorpat. A* 19:1–34

Pojeta, J. 1971. Review of Ordovician pelycypods. *US Geol. Surv. Prof. Pap.* 695. 46 pp.

Pojeta, J., Runnegar, B., Kříž, J. 1973.

Fordilla troyensis Barrande: the oldest known pelecypod. *Science* 180:866–68

Pojeta, J., Runnegar, B., Morris, N. J., Newell, N. D. 1972. Rostroconchia: a new class of bivalved mollusks. *Science* 177:264–67

Rudwick, M. J. S. 1968. Some analytic methods in the study of ontogeny in fossils with accretionary skeletons. *Paleontol. Soc. Mem.* 2:35–59

Seilacher, A. 1954. Ökologie der triassischen Muschel *Lima lineata* (Schloth) und ihrer Epöken. *Neues Jahrb. Geol. Palaeontol. Monatsh.* 4:163–83

Seilacher, A. 1972. Divaricate patterns in pelecypod shells. *Lethaia* 5:325–43

Simpson, G. G. 1953. *The Major Features of Evolution.* New York: Columbia Univ. Press. 434 pp.

Stanley, S. M. 1968. Post-Paleozoic adaptive radiation of infaunal bivalve molluscs—a consequence of mantle fusion and siphon formation. *J. Paleontol.* 42:214–29

Stanley, S. M. 1969. Bivalve mollusk burrowing aided by discordant shell ornamentation. *Science* 166:634–35

Stanley, S. M. 1970. Relation of shell form to life habits of the Bivalvia. *Geol. Soc. Am. Mem.* 125. 296 pp.

Stanley, S. M. 1972. Functional morphology and evolution of byssally attached bivalve mollusks. *J. Paleontol.* 46:165–212

Stanley, S. M. 1973. Effects of competition on rates of evolution, with special reference to bivalve mollusks and mammals. *Syst. Zool.* 22:486–506

Stanley, S. M. 1974. Relative growth of the titanothere horn: a new approach to an old problem. *Evolution* 28:447–57

Thorson, G. 1966. Some factors influencing the recruitment and establishment of

marine benthic communities. *Neth. J. Sea. Res.* 3:267–93

Trevallion, R. R., Edwards, C., Steele, J. H. 1970. Dynamics of a benthic bivalve. *Marine Food Chains,* ed. J. H. Steele, 285–95. Edinburgh: Oliver & Boyd

Trueman, E. R. 1966. Bivalve mollusks: Fluid dynamics of burrowing: *Science* 152:523–25

Van Valen, L. 1971. Adaptive zones and the orders of mammals. *Evolution* 25:420–28

Waller, T. A. 1972. The functional significance of some shell microstructures in the Pectinacea (Mollusca: Bivalvia): *Int. Geol. Congr.* 24(7):48–56

Yonge, C. M. 1936a. Mode of life, feeding, digestion, and symbiosis with zooxanthellae in the Tridacnidae: *Great Barrier Reef Exp. 1928–1929, British Mus. Sci. Rep.* 1:283–321

Yonge, C. M. 1936b. The evolution of the swimming habit in the Lamellibranchia. *Mus. Roy. Hist. Nat. Belg. Mem.* 2, T. 3:77–100

Yonge, C. M. 1950. Life on sandy shores: *Sci. Progr.* 38:430–44

Yonge, C. M. 1953. The monomyarian condition in the Lamellibranchia. *Trans. Roy. Soc. Edinburgh* 62:443–78

Yonge, C. M. 1957. Mantle fusion in the Lamellibranchia: *Publ. Sta. Napoli* 29:25–31

Yonge, C. M. 1962. On the primitive significance of the byssus in the Bivalvia and its effects in evolution. *J. Mar. Biol. Assoc. U. K.* 42:112–25

Yonge, C. M., Campbell, J. I. 1968. On the heteromyarian condition in the Bivalvia with special reference to *Dreissena polymorpha* and certain Mytilacea. *Trans. Roy. Soc. Edinburgh* 68:21–43

AGE PROVINCES IN THE NORTHERN APPALACHIANS

✺10045

R. S. Naylor[1]

Earth and Planetary Science Department, Massachusetts Institute of Technology, Cambridge, Massachusetts 02139[2]

INTRODUCTION

Most of the northern Appalachian region is underlain by well-stratified volcanic and sedimentary rocks of Late Precambrian through Early Devonian age. Even in structurally complex areas, these units exhibit a characteristic stratigraphic and structural continuity, which greatly facilitates study of their age relationships. Locally, these units are underlain by rock sequences which lack this internal continuity. Such sequences are defined as *basement* in this review.

Isotopic dating has proved a valuable technique for studying the history of basement rocks, and the bulk of this paper is devoted to a discussion of basement age provinces, updating a previous review by Isachsen (1964). Age data for the younger rocks have been recently synthesized by Lyons & Faul (1968), hence this topic is limited to comments on data published subsequent to that review. Although its geology is closely related to that of the mainland Appalachians, Newfoundland has been excluded from this review, partly because of the paucity of isotopic dating in that area.

Basement Rocks

The distribution of basement rocks in the Northern Appalachians is shown in Figure 1. Precambrian rocks constitute the basement on both flanks of the range. To first order, isotopic dating demonstrates that these flanking rocks comprise two distinct provinces, described in this paper as the Western Basement and the Eastern Basement. The Western Basement is characterized by high temperature, high pressure (granulite facies) metamorphism at 1100 to 1200 m.y. and is part of the Grenville basement province at the eastern edge of Paleozoic North America. The Eastern Basement is characterized by widespread volcanism and plutonism at about 600 to 650 m.y. and is commonly metamorphosed only to greenschist

[1] The author's studies of basement rocks and Acadian granites in the northern Appalachians have been sponsored by the National Science Foundation.

[2] The author's present address is the Earth Sciences Department, Northeastern University, Boston, Massachusetts 02115.

387

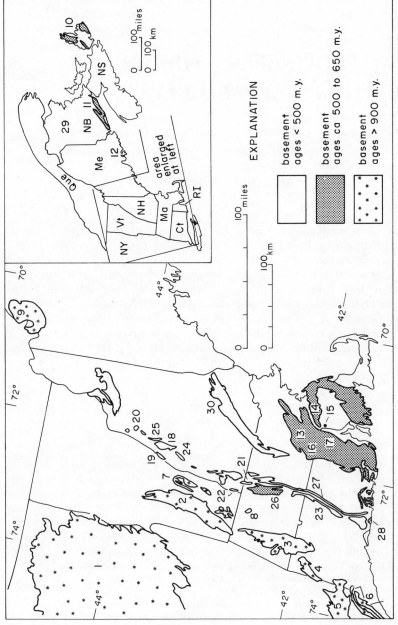

Figure 1 Distribution and isotopic ages of basement rocks in the Northern Appalachians. Numbers refer to localities described in the text.

facies. It appears to have originated as a volcanic and plutonic terrane within the Appalachian geosyncline during the earliest stages of the Appalachian orogenic cycle. Rocks older than 650 m.y. have not been positively identified in the Eastern Basement area, but certain localities which may prove older are suggested for future isotopic dating. It is not known (if indeed they are older) whether these units are basement on which the Eastern Basement rocks were deposited or whether they are separate blocks of older basement caught up in a zone of suture along the western margin of the Eastern Basement terrane.

By the definition given above, certain rocks occurring in the central part of the orogenic belt may also be classified as basement. Isotopic dating indicates that many of these rocks are Ordovician and thus developed as part of the evolution of the Appalachian range. Units older than Ordovician have also been recently identified in this belt. It is not known whether these rocks are closely related to the Eastern Basement province or whether they resulted from contemporaneous events occurring in widely separated parts of the geosyncline.

Techniques

The following decay constants are used in this paper:

$$\lambda Rb^{87} = 1.39 \times 10^{-11} \text{ yr}^{-1}$$
$$\lambda U^{235} = 9.72 \times 10^{-10} \text{ yr}^{-1}$$
$$\lambda U^{238} = 1.54 \times 10^{-10} \text{ yr}^{-1}$$

Zircon and Rb-Sr whole-rock ages are considered to be the most useful in outlining the age provinces. Many of the K-Ar and Rb-Sr mineral ages reflect secondary events and processes, hence no attempt is made to systematically review these data.

AGE PROVINCES

Western Basement

Basement rocks of the Grenville Province extend eastward into the Appalachians. Characteristic ages of these rocks reflect high pressure, high temperature metamorphism between 1100 and 1200 m.y. although, locally, ages possibly reflecting older events have been measured. Also included are a few units with ages younger than 1100 m.y. but which lie unconformably beneath the stratified Appalachian section. The effects of multiple Paleozoic disturbances become more severe eastward, obscuring the original character of the rocks.

ADIRONDACK MASSIF   Suites of zircons from igneous rocks throughout the Adirondack Massif (Figure 1, Locality 1) yield primary concordia ages in the range 1100 to 1200 m.y. (Silver 1969). Similar ages of 1095 to 1165 m.y. are obtained by the rubidium-strontium whole-rock technique (Hills & Gast 1964).

GREEN MOUNTAIN BELT   The Grenville basement rocks reappear to the east in a belt of en-echelon massifs, the largest and most northerly of which is the Green

Mountain Massif (Locality 2). Faul et al (1963) report Pb-alpha ages of 1100 to 1530 m.y. from these rocks. Attempts at Rb-Sr whole-rock dating have been frustrated by generally unfavorable Rb/Sr ratios.

Mineral ages by K-Ar and Rb-Sr range from 310 to 980 m.y. (Faul et al 1963, Naylor 1967), reflecting Paleozoic metamorphism. Rb-Sr dating of muscovite from pegmatite at Buttermilk Falls shows an interesting pattern of discordance. The interior portion of a 1 cm diameter book yields an age of 1070 m.y., but the outer portion yields 1000 m.y. and fine-grained muscovite from the pegmatite, 930 m.y. (Naylor 1967).

The next massif to the south along this belt is the Berkshire Massif (Locality 3) where Ratcliffe & Zartman (1971) report zircon ages of 1060 m.y. from the Tyringham and Washington Gneisses. Similar rocks occur further southwest in the Housatonic Highlands (Locality 4), the Hudson Highlands (Locality 5), and the Manhattan Prong (Locality 6), where zircons from the Fordham Gneiss have Pb^{207}/Pb^{206} ages ranging from 800 to 1300 m.y. (Grauert & Hall 1973).

GNEISS DOMES OF SOUTHWESTERN NEW ENGLAND   Rocks correlated with the Grenville reappear still further to the southeast in a belt of mantled gneiss domes. Low Rb/Sr ratios and severe overprinting by Paleozoic disturbances have hampered geochronologic study of these domes, but Faul et al (1963) report a Pb-alpha age of 900 m.y. from the Chester Dome (Locality 7) in Vermont. Isotopic zircon dates on rocks from the domes closest to the Green Mountain belt will probably yield ages of 900 m.y. or older.

As discussed later, other gneiss domes crop out farther to the east but contain core rocks younger than 600 m.y. Except for the Chain Lakes complex, there is no direct evidence that Grenville-type rocks of the Western Basement extend farther eastward beneath the Appalachians than the more westerly of the gneiss domes. Preliminary data (R. S. Naylor, unpublished) indicate that the Shelburne Falls Dome (Locality 8), west-central Massachusetts, is more closely related to the domes of the Bronson Hill Anticlinorium than to the Grenville basement.

CHAIN LAKES MASSIF   A 30 by 50 km massif of sillimanite grade granulite (Locality 9) is exposed along the Maine-Quebec boundary just northeast of the border with New Hampshire. Naylor et al (1973) report Pb^{207}/Pb^{206} ages of 1500 to 1510 m.y. on zircons from a sample of this rock north of Eustis, Maine.

YOUNGER ROCKS   Rocks younger than 1000 m.y. but older than the stratified cover-rocks have been documented near New York City and may occur elsewhere in the western basement. Long (1969) obtained a Rb-Sr whole-rock isochron age of 530 to 575 m.y. for the Yonkers Gneiss (Locality 6) in the Manhatten prong, and this result has been substantiated by zircon dating (Grauert & Hall 1973). The Yonkers Gneiss lies unconformably beneath the Lowerre Quartzite and Inwood Marble of the stratified cover-sequence. The isotopic results suggest it is intrusive into the older Fordham Gneiss rather than being interstratified meta-arkose or meta-rhyolite as suggested by Hall (1968). The Stamford Granite Gneiss

(Skehan 1961) at the southern end of the Green Mountains (Locality 2) and the Bull Hill Gneiss (Doll et al 1961) of the Chester (Locality 7) and Athens Domes, Vermont, are possible younger basement units, but neither unit has been isotopically dated.

Eastern Basement

A distinctive basement terrane characterized by volcanic and plutonic rocks underlies fossiliferous Lower Cambrian strata in several areas along the southeastern margin of the Northern Appalachians. Except near later plutons, most of the basement rocks are metamorphosed only to the greenschist facies, yet isotopic dating of these rocks has proved surprisingly difficult. Rb-Sr whole-rock dating of many of the volcanic units (Fairbairn et al 1966) yields ages younger than the accepted time scale ages of the overlying Lower Cambrian strata. Dating of the associated plutonic rocks yields more acceptable ages and demonstrates widespread Late Precambrian plutonic and volcanic activity. However, even on the plutonic rocks there is a discrepancy between zircon dates (averaging 600 to 650 m.y.) and Rb-Sr whole-rock dates (averaging 530 to 590 m.y.). Mineral ages (mostly by K-Ar) range from 300 to almost 900 m.y., which suggests combinations of metasomatism, inherited radiogenic argon, and later disturbance. The Eastern Basement is nowhere characterized by granulite facies metamorphism and has nowhere yielded zircon or Rb-Sr whole-rock ages greater than 650 m.y. These features indicate it is a different terrane than the previously characterized Western Basement.

Possible exceptions are several areas of higher grade rocks including quartzite, marble, granite, schist, and gneiss occurring along or near the northwestern boundary of the Eastern Basement terrane (examples include parts of the George River Group, Cape Breton Island, Nova Scotia; the Green Head Group, St. John area, New Brunswick; and parts of the Nashoba Formation, eastern Massachusetts). These units have not been isotopically dated, but they may be significantly older than 650 m.y. Their field relationships are not well established. It is not known whether they are slivers of older basement caught up in a zone of suture or whether they represent a yet older basement on which the Eastern Basement rocks were deposited.

CAPE BRETON ISLAND, NOVA SCOTIA   Cormier (1972) has published Rb-Sr wholerock isochron ages of 530 to 575 m.y. for granitic rocks distributed across the breadth of Cape Breton Island (Locality 10). Mineral ages on these same rocks by K-Ar and Rb-Sr range from 360 to 585 m.y. Wiebe (1974) suggests these granitic rocks are contemporaneous with volcanics of the Forchu group on which Fairbairn et al (1966) determined a Rb-Sr whole-rock age of 509 ± 40 m.y. The Forchu Group, consisting of felsic and intermediate volcanic rocks with minor intercalated sedimentary rocks, is overlain by Lower Cambrian sedimentary rocks. The George River Group, consisting of quartzite, marble, schist and gneiss, and amphibolite (Weeks 1954), is interpreted as a significantly older unit, but these rocks have not been isotopically dated.

SOUTHEASTERN NEW BRUNSWICK AND COASTAL MAINE   Late Precambrian felsic to intermediate volcanic rocks of the Coldbrook Group (Locality 11) underlie a 40 km wide belt along the southeastern coast of New Brunswick (Potter et al 1968). Fairbairn et al (1966) report a Rb-Sr whole-rock isochron age of 495 ± 48 m.y. for the Coldbrook Group, but Rb-Sr whole-rock or zircon ages have not been reported for granitic rocks in this terrane.

The Green Head Group, occurring within this belt, consists of marble, quartzite, schist, and gneiss. Marble horizons locally bear the stromatolite, *Archaeozoon acadiense* (Hayes & Howell 1937). Although these rocks have not been isotopically dated, and their contact with the Coldbrook Group rocks has been variously interpreted as a fault or an angular unconformity, the Green Head Group is regarded by many workers as an older basement complex.

The Precambrian belt of New Brunswick strikes offshore to the south, and correlatives of these rocks have not been identified along the Maine coast. The Ellsworth Formation (Locality 12) east of Penobscot Bay, Maine, is more highly deformed than the overlying units, and some geologists have speculated that it may be Precambrian. Isotopic dates have not been published for the Ellsworth Formation.

MASSACHUSETTS AND RHODE ISLAND   The Milford Granite (Locality 13) of southeastern Massachusetts yields a Rb-Sr whole-rock isochron age of 614 ± 24 m.y. and zircon Pb^{207}/Pb^{206} ages of 630 m.y. R. E. Zartman and R. S. Naylor (in preparation) and Fairbairn et al (1967) report Rb-Sr whole-rock ages of 591 ± 28 m.y. for the Dedham Granodiorite (Locality 14), 514 ± 17 m.y. for the granite at Hoppin Hill (Locality 15; a possible correlative of the Dedham), and 569 ± 4 m.y. for the Northbridge granite-gneiss (Locality 16). The isotopic ages and the relationships of these units to overlying fossiliferous Lower Cambrian strata indicate widespread Late Precambrian plutonic activity in southeastern New England. Crosscutting relationships indicate that part of the Westboro Quartzite and part of the greenstone mapped as Marlboro Formation predate the granitic rocks.

Schist, quartzite, and gneiss of the Blackstone Series (Locality 17) in Rhode Island are possibly part of this Late Precambrian terrane (Quinn 1971).

Basement Rocks of the Central Orgenic Belt

Rocks of the Western Basement have not been identified east of the Chain Lakes Massif and the line of gneiss domes in southwestern New England, and the Eastern Basement assemblage has not been identified west of a series of late stage faults running roughly parallel to the present coastline. Most of the rocks between the two basement terranes belong to the Late Precambrian through Early Devonian Appalachian stratified sequence, but locally rocks occur which may be classified as basement as the term is used in this paper.

In the Bronson Hill Anticlinorium the Ammonoosuc Volcanics (Localities 18–27) are the oldest of the stratigraphically continuous units. Extrapolating from a fossil locality in western Maine (Harwood & Berry 1967) the Ammonoosuc Volcanics in this belt are considered Middle Ordovician, but it is possible that basalts of older age have been mapped as Ammonoosuc in the more highly metamorphosed areas. Naylor (1969) reports a Rb-Sr whole-rock isochron age of 440 ± 30 m.y. for the Ammonoosuc Volcanics in New Hampshire, revising an earlier result of Brookins & Hurley (1965). Brookins & Methot (1971) report a Rb-Sr whole-rock isochron age of 460 ± 15 m.y. for the Middletown Gneiss (equivalent to Ammonoosuc) in Connecticut.

OLIVERIAN SERIES   The Oliverian Domes of the Bronson Hill Anticlinorium have cores of light-colored gneiss lying conformably below amphibolites of the Ammonoosuc Volcanics. The core-gneisses of the domes cannot be dated strati-graphically because they are unfossiliferous and do not connect from dome to dome at the present erosion surface. Naylor (1969) identified two distinct lithologic units in the core of the Mascoma Dome (Locality 18), central-western New Hampshire. The major unit is interpreted as a sequence of massive, weakly stratified, dacitic volcanics. These rocks are cut by a pluton of quartz monzonite. Zircons from both units define a chord with a primary concordia intercept of 450 ± 25 m.y., and Rb-Sr whole rock data for the plutonic unit indicate an age of 440 ± 45 m.y. Similar Rb-Sr ages were measured on granitic rocks from the Lebanon Dome (Locality 19) ten miles to the west. Naylor (1969) concluded that all of these rocks originated from Ordovician volcanic and plutonic activity and could detect no sign of anatexis resulting from kyanite-grade metamorphism during the Acadian Orogeny.

Rocks similar to those of the Mascoma Dome occur elsewhere in the belt and are tentatively correlated. The Owls Head Dome (Locality 20), the Warwick Dome (Locality 21), the Vernon Dome (Locality 22), and the Glastonbury Dome (Locality 23) contain rocks closely resembling both the plutonic and volcanic rocks of the Mascoma Dome. The elongate Croydon (Locality 24) and Smarts Mountain (Locality 25) Domes appear to lack significant bodies of plutonic rock but contain core-rocks similar to the dacitic volcanics. At the north end of the Glastonbury Dome (Locality 23) the core rocks intrude amphibolite mapped as Ammonoosuc (Leo 1974). It is not certain whether this crosscutting relationship dates from the Ordovician or whether it is a result of Acadian anatexis, and it should be noted that Brookins & Hurley (1965) report a Rb-Sr whole-rock age of 355 ± 10 m.y. for gneiss from the Glastonbury Dome.

OLDER ROCK UNITS IN THE BRONSON HILL ANTICLINORIUM   Several domes contain rocks which differ in lithology from the Oliverian Series rocks described above. Because some of these rocks yield older ages, considerable caution is required in correlating core rocks among the domes of the Bronson Hill belt.

Naylor et al (1973) report a Pb^{207}/Pb^{206} age of 575 m.y. for the Dry Hill Gneiss in the core of the Pelham Dome (Locality 26), which lies along the western margin of the Bronson Hill Anticlinorium in north-central Massachusetts. As in the other domes, the Pelham core rocks are mantled by the Ammonoosuc Volcanics but the core rocks contain quartzite, schist, and other rocks not found in the Oliverian Series.

Along the eastern margin of the Bronson Hill Anticlinorium in Massachusetts and Connecticut the Ammonoosuc Volcanics are underlain by the Monson Gneiss (Locality 27), a unit which resembles the Oliverian dacitic metavolcanics except that it shows a greater degree of compositional differentiation and layering. Brookins & Hurley (1965) report a Rb-Sr whole-rock isochron age of 472 ± 15 m.y. for the Monson Gneiss. This determination was revised to 480 ± 15 m.y. by Brookins & Methot (1971).

DOMES IN SOUTHERN CONNECTICUT From the point where the Bronson Hill Anticlinorium extends into Long Island Sound, a terrane of gneiss domes extends eastwards along the Connecticut Coast towards the area underlain by the Eastern Basement. At the southern end of the Bronson Hill Anticlinorium the New London Gneiss, the Mamacoke Formation, and the Plainfield (quartzite) Formation structurally underlie rocks which form the cores of domes further north (Dixon & Lundgren 1968). Hills & Dasch (1972) report a Rb-Sr whole-rock isochron age of 616 ± 78 m.y. for the Stony Creek Granite (Locality 28) which cuts the Plainfield Formation.

MIRAMACHI ANTICLINORIUM, NEW BRUNSWICK The Tetagouche Group of the Miramachi Anticlinorium (Locality 29) in central New Brunswick (Potter et al 1968) contains volcanic rocks resembling those of the Bronson Hill Anticlinorium and its extension into northern Maine. Parts of the group are stratigraphically dated as Ordovician but zircon and Rb-Sr whole-rock ages have not been reported. Local occurrences of gneiss possibly represent inliers in older basement (G. M. Boone and N. Rast, personal communication).

MASSABESIC GNEISS, SOUTHERN NEW HAMPSHIRE A 10 km wide belt of high-grade schist and gneiss (Sriramadas 1966) known as the Massabesic Gneiss (Locality 30) extends from near Berwick, Maine, through Manchester, New Hampshire, into Massachusetts, where it is mapped as part of the Fitchburg "Pluton." Preliminary Rb-Sr whole-rock dating (R. S. Naylor, unpublished data) suggests the gneiss is older than Devonian and may represent older basement. The unit is bounded by steep metamorphic gradients and its structural relationships with the surrounding units are uncertain.

Appalachian Age Province

Stratified, dominantly marine, sedimentary and volcanic rocks of Late Precambrian (?) through Early Devonian age characterize the Appalachian age province. Rb-Sr whole-rock dating may set useful limits on the depositional ages of some units in belts where fossils have not been found (see Lyons & Faul 1968), but on the whole, fossils provide the most precise means for dating the stratified rocks. Fossil localities are abundant in the northern part of the area, and the stratigraphic continuity of these units permits fossil ages to be extrapolated into the highly metamorphosed southern New England area with considerable confidence. Improved Rb-Sr or zircon dating of volcanic units near the base of the section, such as the Tibbit Hill Volcanics, northern Vermont, should provide valuable information on the earliest history of this depositional cycle.

About 20% of this province is underlain by granitic intrusives, mostly of Devonian age. Lyons & Faul (1968) noted that the ages (mostly mineral ages by Rb-Sr and K-Ar) on these granites cluster at about 360 m.y., and suggested this as the best date for the climatic Acadian Orogeny. Rb-Sr dating of coarse muscovite fractions from some of these rocks (Naylor 1971, and unpublished data) yields ages of 380 m.y. and older, suggesting that the clustering at 360 m.y. may represent cooling effects. The granites yielding 380 m.y. muscovite ages in eastern Vermont were intruded after regional kyanite- and sillimanite-grade metamorphism and the formation of large nappes. The closeness of the minimum ages for these late-stage granites to the deposition age of the youngest country rocks indicates that the Acadian Orogeny was abrupt in onset and brief in duration. These data are consistent with hypotheses involving the collision of two plates following an earlier interval of plate closure.

Postorogenic, mostly nonmarine, Middle Devonian through Carboniferous sedimentary rocks were deposited on both flanks of the Appalachians. On the eastern side these sediments contain minor intercalated volcanic units, and show the effects of one or more pre-Triassic disturbances. In southeastern Rhode Island, Upper Carboniferous rocks are metamorphosed to kyanite-staurolite grade (Quinn 1971) and are cut by granite bodies dated at 240 to 250 m.y. (see Lyons & Faul 1968). Over a wide belt in southeastern New England, Rb-Sr and K-Ar dating yields mineral ages of approximately 250 m.y. irrespective of the primary age of the rocks (Faul et al 1963, Zartman et al 1970). Except for the extreme southeastern part of the region, where late orogenic activity is evident, Zartman et al (1970) conclude most of this overprinting is due to burial metamorphism beneath a formerly more widespread cover of Carboniferous rocks.

Triassic and Younger Ages

Triassic red beds and associated basalts occur in the lower Connecticut River Valley and in the area around the Bay of Fundy. Armstrong & Besancon (1970) report K-Ar ages clustering at 200 m.y. for Triassic diabase dikes and basalt sills and flows in the Connecticut Valley area; they conclude that these ages are slightly low due to argon loss during low grade burial metamorphism.

Foland et al (1971) have demonstrated that plutons of the White Mountain Plutonic Series in New Hampshire were emplaced over a 100 m.y. interval from about 230 to about 115 m.y., and other younger ages are cited in the review by Lyons & Faul (1968). These Triassic and younger ages appear more closely related to the opening of the modern Atlantic Ocean than to the earlier processes which built the Appalachian mountains.

DISCUSSION

Tectonic Significance of Basement Rocks

The Precambrian rocks have exerted a strong influence on theories of the evolution of the northern Appalachian mountains. As recently as the 1932 Geologic Map of the United States (Stose 1932) most of the high grade metamorphic rocks of the Appalachians were considered Precambrian. Geologists subsequently determined that most of the well-stratified metasedimentary rocks in the central part of the range are of early Paleozoic age, and eventually located the contact between these rocks and the Precambrian rocks of the Western Basement (Figure 1, Localities 2–6). Knowledge that Precambrian rocks of the Eastern Basement formed the opposite flank of the range has led most geologists mapping in the northern Appalachians to favor ensialic theories for the origin of the mountains, with the central geosyncline viewed as a down-warp floored by a more-or-less continuous continental crust.

Two papers which espoused an opposing or ensimatic theory were those of Drake et al (1959) and Dietz (1963). Both noted similarities between the Early Paleozoic miogeosyncline and eugeosyncline on the western flank of the Appalachians and the modern shelf and continental rise deposits of the Atlantic coast of North America. Both papers attributed genetic significance to these similarities, proposing that the Paleozoic geosyncline was partly ensimatic. At the time, most geologists interpreted this hypothesis as implying progressive accretion of North America into the Atlantic Ocean without ocean-floor spreading or continental drift. On this assumption, the presence of Precambrian basement between the geosyncline and the Atlantic was a serious obstacle to acceptance of the ensimatic hypothesis. Significantly, neither paper mentioned the Eastern Basement.

A solution to this dilemma was proposed by Wilson (1966) who applied the ocean-floor spreading hypothesis to the evolution of the Appalachians. He proposed that the geosyncline was deposited in a Proto-Atlantic Ocean which subsequently closed on itself. Wilson conjectured that the Eastern Basement was a portion of Africa abutted onto North America by closure of the Proto-Atlantic and stranded by later opening of the modern Atlantic to the east.

Improved understanding of the age and distribution of basement rocks is critical to working out the structure of the Appalachians. In applying plate tectonics to theories of the evolution of the range, this knowledge will be helpful in identifying the boundaries and possibly even the origins of the plates but, as discussed below, is unlikely to provide a direct proof of the hypothesis.

Wilson (1966) noted the contrast between the Western and Eastern Basement terranes and cited this as part of the evidence that they comprise separate plates. In its simplest form, the theory of continuous basement beneath the geosyncline would favor the presence of identical basement on both flanks, but there is nothing to prevent the basement from changing character laterally. On the other hand, the simplest model of a narrow ocean opening and reclosing with little motion parallel to its axis would also permit the basement on both flanks to be identical, so the issue is not conclusive.

The observed scarcity of sialic Precambrian basement towards the center of the geosyncline favors the ensimatic theory, but conclusive evidence is elusive. The relatively light density of sialic basement should favor its rise towards the surface during regional metamorphism as in the Vermont gneiss domes. Except for this it is easy to argue that sialic basement is present beneath the geosyncline, only buried beneath the thick sediments. Plate tectonics permits fragments of exotic basement to be rafted in on the various plates, so the discovery of old basement embedded in the central part of the range could be consistent with either theory.

Distribution of Basement Rocks

There seems little doubt that the large massifs (Figure 1, Localities 2–6) of Western Basement represent an eastward extension of the Grenville-Adirondack basement. The massifs occur approximately along the boundary between the mio- and eugeosynclinal sections which Rodgers (1968) proposed marks the shelf edge of Early Paleozoic North America. Although not conclusively proven by isotopic dating, it seems likely that the Vermont gneiss domes (Locality 7) represent a further eastward extension of similar basement, demonstrating that at least the western part of the eugeosyncline is ensialic. The analogy between the eugeosyncline and modern continental rises suggests that the sialic basement thins and wedges out eastward, and this is supported by the observed lack of demonstrable Western Basement east of the line of domes. It is conceivable that this thinning facilitated the tectonic involvement of the basement in doming during later metamorphism.

It seems noteworthy that a prominent belt of ultramafic bodies (Doll et al 1961) occurs between the domes and the massifs, that is, on the landward side of the dome basement. Chidester & Cady (1972) have argued that the ultramafic bodies are intrusives which punched their way through the sialic crust. Analogy with ultramafic bodies along-strike in Newfoundland suggests, on the other hand, that the bodies may be ophiolites transported westward up and over the basement and subsequently dismembered and metamorphosed during the Devonian Acadian Orogeny.

The Chain Lakes Massif (Locality 9) is anomalous in several respects. Its isotopic age and lithology are not typical for Western Basement, and its position lies considerably east of the strike of the Vermont Domes. Ordovician and younger rocks flanking the massif are virtually unmetamorphosed, hence the structure is clearly not an Acadian gneiss dome, as gneiss domes form only in high grade metamorphic terrane. The massif marked a stratigraphic high during the mid Paleozoic as the Silurian and Devonian sediments thin and wedge out towards its axis. It is part of Boucot's (1961) Somerset Island, and may well represent an exotic block of basement rafted in on a non-North American plate.

Isotopic dating has shown the Monson and Oliverian units of the Bronson Hill Anticlinorium (Localities 18–27) to be of early Paleozoic age. Together with the Ammonoosuc Volcanics the rocks in this belt are chiefly volcanics with related intrusives, and various workers have suggested this is the site of an Early Paleozoic volcanic arc. Many island arcs are complex structures with long histories, hence the presence of older rocks in the Pelham Dome (Locality 26) is not inconsistent with this hypothesis. The rocks of the Miramachi Anticlinorium in New Brunswick (Locality 29) probably originated in a similar fashion.

Wilson (1966) implied the Eastern Basement was part of the stable craton from the opposite side of the ocean basin from which the Appalachians developed. The crystalline rocks of the Eastern Basement are mostly greenstone, felsite, and calc-alkaline granodiorite to quartz monzonite—rocks which could readily originate in a volcanic arc or other noncontinental environment. The Meguma belt, with Early Paleozoic geosynclinal sediments nearly 10 km thick, lies to the east of the Eastern Basement, and an additional hundred kilometers or more of unknown terrane, submerged on the continental shelves separates the Eastern Basement from Africa. The rocks are of Late Precambrian age, which corresponds to the earliest stages in the development of the Appalachians. These observations suggest the Eastern Basement rocks are more likely to have developed within the Appalachian geosyncline during its early stages than to represent the stable craton on its eastern flank. The older units mentioned in the description of Localities 10–17 may, however, be fragments of that craton.

The relationship of the Eastern Basement to rocks further west is a major topic for continued study. As mentioned in the locality descriptions, the Eastern Basement wraps around the southeast coast of Connecticut and structurally underlies the sequence of rocks at the southern end of the Bronson Hill anticlinorium (Locality 28). Isotopic dating indicates the two sequences differ in age by almost 200 m.y. but the lithologies are similar, a circumstance which poses problems for field mapping in this high grade metamorphic terrane. One hypothesis is that the Eastern rocks constitute a basement on which the Bronson Hill sequence and the Merrimack Synclinorium units to the east were deposited. An alternate hypothesis, favored by the present author, is that the contact between these units represents a major Appalachian suture.

Isotopic ages of about 600 m.y. are characteristic of the Eastern Basement, and similar ages are also locally evident in western New England. These ages may reflect events related to the opening of the Appalachian geosyncline. As such, they could occur anywhere in the region and need not indicate proximity of the Eastern Basement to western New England 600 million years ago.

In summary, the basement data apparently cannot be used to distinguish between ensialic and ensimatic (including plate-tectonic) hypotheses for the origin of the Appalachian geosyncline. The author favors the plate-tectonic hypothesis for its ability to explain observed crustal shortening without requiring a great bulge of sialic basement beneath the range. Confirmation of the plate-tectonic hypothesis awaits positive identification of suture zones and of incorporated fragments of ocean crust. In the meantime, the hypothesis is of great value in suggesting leads for future investigation.

Literature Cited

Armstrong, R. L., Besancon, J. 1970. A Triassic time-scale dilemma: K-Ar dating of Upper Triassic mafic igneous plutons, eastern U.S.A. and Canada and Post-Upper Triassic plutons, western Idaho, U.S.A. *Eclogae Geol. Helv.* 63:I, 15–28

Boucot, A. J. 1961. Stratigraphy of the Moose River Synclinorium, Maine. *US Geol. Surv. Bull.* 1111E:153–88

Brookins, D. G., Hurley, P. M. 1965. Rb-Sr geochronological investigations in the Middle Haddam and Glastonbury quadrangles, eastern Connecticut. *Am. J. Sci.* 263:1–16

Brookins, D. G., Methot, R. L. 1971. Geochronologic investigations in south-central Connecticut: I: Pre-Triassic basement rocks. *Geol. Soc. Am. Abstr. with Programs* 3:I, 20

Chidester, A. H., Cady, W. M. 1972. Origin and emplacement of alpine-type ultramafic rocks. *Nature Phys. Sci.* 240:27–31

Cormier, R. F. 1972. Radiometric ages of granitic rocks, Cape Breton Island, Nova Scotia. *Can. J. Earth Sci.* 9:1074–86

Dietz, R. S. 1963. Collapsing continental rises: an actualistic concept of geosynclines and mountain building. *J. Geol.* 71:314–33

Dixon, H. R., Lundgren, L. W. 1968. Structure of eastern Connecticut. In *Studies in Appalachian Geology: Northern and Maritime*, ed. E. Zen et al, 219–29. New York: Interscience

Doll, C. G., Cady, W. M., Thompson, J. B., Billings, M. P. 1961. *Centennial Geologic Map of Vermont.* Vermont Geol. Surv.

Drake, C. L., Ewing, M., Sutton, G. H. 1959. Continental margins and geosynclines: the eastcoast of North America north of Cape Hatteras. In *Physics and Chemistry of the Earth*, ed. L. H. Ahrens, F. Press, K. Rankama, S. K. Runkorn, 3:110–98. London: Pergamon

Fairbairn, H. W., Bottino, M. L., Pinson, W. H., Hurley, P. M. 1966. Whole-rock age and initial ⁸⁷Sr/⁸⁶Sr of volcanics underlying fossiliferous Lower Cambrian in the Atlantic Provinces of Canada. *Can. J. Earth Sci.* 3:509–21

Fairbairn, H. W., Moorbath, S., Rama, A. O., Pinson, W. H., Hurley, P. M. 1967. Rb-Sr age of granitic rocks of southeastern Massachusetts and the age of the Lower Cambrian at Hoppin Hill. *Earth Planet. Sci. Lett.* 2:321–28

Faul, H., Stern, T. W., Thomas, H. H., Elmore, P. L. D. 1963. Ages of intrusion and metamorphism in the northern Appalachians. *Am. J. Sci.* 261:1–19

Foland, K. A., Quinn, A. W., Giletti, B. J. 1971. K-Ar and Rb-Sr Jurassic and Cretaceous ages for intrusives of the White Mountain magna series, northern New England. *Am. J. Sci.* 270:321–30

Grauert, B., Hall, L. M. 1973. Age and origin of zircons from metamorphic rocks in the Manhattan Prong, White Plains area, southeastern New York. *Carnegie Inst. Wash. Yearb.* 72:293–97

Hall, L. M. 1968. Times of origin and deformation of bedrock in the Manhattan Prong. In *Studies in Appalachian Geology: Northern and Maritime*, 117–27 New York: Interscience

Harwood, D. S., Berry, W. B. N. 1967. Fossiliferous lower Paleozoic rocks in the Cupsuptic quadrangle, west-central Maine. *US Geol. Surv. Prof. Pap.* 575-D:16–23

Hayes, A. O., Howell, B. F. 1937. Geology of St. John, New Brunswick. *Geol. Soc. Am. Spec. Pap.* 5. 146 pp.

Hills, F. A., Dasch, E. J. 1972. Rb-Sr study of the Stony Creek granite, southern Connecticut: A case for limited remobilization. *Geol. Soc. Am. Bull.* 83:3457–64

Hills, A., Gast, P. W. 1964. Age of pyroxene-hornblende granitic gneiss of the eastern Adirondacks by the rubidium-strontium whole-rock method. *Geol. Soc. Am. Bull.* 75:759–66

Isachsen, Y. W. 1964. Extent and configuration of the Precambrian in northeastern United States. *Trans. NY Acad. Sci. Ser. II* 26:812–29

Leo, G. W. 1974. Metatrondhjemite in the northern part of the Glastonbury Gneiss Dome, Massachusetts and Connecticut. *Geol. Soc. Am. Abstr. with Programs* 5(1):47–48

Long, L. E. 1969. Whole-rock Rb-Sr age of the Yonkers Gneiss, Manhatten Prong. *Geol. Soc. Am. Bull.* 80:2087–90

Lyons, J. B., Faul, H. 1968. Isotope geochronology of the northern Appalachians. In *Studies in Appalachian Geology: Northern and Maritime*, 305–18. New York: Interscience

Naylor, R. S. 1967. A field and geochronologic study of mantled gneiss domes in central New England. PhD thesis. Calif. Inst. Technol., Pasadena, Calif. 123 pp.

Naylor, R. S. 1969. Age and origin of the Oliverian Domes, Central-Western New Hampshire. *Geol. Soc. Am. Bull.* 80:405–28

Naylor, R. S. 1971. Acadian Orogeny: an abrupt and brief event. *Science* 172:558–60

Naylor, R. S., Boone, G. M., Boudette, E. L., Ashenden, D. D., Robinson, P. 1973. Pre-Ordovician rocks in the Bronson Hill and Boundary Mountain anticlinoria, New England, U.S.A. *Trans. Am. Geophys. Union* 50(4):495

Potter, R. R., Jackson, E. V., Davies, J. L. 1968. *Geological Map of New Brunswick.* Dep. Nat. Resources, New Brunswick. Map No. NR-1

Quinn, A. W. 1971. Bedrock Geology of Rhode Island. *US Geol. Surv. Bull.* 1295:1

Ratcliffe, N. M., Zartman, R. E. 1971. Precambrian granitic plutonism and deformation in the Bershire Massuf of western Massachusetts. *Geol. Soc. Am. Abstr. with Programs* 3(1):49

Rodgers, J. 1968. The eastern edge of the North American continent during the Cambrian and Early Ordovician. In *Studies in Appalachian Geology: Northern and Maritime*, ed. E. Zen et al, 141–49. New York: Interscience

Silver, L. T. 1969. A geochronologic investigation of the anorthosite complex,

Adirondack Mountains, New York. In *Origin of Anorthosite and Related Rocks,* ed. Y. W. Isachsen, 466. State Mus. Sci. Serv. Mem. 18

Skehan, J. W. 1961. The Green Mountain Anticlinorium in the vicinity of Wilmington and Woodford Vermont. *Vermont Geol. Surv. Bull.* No. 17. 159 pp.

Sriramadas, A. 1966. Geology of the Manchester Quadrangle, New Hampshire. *N. Hampshire Dep. Resources Develop. Bull.* 2:78

Stose, G. W. 1932. *Geologic Map of the United States.* US Geol. Surv.

Weeks, L. J. 1954. Southeast Cape Breton Island, Nova Scotia. *Geol. Surv. Can. Mem.* 277:112

Wiebe, R. A. 1974. Late Precambrian rocks of Cape Breton Island. *Geol. Assoc. Can. Ann. Meet. Program* 1974:98

Wilson, J. T. 1966. Did the Atlantic close and then re-open? *Nature* 211:676-81

Zartman, R. E., Hurley, P. M., Krueger, H. W., Giletti, B. J. 1970. A Permian disturbance of K-Ar radiometric ages in New England: its occurrence and cause. *Geol. Soc. Am. Bull.* 81:3359-74

METALLOGENESIS AT OCEANIC   ✻10046
SPREADING CENTERS[1]

Enrico Bonatti
Lamont-Doherty Geological Observatory of Columbia University,[2]
Palisades, New York 10964

INTRODUCTION

The frequent association in the geological record of basalts of submarine origin with metal ore deposits (such as massive sulfide ores and ferromanganese sedimentary deposits) had led, even prior to the plate-tectonic era, to the suggestion that submarine volcanism may provide conditions leading to processes of metal segregation and ore formation.

Progress in the methods of sampling the oceanic crust, both by conventional techniques and by deep drilling, and the synthesis achieved by plate-tectonic theory permit a new evaluation of the relationship of metallogenesis in active regions of the ocean floor to tectonism, volcanism, and hydrothermalism in the ocean basins. This new knowledge is also leading to a new understanding of the origins of a wide class of ore deposits on the continents.

This paper attempts to review and evaluate metallogenesis in tectonically active areas of the oceans. I discuss data on metal deposits from the present oceanic crust, specifically from (*a*) the upper crustal layer (layer 1) consisting of sediments, and (*b*) from the basaltic layer (layer 2) beneath the sediment blanket. I also discuss briefly the possibility of metallogenesis in the deeper zone of the oceanic crust and in fracture zones.

Of the metal deposits from the sediment layer, those associated with spreading centers are shown to be caused mainly by hydrothermal systems, ultimately related to igneous intrusion and volcanism, and to the emplacement of new oceanic crust at constructive plate margins. These same processes occurring at spreading centers lead to metallogenesis within the basaltic layer, with formation of metal sulfide deposits.

Data on metal ore deposits from presently emerged fragments of ancient oceanic

---

[1] Contribution No. 2167 from Lamont-Doherty Geological Observatory and from the Rosenstiel School of Marine and Atmospheric Science.

[2] On leave from the Rosenstiel School of Marine and Atmospheric Science, University of Miami.

401

crust are also reviewed; it is shown that such metal deposits are similar to deposits from the modern oceanic crust, and were formed at ancient spreading centers.

METALLIFEROUS SEDIMENTS FROM OCEAN BASINS

Ferromanganese deposits are widespread on the ocean floor, and have been the subject of a voluminous literature. Since the last century (Murray & Renard 1891), two main views have been expressed as to their origin: one favoring slow precipitation of metals from "normal" sea water, the other favoring submarine volcanism as a source of metals. More recent work indicates that both views are essentially valid, although different processes of metal concentration prevail in different regions of the ocean floor.

Bonatti et al (1972c) suggested the following classification of submarine metalliferous sediments, with four main types of deposits, depending on the prevalent source of the metals: hydrogenous, diagenetic, hydrothermal, and halmyrolitic (Figure 1).

HYDROGENOUS DEPOSITS   These are formed by slow precipitation of metals from normal sea water, in which the metals are provided primarily by weathering of the continents. They are widespread in areas of low sedimentation rate, such as abyssal plains in regions remote from continents, or in topographic highs where terrigenous sedimentation is scarce or absent.

DIAGENETIC DEPOSITS   These are produced by remobilization of metals during diagenesis of marine sediments, occurring especially in hemipelagic areas of the ocean floor with relatively high rates of deposition of organic matter. Reduced conditions are established below the top of the sediment column; several metals are dissolved in this reduced zone, mobilized by diffusion and advection, and redeposited in the top oxidized zone.

HYDROTHERMAL DEPOSITS   These are found in tectonically active regions of the ocean floor, such as spreading centers, and are associated with igneous and/or

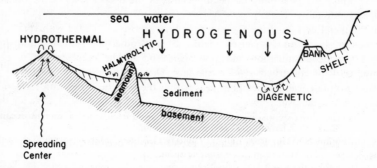

Figure 1   Scheme illustrating the four types of metalliferous sediments from the ocean floor.

hydrothermal activity. These deposits are discussed in detail in the next section of this paper.

HALMYROLITIC DEPOSITS   These deposits are caused by low temperature reactions (halmyrolysis) of solids (mainly basaltic glass) with sea water; such reactions may cause release of metals into sea water and subsequent precipitation of the metal on the sea floor. Halmyrolytic deposits are important only in the limited areas of the ocean floor where submarine basaltic pyroclastics (hyaloclastites) are abundant, such as close to some volcanic seamounts.

HYDROTHERMAL METALLIFEROUS SEDIMENTS FROM SPREADING CENTERS

The following review of data on the distribution and geochemistry of hydrothermal metalliferous sediments is a basis for discussing the processes leading to their formation.

Several years ago it was recognized that sediments enriched in various metals occur along active oceanic ridges, and it was suggested that they originated through submarine hydrothermal activity, often related to volcanism (Skornyakova 1964, Arrhenius & Bonatti 1965, Bonatti & Joensuu 1966, Bostrom & Peterson 1966). Subsequent work has better defined the distribution and geochemistry of these deposits, and has established their general association with spreading centers.

The metal enrichment of ocean ridge deposits may be explained in two ways: (*a*) lack of dilution of hydrogenous metal-rich phases (i.e. phases deposited from normal sea water by slow chemical precipitation) by terrigenous or biogenous sediments; (*b*) a local additional source of metals present in the vicinity of the "anomalous" metalliferous deposits. In this discussion, I show that the second explanation is probably correct, and work with the assumption that a local source of metals at the spreading centers is provided by the discharge of hydrothermal fluids onto the sea floor, in a process closely linked with volcanism and the emplacement of new oceanic crust.

In this discussion, occasional reference is made to metal deposits presently being formed near the discharge of shallow water hydrothermal vents in volcanic areas such as Santorin or Thera (Greece), Banu Wuhu (Indonesia), Vulcano (Tyrrenian Sea), and Matupi Harbor (New Britain, southwest Pacific). These shallow-water deposits have some characteristics in common with the deep-sea deposits from spreading centers; because their growth is accessible to direct observation, they provide useful hints as to processes of metallogenesis at spreading centers.

Areal Distribution

Figure 2 shows the distribution of known metalliferous sediments along modern spreading centers. Because the detailed exploration of oceanic ridges is only at the initial stages, more deposits will undoubtedly be discovered in the future. At some sites, shown in Figure 2, metalliferous deposits have been sampled in which the

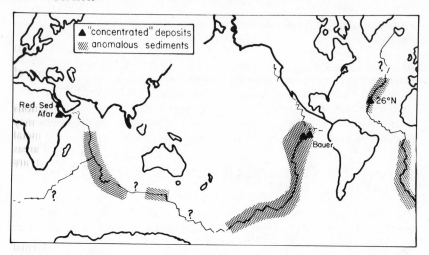

Figure 2   Distribution of metalliferous sediments along spreading centers (modified from Bostrom 1974). Also indicates a few sites where samples of "concentrated" metal deposits were recovered.

metal-containing phases are relatively undiluted by terrigenous and/or biogenous components. Generally the deposits are diluted by such components, and are recognized because of their anomalously high content of various metals.

EAST PACIFIC RIDGE   The portion of the East Pacific Ridge running from the Gulf of California to about 20°S has been explored rather extensively. At some sites rather undiluted metal deposits have been sampled. One is site Amph D2, a basaltic topographic high on the axis of the ridge at about 8°S (Bonatti & Joensuu 1966); another is the Bauer Deep, a basin located between the East Pacific Ridge and the now inactive Galapagos Ridge between 5 and 15°S (Sayles & Bischoff 1973, Anderson & Halunen 1974), where the metalliferous deposits occur throughout the sediment column for a thickness of up to 100 m (Hart et al 1974). In addition to these areas, sediments from the East Pacific Ridge and Pacific-Antarctic Ridge generally show unusually high metal content (Bostrom et al 1969).

INDIAN RIDGE   No sites with "concentrated" metal deposits have been found to date along the active Indian Ridge; however, metal-enriched sediments appear to follow the ridge (Bostrom 1974). Note that anomalous sediments of this type are absent from the aseismic Ninety-East Ridge.

GULF OF ADEN-AFAR-RED SEA RIFTS   The Indian (Carlsberg) Ridge extends into the Gulf of Aden spreading center and farther into a series of emerged spreading centers in the Afar region. Metal-rich deposits, formed in a shallow sea about 200,000 years ago, exist in the Afar Rift (Bonatti et al 1972a). The spreading axis continues along the Red Sea axis, site of the well-known metalliferous hot-brine deposits (see Degens & Ross 1969, Backer & Schoell 1972).

MID-ATLANTIC RIDGE   A "concentrated" deposit has been found within the axial rift valley at 26°N, apparently associated with temperature and chemical anomalies of the bottom sea water (Scott et al 1974, Rona et al 1974). In addition, sediments from the vicinity of the Mid-Atlantic Ridge in the North and South Atlantic show metal enrichment (Bostrom 1974).

Vertical Distribution

During the Deep Sea Drilling Project, metalliferous sediments similar in composition to those from spreading centers were sampled at many sites in the Atlantic, Pacific, and Indian Oceans. The deposits generally constitute a layer several meters thick at the base of the sedimentary column, just above the contact with the basaltic basement; therefore, they were probably originally formed close to a spreading center and subsequently moved laterally with the spreading oceanic crust (Figure 3). This suggests that the processes which create metalliferous deposits close to present spreading centers have operated throughout the history of the major oceans.

Geochemistry

The chemical composition of a number of representative metalliferous sediments from spreading centers is shown in Table 1. The enrichment of various elements such as Fe, Mn, Ni, Co, Cu, Zn, Cr, U, As, and Hg, relative to normal deep sea sediments is best seen by considering the ratio of the element in question to aluminum. The chemistry of some of these elements is examined in more detail next.

IRON AND MANGANESE   These two elements are generally the two principal components of the active ridge metalliferous sediments. The Fe/Mn ratio in these

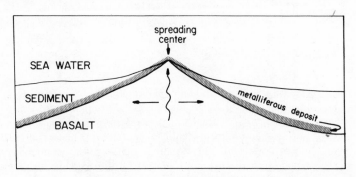

Figure 3   Scheme illustrating the distribution of metalliferous sediments at the base of the sediment column, as inferred from results of deep drilling by the *Glomar Challenger* in the major oceans.

Table 1 Composition of samples from metalliferous "hydrothermal" deposits; all except the last two are from the vicinity of spreading centers. Sources: E. Pacific Ridge, Bostrom et al (1969); E. Pacific Ridge (Amph D2), Bonatti & Joensuu 1966; Bauer Deep, Sayles & Bischoff 1973; Mid-Atlantic Ridge 26°N, Scott et al 1974; Afar, Bonatti et al 1972a; Matupi Harbor, Ferguson & Lambert 1972; Thera, Bonatti et al 1972b. Uranium data, Rydell & Bonatti 1973.

	Per Cent				ppm					
	Si	Al	Fe	Mn	Ni	Co	Cu	Zn	Ba	U
E. Pacific Ridge	6.1	0.5	18.0	6.0	430	105	730	380	–	~4-12
E. Pacific Ridge (Amph D2)	8.2	0.5	32.5	1.94	400	35	74	–	115	~2
Bauer Deep	18.0	1.4	18.2	5.7	950	90	1,100	600	13,000	–
Mid-Atlantic Ridge 26°N	–	–	0.01	39.2	100	18	12	–	–	~16
DSDP (E. Pacific)	–	–	17.5	4.5	535	83	917	358	–	–
Afar (Fe-rich)	14.0	3.7	29.0	0.15	18	17	11	–	135	0.5
Afar (Mn-rich)	<0.2	0.0	0.15	54.2	<10	<5	<5	–	58,000	~4
Matupi Harbor	–	–	44.0	0.034	–	–	47	52	–	–
Thera	11.6	1.2	35.0	0.6	<5	<5	30	–	90	16.1

deposits is highly variable, ranging from >10 to <0.1, in contrast to slowly deposited hydrogenous ferromanganese sediments, whose Fe/Mn ratios range around unity, as seen in Figure 4 (Bonatti et al 1972c).

Such wide range of Fe/Mn ratios in the hydrothermal deposits is probably caused by drastic fractionation of the two elements, both beneath the sea floor before discharge of the solutions and above it during deposition, as is discussed below. Fe in these deposits is generally in the form of very poorly crystallized hydroxides of the limonite-goethite family; in some cases, as in the Bauer Deep, Red Sea, and Afar deposits, a Fe-smectite is also present as a major Fe-containing phase. Mn is generally in the form of poorly crystallized oxides and hydroxides, except when considerable compaction and diagenesis has occurred, as in the Afar deposit, where well-crystallized pyrolusite appears.

NICKEL, COBALT, COPPER, ZINC, CHROMIUM   These elements are generally more abundant in the hydrothermal deposits than in normal pelagic sediments. However, the ratio of Fe+Mn to minor metals is substantially higher in hydro-

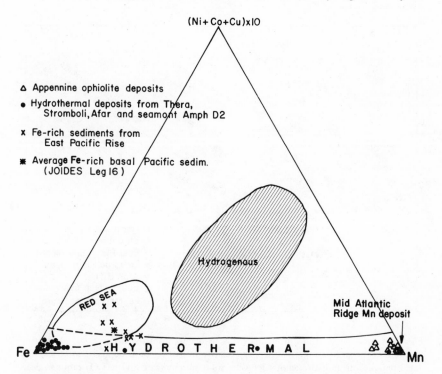

Figure 4   Ratio Fe/Mn(Cu + Ni + Co) in metalliferous sediments from the ocean floor, illustrating the different composition of hydrothermal and hydrogenous deposits. Data from metalliferous sediments from the Mesozoic northern Apennine ophiolites are also shown (modified from Bonatti et al 1972c).

thermal deposits than in hydrogenous ferromanganese deposits (Figure 4). This probably reflects in part the relatively more rapid rate of accumulation of the hydrothermal deposits, which limits the opportunity of "scavenging" of minor metals from sea water by Fe and Mn hydroxides (Bonatti et al 1972c). However, in some instances, as in the Red Sea deposits, discharge of the hydrothermal solutions takes place in a bottom morphology which prevents ready mixing of the solutions with oxygenated sea water and causes the formation of stagnant, reduced brine pools. Deposition of sulfides may then occur, and minor metals with strong affinities for sulfide phases, such as Cu and Zn, may become highly concentrated in the deposits.

SILICON AND ALUMINUM   The Si/Al ratio in the metalliferous deposits from spreading centers tends to be higher than in hydrogenous ferromanganese deposits (Figure 5), where the ratio is generally close to that of deep sea sediments, probably reflecting dilution of the Fe-Mn minerals by terrigenous and/or basaltic silicates. The high Si/Al ratio of hydrothermal deposits suggests

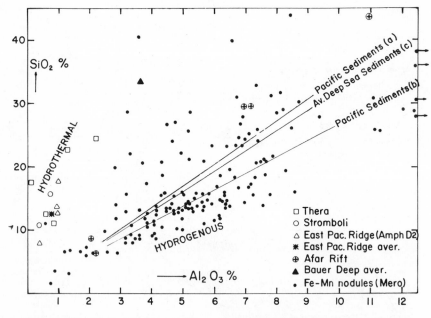

Figure 5 SiO_2/Al_2O_3 ratio in metalliferous sediments from the ocean floor, illustrating the higher ratio in hydrothermal deposits relative to hydrogenous deposits. Mero's (1965) data are largely from hydrogenous deposits. Solid lines indicate SiO_2/Al_2O_3 ratios in deep sea sediments [respectively from (a) Landergren (1964), (b) Goldberg & Arrhenius (1958), and (c) Turekian & Wedepohl (1961)], where the SiO_2/Al_2O_3 ratio is determined mainly by terrigenous or volcanogenous silicates. This graph suggests that a local source of SiO_2 is present in many "hydrothermal" metalliferous deposits.

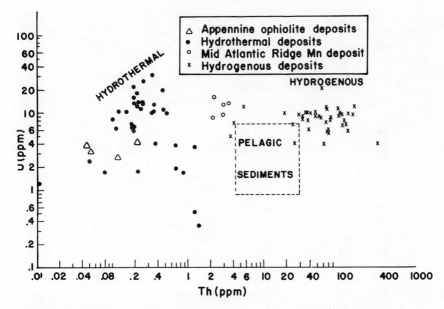

Figure 6   U/Th ratio in metalliferous deposits from the ocean floor (from Bonatti et al 1972c).

a local supply of silica. This view is supported by the direct observation of silica as one of the main components of submarine hydrothermal solutions discharged in areas of basic volcanism such as Thera (Bonatti et al 1972b) and Banu Vukhu, Indonesia (Zelenov 1965), and the Reykjanes area of Iceland (Björnsson et al 1972). Silica in the hydrothermal deposits is generally in the form of noncrystalline opaline phases, and in Fe-smectites when this phase is present.

URANIUM AND THORIUM   The U content of hydrothermal metal deposits is several times higher than that of normal pelagic sediments, and their U/Th ratio is one to two orders of magnitude higher than that of hydrogenous deposits (Figure 6), due primarily to the paucity of Th in the former. The different U/Th ratios in the two types of deposits have been ascribed to the more rapid growth of the hydrothermal deposits which would prevent any significant incorporation of Th from sea water, where this element is contained in exceedingly low concentration (Bonatti et al 1972c).

The U^{234}/U^{238} ratios of young (relative to the half life of U^{234}) hydrothermal deposits have been plotted (Figure 7) vs the U concentrations in the same deposits (from Rydell & Bonatti 1973). The U^{234}/U^{238} ratio in a number of samples is close to that of sea water (~ 1.15), suggesting that much of the U of these deposits derives from sea water. However, a number of deposits display anomalously high U^{234}/U^{238} ratios, suggesting that U in these cases is derived from the basaltic substratum, with preferential leaching of U^{234}. Samples with an anomalous U^{234}/U^{238} ratio tend to have low concentrations of U (Figure 7); these results support a hydrothermal model of metallogenesis, as discussed in detail by Rydell & Bonatti (1973).

RARE-EARTH ELEMENTS   The concentration of rare-earth elements in hydrogenous ferromanganese deposits, when normalized to chondrites, is characterized by a

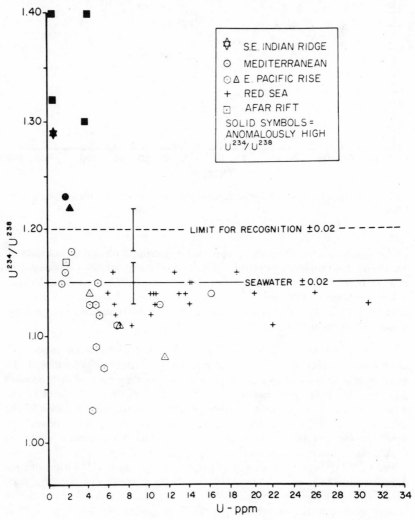

Figure 7 U^{234}/U^{238} ratio versus U concentration in metalliferous "hydrothermal" deposits (from Rydell & Bonatti 1973).

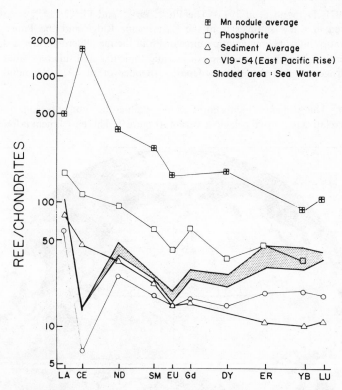

Figure 8 Distribution of rare-earth elements (normalized to chondrite abundances) in a sample of metalliferous deposits from the East Pacific Ridge, compared with phosphorite, "normal" pelagic sediments, sea water, and "hydrogenous" metalliferous deposits (Mn nodule average). Data by R. Kay, from Bender et al 1971.

strong positive anomaly of Ce. The reason for this Ce enrichment is that Ce, alone among the trivalent rare earths, can be oxidized to the relatively insoluble 4^+ valence state in sea water and can precipitate as oxide or hydroxide (Goldberg et al 1963, Piper 1972). Metalliferous deposits from spreading centers show instead a depletion of Ce, with a rare-earth pattern similar to that of sea water (Figure 8) and with a total rare-earth content one to two orders of magnitude lower than hydrogenous deposits (Bender et al 1971, Dymond et al 1973).

OXYGEN, STRONTIUM, AND LEAD ISOTOPES   A few O^{18}/O^{16} data are available for hydrothermal deposits from the East Pacific Ridge and the Bauer Deep; they suggest that the metalliferous sediments were formed in isotropic equilibrium with sea water (Dymond et al 1973). The Sr^{87}/Sr^{86} ratio of the deposits is close to that of sea water, suggesting sea water as the main source of Sr (Bender et al 1971, Dymond et al 1973). The Pb^{207}/Pb^{204} and Pb^{206}/Pb^{204} ratios were measured in a few samples from the East Pacific Ridge and the Bauer Deep. These ratios indicate a magmatic source of Pb in the metalliferous ridge sediments because they are similar to ratios of oceanic tholeiites and different from ratios of hydrogenous ferromanganese deposits (Bender et al 1971, Dymond et al 1973).

BARIUM   This element is highly enriched in metalliferous sediments from spreading centers relative to normal pelagic sediments (Figure 9). This enrichment is frequently

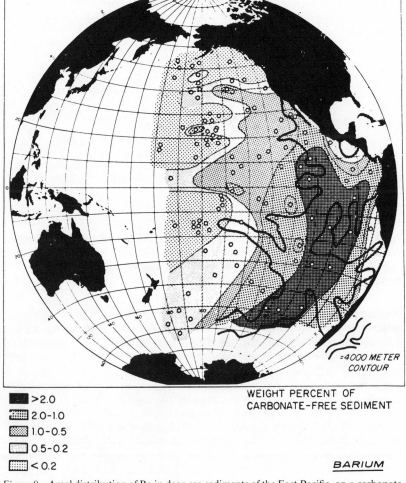

WEIGHT PERCENT OF
CARBONATE-FREE SEDIMENT

>2.0
2.0-1.0
1.0-0.5
0.5-0.2
<0.2

=4000 METER
CONTOUR

BARIUM

Figure 9   Areal distribution of Ba in deep sea sediments of the East Pacific, on a carbonate-free basis (from Bonatti & Arrhenius 1971; see references for sources of data).

paralleled by the appearance of barite in the deposits, as in sediments from the axis of the East Pacific Rise (Arrhenius & Bonatti 1965) and in the metalliferous deposits from the Afar spreading center (Bonatti et al 1972a). Ba appears to coprecipitate with Mn rather than with Fe in these deposits, and is probably among the elements supplied to the sea floor by hydrothermal solutions.

Rate of Accumulation of the Deposits

Metals accumulate in the active ridge deposits at rates one to two orders of magnitude faster than in pelagic sediments and in ferromanganese hydrogenous deposits, as estimated by Bostrom (1970) and as measured directly (Bender et al 1971) in cores from the East Pacific Rise by the excess Th^{230} method. Such relatively rapid accumulation of the active ridge deposits casts some doubts on the concept that their high metal concentration is caused only by lack of dilution of the hydrogenous metals by biogenous or terrigenous sediments. Further doubts arise when one considers that the geochemistry of the deposits from spreading centers is clearly different from that of the hydrogenous deposits, as discussed previously. Thus, the data suggest that in addition to sea water, another source of metals exists along spreading centers, therefore supporting the hydrothermal hypothesis of their origin.

HYDROTHERMAL SYSTEMS AT SPREADING CENTERS

It was suggested several years ago that sub-bottom hydrothermal systems operate along active oceanic ridges (Arrhenius & Bonatti 1965, Elder 1965, Bonatti & Joensuu 1966, Bostrom & Peterson 1966). More recent work has confirmed those earlier suggestions. Sub-bottom hydrothermal circulation has been advocated as a cause of metamorphism of metabasic rocks dredged from the Mid-Atlantic Ridge (Melson et al 1968, Miyashiro et al 1971, Bonatti et al 1974b). Heat flow measurements in the vicinity of spreading centers show (a) a high scatter of values and (b) a distribution whereby even the higher values close to the spreading axis are substantially lower than those estimated for cooling of a lithospheric plate by conduction during spreading. These data indicate that close to a spreading axis a substantial proportion of the heat is dissipated by convection rather than conduction (Palmason 1967, Williams et al 1974, and references therein).

In the convective-hydrothermal circulation model, sea water enters the basaltic oceanic crust and is subsequently driven back up by thermal convection, caused mainly by hot basaltic dykes and intrusions emplaced beneath the spreading center. Considering the distribution of heat flow maxima and minima in the vicinity of the Galapagos spreading center and assuming a Raleigh-type convection, Williams et al (1974) estimated that the convective circulation exists to a depth of 2–3 km beneath the sea floor. Estimates of the permeability of the basaltic upper oceanic crust are consistent with such circulation (Palmason 1967, Lister 1972).

Direct observations in the Reykjanes peninsula of Iceland, where a spreading center emerges, confirm the existence of intense hydrothermal circulation to a depth of several kilometers (Palmason & Semundsson 1974), with a number of hot springs and fumaroles discharging at the surface (Figure 10). Another region where a spreading center emerges is the Afar Rift, also locus of intense geothermal activity (Bonatti et al 1972a and see Figure 11). The importance of hydrothermal systems at spreading centers as a means of chemical exchange between the oceanic crust and sea water can be appreciated if one considers that Spooner & Fyfe (1973) estimated that at present rates the entire present ocean volume could be heated to 300°C during circulation in the oceanic crust in 100 million years.

METAL ENRICHMENT OF HYDROTHERMAL SOLUTIONS BENEATH SPREADING CENTERS

In order to verify the hypothesis that at least some of the metals enriched in the deposits from spreading centers are supplied by hydrothermal solutions, the composition of such solutions should be estimated upon their discharge. This can be done by considering hydrothermal systems where sea water circulates at depth within basalt, and where the resulting solutions are accessible to sampling upon discharge. The composition of solutions from a number of such hydrothermal systems, including those from the Reykjanes spreading center in Iceland, is reported in Table 2. These solutions are clearly enriched in metals and silica relative to sea water; for instance, the solutions discharged at Matupi Harbor are enriched

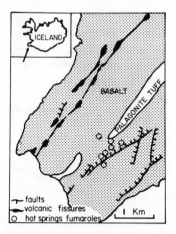

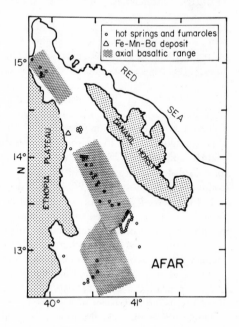

Figure 10 (left) Hot springs and fumaroles in the Reykjanes geo-thermal field, along the axis of the emerged spreading center (from Tomasson & Kristmannsdottir 1972).

Figure 11 (right) Northern Afar (Ethiopia) spreading center, with location of a number of hot springs and fumaroles close to the axial basaltic ranges (modified from CNR-CNRS 1973).

Table 2 Composition of some hydrothermal solutions from volcanic areas compared with sea water. Data for Reykjanes from Björnsson et al (1972); for Matupi Harbor from Ferguson & Lambert (1972); others from White (1968). Values in ppm.

	SiO_2	Mn	Fe	Cu	Zn	SO_4^{-2}	Cl^-
Reykjanes	374	—	485	—	—	72	20,745
Matupi Harbor (a)	—	111	97	0.05	2.53	5,420	22,500
Matupi Harbor (b)	—	20	108	0.06	1.35	5,820	1,760
Salton Sea	400	1,400	2,290	8	540	5.4	155,000
Atlantis II	—	72	59	0.8	2.5	954	163,000
Sea Water	3	0.002	0.003	0.003	0.005	2,600	19,800

relative to sea water by a factor of 6×10^4 for Mn; 3×10^4 for Fe; $\times 16$ for Cu; 5×10^2 for Zn; 3×10^3 for Pb (Ferguson & Lambert 1972, Spooner 1974). In the Reykjanes solutions we see enrichments of $\times 125$ for SiO_2 and $\times 243$ for Fe (Björnsson et al 1972, Spooner & Fyfe 1973). It is notable, as Spooner (1974) pointed out, that metal enrichment is considerable also when the hydrothermal solutions are not significantly more saline than sea water, as, for example, in the Reykjanes and Matupi Harbor geothermal systems.

Whenever solutions of this type are observed to discharge under sea water, metal-rich precipitates are found deposited near their discharge. The composition of the precipitates is similar to that of deposits of deep-sea spreading centers, for example, in Thera (Bonatti et al 1972b), Banu Wuhu (Zelenov 1964), and Matupi Harbor (Ferguson & Lambert 1972).

Metal enrichment in hydrothermal systems at spreading centers can be caused by three distinct processes: (a) leaching from sediment, (b) leaching from basalt, and (c) contribution from deep-seated (mantle) sources.

LEACHING FROM SEDIMENTS   If the hydrothermal waters go through sedimentary strata during their sub-bottom circulation, the more easily soluble elements, including Fe and Mn, can be leached from the sediments. If evaporites are present, the waters will become highly saline, thus increasing their capacity to carry metals through the formation of highly soluble metal-chloride complexes. This situation is exemplified in the hydrothermal brines discharged at the axis of the Red Sea Rift and in the Afar Rift. In the Afar Rift, there are thermal springs discharging saturated $FeCl_2$ solutions.

LEACHING FROM BASALTS   Sediments are generally thin or absent close to the axis of oceanic ridges; therefore, metal-leaching from sediments is not significant in most instances. The sub-sea-floor hydrothermal circulation occurs in basalt, which is probably the main source for the elements enriched in the hydrothermal waters and in the metalliferous deposits. A number of possible reactions between basalt and thermal waters up to about 300°C can be effective in extracting metals and silica from the basalt. For example, in olivine the possibilities range from simple reactions of hydrolysis, with or without serpentinization, such as:

$$\underset{\text{(fayalite)}}{Fe_2SiO_4} + 4H_2O = 2Fe^{2+} + 4OH^- + H_4SiO_4 \qquad\qquad 1.$$

$$\underset{\text{(fayalite)}}{2Fe_2SiO_4} + 3H_2O = \underset{\text{(serpentine or talc)}}{Fe_3Si_2O_5(OH)_4} + Fe^{2+} + 2OH^- \qquad\qquad 2.$$

to redox reactions such as:

$$\underset{\text{(fayalite)}}{Fe_2SiO_4} + \tfrac{1}{2}O_2 = \underset{\text{(hematite)}}{Fe_2O_3} + \underset{\text{(quartz)}}{\tfrac{1}{2}(SiO_2)_{\text{solid}}} + \tfrac{1}{2}(SiO_2)_{\text{solution}} \qquad\qquad 3.$$

$$\underset{\text{(fayalite)}}{11Fe_2SiO_4} + SO_4{}^{2-} + 4H^+ = \underset{\text{(magnetite)}}{7Fe_3O_4} + \underset{\text{(pyrite)}}{FeS_2} + 11SiO_2 + 2H_2O \qquad\qquad 4.$$

As pointed out by Spooner & Fyfe (1973), these redox reactions cause the hydrothermal waters to become reduced, which would further favor the solution of Fe, Mn, and other metals. Some elements, such as U, may actually be lost by the thermal solutions at this stage and added to the basalt (Rydell & Bonatti 1973). The possible formation of sulfides at depth and the resultant fractionation of metals is discussed in the next section of this paper.

Silica can be extracted from the basalt along with the metals, since its solubility at temperatures inferred for the thermal solutions is several orders of magnitude higher than in bottom sea water. Transport of the metals as chloride complexes or by other means is still effective even when chloride concentration of the hydrothermal solutions is not higher than that of sea water, as suggested by data on hydrothermal solutions from Reykjanes peninsula and Matupi Harbor (Spooner 1974).

Basalts which host hydrothermal systems within the oceanic crust will be affected by hydrothermal metamorphism, ranging from zeolite to greenschist facies. One problem not completely clarified, as yet, is that of finding within oceanic basalts and metabasalts the metal-depleted residue after hydrothermal extraction of metals. Corliss (1971) found that the microcrystalline interior of deep-sea pillow basalts is depleted in Fe, Mn, and other metals, relative to the rapidly chilled glass margins, which approximate the original composition of the lava. Corliss ascribes this metal depletion to high temperature deuteric reactions between sea water and the basalt, similar in principle to the hydrothermal reactions assumed to occur in the oceanic crust.

CONTRIBUTION FROM DEEP-SEATED (MANTLE) SOURCES   The possibility that some of the elements enriched in the metalliferous deposits from spreading centers are derived directly from volatile phases in the mantle must be considered.

Degassing of the mantle at spreading centers and in deep crustal fracture zones is suggested by the discovery of excess ^3He in deep ocean water above the East Pacific Ridge (Clarke et al 1969, Craig et al 1974, Lupton & Craig 1974) and in the North Atlantic above the Mid-Atlantic Ridge near the Gibbs fracture zone (Jenkins et al 1972).

Bostrom (1974) has suggested that mantle-derived, CO_2-rich volatile phases, similar in composition to carbonatites, are the main carrier of the elements enriched in the deposits from spreading centers. It is premature to evaluate the extent of the contribution from the deep-seated source to the metalliferous deposits. Leaching of the basaltic crust seems to be an adequate explanation for most of the elements enriched in the hydrothermal solutions. For some elements, for instance Ba, whose concentration in oceanic basalt is exceedingly low, or S, which is highly volatile, a contribution by deep-seated volatile sources may be important.

In any case, mantle-derived volatile components probably travel upwards in the crust through the same system of fissures and fractures that channels the intrusion of basaltic liquid beneath a spreading center; hydrothermal circulation may be the means by which many of these components reach the sea floor. This view is supported by the existence of anomalous maxima in the concentration of Fe, Cu, and other metals in those deep water masses above spreading centers where excess ^{3}He was also found (Brewer et al 1972).

METAL DEPOSITION ON THE SEA FLOOR UPON HYDROTHERMAL DISCHARGE

Discharge of the hydrothermal solutions through the sea floor produces two possible results.

(a) If the discharge occurs in a topographic and hydrologic situation which prevents rapid mixing of the hydrothermal brines with oxygenated sea water, ready oxidation of the metals will not occur. A sulfide-rich assemblage may be deposited, possibly containing high concentrations of those minor metals which have strong affinity for sulfide phases, such as Cu and Zn. This situation (Figure 12) is typified by the Red Sea hot brine deposits.

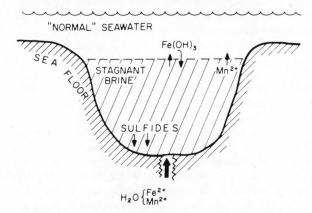

Figure 12 Qualitative and schematic representation of processes occurring when Fe-Mn-bearing hydrothermal solutions debouch in a topographic situation where stagnant conditions can be established, as in the Red Sea deposits (modified from Bischoff 1969). Precipitation of Fe-smectite from the upper zone of the brine has been omitted from the picture.

(b) If the discharge occurs into oxygenated, circulating sea water, as is generally the case along oceanic ridges, gradual oxidation of the "excess" metals will take place during mixing. Some fractionation of the metals will occur at this stage, for instance, fractionation of Fe from Mn, because Fe tends to be oxidized more rapidly and to precipitate faster than Mn (Figure 13). This fractionation may account in part for the extreme range of Fe/Mn ratios found in hydrothermal metalliferous sediments (Figure 4). A clear illustration of this process is in the Matupi Harbor hydrothermal system. At discharge, the hydrothermal solutions contain concentrations of both Fe and Mn several orders of magnitude higher than sea water, with the ratio Fe/Mn roughly equal to unity, or, in some instances, with Mn more abundant than Fe. However, the precipitates deposited close to the discharge of these solutions are Fe-rich but almost free of Mn (Ferguson & Lambert 1972).

Depending on bottom water circulation, on the rate of deposition of hydrothermal metals, and on the rate of deposition of other authigenous, biogenous, and terrigenous sediment components, the hydrothermal deposit may be diluted to a variable extent. Close to the hydrothermal discharge, deposition of metals may be rapid and undiluted metalliferous deposits may be formed, as observed on a small scale in shallow-water systems at Thera, Banu Vukhu, and Matupi Harbor, and as inferred for deep-sea deposits such as Amph D2 from the East Pacific Rise.

Scavenging of elements from sea water by Fe and Mn hydroxide particles may occur during deposition. This process may be efficient for elements such as P (Berner 1973) and U, and may provide a partial explanation of the evidence that some components of the "hydrothermal" metalliferous sediments are at least in part derived from sea water. An additional explanation for such sea-water-derived components is that the fluid involved in the sub-sea-floor hydrothermal circulation is mainly sea water, some of whose original components may be brought back upon discharge; another, of course, is the aforementioned contamination from normal marine sediments.

METAL SULFIDE DEPOSITS AT SPREADING CENTERS

If the hydrothermal model is correct, metal-rich solutions circulate quite deep within the crust beneath spreading centers. This setting is favorable to the formation of metal sulfide deposits.

Isolated veinlets or grains of sulfide minerals are not uncommon in oceanic basalts and metabasalts. More extensive sulfide mineralizations have been reported by Baturin (1971) in rocks from the Indian Ridge, and by Bonatti et al (1974b) in basalts from the equatorial Mid-Atlantic Ridge. In this latter deposit, sulfides are contained in the basalt as "stockwork"-type veinlets and as disseminated mineralizations; the most common sulfide phases are pyrite and chalcopyrite, accompanied by a series of Fe- and Cu-rich alteration minerals.

To date, massive sulfide deposits from modern oceanic ridges have not been sampled, if we exclude the possible example of the Red Sea metalliferous deposits, where a sulfide facies exists. However, various considerations (see Sillitoe 1973),

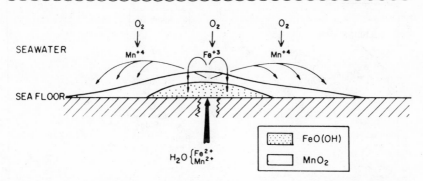

Figure 13   Qualitative and schematic representation of processes of formation of Fe-Mn deposits when Fe-Mn-bearing hydrothermal solutions are discharged into an oxygenated sea bottom.

especially the evidence from presently emerged fragments of ancient oceanic crust (discussed in the next section), suggest that massive-type metal sulfide deposits are created at modern spreading centers and must be present in the oceanic crust, primarily within the basaltic layer.

The formation of sulfides during the sub-sea-floor hydrothermal circulation involves reactions between S^{2-} and metals. The sulfide ion can be supplied by various sources, one being the reduction of sea water SO_4^{2-} during the sub-bottom circulation, through a number of redox reactions exemplified by reaction 4. It is significant in this context that in geothermal systems where sea water circulates within basalt, the SO_4^{2-} content of the solutions at discharge is generally drastically lowered, relative to sea water (see Table 2).

Another source is the sulfur trapped in deep-sea basalt. Ocean-ridge basalts contain an average of about 800 ppm sulfur (Moore & Schilling 1973), some of which can be extracted by the hydrothermal solutions.

Sulfur should also be contained in that volatile fraction which ascends directly from the upper mantle beneath spreading centers without being trapped in the cooling basalt; this fraction may be caught in part by the sub-sea-floor hydrothermal liquids.

The factors determining transport, selective deposition, and fractionation of the various metal species in hydrothermal solutions are complex. These factors are in principle identical to those operating in hydrothermal systems on land, and have been treated extensively in the literature (see, for example, Barnes & Czamanske 1967).

The partition coefficients of the various metals between sulfide solid phase and solution are determined by a number of variables,—among them, the activity of the pertinent ions, temperature, redox conditions, pH, etc. Cu and Zn, in addition to Fe, partition strongly with sulfide phases in a range of conditions reached in hydrothermal systems, as shown both by experimental work and by their abundance

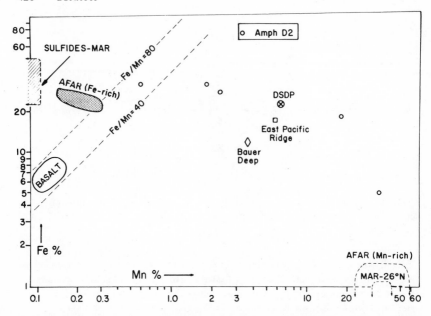

Figure 14   Fe vs Mn concentration in a number of metalliferous sediments from spreading centers, compared with ocean-ridge basalts and with metal sulfide mineralizations from the equatorial Mid-Atlantic Ridge. This plot illustrates the Fe/Mn fractionation in hydrothermal systems, with high Fe/Mn ratio in the sulfide-rich deposits. Field between dashed lines indicates the range of Fe/Mn ratios in oceanic basalts. Data on sulfides are from Bonatti et al 1974a; on Afar, from Bonatti et al 1972a; Bauer Deep (average) from Sayles & Bischoff 1973; DSDP (average) from Cronan 1973; East Pacific Ridge (average of axial samples) from Bostrom et al 1969; Amph D2 from Bonatti & Joensuu 1966; Mid-Atlantic Ridge at 26°N from Scott et al 1974.

in hydrothermal metal sulfide deposits. In contrast, other elements such as Mn are kept in solution during sulfide precipitation. As a result, fractionation of the metals occurs, particularly fractionation of Fe from Mn. This means that in those hydrothermal systems where abundant sulfides have been deposited beneath the sea floor, the solutions at discharge may show substantially lowered Fe/Mn ratios. This may partly explain why many hydrothermal metalliferous sediments contain much lower Fe/Mn ratios than basalt, from which both metals were presumably extracted, as illustrated in Figure 14.

The depth below sea floor at which deposition of sulfides occurs depends on the physical-chemical conditions of each hydrothermal system. Deposition of sulfides just beneath the submarine discharge of hydrothermal springs can be observed directly near the island of Vulcano in the Tyrrenian Sea (Honnorez 1969, Honnorez et al 1973).

METAL RELEASE DURING LOW TEMPERATURE SEA WATER–BASALT REACTIONS

It was shown in the previous discussion that sub-sea-floor reactions of heated sea water with basalt provide a major source of metals to the sea floor. Low temperature sea water-basalt reactions occurring on the sea floor may also result in release of metals and in the formation of "halmyrolytic" deposits (Bonatti et al 1972c). This process is significant only when it involves basaltic glass, which is highly unstable on the sea floor, particularly when it is in the form of comminuted fragments (hyaloclastites).

An example is shown in Figure 15 and Table 3 in which the composition of the fresh core of a basaltic glass grain from a South Pacific seamount is compared with its altered rim. Note the depletion of Mn caused by sea water alteration, and the opposite behavior of Fe, which is left behind during alteration. This is in

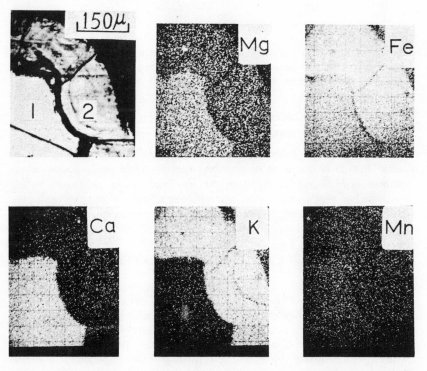

Figure 15 X-ray scanning pictures for Mg, Fe, Ca, K and Mn at the boundary zone between fresh basaltic glass 1 and alteration rim 2, in a grain from seamount Amph D5, South Pacific. For quantitative data refer to Table 3 (from Bonatti 1970).

422 BONATTI

Table 3 Results of electron probe analyses of a grain of basaltic glass from sea mount Amph D5 (see Figure 15). Column 1 = composition of the unaltered glass core. Columns 2a and 2b = composition of two points within the altered rim produced by reaction with sea water (from Bonatti 1970).

Per cent	1	2a	2b
SiO_2	50.3	42.4	45.4
Al_2O_3	14.7	13.1	11.6
FeO	9.9	16.9	18.6
CaO	11.6	5.3	7.0
MgO	7.8	3.2	3.7
K_2O	0.1	2.7	3.4
Na_2O	3.1	1.5	2.2
MnO	0.2	<0.1	<0.1
Total	97.6	85.1	91.9

contrast to high temperature sea water–basalt reactions, as in hydrothermal systems, where both Fe and Mn appear to be lost by the basalt. Mn is presumably released to sea water from which it may subsequently precipitate. The alteration process, if brought to completion, results in the following scheme:

basaltic glass + sea water = zeolite (phillipsite) +
Fe-smectite + Fe-hydroxide + MnO_2

Halmyrolytic deposits of this type are not very abundant along spreading centers, because fissural eruptions of basalt along oceanic ridges produce only minor quantities of hyaloclastites. However, they are important in regions of the sea floor with low sedimentation rates and with central (seamount-type) volcanism, where eruptions producing hyaloclastites by thermal shattering of the lava are common, as in wide regions of the South Pacific (Bonatti 1967).

METALLOGENESIS IN THE OCEANIC LOWER CRUST AND UPPER MANTLE

Metal deposits may be formed in the lower zone of the crust and in the upper mantle during the creation of new oceanic lithosphere at spreading centers. The lower oceanic crust probably consists of a complex mixture of gabbros, metagabbros and serpentinite, with a limited occurrence of cumulitic (layered) rocks towards the lower levels, while the upper mantle consists probably of peridotitic material. There is no direct evidence of metallogenesis in the oceanic lower crust-upper mantle; however, metallogenesis can be inferred there by analogy with exposed oceanic sections, such as some ophiolitic complexes. Thus, we can postulate the presence of chromite mineralizations in the cumulitic gabbre-ultramafic zone, probably formed by magmatic segregation during the slow cooling of lower crustal basic magma chambers; or the presence of Cr, Pt, and Ni mineralizations associated with lower crustal and upper mantle ultramafic bodies.

METALLOGENESIS IN OCEANIC FRACTURE ZONES

Oceanic ridges are offset by fracture zones; the largest, such as those from the equatorial Atlantic, extend through the crust and probably throughout the entire thickness of the oceanic lithosphere. Heat flow in offset zones is relatively high, even though basaltic volcanism is quantitatively scarce. This tectonic setting appears a priori favorable for metallogenesis, with the deep crustal fractures providing channels for upward injection of deep-seated mineralizing fluids. This hypothesis is supported by the discovery of iron sulfide concretions, probably of hydrothermal origin, at the Romanche Fracture Zone (Bonatti et al 1974a), and by data suggesting injection of ^{3}He and metals through the sea floor at the Charlie-Gibbs Fracture Zone (Jenkins et al 1972). The metal-rich Salton Sea geothermal field (White 1968) is probably also associated with a transform zone within the San Andreas system.

Metallogenesis in the large fracture zones is potentially interesting if we consider suggestions that some of the major metal ore deposits on continental margins are aligned along the predrift extension of some of the major oceanic fracture zones (see Kutina 1974). This hypothesis need not be in conflict with plate-tectonic theory, if it is assumed that major fracture zones, such as those of the equatorial Atlantic, developed as rejuvenation of older predrift zones of tectonic weakness (Bonatti 1973).

METAL DEPOSITS FROM ANCIENT SPREADING CENTERS

Ophiolitic bodies, consisting of assemblages of peridotite, gabbro, and basalt overlain by marine sedimentary sequences including cherts and limestones, are widespread in orogenic areas at the margins of large plates. They have been recognized as representing uplifted (obducted) fragments of ancient oceanic crust, produced at former oceanic and/or back-arc basin spreading centers.

Metal deposits are widespread in ophiolite formations; Figure 16 illustrates their characteristics with an example from the Mesozoic ophiolites in the northern Apennines. Figure 17 illustrates schematically the stratigraphy of the northern Apennine ophiolites. Metalliferous beds are frequently present in the chert formation (of Malm age), close to the contact with the basalt unit; thus the stratigraphic position of these metal-rich sediments is identical to that of basal metalliferous deposits from the present oceans. Their geochemistry contrasts with that of hydrogenous metal deposits from the present oceans, and is instead similar to the geochemistry of deposits from modern spreading centers, as summarized in Table 4 and Figures 4 and 6 (Bonatti & Zerbi 1974). Fe and Cu sulfide ore deposits are widespread within the basalt unit of the northern Apennine ophiolites, both as massive ore bodies and as stockwork and disseminated mineralizations. Many of these deposits are and/or have been exploited economically. Pyrite and chalco-

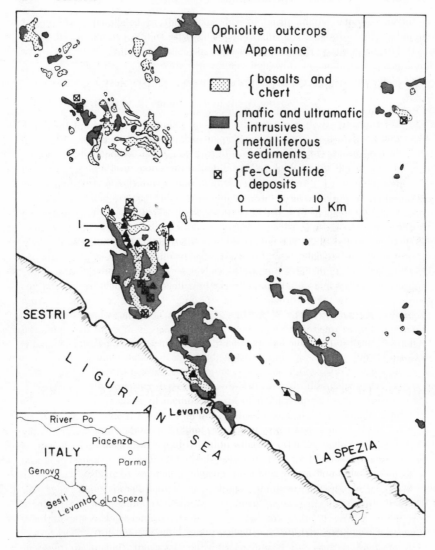

Figure 16 Ophiolitic exposures in the northwest Apennines (Italy). Shows Mn-rich sedimentary deposits, and metal sulfide deposits (from Bonatti & Zerbi 1974).

pyrite are their prevalent constituents, but a series of other minerals, including marcasite, sphalerite, pyrrotite, covellite, hematite, magnetite, geothite, malachite, and melanterite, as well as traces of Ni and Co sulfides, have been reported (see Bonatti & Zerbi 1974 and references therein).

These sulfide deposits and the manganiferous sediments associated with them probably formed through the same metallogenetic processes hypothesized to occur

Table 4 Comparison of Mesozoic metalliferous sediments from the Apennine ophiolites with "hydrogenous" and "hydrothermal" metalliferous sediments from modern oceans, indicating their similarities with the "hydrothermal" deposits (from Bonatti & Zerbi 1974)

Apennine ophiolite metalliferous sediments	"Hydrothermal" submarine metalliferous sediments	"Hydrogenous" submarine metalliferous sediments
Stratigraphically at the base of the sedimentary column, just above basalt unit	Stratigraphically at the base of the sedimentary column, just above basalt basement	Stratigraphically throughout sedimentary column, prevalently at the top
Fe/Mn ratio extremely low (<0.1)	Fe/Mn ratio ranging from extremely low (<0.1) to extremely high (>10)	Fe/Mn ratio ranging between >0.5 and <5
Content of minor metals (Ni, Co, Cu, Zn) low relative to Fe + Mn	Content of minor metals (Ni, Co, Cu, Zn) generally low relative to Fe + Mn	Content of minor metals (Ni, Co, Cu, Zn) high relative to Fe + Mn
U/Th ratio high ($> \sim 1$)	U/Th ratio high ($> \sim 1$)	U/Th ratio low ($< \sim 1$)

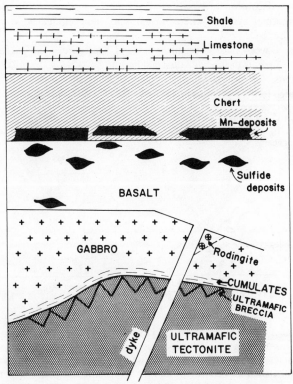

Figure 17 Schematic stratigraphy of the northern Apennine ophiolite complex (from Bonatti & Zerbi 1974).

at modern spreading centers. Note that the Fe/Mn fractionation between Fe-rich, Mn-poor sulfide and Fe-poor, Mn-rich sedimentary deposits suggested earlier for modern spreading centers is verified in the Apennine ophiolite deposits (Figure 18).

A similar association of metal sulfide deposits in pillow basalt, overlain by Fe–Mn-rich sediments, exists in the Troodos Massif of Cyprus, an ophiolitic complex probably created at a Mesozoic spreading center (Moores & Vine 1971). The characteristics of these deposits (Ederfield et al 1972, Costantinou & Govett 1973), as well as of other similar deposits from various parts of the world, including those of Ordovician age from Newfoundland (Duke & Hutchinson 1974), are consistent with the model advocated here for metallogenesis at modern spreading centers. Thus it appears that metallogenesis occurred along constructive plate margins throughout the late history of the Earth.

MODEL OF METALLOGENESIS AT SPREADING CENTERS

Metallogenesis at spreading centers is caused ultimately by the same processes that determine the emplacement of new lithosphere at constructional plate margins. A model of metallogenesis can be summarized as follows (Figures 19 and 20):

1. Intrusion and effusion of basalt at spreading centers favors the development of hydrothermal systems where heated sea water circulates in the oceanic crust (Figure 19).
2. Reactions between hydrothermal liquid and basalt cause enrichment of silica and various metals in the hydrothermal liquid.
3. Components derived by volatile transport from the upper mantle may be incorporated in the hydrothermal liquid.

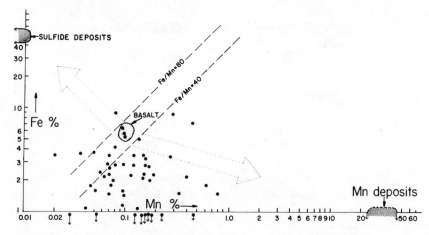

Figure 18   Fe vs Mn plot of metalliferous sediments and metal sulfide deposits from the Appenine ophiolites, illustrating fractionation of Fe from Mn. Dots represent samples of chert and shale associated with the metalliferous sediments (from Bonatti & Zerbi 1974).

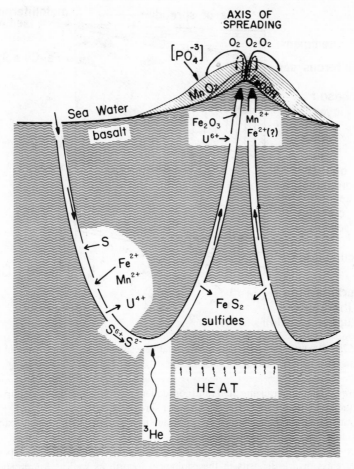

Figure 19 Qualitative scheme illustrating the hydrothermal model of metallogenesis at spreading centers. Fe and Mn stand for the metals leached from the basaltic crust. U stands for elements which may be lost to the basalt, ^{3}He for volatile components supplied by the upper mantle, and $[PO_4]^{-3}$ for elements scavenged from sea water.

4. Metal sulfide mineralizations, including probably massive ores, are deposited by the hydrothermal liquids in the oceanic crust.
5. Metalliferous sediments are deposited on the sea floor upon discharge of the hydrothermal solutions.
6. Chromite mineralizations may be formed in the lower zone of the crust, within cumulate gabbroic and ultramafic rocks (Figure 20), in analogy with observations obtained in ophiolite complexes.
7. Pt, Ni, and Cr mineralizations may be present in the oceanic harzburgite-lherzolite upper mantle, fragments of which are exposed in fracture zones.

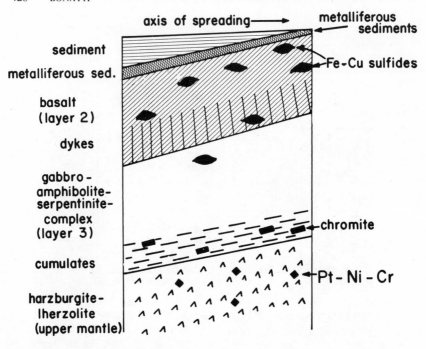

Figure 20 Qualitative scheme illustrating metal deposits in the oceanic crust and upper mantle.

8. Upwelling of mineralizing fluids and consequent metallogenesis may be important in the large fracture zones which intersect spreading centers.
9. Some limited metallogenesis may occur due to low temperature basalt-sea water reactions on the sea floor (halmyrolysis).

In conclusion, metallogenesis at spreading centers may be important in explaining a wide class of ore deposits in the geological record, as well as for prospecting for new deposits. Of equal importance are processes of metallogenesis occurring at "destructive" plate margins, as above subduction zones (see Sillitoe 1973), a subject outside the scope of this paper.

ACKNOWLEDGMENTS
Research supported by the National Science Foundation, International Decade of Ocean Exploration Program, Grant GX-40428, and by the Office of Naval Research Contract N00014-67-A-0201-0013.

Literature Cited

Anderson, R. N., Halunen, A. J. 1974. The implications of heat flow for metallogenesis in the Bauer deep, southeast equatorial Pacific. *Nature.* In press

Arrhenius, G., Bonatti, E. 1965. Neptunism and volcanism in the ocean. *Progr.*

Oceanogr. 3:7–22. London: Pergamon

Backer, H., Schoell, M. 1972. New deeps with brines and metalliferous sediments in the Red Sea. *Nature Phys. Sci.* 240: 153–58

Barnes, H. L., Czamanske, G. K. 1967. Solubility and transport of ore minerals. In: *Geochemistry of Hydrothermal Ore Deposits* (ed. H. L. Barnes), 334–811. New York: Rinehart & Wilson

Baturin, G. N. 1971. Sulfide ores in the Arabic-Indian Ridge. In: *The History of the World Ocean* ed., L. A. Zenkevitch, 259–65. Moscow: Nauka

Bender, M., Broecker, W., Gornitz, V., Middel, U., Kay, R., Sun, S. S., Biscaye, P. 1971. Geochemistry of three cores from the East Pacific Rise. *Earth Planet. Sci. Lett.* 12:425–33

Berner, R. A. 1973. Phosphate removal from sea water by adsorption on volcanogenic ferric oxides. *Earth Planet. Sci. Lett.* 18: 77–86

Bischoff, J. L. 1969. Red Sea geothermal brine deposits: Their mineralogy, chemistry and genesis. In: *Hot Brines and Recent Heavy Metal Deposits,* 368–401. New York: Springer

Björnsson, S., Arnorsson, S., Tomasson, J. 1972. Exploration of the Reykjanes brine area. *Rep. U. N. Symp. Geotherm. Energy* (Pisa). 1–25

Bonatti, E. 1967. Mechanisms of deep sea volcanism in the South Pacific. *Res. Geochem.,* (ed. P. Abelson), 2:459–91. New York: Wiley

Bonatti, E. 1970. Deep sea volcanism. *Naturwissenschaften* 57:379–84

Bonatti, E., Arrhenius, G. 1971. Acidic rocks on the Pacific Ocean floor. In *The Sea,* ed. A. Maxwell, 4:445–64. New York: Wiley

Bonatti, E., Fisher, D. E., Joensuu, O., Rydell, H. S., Beyth, M. 1972a. Iron-manganese-barium deposit from the northern Afar Rift (Ethiopia). *Econ. Geol.* 67:717–30

Bonatti, E., Honnorez, J., Honnorez-Guer-stein, M. 1974a. Metal sulfide mineralizations in the equatorial Mid-Atlantic Ridge. Submitted to *Earth Planet. Sci. Lett.*

Bonatti, E., Honnorez, J., Joensuu, O., Rydell, H. 1972b. Submarine iron deposits in the Mediterranean Sea. In *The Mediterranean Sea,* ed. D. J. Stanley, 701–10. New York: Dowden, Hutchinson & Ross

Bonatti, E., Honnorez, J., Kirst, P., Radicati, F. 1974b. Metagabbros from the Mid-Atlantic Ridge at 6°N: Contact-hydro-thermal-dynamic metamorphism beneath the axial valley. *J. Geol.* In press

Bonatti, E., Joensuu, O. 1966. Deep sea iron deposits from the South Pacific. *Science* 154:643–45

Bonatti, E., Kraemer, T., Rydell, H. 1972c. Classification and genesis of submarine iron-manganese deposits. In *Ferromanganese Deposits on the Ocean Floor,* ed. D. Horn, 149–65. Washington DC: Nat. Sci. Found.

Bonatti, E., Zerbi, M. 1974. Metalliferous deposits from the Apennine ophiolites: Mesozoic equivalent of deposits from modern spreading centers. *Bull. Geol. Soc. Am.* In press

Bostrom, K. 1970. Submarine volcanism as a source for iron. *Earth Planet. Sci. Lett.* 9:348–54

Bostrom, K. 1974. The origin and fate of ferromanganoan active ridge sediments. *Stockholm Contrib. Geol.* 27:149–243

Bostrom, K., Peterson, M. N. A. 1966. Precipitates from hydrothermal exhalations on the East Pacific Rise. *Econ. Geol.* 61:1258–65

Bostrom, K., Peterson, M. N. A., Joensuu, O., Fisher, D. E. 1969. Aluminum-poor ferromanganoan sediments on active oceanic ridges. *J. Geophys. Res.* 74:3261–70

Brewer, P. G., Spencer, D. W., Robertson, D. E. 1972. Trace element profiles from the GEOSECS II test station in the Sargasso Sea. *Earth Planet. Sci. Lett.* 16: 111–16

Clarke, W. B., Beg, M. A., Craig, H. 1969. Excess [3]He in the sea: Evidence for terrestrial primordial helium. *Earth Planet. Sci. Lett.* 6:213–17

CNR-CNRS Afar Team. 1973. Geology of northern Afar. *Rev. Geogr. Phys. Geol. Dynam.* 15:443–90

Corliss, J. B. 1971. The origin of metal-bearing submarine hydrothermal solutions. *J. Geophys. Res.* 76:8128–38

Costantinou, G., Govett, G. J. S. 1973. Geology, geochemistry and genesis of the Cyprus sulfide deposits. *Econ. Geol.* 68: 843–58

Craig, H., Clarke, W. B., Beg, M. A. 1974. Excess [3]He in deep water on the East Pacific Rise. *Earth Planet. Sci. Lett.* In press

Cronan, D. S. 1973. Basal ferruginous sediments cored during Leg 16, Deep Sea Drilling Project. *Initial Reports Deep Sea Drilling Project,* 16:601–04

Degens, E. T., Ross, D. A. 1969. *Hot Brines and Recent Heavy Metal Deposits in the Red Sea.* New York: Springer. 600 pp.

Duke, N. A., Hutchinson, R. W. 1974. Geological relationships between massive sulfide bodies and ophiolitic volcanic rocks neaᵢ York Harbour, Newfoundland. *Can. J. Earth Sci.* 11:53–69

Dymond, J., Corliss, J. B., Heath, G. R., Field, C. W., Dasch, E. J., Veeh, H. H. 1973. Origin of metalliferous sediments from the Pacific Ocean. *Geol. Soc. Am. Bull.* 84:3355–72

Ederfield, H., Gass, I. G., Hammond, A., Bear, L. M. 1972. The origin of ferromanganese sediments associated with the Troodos Massif of Cyprus. *Sedimentology* 19:1–19

Elder, J. W. 1965. Physical processes in geothermal areas. *Am. Geophys. Union Monogr.* 8:211–39

Ferguson, J., Lambert, I. B. 1972. Volcanic exhalations and metal enrichments at Matupi Harbour, New Britain. *Econ. Geol.* 67:25–37

Goldberg, E. D., Arrhenius, G. 1958. Chemistry of Pacific pelagic sediments. *Geochim. Cosmochim. Acta* 13:153–212

Goldberg, E. D., Koide, M., Schmitt, R., Smith, R. 1963. Rare-earth distributions in the marine environment. *J. Geophys. Res.* 68:4209–17

Hart, S. R., et al. 1974. Report on Leg 34 of Deep Sea Drilling Project. *Geotimes* 19:20–24

Honnorez, J. 1969. La formation actuelle d'un gisement sous-marine de sulfures fumerolliens a Vulcano (Mer. Tyrrhenienne): Partie 1. *Mineral. Deposita* 4:114–31

Honnorez, J., Honnorez-Guerstein, M. B., Valette, J., Wauschkuhn, A. 1973. Present day formation of an exhalative sulfide deposit at Vulcano (Thyrrenian Sea): Part 2. In: *Ores in Sediments*, 139–66. Berlin: Springer

Jenkins, W. J., Beg, M. A., Clarke, W. B., Wangersky, P. J., Craig, H. 1972. Excess ³He in the Atlantic Ocean. *Earth Planet. Sci. Lett.* 16:122–26

Kutina, J. 1974. Structural control of volcanic ore deposits in the context of global tectonics. *Bull. Volcanol.* In press

Landergren, S. 1964. On the geochemistry of deep sea sediments. *Rep. Swed. Deep Sea Exped.* 10:57–70

Lister, C. R. B. 1972. On the thermal balance of a mid-ocean ridge. *Geophys. J. Roy. Astron. Soc.* 26:515–35

Lupton, J. E., Craig, H. 1974. Excess ³He in oceanic basalts: Evidence for terrestrial primordial helium. *Earth Planet. Sci. Lett.* In press

Melson, W. G., Thompson, G., Van Andel, T. H. 1968. Volcanism and metamorphism in the Mid-Atlantic Ridge, 22°N. *J. Geophys. Res.* 73:5925–41

Mero, J. L. 1965. *The Mineral Resources of the Sea.* New York: Elsevier. 312 pp.

Miyashiro, A., Shido, F., Ewing, M. 1971. Metamorphism in the Mid-Atlantic Ridge near 24° and 30°N. *Phil. Trans. Roy. Soc. London* 268:589–603

Moore, J. G., Schilling, J. G. 1973. Vesicles, water and sulfur in Reykjanes Ridge basalt. *Contrib. Mineral. Petrol.* 41:105–18

Moores, E. M., Vine, F. J. 1971. The Troodos Massif, Cyprus, and other ophiolites as oceanic crust. *Phil. Trans. Roy. Soc. London* 268:443–66

Murray, J., Renard, A. F. 1891. Deep sea deposits. *Challenger Exped. Rep.* 3. London

Palmason, G. 1967. On heat flow in Iceland in relation to the Mid-Atlantic Ridge. In: *Iceland and Mid-Ocean-Ridges.* Soc. Sci. Island. 38:111–27

Palmason, G., Semundsson, K. 1974. Iceland in relation to the Mid-Atlantic Ridge. *Ann. Rev. Earth Planet. Sci.* 2:25–46

Piper, D. Z. 1972. Rare-earth elements in manganese nodules from the Pacific Ocean. In: *Ferromanganese Deposits on the Ocean Floor,* ed. D. Horn, 123–30

Rona, P. A., McGregor, B. A., Betzer, P. R., Krause, D. C. 1974. Anomalous water temperatures over Mid-Atlantic Ridge crest at 26°N. *Am. Geophys. Union. Trans.* 55:293

Rydell, H. S., Bonatti, E. 1973. Uranium in submarine metalliferous deposits. *Geochim. Cosmochim. Acta* 37:2557–65

Sayles, F. L., Bischoff, J. L. 1973. Ferromanganoan sediments in the equatorial East Pacific. *Earth Planet. Sci. Lett.* 19:330–36

Scott, R. B., Rona, P. A., McGregor, B. A., Scott, M. R. 1974. The TAG hydrothermal field. *Nature* 251:301–02

Sillitoe, R. H. 1973. Environment of formation of volcanogenic massive sulfide deposits. *Econ. Geol.* 68:1321–26

Skornyakova, I. S. 1964. Dispersed iron and manganese in Pacific Ocean sediments. *Int. Geol. Rev.* 7:2161–74

Spooner, E. T. C. 1974. Sub-sea-floor metamorphism, heat and mass transfer: An additional comment. *Contrib. Mineral. Petrol.* 45:169–73

Spooner, E. T. C., Fyfe, W. S. 1973. Sub-sea-floor metamorphism, heat and mass transfer. *Contrib. Mineral. Petrol.* 42:287–304

Tomasson, J., Kristmannsdottir, H. 1972. High temperature alteration minerals and thermal brines, Reykjanes, Iceland.

Contrib. Mineral. Petrol. 36:123–34

Turekian, K. K., Wedepohl, K. H. 1961. Distribution of the elements in some major units of the earth's crust. *Bull. Geol. Soc. Amer.* 72:175–92

White, D. F. 1968. Environments of generation of some base metal ore deposits. *Econ. Geol.* 63:301–35

Williams, D. L., Von Herzen, R. P., Sclater, J. G., Anderson, R. N. 1974. Lithospheric cooling and hydrothermal circulation in the Galapagos spreading center. *Geophys. J. Roy. Astron. Soc.* In press

Zelenov, K. K. 1964. Iron and manganese in exhalations of the submarine Banu Vukhu Volcano (Indonesia). *Dokl. Akad. Nauk. SSSR* 155:1317–20

CHEMISTRY OF INTERSTITIAL WATERS OF MARINE SEDIMENTS

✣10047

Joris M. Gieskes

Scripps Institution of Oceanography, La Jolla, California 92037

INTRODUCTION

In recent years much interest has developed in the processes that have led to the present chemical composition of the oceans. Among the pioneering efforts were those of Goldschmidt (1933) and Conway (1943). These authors made up approximate mass balances and calculated the amounts of primary rocks that must have weathered to account for the chemical composition of the ocean and the sediments. This work was later amplified by Horn & Adams (1966). In the mass balances considered by these various authors, no serious attention was given to the possible recycling of materials, nor to the ocean as just one geochemical reservoir through which the elements pass on to a subsequent reservoir, namely the sediments. Considerations of this type were introduced by Barth (1952) and Goldberg & Arrhenius (1958). The latter authors found good agreement between measured rates of input of some elements into the ocean, via the rivers and the atmosphere, and the rate of output into the sediments. Such balances suggest that a steady state exists in the overall geochemical cycle.

In a different approach Sillén (1967) used a simple chemical equilibrium model to account for the composition of sea water. He assumed an equilibrium between the ocean and the marine sediments, particularly with the alumino-silicate and the carbonate phases. Broecker (1971) criticized this model, especially because it tends to discourage any inquiry into the reactions that may take place in the various geochemical reservoirs. Broecker specifically cites examples that could imply variations in the ionic composition of sea water with time, and he advocates a kinetic approach to the problem, which would then lead to a search for the processes that control the apparent steady state as it exists today.

Garrels & Mackenzie (1971) developed a steady-state model based on the concept of chemical uniformitarianism, in which the total mass and the chemical composition of sedimentary rocks, as well as the chemical composition of the deposited materials, have remained constant with time, that is, at least during the last two billion years. These authors (Garrels & Mackenzie 1972, Garrels & Perry 1974) have developed this concept into a quantitative model involving both estimates of overall fluxes and residence times of chemical elements in various geochemical

433

reservoirs. Some of the criticisms raised by Broecker (1971) against equilibrium models could be raised equally well against these dynamic models, since little information is obtained on possible short-time variations in the apparent steady state.

Holland (1972) recently made an important analysis of the possible magnitude of deviations from the present-day average composition of sea water, using known chemical and mineralogical information on evaporite deposits. He concluded that, at least from Permian time, only small variations may have occurred.

Important clues as to the possible variation in the chemical composition of sea water, at least over the last 150 million years, may be contained in the present-day marine sediments, particularly in the long cores obtained by the Deep Sea Drilling Project. The time limitation, of course, is set by the lack of older submerged marine sediments, owing to their removal by subduction into trenches or their uplift to subaerial exposure.

In any study of the chemistry of marine sediments, one of the important aspects is the composition of the dissolved salts in the interstitial waters. For instance, if Sillén (1967) were correct in his assumption of an equilibrium state between the sediments and the ocean, the chemical composition of the interstitial fluids should reflect this. On the other hand, in steady-state considerations, accurate estimates of the fluxes of materials across the reservoir boundaries are needed. In addition, any changes in the chemistry of the sediments due to authigenic reactions may leave an imprint on the chemical composition of the connate interstitial waters.

In the present paper I explore some aspects of the study of the chemistry of interstitial waters, with particular reference to some of the problems mentioned above. Attention is given to sampling and analytical procedures, to the problems of ionic transport in the interstitial waters, and to the significance of the results hitherto obtained, especially with regard to their implications in problems of diagenesis and fluxes of materials through the sediment column.

SAMPLING AND ANALYTICAL PROCEDURES

One of the most important problems in the study of the chemistry of interstitial waters is the validity of the sampling procedures. The sediment-interstitial water system is a complex combination of an ion exchange matrix (mainly clay minerals and zeolites) and other solid material that may be in equilibrium with the interstitial fluids. If this system is indeed an equilibrium system, or even if it is partly so, recovery of the interstitial fluids under conditions of temperature and pressure different from those existing in situ may lead to results different from the real situation. Among the first workers to recognize this problem were Mangelsdorf et al (1969) with reference to anomalous distributions of potassium ions in piston cores. Bischoff et al (1970) subsequently demonstrated that, upon warming of a core, monovalent cation concentrations (Na^+, K^+) increased in the interstitial waters, whereas those of divalent cations (Ca^{2+}, Mg^{2+}) decreased. For dissolved silica, Fanning & Pilson (1971) showed that large increases occur when sediments are warmed before extraction of the interstitial fluids.

This work was confirmed in a variety of sedimentary environments during Legs 15 and 25 of the Deep Sea Drilling Project (Sayles et al 1973a, Gieskes 1973, 1974). These latter studies also showed that dissolved anion concentrations (Cl^-, SO_4^{2-}) are essentially not affected by temperature, but that the alkalinity generally rises with temperature, with a simultaneous drop in pH. These various results can best be explained in terms of cation exchange with the clay matrix (Na^+, K^+, Ca^{2+}, Mg^{2+}), whereas the temperature effect on dissolved silica is probably related to chemisorption of dissolved silica on clay surfaces. Evidence for this is circumstantial, but it appears that the magnitude of the temperature effect on dissolved silica is about the same (a concentration increase of about 150 ± 50 $\mu mol/liter$ upon warming from 5 to 25°C) regardless of the total concentration. However, in sediments with a small clay component, the temperature effect decreases (Gieskes 1974). The effect of temperature on alkalinity is probably a result of both ion exchange phenomena (H^+ from clay matrix) and carbonate solubility phenomena (Gieskes 1973, 1974, Hammond 1973).

Recent workers have realized these problems and have made efforts to obtain the interstitial water samples at temperatures as close to the in situ temperature as possible. Pressure effects are usually ignored and appear to be small for the major cations, but they could be important for alkalinity if carbonate solubility equilibria are involved. If the latter is the case, a squeezing pressure smaller than that prevailing in situ would yield too low an alkalinity. Thus, pressure-induced effects may offset temperature effects to some degree if squeezing is performed at room temperatures. In surface sediments, a good measure of the validity of the analysis with regard to the in situ situation is the comparison between the composition of supernatant sea water obtained by extrapolation and the true sea water composition. Under any circumstances, the extraction of the interstitial fluids should be accomplished as soon as possible after retrieval of the sediment, preferably at the in situ temperature.

Of course, the best way to assess true in situ chemical concentration gradients is by means of the in situ extraction of the interstitial fluids. Such apparatus is presently in use for the analysis of interstitial water ionic components (Sayles et al 1973b) as well as for the in situ extraction of dissolved gases (Barnes 1973). Such apparatus could best be used in conjunction with the more traditional box cores and/or piston and gravity cores. For the analysis of trace metals in anaerobic sediments, extra precautions must be taken to work in an inert atmosphere, in order to prevent oxidation processes (cf Bray et al 1973).

Effects of temperature and possibly of pressure are a serious matter in surface cores because precise information is needed on the magnitude of concentration gradients. Changes of 10% in the potassium concentration or of -2% in the magnesium concentration (5 to 25°C) cannot be tolerated there. However, in very long cores, such as those obtained in the Deep Sea Drilling Project, concern about temperature effects is much less important. Work done during Legs 15 and 25 (Sayles et al 1973a, Gieskes 1973, 1974) shows clearly that only small changes in the overall concentration gradients occur. In addition, if serious attention were given to obtaining samples at the in situ temperature, a knowledge of the local heat flow would be necessary, as one deals here with sediment thicknesses of up to 1000 m. Concentration gradients in Deep Sea Drilling Project cores serve as qualitative indicators of fluxes and reactions in the sediments, and detailed considerations of temperature effects appear unnecessary. However, some of the validity of present work will be checked in the near future with an in situ sampler designed to retrieve interstitial waters in deep drill holes.

Another serious problem in the study of interstitial water chemistry is sample preservation. Especially in cores affected by in situ reducing conditions, some of the components are very labile. For instance, values of dissolved ammonia have been shown to decrease upon storage (Gieskes 1974), especially if not stored at low pH or in a deep frozen state. Similarly, dissolved silica may be affected to some degree by biological activity if improperly stored (best storage is $-20°C$ in polyethylene). For the storage of samples intended for trace metal analysis, storage at low pH is appropriate, because under those circumstances the reduced valence states are stabilized. Storage of high alkalinity samples is not a recommended practice, as in such cases calcium carbonate may precipitate upon storage (Berner et al 1970, Gieskes 1973). I prefer to perform as many analyses as possible aboard ship, but if this is not possible, to titrate a portion of the samples for alkalinity (Gieskes & Rogers 1973) and then to store the titrated sample for further analysis. Nutrient data can best be obtained from frozen samples ($-20°C$) stored in glass for inorganic phosphate and ammonia and in polyethylene for dissolved silica.

Various methods are available for the analysis of interstitial waters of marine sediments, but this brief review does not concern itself with these methods. Some recent publications give detailed descriptions of analytical procedures (Presley 1971, Nissenbaum et al 1972, Sholkovitz 1973, Gieskes 1973, 1974). Fortunately, there are no standard methods available and investigators should use well-tested methods of their own preference.

One additional problem is the proper sampling procedure for obtaining sediment cores. For surficial sediments, three major techniques are available: gravity coring, piston coring, and box coring. In all these techniques it is imperative to take great precautions to obtain the surface of the sediment. This is often a major problem but, especially in nearshore sediments with high accumulation rates, the surface sediments are of great interest because they may contain a large amount of man-produced trace metal components (Chow et al 1973). In such cases, use can be made of radio tracers, especially Pb-210 and Th-232, to check for the presence of most recently deposited material (Koide et al 1973). A further problem is the possible compaction of sediments during the coring procedure. A compacted core may show concentration gradients steeper than those that actually exist. Here again, an in situ sampler, though perhaps providing a less detailed concentration-depth profile, may serve as a reference to which interstitial water concentration profiles obtained from shipboard extractions can be adjusted.

In summary, it is now possible, with proper precautions in sampling procedures, sample preservation, and sample storage, to obtain high quality data on the chemical composition of interstitial waters of recent sediments, as well as of those obtained from deep drill holes. These data can then be used to assess the relative importance of processes, chemical or physical, that have led to any observed changes in the chemical composition of the interstitial waters.

RECENT SEDIMENTS

In this section I discuss in greater detail the chemistry of the interstitial waters of the first few meters of marine sediments, as obtained by box, gravity, or piston-core devices. The aims of interstitial water studies on such sediments are threefold:

1. To establish criteria that define depositional conditions, in particular the reduction-oxidation states, that in turn affect the distribution and mobility of chemical species in the interstitial fluids and the solid phases.
2. To study diagenetic processes occurring within a reasonably short time span after deposition of the sediment. Such processes include both the diagenesis of organic matter by biological action, usually referred to as early diagenesis, and the formation of authigenic minerals by the inorganic transformation of deposited material, e.g. clay mineral diagenesis, alteration of volcanic ash, and precipitation of barite.
3. The study of possible fluxes of chemical species into or out of the sediment column through the interstitial fluids.

Depositional Conditions

Most investigations of recent marine sediments have not included detailed work on the chemistry of the interstitial fluids. At best, depositional conditions were estimated from direct measurements of reduction-oxidation potentials by means of electrometric techniques (van der Weijden et al 1970). Such studies usually serve to establish whether conditions are "oxidizing" or "reducing" but do not give extensive information on what processes control any differences in redox states or on what species do get mobilized. Investigations can best be carried out on the basis of analyses of the chemistry of both the solids and the interstitial waters. A very illustrative example of what information can be gained from the solids is the study of Bonatti et al (1971) in the eastern equatorial Pacific (Galapagos Rise). They investigated the distribution of various easily oxidizable or reducible species, including manganese, iron, cobalt, uranium, and chromium. Such studies should be complemented by detailed investigations of the chemical composition of the interstitial waters.

The changes in redox conditions in marine sediments can be traced back to the biochemical decomposition of organic matter. In areas of low deposition rates, the organic carbon content is usually low, and most of this material is oxidized in the upper few centimeters by organisms utilizing dissolved oxygen as their main oxidation agent. This is sometimes reflected in the interstitial waters by an increase in dissolved nitrate (Hartmann et al 1973b). If dissolved oxygen is completely or nearly depleted, nitrate reduction or denitrification may occur, as Arrhenius has demonstrated (1963) in the South Pacific. Upon total depletion of dissolved oxygen and nitrate, the process of sulfate reduction becomes predominant. Finally, upon depletion of dissolved sulfate, the reduction of carbon dioxide to methane, as well as fermentation processes, becomes important (Nissenbaum et al 1972). All the above involve biochemical processes, probably closely related to those observed in open ocean water and enclosed basins (Richards 1965, Gieskes & Grasshoff 1969, Brewer & Murray 1973). Theoretical predictions of these processes have been made by Thorstenson (1970) and, for sediments, by Gardner (1973). These theoretical calculations do in fact agree with the field observations, i.e. as far as the combustion of organic matter is concerned.

Manheim (1961) showed that the main change in the redox potential of sea water occurs after the complete removal of the dissolved oxygen and nitrate in the zone where sulfate reduction processes begin to prevail. Recent studies on the nitrate-nitrogen and iodate-iodide "redox couples" in sea water appear to confim this observation (Liss et al 1973). However, as is often the case, most chemical species constituting redox couples in sea water are controlled by biological processes and hence may not be in true equilibrium. There seems no doubt that once a zone has been reached in which bacterial sulfate reduction processes do occur, as evidenced by decreases in dissolved sulfate and possible increases in dissolved sulfide and/or the formation of pyrite, reducing conditions are established (Berner 1970, Presley et al 1972). The term reducing is used because under these circumstances some elements of geochemical interest, e.g. manganese, uranium, and chromium, can change their oxidation state. Often terms such as "slightly reducing," "slightly oxidizing," or "not reducing enough" are used in the literature. Such terminology is often unfounded and misleading. Of course, oxidation-reduction processes are mainly dependent on the possibility of electron transfer. When such processes are less likely to occur, whether from lack of organisms or lack of substrate, one could talk about "slightly reducing," but such a definition would be relatively meaningless. Redox reactions are very important, and investigations of the detailed circumstances under which such reactions occur in the sediment column are among the major objectives of interstitial water studies.

In summary, in marine sediments, particularly in nearshore sedimentary environments, the biological decomposition of organic matter is of great importance. These processes, usually controlled by bacterial degradation processes, are of primary importance in establishing redox conditions in the sediments. Usually sediments are classified as "oxidizing" when oxygen is present and "reducing" when sulfate reduction processes have become important. The study of the distribution of dissolved nutrients (NO_3^-, NH_4^+), as well as of such dissolved species as sulfate, sulfide, and manganese in the interstitial waters, is of great use in more closely defining the depositional conditions that characterize a particular sediment.

A typical example of the importance of the investigation of redox conditions in sediments is the problem of manganese enrichment in core tops in the eastern equatorial Pacific Ocean (Lynn & Bonatti 1965, Bonatti et al 1971). On the Galapagos Rise, the first few centimeters of the carbonate cores were observed to show a dark coloration caused by an enrichment in manganese as manganese dioxide. Bonatti et al (1971) showed that, below 7 cm, reducing conditions prevailed

Observed increases are often due to the dissolution of underlying salt deposits (Manheim & Bischoff 1969, Michard et al 1974).

SULFATE   Variations in dissolved sulfate are often observed, especially in nearshore sediments, where sulfate reduction processes are important and occur in the sediment actually in contact with the interstitial water, as evidenced by the presence of dissolved sulfide and/or pyrite (Berner 1964, Sholkovitz 1973, Goldhaber 1974). The latter observations are particularly applicable to rapidly accumulating sediments in nearshore areas and relatively isolated marine basins. In many other sedimentary environments, sulfate depletions have been observed, but, as demonstrated below, these depletions can be caused by processes occurring at greater depths than those cored by traditional piston cores.

SODIUM AND POTASSIUM   There are many observations available on decreases in potassium with depth into the sediment column. Particularly the in situ work of Sayles et al (1973b) leaves no doubt as to the reality of this observation. In addition, the very careful work of Shishkina (1958) demonstrates that small depletions of sodium do occur. The nature of the sink for these constituents has not been satisfactorily determined. Suggestions of potassium uptake into authigenic products such as illites are most attractive (Weaver 1967), but such processes are most likely to occur after deep burial of the sediments, e.g. Perry & Hower (1970). In addition, possible sinks for sodium and potassium can be authigenic feldspars (Kastner 1974). Also, particularly in nearshore sediments, in which large depletions in these elements are often observed, uptake by detrital minerals, particularly chlorite, has been suggested by Griffin & Goldberg (1963). Such uptake, however, is very difficult to detect. It appears, therefore, that gradients in concentrations of K^+ and Na^+ are of some interest in problems of reservoir exchange. These concentration changes do not give us information on reactions taking place in the sediments unless such reactions can be really confirmed.

CALCIUM   Changes in dissolved calcium concentrations to either higher or lower values have often been observed in interstitial waters. In nearshore sediments with high rates of sedimentation and large sulfate depletions, one usually also observes large increases in the alkalinity of the interstitial waters. In such cases, precipitation of calcium carbonate can occur (Presley & Kaplan 1968, Sholkovitz 1973). Nissenbaum et al (1972) suggested that strontium may be removed by co-precipitation with the calcium carbonate. Even though mineralogical investigations have not detected such a carbonate phase, some isotopic information on the $^{13}C/^{12}C$ ratio of the carbonate has been interpreted as evidence for this precipitation of $CaCO_3$ (Presley & Kaplan 1968). In areas of more slowly accumulating sediments, however, increases in dissolved calcium have often been observed (Bischoff & Sayles 1972, Sayles et al 1973b). Here again, especially in the absence of a detailed investigation of the solid phases, little can be said about the cause of such con-centration increases. Again in this case the source for the dissolved calcium may be located deeper in the sediments.

in these sediments. In the deeper part of these cores, manganese dioxide coul
be reduced to manganese (II), which would then dissolve to a certain extent in tl
interstitial waters. The concentration of this dissolved manganese would be limit
by the presence of a mixed manganese carbonate. The increased manganese co
centration in the interstitial waters would finally cause a diffusion gradient in
the upper "oxidizing" layers. This postulate is in qualitative agreement w
observations made on interstitial waters in other areas (Hartmann 1964, Li et
1969, Bischoff & Ku 1971, Bischoff & Sayles 1972). Indeed, some preliminary w
in the author's laboratory (C. W. Rogers, unpublished data) indicates that
cores obtained at the same location as those of Lynn & Bonatti (1965), in
sediment immediately below the black surface layer, the interstitial mangar
concentration increases to about 5 ppm and that decreases in the sulfate
increases in the ammonia concentrations occur. Various models have been propc
to explain this migration of manganese to the top of the core. The most sophistic;
model is that proposed by Michard (1971). This model, although it describes
distribution of the solid manganese phases accurately, has some less attrac
aspects. The assumption of a zone in which "conditions are not reducing enc
to reduce manganese" is particularly disturbing because it is untested. Here,
detailed interstitial water studies will be of great value in establishing the condit
under which this mobilization and migration of manganese occurs. Enrichn
in manganese cannot be explained in all areas of the ocean floor by up
migration of manganese, as has been demonstrated for East Pacific Rise sedin
by Bender (1971).

Diagenesis

In the recent literature, many reports have appeared on variations in the
position of interstitial waters of recent marine sediments. In the majority,
emphasis was put on the determination of the so-called nutrient componen
constituents directly attributable to the diagenesis of organic matter (NO_3^-,
NH_4^+, PO_4^{3-}), except in those studies dealing with nearshore sediments. In
investigations, however, detailed attention has been given to the major ioni
stituents of sea water, i.e. chloride, sulfate, sodium, potassium, calciun
magnesium, as well as some of the minor constituents, particularly stro
manganese, bicarbonate, and dissolved silica.

In this section, some of the variations in concentrations of these interstitia
constituents are discussed in detail, although the discussion cannot be cc
within the scope of this short review. In a later section, some generalization
observations in interstitial waters of Deep Sea Drilling Project cores are pr
At this stage, however, I discuss only the observations made on interstitial
obtained from sediment cores no longer than perhaps 15 m.

CHLORIDE   In most recent sediments, especially within the first few mete
little, if any, change has been observed in the chloride concentration. Exc
of course, are areas in which fresh water input from the lower strata is a
For instance, Manheim & Chan (1974) made such observations in the Bl

MAGNESIUM   Dissolved magnesium (unless it is above an underlying salt dome) generally decreases in concentration with depth. One must distinguish at least two types of sediments. In nearshore areas, especially when sedimentation rates are high and sulfate reduction is important, large decreases in magnesium have been observed. Sometimes such decreases have been interpreted in terms of diagenetic processes involving the exchange of iron for magnesium in nonexchangeable positions in detrital clays, and the subsequent precipitation of the mobilized iron as pyrite (Drever 1971). Such observations, again, must be quantified by experimental observations. In certain cases it appears that depletions in the magnesium concentration may be explained by an increase in exchange capacity of the clay minerals upon burial, possibly related to the removal of iron oxide interlayers (Sholkovitz 1973, Clancy 1973). In areas where accumulation rates are much slower, magnesium depletions may be due to entirely different processes, not necessarily occurring in the adjacent sediments. This is discussed in more detail in a subsequent section.

STRONTIUM   In rapidly accumulating sediments, decreases have sometimes been observed which are probably related to precipitation of carbonate phases, as discussed previously (Nissenbaum et al 1972). Often, however, increases are observed that are probably related to carbonate recrystallization processes. This is discussed in greater detail under the section on Deep Sea Drilling Project results.

MANGANESE   The mobilization of manganese under reducing conditions is discussed in a previous section.

ALKALINITY   In many nearshore environments, large increases in the interstitial water alkalinity values have been observed (Berner et al 1970, Sholkovitz 1973, Goldhaber 1974). Such increases can be accounted for by the bacterial sulfate reduction processes discussed previously. Also, these increases in alkalinity have been considered the principal reason for the depletion of calcium by calcium carbonate precipitation. An important observation is that in such depositional environments, even the most delicate foraminiferal tests are protected from dissolution (Berger & Soutar 1970).

In open ocean environments, observed changes in alkalinity are small, and some authors have reported slight minima in the alkalinity just below the sediment-water interface (Hartmann et al 1973b; J. Morse, personal communication). As no obvious changes in other constituents were observed, the possibility of an artifact cannot be ruled out. Precisely because of such problems, the use of an in situ sampler seems imperative.

DISSOLVED SILICA   The distribution of dissolved silica has received special attention in the recent literature (Bischoff & Sayles 1972, Hurd 1973, Schink et al 1974). From observations it is apparent that dissolved silica in interstitial waters can range in value from 100 to 1200 μmol/liter, depending on the type of sediment. Both the nature of the detrital silicate components (clay minerals and quartz) and the amount of opaline silica in the form of biogenous silica are of great importance in determining the interstitial dissolved silica value. From the observations it appears that some steady-state value is maintained. This steady state probably involves the dissolution of biogenous silica and reactions with clay minerals. That such reactions may indeed occur has been shown by MacKenzie & Garrels (1965) and Siever & Woodford (1973). These authors, however, were not able to deduce the nature of such reactions or whether, in fact, new mineral phases were created. As long as such information is not available, arguments that these postulated reactions of dissolved silica with clay minerals are important in controlling the chemistry of the oceans must be rather speculative (Schink et al 1974).

From the above description of actual observations, it is clear that a rather great variability in the data occurs. Among the major objectives of interstitial water studies, however, is the detection of possible authigenic reactions occurring in the sediments. Regular concentration changes with depth in any of the above components can indicate which components are taken up in such reactions and which are released. In the absence of knowledge of the reaction or reactions actually occurring, such an exercise can be only academic. The shape of the concentration gradients, however, is most helpful both in estimating the nature of the reactions and in establishing the most likely location for their occurrence. For the latter purpose, we must have a detailed knowledge of the transport properties of the sediment-interstitial water system, particularly with regard to its diffusional characteristics. In this respect it is useful to consider the problem of how far diffusional communication with overlying sea water is possible in a system with an upward-moving boundary, i.e. the sediment-water interface. For these purposes, we can calculate for a certain fixed time, t, the mean diffusion path, i.e. the path across which an original concentration gradient would be eliminated by more than 90% $[z_M = (Dt)^{1/2}$, where z_M = depth, D = diffusion coefficient, t = time]. This path length can then be compared with the thickness of the accumulated sediment in the same time, t. When the latter exceeds the mean diffusion path length, the underlying sediment becomes essentially closed to diffusional communication with overlying sea water. Results of such calculations are presented in Table 1, in which a range of possible diffusion coefficients and sedimentation rates has been considered. There is still much controversy about the values of diffusion coefficients in interstitial waters, but some values can be considered trustworthy enough for our present considerations. For dissolved sulfate ions, Berner (1964) found 3×10^{-6} cm²/sec,

Table 1 Depth (meters) into the sediments to which diffusional communication is possible with overlying sea water

Diffusion coefficient (cm²/sec)	Sedimentation rate (cm/1000 yr)			
	500	50	5	1
5×10^{-6}	3	30	300	1500
3×10^{-6}	2	18	180	900
1×10^{-6}	0.6	6	60	300
1×10^{-7}	<0.1	0.6	6	30

whereas Goldhaber (1974) preferred a value of 1×10^{-6} cm^2/sec. Li & Gregory (1974) gave a value for dissolved sulfate in red clay at 5°C of 3×10^{-6} cm^2/sec and for calcium of 0.8×10^{-6} cm^2/sec. Michard et al (1974) found a value of 3×10^{-7} cm^2/sec at 12°C from magnesium gradients in a Mediterranean core. Also, Fanning & Pilson (1974) gave a value of 3.3×10^{-6} cm^2/sec at 5°C for dissolved silica. Thus it can be safely assumed that diffusion coefficient values are probably equal to 3×10^{-6} cm^2/sec or less. It follows, then, that for nearshore sediments with accumulation rates of 50–500 cm/1000 yr, piston cores will indeed core intervals that may encompass, in part, sediments that are not in communication with overlying sea water. This computation ignores any upward advection due to compaction and thus gives maximum depths. In addition, effects of bioturbation are necessarily not included.

Interpretations of concentration gradients in interstitial waters in nearshore sediments, therefore, can be made with some degree of confidence as the system becomes closed to communication with the overlying sea water within the cored interval. In open ocean sediments, however, piston cores sample only a relatively small section of the "open system." It seems appropriate to discuss only nearshore sediments in this section and to return to observations in deep sea sediments under the discussion of Deep Sea Drilling Project results.

For nearshore sediments, several diagenetic models relating changes in pH, alkalinity, nutrients, sulfate, calcium, and/or magnesium have been proposed (Berner et al 1970, Sholkovitz 1973, Hartmann et al 1973a, Gardner 1973). Such models are indeed most useful in relating observed concentration changes, but so far these models have not been verified or made useful in terms of detecting the actual products of the suggested diagenetic reactions. The exception to this is the case of sedimentary pyrite formation (Berner 1970, Goldhaber 1974).

Environmental conditions can have great effects on processes of early diagenesis and related processes. This was, for instance, demonstrated by Sholkovitz (1973) in the sediments of the Santa Barbara Basin. He found large concentration gradients in the varved sediments underlying the quiescent, almost anaerobic basin waters, whereas in the sediments underlying more oxygenated water, notwithstanding similar sedimentation rates, he observed only small concentration gradients. This was probably due to a much faster breakdown of reactive organic matter by bottom-dwelling organisms, as well as to possible bioturbation. Goldhaber (1974) also inferred bioturbation as a mechanism in his observations on anomalous changes in sulfate depth profiles in sediments of the Gulf of California.

A final observation should be made here. In some sediments, particularly deep Pacific clays, authigenic minerals, e.g. zeolites and montmorillonites, are presently being formed; yet no clear concentration gradients in interstitial water constituents are observed here. Many authigenic reactions may indeed involve the interstitial fluids as medium for ionic transport during the reactions, but the total stoichiometry of the reactions may involve no change in the composition of the interstitial fluids. Thus the formation of authigenic minerals per se does not necessitate a concentration gradient of an interstitial water component, nor does a gradient imply that such observed concentrations are related to the formation of a particular authigenic mineral. For this, careful mineralogical work, as well as consideration of the mass balance between reactants and products, is necessary.

Fluxes Across the Sediment-Water Interface

As discussed in the previous section, in many parts of the ocean floor, both near shore and in the pelagic realm, concentration gradients with depth of several interstitial water components have been observed. For many of these components it is conceivable that appreciable diffusional transport will occur along these concentration gradients. In order to estimate possible fluxes of material, a primary requirement is a good knowledge of the core on which observations have been made. The necessary information includes data on porosity, mineralogical composition, possible compaction as a result of the coring operation, and rate of sedimentation. Finally, it is imperative that the top of the core, i.e. the sediment-water interface, be recovered. For this type of study, the simultaneous use of the in situ sampling technique (Sayles et al 1973b, Barnes 1973) is invaluable. Used by itself, this technique will yield correct gradients but will not yield enough information on sediment type.

Another important requirement for diffusional flux calculations is the possession of precise estimates of the various diffusion coefficients, as they are applicable in sediment-interstitial water systems. Only a few direct estimates of such coefficients are available (Duursma & Bosch 1970, Li & Gregory 1974, Fanning & Pilson 1974). Most estimates, however, have been obtained from best fits to diffusion-model-derived equations (Berner 1964, Lerman & Weiler 1970, Goldhaber 1974, Michard et al 1974). There still appears to be considerable uncertainty about the exact values of most diffusion coefficients.

In sediments, a simple steady-state model is often used in which diffusion is considered to occur in a system of moving coordinates (the sediment-water interface is the moving boundary) and in which terms involving rates of change in concentration due to chemical reaction are considered zero or first order:

$$\frac{\delta c}{\delta t} = D\frac{\delta^2 c}{\delta z^2} - u\frac{\delta c}{\delta z} + J = 0$$

where D is the diffusion coefficient, c is the concentration, z is a depth parameter, u is the sedimentation rate, and J is the production or consumption term. This equation is an oversimplification because it does not account for upward advection due to compaction, nor does it account for a variable diffusion coefficient as a function of porosity and other factors. Various models have been proposed that take some or all of such effects into consideration (Goldberg & Koide 1962, Berner 1964, Anikouchine 1967, Tzur 1971, Lerman & Jones 1973, Goldhaber 1974). In each case, the relative importance of the various factors affecting the above simplified diffusion equation must be carefully considered. In general it is assumed that the diffusion coefficient is constant with depth, which may be approximately true in surface sediments, provided only small changes in water content or porosity occur. Often the upward advection of water is ignored, which again must be considered carefully, especially in sediments with very high sedimentation rates. Finally, the sedimentation rate is also often ignored, so that diffusion processes are essentially considered to occur in a fixed frame of reference. This is probably a justifiable assumption in the consideration of diffusion gradients over a few tens of centimeters in slowly accumulating sediments, for instance, in the problem of silica diffusion out of the sediments (Fanning & Schink 1969, Hurd 1973, Fanning & Pilson 1972, 1974). Often, dissolved silica gradients have been observed to depths of 20–40 cm into the sediment. The problem with the dissolved silica profiles is that at this stage we still have very little knowledge of what reactions are responsible for the apparent steady-state concentration gradients in the sediments in the lower sedimentary layers, i.e. in the first few meters below the "diffusion layers." We need such knowledge to better understand the nature of the reaction term in the above diffusion equation. In his analysis, Hurd (1973) considered only the dissolution rate of opal, which was possibly hindered by overgrowth of silicates. Perhaps a more complex mechanism is at work, in which case evaluation by a careful analysis of the solids in the upper few centimeters below the sediment-water interface is needed. Even so, diffusion of dissolved silica from the sediment contributes only a minor part, at best a few tenths of a percent, of the dissolved silica input into the deep ocean, which is mainly due to the dissolution of siliceous organisms from the sediment-water interface (Berger 1970, Edmond 1973, 1974, Hurd 1973).

In a previous section we discussed in some detail the work done on manganese migration. However, much quantitative work on the importance of manganese diffusion out of the sediments is not yet possible, even in areas where extensive deposits of manganese nodules have been observed. It appears that, especially in carbonate sediments, manganese migration can be important (Lynn & Bonatti 1965, Li et al 1969, Michard 1971), but little quantitative work has been done.

For other elements, no serious attempts have been made to calculate overall fluxes into or from the sediments, usually because of a lack of knowledge of diffusion gradients in different areas of the ocean. Such calculations, when they become available, will be very important. For instance, it is not yet possible to estimate the flux of magnesium (both via solid phases and through the interstitial waters) into the sediments of the ocean. In his study of the magnesium problem, Drever (1974) suggested that for magnesium a true steady state may not exist in the ocean, and that at present the ocean may be increasing its magnesium content (see also Broecker 1971).

DEEP SEA DRILLING CORES

As discussed above, data on chemical concentration gradients obtained from piston-coring devices penetrating only a few tens of meters into slowly accumulating sediments may not reveal much information on diagenetic processes taking place in the cored sediment section, especially with regard to some of the major constituents of sea water. The reactions or processes that are the cause of these gradients may be situated deeper in the sediment column. With the onset of the Deep Sea Drilling Project, a unique opportunity arose to study the distribution of the various chemical components of the interstitial waters in long sediment sections, which often extend down to oceanic basement. Much of the pioneering work done by Kaplan, Manheim, Presley, and Sayles in this area has recently been summarized by Manheim & Sayles (1974). In general, a broad classification of observed concentration changes is possible, especially if reference is made to the depletion in sulfate with depth. This was done by Broecker (1973) in his introduction to the extensive geochemical studies carried out as part of Leg 15 of the "Glomar Challenger" in the Carribean Sea. It appears that a primary factor in such a classification is the sedimentation rate. This is not surprising if we consider the data given in Table 1. At low rates of sedimentation, diffusional communication with the overlying sea water becomes feasible to great depths. In such sediments, usually small depletions, if any, are observed in most dissolved constituents. Of course, exceptions always occur and can be related to deposits with high organic carbon contents in the lower sedimentary strata (e.g. Site 214, Manheim et al 1974). With progressively increasing sedimentation rates, the communication depth decreases and the organic carbon content increases; thus, larger depletions in sulfate will occur. In such sediments, gradients in various cations are commonly observed (depletions in Mg^{2+} and K^+, increases in Ca^{2+} and sometimes in Sr^{2+}). When sedimentation rates become higher than 4–5 cm/1000 yr, sulfate depletions can become large, and in sediments with very fast sedimentation rates, complete depletions in dissolved sulfate occur. These depletions are functions not only of increased sedimentation rates, but also of increased organic carbon content. In sediments with similar sedimentation rates, the largest sulfate depletions, of course, occur in the sediments with the highest organic carbon content (cf Sites 241 and 242, Gieskes 1974). In holes with high sedimentation rates (>5 cm/1000 yr), maxima in alkalinity (HCO_3^-) and minima in dissolved calcium have also been observed (Manheim & Sayles 1974, Gieskes 1973, 1974). It should be emphasized that this observation is not a general rule, as obvious exceptions have been observed, for instance in Antarctic sediments (Mann and Gieskes, in preparation). In these sediments, large depletions in sulfate are accompanied by constant alkalinities and increased calcium concentrations. There appears to be a good correlation between the depth at which the alkalinity and calcium extrema occur and the sedimentation rates: 6–8 m in Site 147 with 50 cm/1000 yr; 70 m in Site 148 with 8 cm/1000 yr; 100–125 m in Site 242 with 5 cm/1000 yr (Gieskes 1973, 1974). With diffusion coefficients between 1×10^{-6} cm²/sec and 3×10^{-6} cm²/sec, these observations compare well with the predicted communication depths of Table 1. Processes involving increases in the calcium concentration and decreases in alkalinity in the deeper parts of these sites are responsible for these observed extrema.

A careful analysis of the observed gradients, as well as supplemental information on sedimentation rates, porosities, mineralogies, and organic carbon contents, usually obtained on a routine basis, can lead to an evaluation of the processes that may be involved and also to discovery of the location where such processes are likely to occur. The task remains, of course, to identify the reactants and the products. For such purposes, an invaluable tool is the study of the oxygen isotope ratio of the interstitial water. Recent investigations of this ratio have led Lawrence (1973, 1974) to suggest that observed large depletions in the $^{18}O/^{16}O$ ratio are the results of authigenic reactions involving the alteration of volcanic ash and/or basalt. Of great importance also is the study of the electrical conductivity of the sediments, which yields information on relative changes in diffusion coefficients in compacting and cementing sediments (Manheim 1970, Manheim & Waterman 1974). Such information is particularly useful in evaluating the nature of concentration–depth profiles. A decrease in bulk sediment diffusion coefficients with depth would imply increased curvature in a gradient, particularly if the gradient is a result of deep-seated reactions.

Rather than attempt a detailed description of various observations made on over 200 Deep Sea Drilling Project sites, I present data obtained during Leg 25 of the "Glomar Challenger" in the Indian Ocean to demonstrate the type of observations made. Site 242 is characterized by continuous sedimentation and fairly high accumulation rates, and consists of hemipelagic clayey nanno-oozes and nanno chalks; Site 249 is typified by a very large sedimentary hiatus at about 170 m depth (approximately 60 m.y.). Detailed information on these holes can be found in Gieskes (1974).

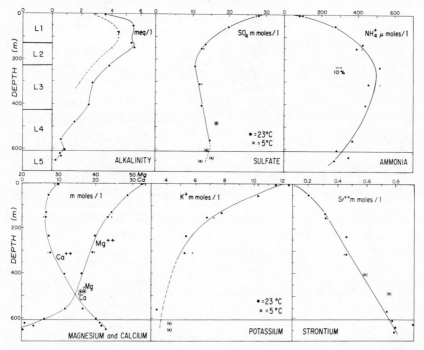

Figure 1 Composition of interstitial waters at Site 242: 15°50.65'S, 41°49.23'E (Gieskes 1974). *L1* = Gray foram-rich nanno-ooze; *L2* = gray clay and foram-bearing nanno-ooze; *L3* = Gray clay and foram-bearing nannochalk; and *L4* and *L5* = brown clay and nanno-chalk separated by acoustic reflector. Numbers in brackets obtained from very small sample sizes. Circled point could be due to sea water contamination.

Site 242 (Figure 1) shows sedimentation rates of 5 cm/1000 yr in the upper 300 m and, below that, much lower rates of about 2 cm/1000 yr. Alkalinity (HCO_3^-) and calcium show extrema at about 100–150 m, and sulfate and magnesium decrease rapidly in the first 200 m. The shapes of all concentration depth profiles imply reactions for all these components, as they are all concave upwards. Below about 200 m, reversals occur not only in alkalinity and Ca^{2+}, but also in SO_4^2 and NH_4^+. Whereas extrema in calcium and alkalinity are commonly observed in rapidly accumulating sediments and appear to be due to secondary processes in the deeper parts of the sedimentary column (cf Site 148, Gieskes 1973), the reversals in the sulfate and ammonia profiles are related to the lower sedimentation rates in the deeper parts of the hole. During these periods of lower sedimentation rates, sulfate depletions were also smaller, due to the greater importance of diffusional replenishment of sulfate from the overlying sea water. For magnesium, there appears to be another reaction site deeper in the hole, probably in the uncored section. The decrease in potassium is large, as is always the case in detrital sediments

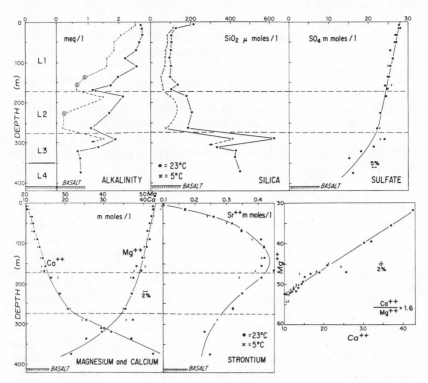

Figure 2 Composition of interstitial waters at Site 249: 29°56.99'S, 36°04.62'E (Gieskes 1974). *L1* = Gray foram-rich nanno-ooze; *L2* = clays and foram-bearing nannochalks; *L3* = gray volcanic ash and siltstones; *L4* = clayey siltstones and silty clay stones. Horizontal broken lines indicate sedimentation hiati.

(Manheim & Sayles 1974). Data on dissolved strontium suggest a reaction through-out the hole, probably related to carbonate recrystallization processes.

Site 249 (Figure 2) shows an entirely different picture. The most important observation in this site was the very large sedimentary hiatus (60 m.y.) at 170 m and another one (15 m.y.) at 270 m. The sulfate, calcium, and magnesium gradients especially are linear in the upper sedimentary unit, slightly curved in the second nannochalk unit, and steeper in the lower volcanogenic unit. I interpret these gradients as being mainly caused by diffusion in the upper two units, whereas reactions involving these ions occur in the deeper parts. Indeed, organic carbon content is high at great depth, and thus it is likely that sulfate reduction is still an ongoing process. Correlations between calcium and magnesium are strong and suggest a reaction involving uptake of magnesium and release of calcium at depth (alteration of ash or basalt?). Dissolved strontium does not correlate well with calcium and is related to carbonate recrystallization processes. Indeed, strontium contents in the carbonate fraction are higher in the upper nanno-ooze unit than in the underlying nannochalks (Marchig & Vallier 1974).

Both these sites demonstrate that, particularly for the major cations in the interstitial waters of marine sediments, piston cores would reveal relatively little information on processes occurring in the lower sediment layers. In fact, in Site 249, concentration gradients in the upper sediments are the result of reactions occurring either in the 300-m lower volcanogenic unit or the underlying basalt. It appears that observations made on concentration gradients in the interstitial waters of piston core-derived sediments in the deep sea can be made more useful if information is available on nearby drill holes. This applies particularly to "less reactive" com-ponents, such as calcium and magnesium.

It is most important to investigate the possibility that alteration of volcanic ash and/or basalt in the basal layers of the sediments can have a pronounced influence on the distribution of the dissolved constituents of the interstitial waters of the overlying sediments (Lawrence 1974). Remarkably little work has been done on such alteration processes, particularly with regard to the relatively low temperature alteration that occurs after burial of the ash or basalt below a relatively thick layer of sediments. That this process does indeed occur seems certain from the observations in Site 245 on the north side of the southwest Indian Ridge (Gieskes 1974, Warner & Gieskes 1974). Here a relatively young basaltic sill (28 m.y.) intrudes on an iron- and manganese-rich basal sediment of Paleocene age. The basalt shows signs of minor alteration in the form of chlorite, chlorophaeite, and calcite (Erlank & Reid 1974). Increases in dissolved calcium and decrease in dissolved magnesium, sodium, and alkalinity may be explained, at least in part, by such alteration reactions. Of course, much careful work needs to be done on the mineralogy and chemistry of the solids before this suggested mechanism can be verified. It can be seen, however, that the above processes must be considered as a possible origin of concentration gradients, even in sediments with high sedimentation rates, such as those of Site 242 (Figure 1).

Thus, careful analysis of concentration gradients observed in marine sediment interstitial waters, particularly in Deep Sea Drilling Project cores, will serve as a most useful guide in determining what possible reactions may occur. More importantly, observed gradients, in conjunction with studies of the diffusional properties of the sediments, can indicate where such reactions may occur in the sediments (or in the underlying basalt). Also, with a knowledge of the diffusion coefficients, estimates can be made of the fluxes involved for the constituents for which concentration gradients have been determined.

SUMMARY AND CONCLUSIONS

Sampling techniques for interstitial waters of marine sediments have advanced a great deal in recent years. Reliable data can be obtained on the chemical composition of the dissolved components of these interstitial waters. In the sampling, especially if the sediments are extruded aboard ship, attention should be given to maintaining the sediment at in situ temperatures, as temperature greatly influences the chemical composition of the interstitial water due to ion exchange, sorption, and solubility equilibria. Especially in studies that aim at evaluating the exchange of dissolved material between the sediment and the overlying sea water, the use of in situ apparatus has been shown to be a great advantage.

Interstitial water studies can often serve to characterize the depositional conditions. For these purposes, especially, investigation of dissolved nutrients (NO_3^-, NO_2^-, NH_4^+, PO_4^{3-}) and dissolved sulfate and sulfide, as well as of some trace elements such as Mn^{2+}, is most useful. A combination of these studies with a detailed investigation of the solids, particularly with respect to elements that can have more than one valence state, will enable one to uniquely define the reduction-oxidation conditions of the sediment-interstitial water system. Only a combination of these studies can serve these purposes, because, for instance, concentration changes in sulfate alone could be due to reactions occurring at a level deeper than that sampled by a piston-, gravity-, or box-coring device.

In any study of the concentration-depth variations of interstitial water components, it is important to know the rate of sediment accumulation. It has been shown that the depth to which diffusional communication with the overlying sea water is possible is small in rapidly accumulating sediments (50–500 cm/1000 yr) on continental shelves, but can be large in slowly accumulating sediments. Thus, cores extending to only 15 m or less below the sediment-water interface often sample a sediment section that is appreciably shorter than the depth to which diffusional communication is possible. Any concentration-depth profiles, particularly when they are smooth and continuous, may originate at levels deeper than that which was sampled. Only in nearshore sediments do any concentration changes have to be explained by processes occurring in the sampled interval. Possible exceptions to this are observed concentration-depth profiles of dissolved manganese, dissolved silica, and some nutrient components.

In slowly accumulating sediments (<2 cm/1000 yr), concentration-depth profiles of calcium, magnesium, sodium, sulfate, and other major sea water constituents can be understood only from information on drill holes, preferably ones that have been drilled to oceanic basement. Then, using information on possible changes with the depth of the diffusional transport properties of the sediments, an analysis can be made of the probable reaction sites in the sediments and the elements involved in such reactions. Authigenic material in the sediments does not necessarily result from the reactions suggested by observed gradients and, conversely, suggested reactions must be verified by actual identification in the sediments. A powerful tool in these studies is the examination of the oxygen isotope ratio of the interstitial waters. Those studies have led to the conclusion that observed $^{18}O/^{16}O$ changes are probably related to alteration of volcanic ash or basalt. One of the important goals in the geochemical studies of marine sediments is the verification of this latter suggestion.

Interstitial water studies to date have neither verified nor refuted the validity of Sillén's equilibrium ocean model, because little or no information is available on the actual equilibrium state between interstitial waters and the solid sediments. Smooth concentration gradients, however, suggest that such an equilibrium state may not exist and that even the alteration of deeply buried volcanic rocks makes a contribution of dissolved material to the ocean, at least for the element calcium. This alteration process may also form a sink for magnesium and sodium, and possibly potassium and carbon dioxide. The significance of the fluxes involved is probably relatively small, but it cannot as yet be evaluated, because of a lack of reliable data on diffusion coefficients, as well as the paucity of geographical data.

It appears that for any study of the geochemistry of deep sea sediments a study of the interstitial fluids is necessary. However, it is equally important to gain quantitative data on the chemical composition and the mineralogy of the solid phases. With that information it will become possible to completely describe the geochemical processes in these sediments. Such studies could also lead to conclusions about possible changes in sea water composition with time.

ACKNOWLEDGMENTS

The author wishes to acknowledge the many discussions on the importance of interstitial water studies with Drs. Miriam Kastner and James Lawrence. Financial support came generously from Grant NSF GA-33229.

Literature Cited

Anikouchine, W. A. 1967. *J. Geophys. Res.* 72:505–9

Arrhenius, G. 1963. *The Sea*, ed. M. N. Hill, 3:655–726. New York: Wiley-Interscience. 963 pp.

Barnes, R. O. 1973. *Deep Sea Res.* 20:1125–28

Barth, T. F. W. 1952. *Theoretical Petrology.* New York: Wiley. 369 pp.

Bender, M. L. 1971. *J. Geophys. Res.* 76: 4212–15

Berger, W. H. 1970. *Geol. Soc. Am. Bull.* 81:1385–1402

Berger, W. H., Soutar, A. 1970. *Geol. Soc. Am. Bull.* 81:275–82

Berner, R. A. 1964. *Geochim. Cosmochim. Acta* 28:1497–1503

Berner, R. A. 1970. *Am. J. Sci.* 268:1–23

Berner, R. A., Scott, M. R., Thomlinson, C. 1970. *Limnol. Oceanogr.* 15:544–49

Bischoff, J. L., Greer, R. E., Luistro, A. O. 1970. *Science* 167:1245–46

Bischoff, J. L., Ku, T. L. 1971. *J. Sediment. Petrology* 40:960–72

Bischoff, J. L., Sayles, F. L. 1972. *J. Sediment. Petrology* 42:711–24

Bonatti, E., Fisher, D. E., Joensuu, O., Rydell, H. S. 1971. *Geochim. Cosmochim. Acta* 35:189–201

Bray, J. T., Bricker, O. P., Troup, B. N.

1973. *Science* 180:1362–64

Brewer, P. G., Murray, J. W. 1973. *Deep Sea Res.* 20:803–18

Broecker, W. S. 1971. *Quaternary Res.* 1: 188–207

Broecker, W. S. 1973. *Initial Reports of the Deep Sea Drilling Project,* ed. B. C. Heezen et al, 20:751–56. Washington, DC: GPO. 958 pp.

Chow, T. J. et al 1973. *Science* 181:551–52

Clancy, J. 1973. *Cation exchange capacity of fine sediments.* MS thesis. Univ. Southern Calif., Los Angeles. 85 pp.

Conway, E. J. 1943. *Proc. Roy. Irish Acad.* 48B:161–212

Drever, J. I. 1971. *Science* 172:1334–36

Drever, J. I. 1974. *The Sea,* ed. E. D. Goldberg, 5:337–57. New York: Wiley-Interscience. 895 pp.

Duursma, E. K., Bosch, C. J. 1970. *Neth. J. Sea Res.* 4:395–469

Edmond, J. M. 1973. *Nature* 241:391–93

Edmond, J. M. 1974. *Deep Sea Res.* 21: 455–80

Erlank, A. J., Reid, D. L. 1974. *Initial Reports of the Deep Sea Drilling Project,* ed. E. S. W. Simpson et al, 25:542–52. Washington DC: GPO. 884 pp.

Fanning, K. A., Pilson, M. E. Q. 1971. *Science* 173:1228–31

Fanning, K. A., Pilson, M. E. Q. 1972. *Deep Sea Res.* 19:847–63

Fanning, K. A., Pilson, M. E. Q. 1974. *J. Geophys. Res.* 79:1293–97

Fanning, K. A., Schink, D. R. 1969. *Limnol. Oceanogr.* 14:59–68

Gardner, L. R. 1973. *Geochim Cosmochim. Acta* 37:53–68

Garrels, R. M., MacKenzie, F. T. 1971. *Evolution of Sedimentary Rocks.* New York: Norton. 397 pp.

Garrels, R. M., MacKenzie, F. T. 1972. *Mar. Chem.* 1:27–41

Garrels, R. M., Perry, E. A. 1974. *The Sea,* ed. E. D. Goldberg, 5:303–36. New York: Wiley-Interscience. 895 pp.

Gieskes, J. M. 1973. *Initial Reports of the Deep Sea Drilling Project,* ed. B. C. Heezen et al, 20:813–29. Washington DC: GPO. 958 pp.

Gieskes, J. M. 1974. *Initial Reports of the Deep Sea Drilling Project,* ed. E. S. W. Simpson et al, 25:361–94. Washington DC: GPO. 884 pp.

Gieskes, J. M., Grasshoff, K. 1969. *Kieler Meeresf.* 25:105–32

Gieskes, J. M., Rogers, W. C. 1973. *J. Sediment. Petrology* 43:272–77

Goldberg, E. D., Arrhenius, G. 1958. *Geochim. Cosmochim. Acta* 13:153–212

Goldberg, E. D., Koide, M. 1962. *Geochim.* *Cosmochim. Acta* 26:417–50

Goldhaber, M. B. 1974. *Equilibrium and dynamic aspects of the marine geochemistry of sulfur.* PhD thesis. Univ. California, Los Angeles. 397 pp.

Goldschmidt, V. M. 1933. *Fortschr. Mineral. Kristallogr. Petrogr.* 17:112–56

Griffin, J. J., Goldberg, E. D. 1963. *The Sea,* ed. M. N. Hill 3:728–41. New York: Wiley-Interscience. 963 pp.

Hammond, D. E. 1973. *Initial Reports of the Deep Sea Drilling Project,* ed. B. C. Heezen et al, 20:831–50. Washington DC: GPO. 958 pp.

Hartmann, M. P. 1964. *Meyniania* 14:3–20

Hartmann, M., Koegler, F. C., Mueller, P., Suess, E. 1973b. *Preliminary Results of Geochemical and Soil Mechanical Investigations on Pacific Ocean Sediments.* Presented at Symp. Origin and Distribution of Manganese Nodules, Hawaii.

Hartmann, M., Mueller, P., Suess, E., van der Weijden, C. H. 1973a. *"Meteor" Forschungsergebnisse* C12:74–86

Holland, H. D. 1972. *Geochim Cosmochim. Acta* 36:637–51

Horn, M. K., Adams, J. A. S. 1966. *Geochim. Cosmochim. Acta* 30:279–97

Hurd, D. C. 1973. *Geochim. Cosmochim. Acta* 37:2257–82

Kastner, M. 1974. *Geochim. Cosmochim. Acta* 38:650–53

Koide, M. Bruland, K. W., Goldberg, E. D. 1973. *Geochim. Cosmochim. Acta* 37: 1171–87

Lawrence, J. R. 1973. *Initial Reports of the Deep Sea Drilling Project,* ed. B. C. Heezen et al, 20:891–900. Washington DC: GPO. 958 pp.

Lawrence, J. R. 1974. *Initial Reports of the Deep Sea Drilling Project,* ed. R. B. Whitmarsh et al, 23:939–42. Washington DC: GPO. 1180 pp.

Lerman, A., Jones, B. F. 1973. *Limnol. Oceanogr.* 18:72–85

Lerman, A., Weiler, R. R. 1970. *Earth Planet. Sci. Lett.* 10:150–56

Li, Y-H., Bischoff, J., Mathieu, G. 1969. *Earth Planet. Sci. Lett.* 7:265–70

Li, Y-H., Gregory, S. 1974. *Geochim Cosmochim. Acta* 38:703–14

Liss, P. S., Herring, J. R., Goldberg, E. D. 1973. *Nature* 242:108–9

Lynn, D. C., Bonatti, E. 1965. *Mar. Geol.* 3:457–74

MacKenzie, F. T., Garrels, R. M. 1965. *Science* 150:57–58

Mangelsdorf, P. C., Wilson, T. R. S., Daniell, E. 1969. *Science* 165:171–73

Manheim, F. T. 1961. *Stockholm Contrib. Geol.* 8:27–36

Manheim, F. T. 1970. *Earth Planet. Sci. Lett.* 9:307–9

Manheim, F. T., Bischoff, J. L. 1969. *Chem. Geol.* 4:63–82

Manheim, F. T., Chan, K. M. 1974. *Am. Assoc. Petrol. Geol.* Memoir 20:155–80

Manheim, F. T., Sayles, F. L. 1974. *The Sea,* ed. E. D. Goldberg, 5:527–68. New York: Wiley-Interscience. 895 pp.

Manheim, F. T., Waterman, L. S. 1974. *Initial Reports of the Deep Sea Drilling Project,* ed. C. C. von der Borch et al, 22:663–70. Washington DC: GPO. 890 pp.

Manheim, F. T., Waterman, L. S., Sayles, F. L. 1974. *Initial Reports of the Deep Sea Drilling Project,* ed. C. C. von der Borch et al, 22:657–62. Washington DC: GPO. 890 pp.

Marchig, V., Vallier, T. L. 1974. *Initial Reports of the Deep Sea Drilling Project,* ed. E. S. W. Simpson et al, 25:405–16. Washington DC: GPO. 884 pp.

Michard, G. 1971. *J. Geophys. Res.* 76: 2179–86

Michard, G., Church, T. M., Bernat, M. 1974. *J. Geophys. Res.* 79:817–24

Nissenbaum, A., Presley, B. J., Kaplan, I. R. 1972. *Geochim. Cosmochim. Acta* 36: 1007–28

Perry, E., Hower, J. 1970. *Clays Clay Minerals* 18:165–77

Presley, B. J. 1971. *Initial Reports of the Deep Sea Drilling Project,* ed. E. L. Winterer et al, 7:1749–55. Washington DC: GPO. 1757 pp.

Presley, B. J., Kaplan, I. R. 1968. *Geochim. Cosmochim. Acta* 32:1037–48

Presley, B. J., Kolodny, Y., Nissenbaum, A., Kaplan, I. R. 1972. *Geochim. Cosmochim. Acta* 36:1073–90

Richards, F. A. 1965. *Chemical Oceanography,* ed. J. P. Riley, G. Skirrow, 1:611–45. New York: Academic. 712 pp.

Sayles, F. L., Manheim, F. T., Waterman, L.S. 1973a. *Initial Reports of the Deep Sea Drilling Project,* ed. B. C. Heezen et al, 20:783–804. Washington DC: GPO. 958 pp.

Sayles, F. L., Wilson, T. R. S., Hume, P. H., Mangelsdorf, P. C. 1973b. *Science* 181: 154–56

Schink, D. R., Fanning, K. A., Pilson, M. E. Q. 1974. *J. Geophys. Res.* 79:2243–50

Shishkina, O. V. 1958. *TIOAN, SSSR* 26: 109–80

Sholkovitz, E. R. 1973. *Geochim Cosmochim. Acta* 37:2043–73

Siever, R., Woodford, N. 1973. *Geochim Cosmochim. Acta* 37:1851–60

Sillén, L. G. 1967. *Science* 156:1189–97

Thorstenson, D. C. 1970. *Geochim Cosmochim. Acta* 34:745–70

Tzur, Y. 1971. *J. Geophys. Res.* 67:4208–11

Van der Weijden, C. H., Schuiling, R. D., Das, H. A. 1970. *Mar. Geol.* 9:81–99

Warner, T. B., Gieskes, J. M. 1974. *Initial Reports of the Deep Sea Drilling Project,* ed. E. S. W. Simpson et al, 25:395–403. Washington DC: GPO. 884 pp.

Weaver, C. E. 1967. *Geochim. Cosmochim. Acta* 31:2181–96

AUTHOR INDEX

CUMULATIVE INDEXES

CONTRIBUTING AUTHORS VOLUMES 1-3

CHAPTER TITLES VOLUMES 1-3

465